New Light on Dark Stars
Red Dwarfs, Low-Mass Stars, Brown Dwarfs

Springer
London
Berlin
Heidelberg
New York
Barcelona
Hong Kong
Milan
Paris
Santa Clara
Singapore
Tokyo

I. Neill Reid and Suzanne L. Hawley

New Light on Dark Stars

Red Dwarfs, Low-Mass Stars, Brown Dwarfs

Springer

Published in association with
Praxis Publishing
Chichester, UK

Dr I. Neill Reid
Formerly of Edinburgh University
Royal Observatory, Edinburgh
Scotland

Dr Suzanne L. Hawley
Associate Professor, Department of Physics and Astronomy
Michigan State University
East Lansing, Michigan, USA

SPRINGER–PRAXIS BOOKS IN ASTROPHYSICS AND ASTRONOMY
SUBJECT *ADVISORY EDITOR*: John Mason B.Sc., Ph.D.

ISBN 1-85233-100-3 Springer-Verlag Berlin Heidelberg New York

British Library Cataloguing-in-Publication Data
  Reid, I. Neill
    New light on dark stars : red dwarfs, low-mass stars, brown
    dwarfs. – (Springer-Praxis books in astrophysics and astronomy)
    1. Red dwarfs   2. Brown dwarfs   3. Stars – Luminosity function
    I. Title   II. Hawley, Suzanne L.
    523.8'8

Library of Congress Cataloging-in-Publication Data
  Reid, I. Neill, 1957–
    New light on dark stars : red dwarfs, low-mass stars, brown dwarfs/I. Neill Reid
    and Suzanne L. Hawley.
       p.   cm. – (Springer–Praxis books in astrophysics and astronomy)
    Includes bibliographical references and index.
    1. Dwarf stars.   2. Dark matter (Astronomy)   I. Hawley, Suzanne L., 1960–
  II. Title.   III. Series.
    QB843.D9 R45 2000
    523.8'8 – dc21   99-049519

Copy editor: R. A. Marriott
Cover design: Jim Wilkie
Typesetting: Originator, Great Yarmouth, Norfolk, UK

Printed on acid-free paper supplied by Precision Publishing Papers Ltd, UK

In memory of Iain Reid

Table of contents

Preface

Perhaps the most common question that a child asks when he or she sees the night sky from a dark site for the first time is: 'How many stars are there?' This happens to be a question which has exercised the intellectual skills of many astronomers over the course of most of the last century, including, for the last two decades, one of the authors of this text. Until recently, the most accurate answer was 'We are not certain, but there is a good chance that almost all of them are M dwarfs.' Within the last three years, results from new sky-surveys – particularly the first deep surveys at near-infrared wavelengths – have provided a breakthrough in this subject, solidifying our census of the lowest-mass stars and identifying large numbers of the hitherto almost mythical substellar-mass brown dwarfs.

These extremely low-luminosity objects are the central subjects of this book, and the subtitle should be interpreted accordingly. The expression 'low-mass stars' carries a wide range of meanings in the astronomical literature, but is most frequently taken to refer to objects with masses comparable with that of the Sun – F and G dwarfs, and their red giant descendants. While this definition is eminently reasonable for the average extragalactic astronomer, our discussion centres on M dwarfs, with masses of no more than 60% that of the Sun, and extends to 'failed stars' – objects with insufficient mass to ignite central hydrogen fusion. From this perspective, the Sun is a high-mass star.

The physical quantity underlying our curious child's question is the contribution made by stars (and brown dwarfs) to the mass density in the Galaxy. This quantity is derived by integrating the stellar and substellar mass function – the number of objects per unit volume and unit mass. Rephrasing matters, a crucial question is whether low-mass stars and/or brown dwarfs can contribute enough dark matter to solve any of the missing-mass problems present in current astronomy. The answer happens to be 'no', but the question is sufficiently important, and the route to the answer, pitfalls and all, is sufficiently interesting to merit more than the brief discussion afforded these objects in most textbooks.

M dwarfs are also of interest in their own right, with low-temperature atmospheres dominated by complex molecular opacities, semi-degenerate but fully convective interiors, and a tendency toward violent activity in their outer chromospheres and coronae. Beyond spectral class M are the cooler L dwarfs, comprising both the lowest-mass stars and brown dwarfs. Identified and recognised as a separate class only within the last two years, they exhibit quite remarkable spectra induced by the combined effects of dust formation and 'weather' within their atmospheres. At even lower temperatures, brown dwarfs of spectral class T are closer cousins to the planetary gas giants of our Solar System than to the Sun.

Our text is written at a level comprehensible to an advanced undergraduate physics major, and is intended as a reference book for post-graduate students and researchers – including the authors themselves. In these chapters we consider low-mass stars and brown dwarfs both as individual objects and as representatives of the parent Galactic populations. Our perspective is that of the observational astronomer rather than the theorist. Although theory plays an indispensible role both in interpreting the available data and in suggesting new avenues for future work, this subject has been driven empirically. With that consideration in mind, the first chapter provides a broad introduction to the range of observational techniques employed in studying these systems. It also happens to collect together most of the formulae (normally dispersed in a variety of sources) which the authors have found useful in preparing for their own observations.

The succeeding four chapters discuss the intrinsic properties of low-mass stars and brown dwarfs. Chapter 2 sets the stage with a thorough discussion of their observational characteristics; Chapter 3 outlines their internal structure; Chapter 4 reviews analyses of their complex atmospheres; and Chapter 5 describes observations and theoretical models of the magnetically-driven activity generated in the high-temperature chromosphere and coronal regions.

Later chapters consider M, L and T dwarfs as members of the Galaxy. Chapter 6 provides the context, with an overview of our current understanding of Galactic structure. Chapters 7 and 8 address the numbering and weighing of stars within the Galactic disk population. Chapters 9 and 10 extend discussion to the search for brown dwarfs and for planetary-mass companions to main-sequence stars – two long-running quests which finally met with success in the closing years of the twentieth century. Finally, Chapter 11 describes the nature and number of the low-mass stars in the metal-poor Galactic halo.

Our aim throughout is to provide a thorough discussion of current knowledge in each of these areas, and to that end each chapter includes an historical overview of the development of that knowledge. Some topics must necessarily receive only cursory discussion within the limits of a single volume. We have, however, appended to each chapter an extensive, although not exhaustive, list of references to papers in the scientific literature which can serve as a starting point for more detailed bibliographic research. Of necessity, these references – and the theories and observations outlined in the book – are current to early 1999. Given the explosive rate of discoveries in brown dwarf and extrasolar planet research between 1995 and the present date, it would be rash to predict what may emerge in the early years of the twenty-first century. The interested reader will be well advised to keep a weather eye on contemporary reviews.

Acknowledgements

Most of the writing of this book was completed while Reid was at the California Institute of Technology, Pasadena, and Hawley was at the Space Sciences Laboratory, University of California, Berkeley, and at the University of Washington, Seattle, and we acknowledge access to the facilities at those establishments. We would also like to thank the staffs at the Palomar and Keck Observatories for their able assistance in acquiring many of the observations discussed here. Particular credit goes to the Palomar Sky Survey team, responsible for completing the Second Palomar Sky Survey (POSS II). One of the authors (Reid) has been fortunate enough to be a participant in a Core Science project associated with the 2MASS sky-survey, undertaken by the University of Massachusetts and the Infrared Processing and Analysis Center (IPAC) at Caltech. We acknowledge the efforts of all associated with this very successful project, which (as Chapter 9 illustrates) has had a very substantial impact on our understanding of both low-mass stars and brown dwarfs.

During the course of this undertaking we have profited from discussions with many of our colleagues. Amongst those whom we would like to particularly thank are France Allard and Peter Hauschildt, for predictions, diagrams and discussions about their recent grids of model atmospheres; John Gizis, for comments, criticism and many figures, particularly in Chapter 11; Davy Kirkpatrick, Jim Liebert, Dave Monet, Adam Burgasser and Conard Dahn of the 2MASS ROT (Rare Objects Team) for stimulating discussions and new discoveries in the low-mass star/brown dwarf field; David Koerner, for reviewing Chapter 10; Lynne Hillenbrand, for figures and comments on Chapters 7, 8 and 9; Manuela Zoccali, for discussions on globular cluster luminosity functions; Bill Abbett, George Fisher, Jeff Valenti and, especially, Chris Johns-Krull, who also provided several figures, for consultations on stellar activity; Adam Burrows, for supplying data from theoretical models and for related discussions; Jeremy Mould, for stimulating interest in M dwarfs, and general joviality; Susana Deustua and Jamie Schlessman for comments and helpful criticism. We would also like to thank the following scientists for providing figures and pre-publication results: Sallie Baliunas, Tom Fleming, Hugh Jones, Anita Krishnamuthi, Greg Laughlin, Jenny Patience, Don Terndrup, and Chris Tinney.

As a technical credit, NASA's Astrophysics Data System proved an invaluable resource for identifying and locating relevant literature. Needless to say, remaining errors and omissions are the responsibility of the authors.

We are grateful to Clive Horwood (Chairman of Praxis) for his patience during the three-year germination of this manuscript, and to Bob Marriott for ably editing the text. Finally, we would like to thank our respective families: Helen and Dwight Hawley, Jim, Rosa and Tom Hughes, on the one side; and Elizabeth Reid, Jamie and the four-legged ones, Ruaraidh, Gwen and Daisy, on the other. They provided tolerance and support during the completion of this manuscript.

List of illustrations, colour plates and tables

1

Astronomical concepts

1.1 INTRODUCTION

M dwarf stars rank among the least spectacular constituents of the local stellar population, yet they represent almost half of the total mass locked up in stars within the Milky Way galaxy. As a result, apart from being interesting in their own right, the global characteristics of these stars have wider implications for studies of Galactic structure and star formation both within our own Galaxy and elsewhere. The main aim of this book is to summarise our current understanding of the physical structure and properties of these stars. That knowledge has been garnered through applying a variety of commonly-used astronomical techniques. This introductory chapter provides an outline of the basic concepts involved, placing a particular emphasis on practical observation.

1.2 THE ELECTROMAGNETIC SPECTRUM

Most astronomical observations are made through the Earth's atmosphere, which is not perfectly transparent. Radiation is absorbed and scattered with various degrees of efficiency at different wavelengths (Figure 1.1). The relevant physical process depends on the energy of the incoming radiation. Wavelength, frequency and energy are related by the equation

$$E = h\frac{c}{\lambda} = h\nu \tag{1.1}$$

where ν is the frequency, λ is the wavelength, c is the speed of light, and h is Planck's constant.

At the longest wavelengths, $\lambda > 100\,\mathrm{m}$, radiation incident on the Earth's atmosphere is completely reflected by the ionosphere. Reflection from the underside of the same layer allows long-wavelength radio stations to broadcast beyond their local horizon. The atmosphere is relatively transparent to higher-frequency radio waves,

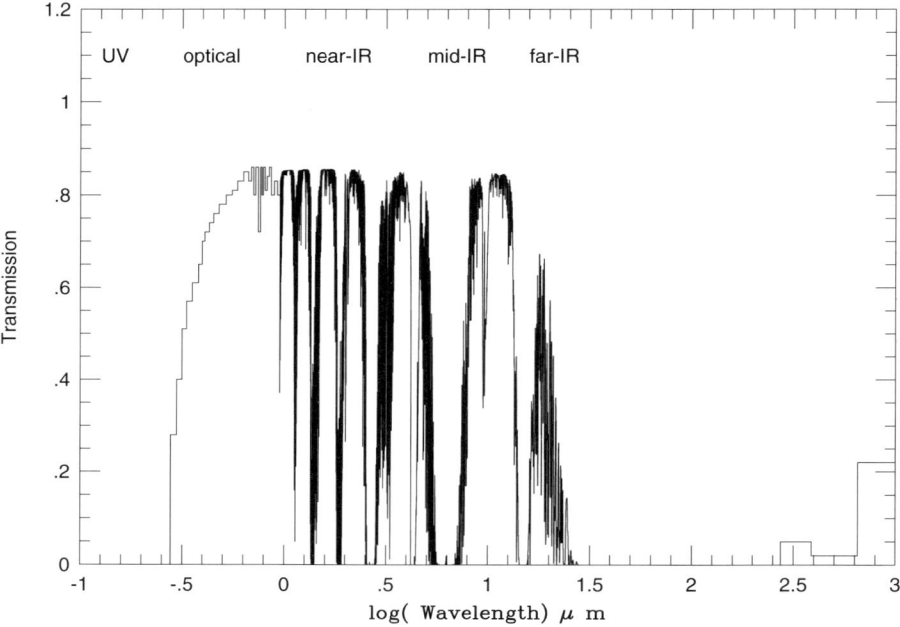

Figure 1.1. Atmospheric transmission at an altitude of 4 km, equivalent to the altitude of Mauna Kea Observatory. (Long-wavelength data courtesy of Peter Stockman.)

spanning wavelengths between 100 m and 1 cm. At wavelengths shorter than 1 cm, electromagnetic radiation has sufficient energy to excite vibrational motion in molecules such as CO, OH and, particularly, H_2O. Those molecules absorb a substantial fraction of the incident millimetre and infrared (1 cm to 1 μm) radiation, although there are regions of moderate transparency, notably at $\sim$5, 3.4, 2.2, 1.6 and 1.25 μm.

Shortward of 1 μm, the OH radical contributes a number of absorption bands (as well as significant emission) and there are two strong O_2 bands at $\lambda \sim$ 7500 Å (the A-band) and $\lambda \sim$ 6800 Å (B-band) (see [R2]). However, by $\lambda \sim$ 7000 Å (7 × 10^{-7} m), the photon energy exceeds that of almost all molecular vibrational transitions. As a result, there is a second highly-transparent window between that wavelength and $\sim$3250 Å – the visible region of the spectrum. This matches the peak in the solar energy distribution, so it is no coincidence that our own (and most other terrestrial) eyes are most sensitive to light at those wavelengths.

The optical window is not completely transparent, since the radiation has a wavelength comparable to the size of aerosol particles in the atmosphere. As a result, scattering occurs, particularly at blue wavelengths, since Rayleigh scattering has a $\lambda^{6/5}$ dependence. The latter wavelength dependence accounts for the fact that the sky is blue. The proportion of the incident radiation that is scattered depends on the total path-length through the atmosphere. Thus, an object appears both fainter and redder (since blue light is scattered preferentially) when near the horizon.

At shorter wavelengths, the high-energy radiation is capable of ionising atoms, and is absorbed completely. Thus, the atmosphere is totally opaque to ultraviolet radiation, $\lambda \sim 3250$ to $120\,\text{Å}$ (100 eV) and soft X-rays ($\lambda > 0.25\,\text{Å}$ or E $< 50\,$keV). At even higher energies, Compton scattering prevents the radiation from penetrating the atmosphere, although γ-rays (E $>1\,$MeV) can be detected from high-energy particle showers as they mutually annihilate through pair-production in the upper atmosphere.

1.3 POSITIONAL ASTRONOMY

A well-defined co-ordinate system is an essential requirement if individual objects are to be identified in a unique manner. In astronomy, objects are observed on the celestial sphere with positions defined in spherical co-ordinate systems, measured in angular units (degrees, radians). The latter are directly analogous to the system of longitude and latitude used to determine positions on the surface of the Earth; indeed, the primary celestial system is a direct projection of the geographic system.

There are four main celestial systems, each defined with reference to a fundamental reference plane which passes through the centre of the sphere (the observer). The circle defined by where this plane intersects the celestial sphere (AFBC in Figure 1.2) is a great circle: defining the diameter of the sphere as r = 1 unit, the length of this circle is 2π, the maximum possible. Any circle defined by a plane which does not pass through the centre of the sphere (such as DSE, parallel to ABC, in Figure 1.2) has a smaller circumference and is known as a small circle. Each great circle has two poles – the two diametrically-opposed points on the celestial sphere which lie 90° from every point on the great circle. In Figure 1.2, P and Q identify the two poles of the great circle ABC.[1]

Any circle drawn through both poles is a great circle, intersecting the reference circle ABC at an angle of 90°. Choosing one 180° segment of a polar circle (for example, PBQ in Figure 1.2) as a second reference then defines an orthogonal co-ordinate system. Consider the point S in Figure 1.2. This point is defined uniquely by two angles: the angle between the reference polar circle, PBQ, and the polar circle passing through S (the angle BPS, which is also the angular length of the arc BF); and by the angular distance PS. In most systems, the latter angle is defined with respect to the fundamental circle, so the angle becomes FS = 90° − PS. In the case of the terrestrial system of latitude and longitude, the fundamental circle is the equator, while the Greenwich meridian defines the reference polar circle. The angle BPS then defines the longitude of a given location on the Earth's surface, while FS gives the latitude.

---

[1] Since P and Q are the poles of only one great circle, ABC, it is clear that a given co-ordinate system can be defined unambiguously either by the position of the reference great circle or by the position of the poles of that great circle.

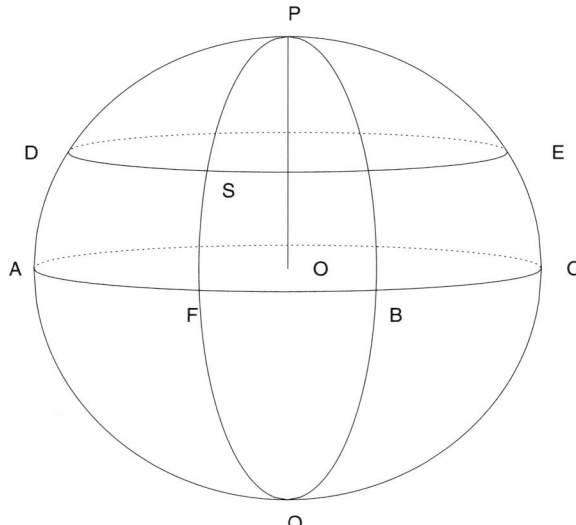

Figure 1.2. Great circles and small circles.

Two other properties of spherical geometry are worth bearing in mind. First, there is only one great circle which passes through any two points on a sphere, unless the two points form the poles of a great circle. Second, just as a straight line is the shortest distance between two points on a flat surface, the shortest distance between two points on a sphere is measured along the great circle joining those two points.

1.3.1   Co-ordinate systems

The first three systems

The alt-azimuth system   The first astronomical co-ordinate system is the altitude/azimuth system. The fundamental plane, the horizon, is defined by the poles: the zenith, the point directly overhead, and the nadir, the diametrically opposed pole, directly underneath the observer. Semi-great circles passing through the zenith are known as vertical circles, and the vertical circle passing through the horizon due north of the observer is the principal vertical circle, which defines the zero point of the azimuth scale. The prime verticals are the vertical circles due east (azimuth = 90°) and due west (azimuth = 270°), and the altitude of a given object is the angular distance above the horizon, measured along the appropriate vertical.

The zenith distance, z, is the complement of the altitude. This angular distance provides a measure of the path-length through the Earth's atmosphere, a quantity important both for determining accurate fluxes (see section 1.5.2) and the degree of refraction. The latter is approximately represented by

$$\zeta = R \tan(z) \tag{1.2}$$

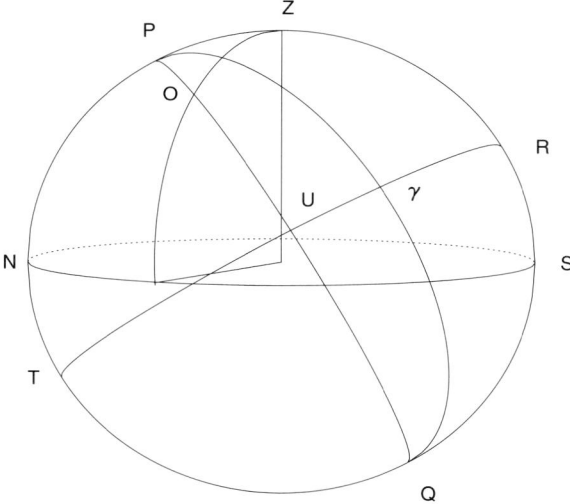

Figure 1.3. Alt-azimuth and equatorial co-ordinate systems.

where ζ is the angle of refraction and R is the constant of refraction, usually expressed in arcseconds. Since blue light is refracted to a larger extent than red light, R is wavelength dependent, with $R \sim 59''4$ at 3500 Å, $R \sim 58''2$ at 5000 Å, $R \sim 57''7$ at 6500 Å and $R \sim 57''5$ at 9000 Å [F1]. This variation leads to differential chromatic refraction, with a point source dispersed into a mini-spectrum at low altitudes.[2]

The equatorial system   All celestial objects rise and set, and the trajectory of their motions depends on the (terrestrial) latitude of the observer. Thus, altitude and azimuth are both time- and location-dependent, and while the alt-azimuth system is used operationally by most large, modern telescopes, it cannot provide a universal reference system. The equatorial system fulfils that purpose by taking the reference frame as the projection of the Earth's equator onto the celestial sphere. The poles of the latter system are therefore the projection of the north and south terrestrial poles.

   If we consider observations made from a given terrestrial location, latitude ϕ, then the zenith distance of the pole, PZ in Figure 1.3, is $(90° - \phi)$. As the Earth rotates, an object, O, describes a circle at constant angular distance, PO, from the pole P. The complement of this angle, the angular distance of the star from the celestial equator as measured along a great circle (angle OU in Figure 1.3), defines the declination, the celestial equivalent of latitude.

   A second reference circle is required to specify the position of O on the celestial sphere. This is provided either by the observer's meridian, the semi-great circle

---

[2] This effect was actually used to identify quasars on Palomar Sky Survey blue plates taken at moderate to large zenith distances: the strong ultraviolet continuum in low-redshift QSOs led to their having more extended images than stars.

PZRSQ (where S is the southern point on the observer's horizon), or by the semi-great circle PγQ. In the former case, the east/west reference line is fixed relative to the observer, and angle ZPO defines the hour angle (HA), measured east or west of the meridian in either degrees or, more usually, hours, minutes and seconds of time. West of the meridian, the HA is the time since the object crossed the meridian (transit); east of the meridian, the HA is the time until transit. Thus, the hour angle of an object is changing continuously.

In contrast to the circle PSQ, PγQ is a fixed reference circle on the celestial sphere. γ is a reference point on the celestial equator, the first point of Aries (see below). As a result, the right ascension (RA) of O, the angle γPS, is constant for the equatorial system RγUT, and defines an unambiguous position. As with hour angle, right ascension is usually measured in units of time, increasing as one moves east of γ. Right ascension and declination are notated as α and δ, respectively.

The ecliptic system   The third reference system is defined by the plane of the Earth's orbit, traced out by the ecliptic, the apparent motion of the Sun on the celestial sphere. The angular separation of the ecliptic and equatorial poles is 23° 27′, and ecliptic co-ordinates are specified in latitude (β) and longitude (λ). The ecliptic intersects the celestial equator at two points (nodes) 180° degrees apart: the vernal equinox where the Sun moves from south to north of the equator, and the autumnal equinox where the Sun returns to the southern celestial hemisphere. The former node is the first point of Aries, denoted γ as mentioned above (Figure 1.3). This point serves as the reference for both ecliptic longitude and right ascension.

Precession   The equatorial and ecliptic systems are not invariant in time. Since the Earth's rotational axis is inclined with respect to both the poles of the ecliptic and the poles of the lunar orbit, and since the Earth is not a perfect sphere, an imbalance of gravitational forces due to both the Moon and the Sun (an applied torque) leads to precession. The main result of luni-solar precession is that the Earth's pole describes a small circle, radius $\sim 23°$, about the ecliptic pole, with a period of $\sim 26,400$ years.[3] The vernal equinox moves backwards (towards negative longitude) along the ecliptic by $\sim 50''\,\mathrm{yr}^{-1}$. Thus, when giving a position in either equatorial or ecliptic co-ordinates, the reference equinox (by year) must be specified. Most catalogues list positions for equinox 1900, 1950 or 2000.

Confusion is often encountered in the astronomical literature over the meaning of the terms equinox and epoch. Equatorial (and ecliptic) co-ordinates are specified with reference to the position of the vernal equinox on a given date; hence, they are cited as, for example, equinox 1950 co-ordinates. However, stars in the immediate solar neighbourhood can possess appreciable angular motion (proper motion) due to their velocity relative to the Sun. One can allow for those motions and correct the position to a given date (epoch), but still maintain the same co-ordinate system

---

[3] There are additional shorter-term effects, notably nutation, which has a period of ~ 18.6 years. Full details are given in [M1] and [S1].

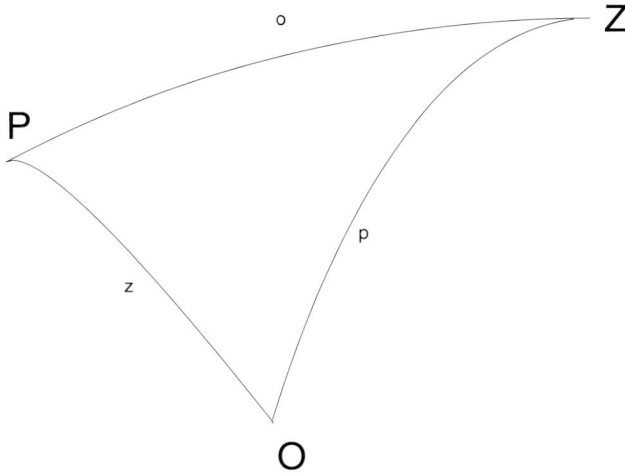

Figure 1.4. Spherical triangles.

(equinox). In other words, the equinox listed for a given object specifies the position of the reference system and the epoch specifies the date when the object was at the stated co-ordinates.

Spherical trigonometry

Before defining the fourth co-ordinate system, it is useful to outline the basic formulae of spherical trigonometry which allow one to manipulate co-ordinates and transform between different systems. The full derivation of these formulae is given in [S2], and we simply state the results here. Consider the spherical triangle PZO (Figure 1.4). We shall denote the inner angles as Z, P and O, while the angular lengths of the three sides are designated z, p and o. The two most useful formulae are the sine rule

$$\frac{\sin O}{\sin o} = \frac{\sin P}{\sin p} = \frac{\sin Z}{\sin z} \qquad (1.3)$$

and the cosine rule

$$\cos o = \cos z \cos p + \sin z \sin p \cos O \qquad (1.4)$$

There are comparable expressions for $\cos z$ and $\cos p$. Also occasionally useful are the mixed formula

$$\sin z \cos O = \cos o \sin p - \sin o \cos p \cos Z \qquad (1.5)$$

and the four-parts formula

$$\cos z \cos O = \sin z \cot p - \sin O \cot P \qquad (1.6)$$

Thus, if we know the hour angle and declination of a given star, we can determine the altitude using the cosine rule, since z = 90°–δ; o = 90°–φ; **P** is the hour angle; and p = 90° – altitude. Given p, the azimuth (**Z**) can be determined by applying the sine rule. Angular distances between sources can also be determined using these formulae.

Galactic co-ordinates

Our Galaxy is a disk system (Chapter 6), and fortified by trigonometry, we can now define the fourth reference system. Galactic longitude and latitude (b, l) are measured with respect to the plane of the Milky Way. The present system (l^{II}, b^{II}) dates from 1958, and replaces the (l^{I}, b^{I}) system defined by Ohlsson [O1]. Under the new system, the north Galactic pole is defined (by the International Astronomical Union) to be at position $\alpha = 12h\,49m$, $\delta = +27°\,24'$ (1950 equinox), while the reference great circle at $l^{II} = 0°$ is defined to lie at a position angle of 123° from the great circle through the north Galactic and north celestial poles. These criteria place the origin of the (l^{II}, b^{II}) co-ordinate system at $\alpha = 17h\,42m\,26s.603$, $\delta = -28°\,55'\,0''.445$ for equinox 1950 [L4].[4] The Galactic plane intersects the celestial equator at $l = 33°, 23°$. Transforming (α, δ) to (l, b) is more complicated than moving between equatorial and ecliptic co-ordinates, since the two systems do not share the same zero-point. However, the following formulae can be applied:

$$\cos(b)\cos(l - 33°) = \cos(\delta)\cos(\alpha - 282°.25)$$

$$\cos(b)\sin(l - 33°) = \cos(\delta)\sin(\alpha - 282°.25)\cos(62°.6) + \sin(\delta)\sin(62°.6)$$

$$\sin(b) = \sin(\delta)\cos(62°.6) - \cos(\delta)\sin(\alpha - 282°.25)\sin(62°.6)$$

$$\cos(\delta)\sin(\alpha - 282°.25) = \cos(b)\sin(l - 33°)\cos(62°.6) - \sin(b)\sin(62°.6)$$

$$\sin(\delta) = \cos(b)\sin(l - 33°)\sin(62°.6) + \sin(b)\cos(62°.6) \quad (1.7)$$

These equations allow co-ordinates to be transformed between the equatorial and Galactic systems.

1.3.2 Stellar astrometry

Measuring accurate positions of celestial objects is the astronomical subdiscipline with the longest pedigree. Indeed, astrometry *was* astronomy until the latter half of the nineteenth century. From the perspective of studying low-mass dwarfs, two astrometric parameters are of prime importance: parallax and proper motion.

Stellar parallax is the apparent motion of a star due to our changing perspective as the Earth orbits the Sun. Measured against a reference frame of more distant objects, the target star describes an ellipse, the semi-major axis of which is the parallax angle, π, and the semi-minor axis $\pi \cos(\beta)$, where β is ecliptic latitude. The

---

[4] Sagittarius A*, the radio source associated with the Galactic centre, is actually at $\alpha = 17h\,42m\,02s$, $\delta = -28°\,47'\,6''.0$ (equinox 1950), or $l = 0°.06$, $b = +0°.15$.

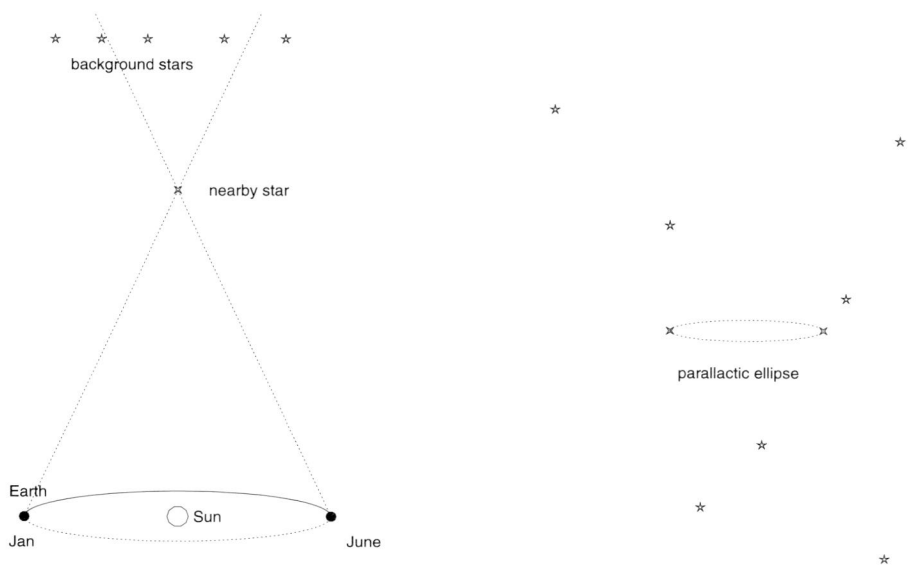

Figure 1.5. Stellar parallax. (*Left*) the change in perspective of an Earth-bound observer; (*right*) the parallactic ellipse traced on the sky by an idealised nearby star (four-point star) against the reference frame of (five-point) background stars. In reality, the motion would include a linear component due to proper motion.

ellipse is the projection of the Earth's orbit (Figure 1.5). Thus, a star at the ecliptic pole describes a near-circular path, while a star in the ecliptic (the plane of the Earth's orbit) exhibits linear motion.

The semi-major axis of the Earth's orbit is 149,597,870 km, defined as one astronomical unit (AU) [L5]. Since this distance, a_E, is much smaller than the distance to even the nearest star, the parallax of a star at distance r can be written using the small angle approximation as

$$\pi = \frac{a_E}{r} \tag{1.8}$$

The units used in astronomical measurement follow from this equation: a star with $\pi = 1$ arcsec is defined to be at a distance of 1 parsec, equivalent to 3.2616 light years or 206,265 AU. The first successful measurements of stellar parallaxes were by Bessel (61 Cygni, 1838), Henderson (α Centauri, 1839) and Struve (Vega, 1840). The nearest known star to the Sun, Proxima Centauri, an M dwarf companion of α Centauri, has a parallax of only $0\rlap{.}''772$. The current limit of precision for parallax measurement is 0.5–1 milliarcseconds (mas).

An important point to note is that *all* objects in a given direction possess shared parallactic motion, but with amplitudes which depend on the reciprocal of their distance. Parallax measurements are generally made from time-series of direct images, taken either by photographic plates or, more recently, by digital detectors such as CCDs (see Section 1.4.2 and [M1]). Reference stars are limited to separations of less than a degree from the target. Their own parallactic motion, in phase with the target, leads to the measured apparent parallax of the target, π, being an under-estimate of the true parallax, π_o. The offset can be corrected if other techniques are used to estimate distances to the reference stars, which typically lie at distances of 500–3,000 parsecs. The recent Hipparcos astrometric space mission [E2] circum-vented this problem by using a compound telescope to project two 1 square degree areas of sky, separated by an angle of $\sim 58°$, onto the same focal plane. This technique – rendered impossible in ground-based observations due to atmospheric distortions – allows measurement of the angular separations of stars with very different parallactic motions, so that absolute astrometric parameters can be determined [V1], [K3].

Parallax is a cyclical motion. Proper motion, in contrast, is secular and cumulative. Stellar proper motions reflect the changing relative positions of the Sun and the target star due to their changing position in the Galaxy. Writing the annual proper motion as μ, then

$$\mu = \frac{V_T}{\kappa r} \ \text{arcsec yr}^{-1} \tag{1.9}$$

where V_T is the heliocentric transverse velocity of the target in km s^{-1}, r is the distance in parsecs and $\kappa = 4.74$. As with parallax, proper motions are generally measured with respect to a reference grid of more distant stars which are typically members of the Galactic disk, and a small correction is usually required to transform the target measurement to an absolute reference frame. However, unlike parallax measurements, faint (15th–18th magnitude) galaxies can be used to define the reference frame for measurements, and absolute motions determined directly [V2].

Many of the nearest stars have proper motions of 1 arcsec or more per year. Figure 1.6 shows an example, the low-luminosity M dwarf LHS 2924, photographed on 20 May 20 1950 by the Oschin Schmidt telescope as part of the original Palomar Observatory Sky Survey, POSS I, and again on 2 May 1992 in the course of the second Palomar Sky Survey, POSS II. With $\mu = 0\rlap{.}''802 \ \text{yr}^{-1}$, the displacement is obvious. Proper-motion surveys based on this type of plate material have been responsible for identifying many of the lowest-luminosity dwarfs currently known to lie in the vicinity of the Sun.

1.3.3 Stellar kinematics

The average space velocity and the velocity dispersions of a group of stars provide a measure of, respectively, the mean rotational velocity and the fraction of kinetic energy resident in random, rather than ordered, motion. Proper motions measure

Figure 1.6. Images of a 5 × 5 arcmin region from the STScI digitisation of POSS I (*left*, 1950) and POSS II (*right*, 1992) photographic plates. The low-luminosity proper motion star LHS 2924 is centred in the POSS II image.

tangential displacement, while radial velocities supply the third co-ordinate. The radial velocity observed at any epoch includes the projected contribution from both the orbital motion of the Earth around the Sun (or, rather, the combined motion of the Earth around the Earth–Moon barycentre and the motion of the latter around the barycentre of the Solar System) and the Earth's diurnal rotation. In round numbers, the former amounts to $\pm18.5\,\mathrm{km\,s^{-1}}$, while the latter is no more than $\pm450\,\mathrm{m\,s^{-1}}$. Both effects must be taken into account in precision velocity work, such as searches for planetary-mass companions (see Chapter 10).

Proper motions are usually measured in the (α, δ) equatorial system. However, it is more useful for statistical purposes to transform these to the Galactic (b, l) system

$$\mu_l = \mu_\alpha \cos(\phi) + \mu_\delta \sin(\phi),$$
$$\mu_b = -\mu_\alpha \sin(\phi) + \mu_\delta \cos(\phi) \tag{1.10}$$

where ϕ is the angle between the direction towards the north celestial pole and the north Galactic pole. Simple application of spherical trigonometry shows that

$$\cos(\phi) = \frac{\cos(\delta)\tan(\delta_0) - \sin(\delta)\cos(\alpha - \alpha_0)}{\sin(\alpha - \alpha_0)} \tag{1.11}$$

where (α, δ) are the equatorial co-ordinates of the star in questions, and (α_0, δ_0) are the equatorial co-ordinates of the north Galactic pole.

If the distance of the programme object is known, then the proper motions can be transformed to velocities, giving the heliocentric velocity triad (V_l, V_b, V_r). For Galactic structure studies it is more useful to transform these velocities to a (right-handed) co-ordinate system with orthogonal axes directed towards the Galactic centre (defined as the U velocity), in the direction of Galactic rotation ($l = 90°$,

$b = 0°$, V velocity) and perpendicular to the Galactic plane (W velocity). The choice of axes is dictated by the fact that the Galactic disk is a flattened, nearly axisymmetric system, with ordered rotational motion (see Chapter 6). The observed (V_l, V_b, V_r) are transformed to (U, V, W) using

$$U = V_r \cos(l) \cos(b) - V_b \cos(l) \sin(b) - V_l \sin(l)$$

$$V = V_r \sin(l) \cos(b) - V_b \sin(l) \sin(b) + V_l \cos(l)$$

$$W = V_r \sin(b) + V_b \cos(b) \tag{1.12}$$

In determining stellar kinematics, stars are usually grouped together based on similarity in properties such as broadband colours or spectral type. The mean heliocentric velocities in U, V and W measure the solar motion, while the velocity distribution is usually characterised as orthogonal Gaussian dispersions (the Schwarzschild velocity ellipsoid). Traditionally, velocity dispersions have been measured for the (U, V, W) axes as $(\sigma_U, \sigma_V, \sigma_W)$. However, in some cases, notably for young stars, the best-fit velocity ellipsoid is misaligned with the (U, V, W) axes. The offset from the radial (U) axis is described as the 'vertex deviation'. Chapter 6 presents recent observational analyses of Galactic stellar kinematics.

1.4   TELESCOPES AND DETECTORS

1.4.1   Telescopes

The two main properties of a telescope are the diameter of the primary optical element (D), which defines both the light grasp and the limiting angular resolution, and the focal length, f, which determines the plate-scale, the number of arcsec per millimetre in the focal plane. The angular resolution, Θ, is set by the diffraction limit:

$$\Theta = \frac{\lambda}{D} \times 206{,}265 \text{ arcsec} \tag{1.13}$$

where λ is the wavelength of the observations, and 206,265 is the number of arcseconds in a radian. Thus, the human eye, with an effective aperture of 5 mm, has an angular resolution of ~ 20 arcsec; an 8-inch (20-cm) telescope can resolve double stars separated by 1 arcsec; and the 200-inch (5-metre) Hale telescope at Mount Palomar has a diffraction limit of 0.02 arcsec at visual wavelengths, although atmospheric turbulence (seeing) prevents this resolution being attained in direct observations.

The focal length, f, of a telescope is the distance between the primary optical element and the principal focus. The plate-scale at that focus is given by

$$p = \frac{206{,}265}{f} \text{ arcsec mm}^{-1} \tag{1.14}$$

The focal ratio of the system is the ratio of the focal length and the diameter of the primary, $R_f = f/D$. Thus, the 200-inch Hale telescope has a primary mirror with a focal ratio of 3.3 (written f/3.3), corresponding to a focal length of 55 feet (16.76 metres) and a plate-scale of 12.3 arcsec mm^{-1}.

The majority of large telescopes built for astronomical research (and, indeed, for amateur work) are reflecting telescopes. The main optical element is a mirror made of low-expansion glass and coated with a thin, highly-reflective layer of aluminium or gold. The primary advantages of a reflector are ease of construction, and cost. The objective lens in a refracting telescope must be edge-supported, and therefore must be sufficiently rigid to minimise deformation as the telescope is slewed. The structural strength of glass therefore sets an upper limit of ~ 1 m to the diameter of refractors. In contrast, with reflecting optics the primary mirror is back-supported, and it is therefore possible to make larger-aperture telescopes – up to 8 m for monolithic mirrors (the Gemini telescopes, the Magellan telescopes, and the ESO Very Large Telescopes) and 10 m for segmented-mirror telescopes (the Keck telescopes on Mauna Kea).

Not only is it substantially more expensive to fabricate high-quality transmitting optics than to make reflecting optics of the same aperture, but mirrors can be ground to faster optical ratios, leading to shorter focal-length telescopes, and hence smaller enclosures (domes). Each of the Keck 10-m telescopes, for example, has a primary mirror with a focal ratio f/2, producing a focal length of only 20 m, and allowing the telescope to be accommodated within a dome little larger than that of the Palomar 200-inch. On the other hand, the largest refractor constructed, the 40-inch Yerkes telescope at Williams Bay, Wisconsin, has a focal ratio of f/19 and a focal length of 63 feet – 8 feet longer than the 200-inch telescope. Very few large refracting telescopes have been constructed since the end of the nineteenth century.

Until recently, most telescopes were mounted equatorially: that is, a rotation axis is aligned with the Earth's polar axis, so motion due to diurnal rotation can be eliminated by tracking in only one co-ordinate (RA). However, it is simpler mechanically to construct an altazimuth telescope, in which the rotational axis is aligned with the local gravitational field. Advances in computer technology now permit accurate tracking in two co-ordinates, taking full account of effects such as telescope flexure and refraction, and most telescopes built in the last 10 years have altazimuth mounts.

The standard telescope design is based on the Cassegrain system: a secondary mirror (B) brings the light to a focus (C) at a point below the primary (A) (Figure 1.7). This provides a stable mount for moderate-sized instruments. In conventional telescopes, the distance AB is close to the focal length of the primary mirror, principally to minimise the diameter of the secondary mirror, and hence the size of the obstruction of the primary beam. This positioning requires that the secondary be suitably figured, which changes the effective focal ratio. Typical values are f/8, although ratios as high as f/400 are possible, with a consequent increase in the plate-scale. Alternatively, in many telescopes the secondary can be removed, and observations undertaken directly at the prime focus, P, at the primary focal ratio of f/2.5 or f/3. The latter focal station gives the largest plate-scale (most arcsec/mm), and thus provides easy coverage of large solid angles, but only limited space for

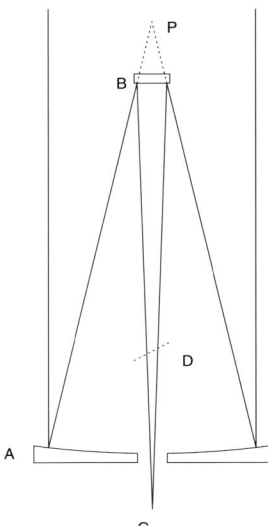 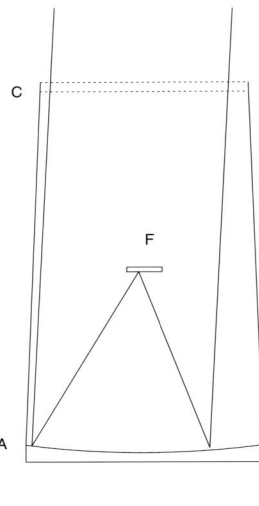

Figure 1.7. Schematics of optical telescopes: *left*, a conventional Cassegrain reflector; *right*, a Schmidt telescope.

mounting instruments. Large instruments obstruct the primary mirror and require a mechanically rigid telescope mount.

The classical Cassegrain design has a parabolic primary mirror. This has the disadvantage of producing off-axis images with significant optical aberrations: coma and spherical aberration [J1], [W2]. These can be corrected to some extent using a lens (corrector plate), but for only a limited field of view (~0°3 in diameter). The Ritchey–Chrétien modified Cassegrain design has a hyperbolic primary which, combined with the appropriate secondary, results in less distortion and a larger useable field of view (up to 1 square degree). Most telescopes built since 1950 have Ritchey–Chrétien primary mirrors (see [W2] for an historical review and thorough discussion of contemporary designs).

Two other foci deserve mention. Both require the introduction of a third mirror at D (usually with additional optics) and direct the beam to larger instruments placed at stationary locations. In equatorial telescopes, the tertiary mirror sends the beam down the polar axis to the coudé room. With a typical focal ratio of f/45, the plate-scale is usually ~1″/mm, enabling very high spectral resolution (see Section 1.6). In altazimuth telescopes, the equivalent focal positions are the Nasmyth platforms, mounted alongside the altitude rotational axis.

Conventional telescopes, even with Ritchey–Chrétien primaries, provide only a small field of view, covering less than 0.0025% of the sky in a single exposure. Such telescopes are ill-suited to surveys of substantial fractions of the celestial sphere. Specialised wide-field telescopes have been designed for that purpose, however, and the most successful of these is the Schmidt telescope, named after Bernhard Schmidt [O3]. The Schmidt primary (A in Figure 1.7, right) has a spherical surface, which would normally introduce spherical aberration. However, a corrector lens (C) placed at the centre of curvature of the primary, can 'pre-correct' the beam, leading to high-quality images over a field of view several degrees in diameter. The focal plane, F, is a spherically curved surface, and detectors (principally photographic plates) must be shaped to conform with that surface.

The two Schmidt telescopes which have had the largest impact on astronomical research are the 48-inch[5] Oschin Schmidt on Palomar mountain, California, and its southern counterpart, the 48-inch UK Schmidt telescope at Siding Spring, New South Wales. Both have an unvignetted field of view of $\sim 3°$ radius at a scale of $67''1/$mm, approximately 34 cm in diameter. This requires a correspondingly large detector which, at present, can be supplied only by photographic plates. The Palomar Schmidt was used to make the first deep survey of the northern skies, the Palomar Observatory Sky Survey (POSS I) undertaken between 1949 and 1957, extending to fainter than 20th magnitude in both blue and red passbands, and covering declinations north of $-33°$. This original atlas has been supplemented by new surveys utilising more sensitive plate material, undertaken by the UK Schmidt in the south (from 1975 to present day) and by Palomar in the north (POSS II – 1985 to 1999). Combined, these newer surveys provide coverage of the whole sky to 22nd magnitude, while matching recent plate material against POSS I plates allows searching for objects which have changed in position or magnitude.

Finally, with the advent of satellite observatories in the early 1960s, it has become possible to survey the sky at wavelengths which are absorbed or scattered by the Earth's atmosphere. Longer wavelength (infrared, far-infrared) and ultraviolet telescopes (such as IRAS, ISO and IUE) are similar in design to their ground-based optical counterparts. X-rays, however, are not reflected when they strike a mirror at normal incidence (perpendicular to the plane of the mirror) but *are* reflected when they strike at grazing incidence (that is, angles of only 1 or 2 degrees to the surface). If the mirrors have suitable parabolic and/or hyperbolic figures, then the radiation can be collected and focused as in an optical telescope [C2], [T1]. Unlike optical telescopes, these mirrors are often made from (or at least coated with) high atomic-number metals. Thus, the primary reflector in an X-ray telescope consists of a metal sleeve at the top end of the telescope (Figure 1.8), rather than a conventional mirror. Recent X-ray missions include Einstein, ROSAT and Chandra.

1.4.2   Detectors

Astronomical observations require the detection of light at extremely low intensity levels. This in turn demands detectors which have both high quantum efficiency (QE)[6] and low noise characteristics, so that a high fraction of the photons striking the detector are recorded, and the detector does not degrade the signal significantly

---

[5] The aperture of a Schmidt telescope is the diameter of the corrector plate, not the primary mirror.

[6] The QE of a detector is the fraction of incident photons which are detected as photoelectrons. Thus, if 100 photons fall on a detector and only five are detected, the QE is 5%.

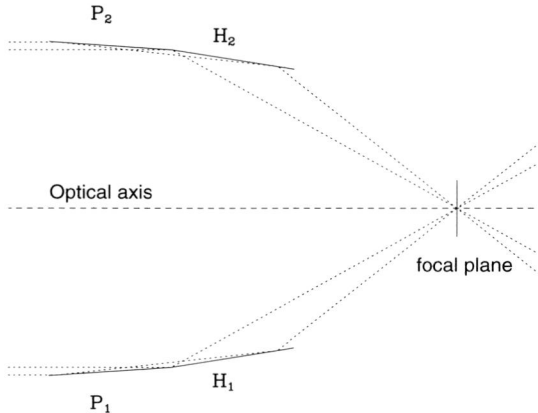

Figure 1.8. A schematic of an X-ray telescope. P_1 and P_2 are grazing-incidence paraboloidal mirrors, and H_1 and H_2 are hyperboloids.

in the act of recording. Photography remained a mainstay of astronomical imaging and spectroscopy for more than 150 years after its invention. Photographic emulsions are sensitive primarily to radiation in the wavelength régime 2000–9000 Å, while photographic plates can be manufactured at a size large enough to take full advantage of the wide field of view offered by Schmidt survey telescopes. The overall QE, however, is only $\sim$2–5%, and over the last two decades other more sensitive detectors have superseded photography in essentially all other roles.

Most modern optical and infrared instruments use semiconductors as detectors. These devices rely on the photoelectric effect to detect radiation. As mentioned in Section 1.2, a photon of frequency ν has energy $h\nu$. If this energy exceeds the binding energy of an electron of an atom in the semiconductor, then the electron is released. Counting the number of free electrons (photoelectrons) emitted within a given time provides a measure of the number of photons striking the photosensitive material; that is, the brightness of the light source. In the original photometers, a voltage was placed across the detector, and the signal detected as a current. Even the brightest sources generate very low currents, however. At optical wavelengths, the signal can be amplified using a photomultiplier tube. A photosensitive material, such as Cs_3 Sb or GaAs, held at a high negative potential, serves as the detector, the photocathode. The potential gradient (typically $\sim 1\,kV$) leads to photoelectrons striking a series of electrodes (dynodes) in a vacuum tube (Figure 1.9), releasing many secondary electrons at each strike and producing a pulse at the anode. This amplification results in a detectable current spike.

Proportional counters operate in a similar fashion in detecting X-rays, although the primary detector is a high atomic-number gas rather than a photocathode. Again, a high-energy photon interacts with an atom, leading to the emission of a photoelectron, which is accelerated in a potential field, colliding with other atoms to produce a cascade of secondary electrons. Unlike the optical photomultiplier, proportional counters can provide some information on the frequency of the

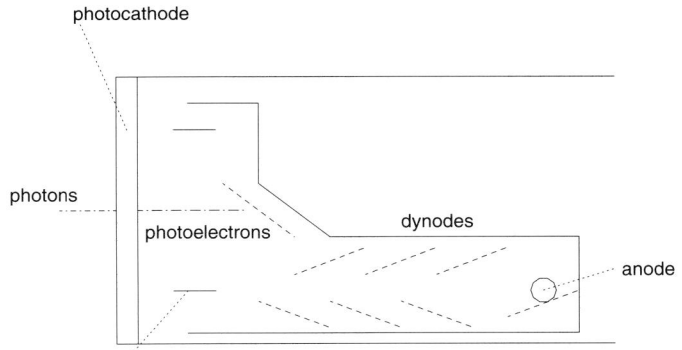

Figure 1.9. A schematic of a photon-counting detector.

photons detected. The higher the frequency, the greater the energy and the larger the number of photoelectrons produced. Comparable detectors, with limited spectral resolution, are being developed for work at optical wavelengths.

The first photon-counting detectors provided no positional information, only the total photon rate within an area of sky defined by either the field of view of the telescope or by an aperture mask in the focal plane. Two-dimensional imaging detectors were developed in the 1970s: notably, the Imaging Photon Counting System [B5], but also reticon arrays and multichannel-plate detectors [E1]. These provided spatial information either by accurate timing or by subdividing the photocathode into separate regions, each with its own photomultiplier chain. As with single-channel devices, each photon is detected separately, and these devices are very efficient in working at very low light-levels, particularly at blue wavelengths where the photocathodes are most efficient (20–30%). However, detecting each pulse of secondary electrons takes a finite time (typically a few microseconds), and the system cannot identify separate pulses due to other (coincident) photons arriving within that deadtime. Thus, the photon count-rate for bright sources is under-estimated; indeed, if the flux level is too high, the resultant high current can damage the detector. This coincidence problem means that photon-counting devices are ill-suited for broadband, wide-field imaging, where multiple sources inevitably lead to high photon-rates, while the low efficiency of photocathodes at wavelengths longer than $\sim 6000\,\text{Å}$ limits their utility at red spectral regions. Given these limitations, recent instrumental development has centred on charge-coupled devices (CCDs) – array detectors which possess neither of these shortcomings, and which have become the standard detector in optical and infrared astronomical equipment.

A CCD consists of an insulating material bonded onto a doped silicon semiconductor (Figure 1.10). A grid of electrodes is embedded within the insulator, with each electrode held at a positive potential. When the device is exposed to light, freely-moving photoelectrons are produced in the doped semi-conductor, and these (or the positive 'holes') migrate towards the local electrode. The efficiency is increased, and the noise contribution from random motion of

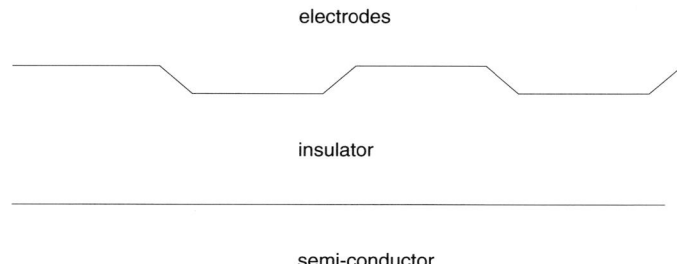

Figure 1.10. A schematic of a CCD array.

electrons (dark current) reduced, if the device is cooled to temperatures of ~ 100–$170\,\text{K}$, usually using liquid nitrogen or liquid helium as a refrigerant. The total charge collected by each electrode is a measure of the photon flux incident on that 'pixel' (picture element) of the detector. By manipulating the voltages appropriately, the charge collected in each pixel can be read out at the end of an exposure, and hence the image is reconstructed. Reading out the CCD chip contributes additional noise to the counts detected in each pixel (readout noise), but this is a small price to pay for quantum efficiencies of 30–40% at blue wavelengths and 80–90% at 6000–9000 Å.

The largest optical CCDs currently manufactured consist of arrays of $4{,}096 \times 4{,}096$ pixels, with individual pixels 12–15 μm in size; 2,048-square optical devices with 12–25 μm pixels are also in use on many telescopes, while the largest infrared arrays have 1,024 pixels to a side. None of the individual chips is much larger than $\sim 5.5 \times 5.5$ cm, so a single-CCD camera cannot provide the same areal coverage as a single photographic plate. However, the solid angle surveyed can be increased by constructing arrays of many CCDs, as has been done at the Japanese Kiso Schmidt [S1], and in the main camera used in the Sloan Digital Sky Survey [G1]. The latter camera includes 30 $2{,}048^2$ CCDs, which sparsely sample the sky in a 6×5 grid on the focal plane, scanning a strip of sky $\sim 2\overset{\circ}{.}5$ wide. With improvements of nearly a factor of ten in quantum efficiency over photographic plate material, these arrays can detect objects substantially fainter than the limiting magnitude of the large-scale photographic sky surveys.

The response of optical CCDs declines significantly beyond 9000 Å, but specialised infrared-array detectors can be used to cover the 1–5 μm region of the spectrum. These devices are fabricated from semiconductors such as InSb and HgCdTd (mercadtelluride), and the technology is less advanced than is the case for the visual-wavelength detectors, with few $1{,}024^2$ arrays having been completed. PbS and SiAs detectors are used at wavelengths beyond 5 μm, where low atmospheric transmission and the high background due to thermal emission make observations extremely difficult. For highest efficiency, the detectors have to be cooled to substantially lower temperatures than optical devices, usually requiring liquid helium as a coolant.

1.5   STELLAR PHOTOMETRY

The telescopes and detectors discussed in the previous section are used astronomically for two main purposes: first, direct imaging, usually for photometry (measuring the brightness of individual objects); and second, spectroscopy, determining the energy distribution as a function of wavelength. The following sections outline the fundamental principles involved in these observational measurements.

1.5.1   The magnitude system

The apparent brightness of astronomical objects is usually measured in units of *magnitude*. The system originated with Hipparchus' division of the naked eye stars into six subgroups, with the brightest stars grouped together in the 'first magnitude' and the faintest stars visible to the naked eye described as being of the 'sixth magnitude'. The human brain/eye combination tends to judge brightness differences as ratios, rather than linear differences. If there are three light sources, A, B and C, where B is twice as bright as A and C twice as bright as B, a visual observer will estimate the difference between A and B as the same as that between B and C, although in linear terms, the relative brightnesses are 1, 2 and 4, respectively. The result is that the magnitude scale is logarithmic, rather than linear, and a given difference in magnitude corresponds to a particular brightness ratio.

Pogson [P2] quantified Hipparchus' original qualitative scale into a system where a difference of five magnitudes is equivalent to a factor of 100 in apparent magnitude, retaining the convention of numerically-increasing magnitudes with decreasing intensity. Hence, magnitude is defined as

$$m = -2.5 \log_{10}(f) + \text{constant} \qquad (1.15)$$

where f is the apparent flux (in $\text{Watts}\,\text{m}^{-2}\,\text{Hz}^{-1}$, $\text{erg}\,\text{sec}^{-1}\,\text{cm}^{-2}\,\text{Å}^{-1}$ or equivalent units). One of the striking advantages of this convention is that the enormous brightness difference of 10^{21} between the apparent magnitude of the Sun (magnitude -26) and the faintest object detectable by the Hubble Space Telescope (magnitude 29) spans only 55 magnitudes. Thus, the magnitude system expresses large brightness differences in a compact, and widely understood, form. It is primarily for this reason that, despite the rumblings of some astrophysicists (for example, [L9]), the system remains in common use today.

The magnitude scale is defined as

$$m_p = -2.5 \log_{10} \frac{f_p}{F_0} \qquad (1.16)$$

where f_p is the measured flux emitted by the source within a particular wavelength region (passband), usually defined by optical filters, and F_0 is the flux density produced by a star which has magnitude 0 in that passband. The latter zero-point is arbitrary, but is usually set to give an A0 star (comparable to Vega) equal magnitudes at all wavelengths. The main exception to this convention is the Gunn

Table 1.1. Broadband filter characteristics.

Band[1]	$\Delta\lambda$ μm	λ_{eff} A0	λ_{eff} M0	λ_{eff} M2	λ_{eff} M6.5	F_0 Janskys	Reference[2]	m_{sky}^3
U	0.325–0.395	0.366	0.366			1181	J2/B2	22.05
B	0.39–0.49	0.44	0.44			4520	J2/B2	22.4
V	0.50–0.59	0.542	0.551			3711	B3/L7	21.3
R	0.565–0.725	0.638	0.722			3180	B3/L7	20.6
I	0.73–0.88	0.787	0.800	0.809	0.818	2460	L7	18.8
J_{CIT}	1.16–1.35	1.22		1.257	1.256	1568	L7	15.5
H_{CIT}	1.49–1.80	1.63		1.635	1.633	1076	L7	14.5
K_S	2.00–2.30	2.15		2.15		650	P1	13.3
K_{CIT}	2.02–2.43	2.19		2.211	2.209	674	L7	13.0
L_{CIT}	3.24–3.73	3.45			3.472	281	B4	8.0
L'	3.52–4.12	3.80			3.80	235	B4	
M	4.5–5.05	4.75	4.75			154	B4	

[1] These passbands are the Cousins UBVRI system as characterised by Bessell ([B2], [B3]) and the Caltech (CIT) version of the Johnson near-infrared JHKL system. The K_S (K-short) passband is truncated at long wavelengths to minimise thermal background radiation. This passband is used in the DENIS and 2MASS near-infrared sky surveys (Chapter 9). Similarly, the L' passband avoids the worst atmospheric absorption in the CIT L-band.
[2] References are for flux zero-points in Janskys – units of $10^{-26}\,\mathrm{Watts\,m^{-2}\,Hz^{-1}}$.
[3] m_{sky} is the sky brightness in magnitudes arcsec^{-1} at a dark-sky site, such as Mauna Kea. Sky brightness longward of $\sim 2.3\,\mu m$ is dominated by thermal radiation from the atmosphere, and is therefore highly variable on short time-scales.

uvgriz system, used with the Sloan Digital Sky Survey (Section 9.63), which adopts a uniform zero-point in all passbands. Magnitudes in the Gunn system are defined as

$$m = -2.5\log_{10} f_\nu - 48.60 \qquad (1.17)$$

where f_ν is the flux in $\mathrm{erg\,cm^{-2}\,s^{-1}\,Hz^{-1}}$.

Broadband photometric systems, with filter-defined passbands of full-width at least 500 Å, are used to map the overall spectral energy distribution of celestial objects; narrowband (< 100 Å) filters are designed generally to examine features in the energy distributions of particular types of star. The most frequently used broadband photometric system is Johnson/Cousins UBVRIJHKLM [J2], [C1], spanning the wavelength range 3000 Å to 5 μm with each passband having a width of ~ 1000 Å (Table 1.1). Within each passband, the measured flux density corresponds to the stellar flux at the effective wavelength of the filter. The latter quantity is found by convolving the spectral energy distribution of the star ($S(\lambda)$) with the shape of the filter bandpass ($B(\lambda)$):

$$\lambda_{eff} = \frac{\int \lambda S(\lambda) B(\lambda) d\lambda}{\int S(\lambda) B(\lambda) d\lambda} \qquad (1.18)$$

The effective wavelength can vary depending on the spectral energy distribution of the target (the spectral type of the star observed). For example, decreasing temperature moves the peak in the emergent energy distribution towards longer wavelengths, steepening the spectral slope at optical wavelengths and moving the effective wavelength to the red. This effect is particularly important in the broad Cousins R-band, where λ_{eff} changes from 6380 Å for an A0 star to 7220 Å at spectral type M0 [B2]. Other passbands are less affected, with typically 200 Å differences in λ_{eff} between spectral types A0 and M0.

Photometric colours are defined as a magnitude difference; for example, $(B - V)$. From the definition of magnitude (equation 1.15), a 'colour' therefore measures the flux ratio in the two passbands:

$$\text{mag}_1 - \text{mag}_2 \equiv \frac{\text{flux}_1}{\text{flux}_2} \qquad (1.19)$$

Traditionally, colours are expressed as the shorter wavelength magnitude *minus* the longer wavelength, so a negative (blue) colour implies f_{short}/f_{long} is high, and a positive (red) colour indicates that f_{short}/f_{long} is low. As discussed further in Chapter 2, colours provide a means of estimating stellar temperatures.

1.5.2 Measuring magnitudes

Observationally, the standard photometric systems were defined using photoelectric photometers and aperture photometry techniques, but most current measurements are made with array detectors. In the former case, a photomultiplier tube was used to measure the brightness of a given source through a circular aperture (usually 5–10 arcsec diameter); in the latter case, software techniques are used to measure the flux within a given radius centred on the star. In either case, since the aim is to measure a large fraction of the stellar flux, the effective aperture size chosen depends on the prevailing atmospheric seeing. However, the underlying night sky also contributes to this measurement and this contribution must be subtracted, either by obtaining a separate offset measurement in aperture photometry, or from the 'sky' pixels immediately adjacent to the object in array photometry.

Typical surface brightness values for a dark-sky site are listed in Table 1.1. Airglow (primarily emission from OH and OI; see [R2]) is a strong contributor shortward of 2 μm, and the sky brightness at these wavelengths is well-correlated with the solar cycle, being higher when the Sun is more active [L6]. Volcanic eruptions can also affect m_{sky} if dust is introduced into the upper atmosphere, where it gradually diffuses around the Earth [L8]. The eruptions of the Mexican volcano El Chichón (1982) and the Philippine volcano Pinatubo (1990) not only produced spectacular sunsets, visible for several months afterwards in North America and Europe, but also raised the night sky brightness (and the optical extinction) for more than two years. At longer wavelengths, thermal radiation from the atmosphere and telescope dominates the background.

The signal-to-noise of an observation is given (following [L9]) by

$$\frac{S}{N} = \frac{A_{eff} N_\nu \Delta \nu t}{[A_{eff} N_\nu \Delta \nu t + \Omega A_{eff} S_\nu \Delta \nu t + Dt + R^2]^{0.5}} \qquad (1.20)$$

where t is the integration time; $\Delta\nu$, the band (Hz); A_{eff}, effective area of telescope, in m^{-2}; N_ν, the source flux density, in photons, $m^{-2}\ s^{-1}\ Hz^{-1}$; S_ν, sky brightness, in photons, $m^{-2}\ s^{-1}\ Hz^{-1}\ arcsec^{-1}$; Ω, solid angle of effective aperture (either the physical diameter of the aperture in the photoelectric photometer, or the circle of integration used in analysing the array photometry); D, dark current; and R, readout noise (zero for aperture photometry).

Modern detectors have both low dark-current and low read-noise (3–7 electrons is typical for a CCD working at liquid nitrogen temperatures). Consequently, photon statistics in the sky background level constitute the dominant source of noise in photometry of faint objects. In aperture photometry, with a photomultiplier tube, the sky measurements are made separately from the object (+sky) observations. One of the major advantages of array photometry is that the sky level can be determined from the same exposure used to measure the source. Moreover, a 2,048-square optical CCD array covers a typical solid angle of at least 150 square arcmin, encompassing many stellar (and non-stellar) objects in a deep exposure on even a moderate-sized telescope. Since these objects are all observed simultaneously, high accuracy relative photometry is possible even during inclement conditions. A caveat is that individual CCD pixels can have slightly different sensitivities, so the intensity levels in each frame must be normalised using a flat-field exposure – an image made by illuminating the CCD with a diffuse, uniform light-source.

Once CCD images have been normalised, a sophisticated profile-fitting technique can be used to determine the relative flux of each object, minimising the contribution from sky-noise. Clearly, the more concentrated the stellar profile (the better the seeing), the smaller the solid-angle for profile-fitting, the lower the contribution from sky-noise and the fainter the limiting magnitude attained in a given exposure time. The instrumental flux measurements themselves are calibrated through observations of standard stars with well-determined magnitudes on particular photometric systems. Extensive lists are provided by Landolt [L2], [L3] for the frequently-used Johnson/Cousins optical system (Table 1.1), and Persson *et al.* [P1] provide standard-star lists for the near-infrared.

In general, an instrumental magnitude is measured for each source, defined by (taking the V-band as an example)

$$v = -2.5\log\left(\frac{N_V}{t}\right) \tag{1.21}$$

where N_V is the total number of counts measured for the source in an integration time of t seconds. Observations of standard stars are used to solve for the constants in an equation matching the form:

$$V = v + k_v \times \sec(z) + C_v \times (V - I) + Z_v \tag{1.22}$$

where V is the magnitude on the standard system; v, the instrumental magnitude; z, the angular distance from the zenith; k_v, the extinction coefficient; C_v, the colour term; and Z_v, the zero-point. The extinction term corrects for absorption through the Earth's atmosphere ($\sec(z)$ is known as the 'airmass' of an observation) using a plane-parallel approximation for atmospheric depth along the line of sight. The colour term (which could also be (B–V) or (V–R) in this case) allows for a potential mismatch between the effective wavelength of the reference system and that used in the observations. Defining these terms accurately requires repeated observations of a reasonable number (15–20) of standard stars, well distributed in both colour and airmass. A more thorough discussion of photometric techniques is given by Henden and Kaitchuck [H2].

The observed magnitude of a star is termed the apparent magnitude. The absolute magnitude is defined as the apparent magnitude that a source has at a distance of 10 parsecs. Hence, since brightness decreases with the square of the distance:

$$M = m + 5 \times \log_{10}(r) - 5 \tag{1.23}$$

The quantity $(m - M)$ is known as the distance modulus, a term often used in assigning the distances to star clusters.

1.5.3   Bolometric magnitudes and effective temperatures

Summing the total energy emitted at all wavelengths by a star determines its *bolometric* magnitude,

$$m_{bol} = -2.5 \log_{10}(f_{tot}/F_0) \tag{1.24}$$

Given a known distance, this can be converted to the absolute bolometric magnitude which, in turn, can be expressed as a luminosity, usually in solar units. Based on absolute measurements of the energy distribution, primarily from satellite data, the Sun has an absolute visual magnitude of $M_V \simeq 4.79$ [L5]. The bolometric correction is approximately -0.12 magnitudes [B7], so $M_{bol}(\odot) \simeq 4.67$, corresponding to $L_\odot \simeq 3.83 \times 10^{26}$ Watts. Stellar luminosities are then given by

$$-2.5 \log_{10}\left(\frac{L}{L_\odot}\right) = M_{bol}(\text{star}) - 4.67 \tag{1.25}$$

The luminosity of a star is also used to define the quantity known as the effective temperature. An ideal radiator of temperature T produces a pure continuum spectrum, with neither absorption nor emission features, and a spectral energy distribution that is described by the Planck formula. This black-body distribution can be written as a function of frequency:

$$F_\nu \delta\nu = \frac{2\pi h}{c} \frac{\nu^3 \delta\nu}{e^{\frac{h\nu}{kT}} - 1} \tag{1.26}$$

where h is the Planck constant, k is the Boltzmann constant, c is the velocity of light, T is the surface temperature and ν is the frequency. In this case the usual units for the flux, F_ν, are Janskys, where 1 Jansky is 10^{-26} Watts m^{-2} Hz^{-1} (MKS units). Alternatively, the Planck curve can be written in wavelength units:

$$F_\lambda \delta\lambda = \frac{2\pi hc^2}{\lambda^5} \frac{\delta\lambda}{e^{\frac{hc}{\lambda kT}} - 1} \tag{1.27}$$

where the usual units are erg cm^{-2} Å^{-1} or erg cm^{-2} μm^{-1} (c.g.s. units). Integrating this distribution produces Stefan's law, which states that the total energy emitted is proportional to the product of the surface area and the fourth power of the temperature. This leads to the definition of the effective temperature of a star:

$$L = 4 \pi R^2 \sigma T_{eff}^4 \tag{1.28}$$

where σ is Stefan's constant, $\dfrac{2\pi^5 k^4}{15 c^2 h^3}$. The effective temperature is that required for a black-body, with the same radius as the star, to radiate the same total energy.

The black-body distribution peaks at a wavelength, λ_{max}, whose value varies inversely with the temperature. The result is Wien's law, which can be derived by determining when the derivative of equation (1.27) is zero. The solution is

$$\lambda_{max} = \frac{2898}{T} \, \mu m \tag{1.29}$$

Hence, the Sun, with $T_{eff} = 5,777\,K$ [L4], has an emergent spectrum which peaks at $0.502\,\mu m$. In contrast, emission from the Earth's atmosphere at $T \sim 300\,K$ peaks at close to $10\,\mu m$.

At wavelengths well longward of λ_{max}, the quantity $hc/\lambda kT$ is small, and $e^{\frac{hc}{\lambda kT}} \approx 1 - hc/\lambda kT$. Thus, equation (1.27) becomes

$$F_\lambda \delta\lambda \propto T\lambda^{-4} \tag{1.30}$$

This is the Rayleigh–Jeans approximation, and the black-body spectrum at $\lambda > \lambda_{max}$ is referred to as the Rayleigh–Jeans tail.

1.5.4   Interstellar absorption

Dust particles are found in higher-density regions of the interstellar medium, which are largely confined to latitudes within 5–10 degrees of the Galactic plane. Just as in the terrestrial atmosphere, these particles modify the spectral energy distribution of background stars, preferentially scattering shorter-wavelength radiation and changing the overall spectral energy distribution. This interstellar reddening is usually quantified by estimating the 'colour excess' of a star – the difference between the observed colour, usually (B–V), and the intrinsic colours, estimated from the spectral type (see Chapter 2). Identifying the latter with the subscript 0, the colour excess is

$$E_{B-V} = (B - V) - (B - V)_0 \tag{1.31}$$

Dust grains in most regions of the interstellar medium have similar properties, so the colour excess at other wavelengths can be scaled to the (B–V) colour excess [B4]. Thus,

$$E_{V-I} = 1.25 \times E_{B-V}, \quad E_{V-K} = 2.78 \times E_{B-V} \tag{1.32}$$

and the total absorption is given by

$$A_V = V - V_0 = 3.12 \times E_{B-V}, \tag{1.33}$$

$$A_K = K - K_0 = 0.34 \times E_{B-V}, \tag{1.34}$$

The quantity $\dfrac{A_V}{E_{B-V}}$ is known as the ratio of total to selective absorption, and this parameter can increase in high-density regions, probably due to the grains having a different composition and different size distribution [W1]. The above equations clearly demonstrate the advantages of using infrared observations to probe dusty regions near the plane of the Galaxy.

1.6 SPECTROGRAPHS AND SPECTROSCOPY

Photometry allows estimation of stellar luminosity (given a known distance) and surface temperatures; spectroscopy permits measurement of individual spectral lines, which can be used to determine composition, temperature and motion toward or away from the observer (the radial velocity). Astronomical spectroscopy has its origins in Newton's optical experiments in the 1660s. Placing a glass prism in a beam of sunlight passing through a hole in the window shutters of an otherwise darkened room, Newton discovered that 'white' sunlight was actually compounded of a mixture of different colours. Relatively little use was made of this discovery until the early nineteenth century, when first Wollaston (1802) and then Fraunhofer (1815) discovered that the solar spectrum was not a continuous band of colours, but exhibited a number of dark bands. The nature of those bands, whether intrinsic to the Sun or terrestrial in origin, remained controversial until 1861, when Kirchhoff [K2] identified the Fraunhofer D-line with the element sodium. Once it became clear that the individual spectral features provided signatures of the presence of different atomic and molecular species within the solar and stellar atmospheres, it became possible to devise stellar classification schemes (see [H1] for a thorough historical review). Spectral classification led to the recognition of underlying patterns of behaviour, eventually formalised as the Hertzsprung–Russell diagram (discussed in more detail in Chapter 2).

The first spectrographs used prisms to disperse the incident starlight, but modern instruments generally use ruled gratings as the principal dispersive optical element. Instruments designed to work at ultraviolet, visual and near-infrared wavelengths – the wavelengths of most interest for M dwarf studies – follow the same basic optical design, although the optical elements in infrared spectrographs are usually cooled to minimise their contribution to the thermal background. The general principles of spectrographic design are outlined by Bowen ([B6], and see also [J1]). Figure 1.11 shows the schematic design of a typical optical spectrograph, consisting of a slit, placed in the focal plane of the telescope, collimator, diffraction grating, camera and detector.

The spectrograph slit isolates the light from a particular astronomical target and minimises the contribution from the sky background. The standard single-object, long-slit assembly is usually both tilted to the normal to the optical axis of the telescope and polished, so that the field can be viewed with an acquisition TV camera (Figure 1.11). The collimating lens has the same focal ratio as the telescope. Placing that lens at the appropriate place behind the slit transforms the diverging beam into

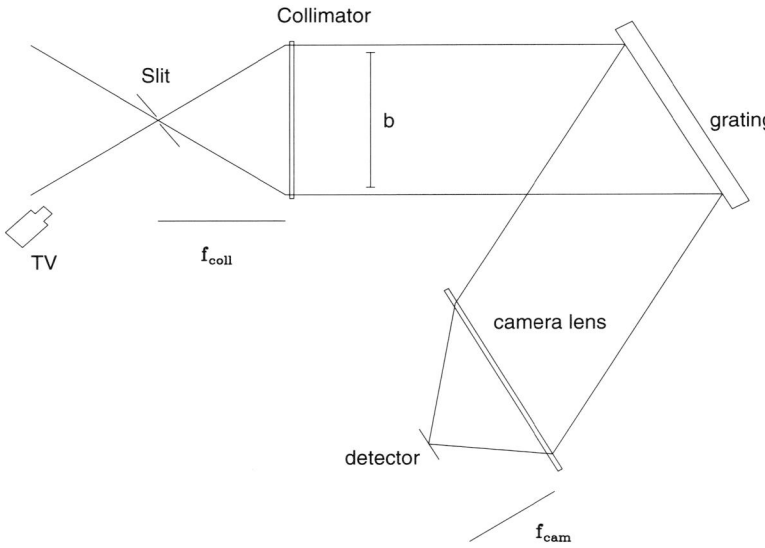

Figure 1.11. A schematic of a diffraction-grating spectrograph.

a parallel beam, which is directed towards the diffraction grating or prism. After dispersion, the beam is imaged onto a detector by the camera lens. Without the dispersing element, this is simply a re-imaging system, with the magnification given by the ratio between the focal length of the camera (f_{cam}) and the focal length of the collimator (f_{coll}). The inverse of this ratio is known as the scale factor. A length, δl, at the focal plane projects to a size δx, given by

$$\delta x = \delta l \frac{f_{cam}}{f_{coll}} \tag{1.36}$$

Generally, the magnification f_{cam}/f_{coll} is chosen so that a point source viewed under typical seeing conditions (see Section 1.7) projects to ~ 2 pixels at the detector. Thus, for the Hale 200-inch telescope, the plate-scale at Cassegrain focus (f/15.7) is 2.59 arcsec mm^{-1}. Given 0.8 arcsec (0.31 mm) as an estimate of the best likely seeing, and a CCD pixel size of $\sim 15\,\mu$m, scale factors of 10 are required to avoid image degradation.

Prisms act as dispersing elements because the angle of refraction decreases as a function of wavelength – the same phenomenon responsible for differential chromatic atmospheric refraction. The dispersion formula is written as

$$\frac{d\theta}{d\lambda} = \frac{B}{b}\frac{dn}{d\lambda} \tag{1.36}$$

where b is the width of the beam (Figure 1.11), B is the base-width of the prism, and n is the refractive index of the prismatic material. The wavelength dependence of n

follows Cauchy's equation:

$$\frac{dn}{d\lambda} \propto \lambda^{-3} \tag{1.37}$$

Thus, prisms have the significant disadvantage that the dispersion, and hence the spectral resolution, decreases rapidly towards longer wavelengths. Moreover, a high dispersion requires that the width of the prism is large compared with that of the beam, which in turn demands a large, expensive and difficult to fabricate optical element. Several prisms can be combined to increase the overall dispersion, but at the potential expense of additional light loss at the air/glass surfaces. As a result of these constraints, diffraction gratings, rather than prisms, are preferred as the primary dispersive element in modern spectrographs.

A diffraction grating consists of a block of optical glass with a series of grooves ruled at regular intervals. The grating can be either reflecting or transmitting. The general properties were originally discovered in 1819 by Fraunhofer (hence Fraunhofer diffraction), who used a series of finely-spaced metal wires to form a grating. Technical difficulties limited their initial usefulness for astronomical purposes, but with the development of more precise ruling techniques it became possible to manufacture large, accurate reflection gratings.

Figure 1.12 illustrates the main properties of a diffraction grating. If we consider a parallel beam incident on a series of apertures, constructive interference between the individual diffraction patterns at each wavelength leads to a series of intensity maxima (orders) in the re-imaged beam. If the spacing between grooves is d, and the angle of incidence is i, then the grating equation

$$d(\sin(i) + \sin(\theta)) = m\lambda \tag{1.38}$$

gives the angle of diffraction, θ, for order m. In most spectrographs, the angle between the optical axes of the collimator and camera, δ_{CC}, is fixed. θ is measured in the same sense as i, so from Figure 1.11 it is clear that

$$\delta_{CC} = i - \theta \tag{1.39}$$

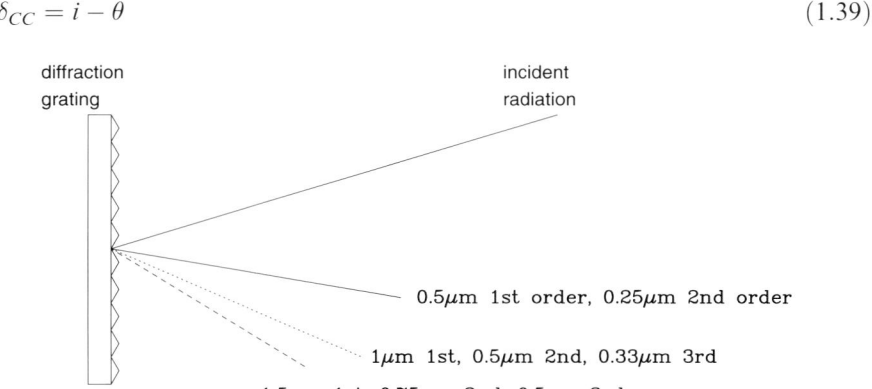

Figure 1.12. Spectral orders produced by a diffraction grating.

Thus, the central wavelength on the detector can be set by changing the tilt of the grating with respect to the optical axis of the collimator. The grating tilt is measured with respect to the normal to that optical axis, and is therefore equal to i in equations (1.36) and (1.37). As an example, consider the red camera on the Hale telescope double spectrograph [O2]. In this case, $\delta_{CC} = 35°$, so if a 600 line/mm grating ($d = 1.67\,\mu m$) is used, and the desired central wavelength is $\lambda = 6000\,\text{Å}$ in first order, then the required grating tilt is $\sim 28°25$. At that angle, second-order radiation of wavelength $3000\,\text{Å}$ (and third order at $2000\,\text{Å}$) is centred on the detector. Since the atmosphere is opaque to radiation shortward of $3250\,\text{Å}$ (Section 1.1.2), this is not a problem for this particular set-up. Indeed, a dichroic is usually employed in the double spectrograph to divert blue light to a separate camera. However, there are situations where a blocking filter is required to isolate the particular wavelength region of interest.

The angular dispersion for a given order can be derived by differentiating the grating equation to produce

$$\frac{\Delta\theta}{\Delta\lambda} = \frac{m}{d\,\cos\theta} \tag{1.40}$$

This can be expressed as the reciprocal linear dispersion, in $\text{Å}\,mm^{-1}$ at the detector, by inverting the above equation and dividing by the focal length of the camera:

$$\frac{\Delta\lambda}{\Delta l} = \frac{d\,\cos\theta}{m f_{cam}} \tag{1.41}$$

A diffraction grating redistributes the incident light amongst several orders, with the intensity distribution dependent on the shape of the grooves. For a simple, square-wave grating, most of the energy resides in the undispersed, zero-order image. However, the shape of the grooves can be adjusted to concentrate up to 90% of the flux within a relatively narrow range of θ. Following the grating equation, for any given wavelength this places most of the intensity within a specific order. These gratings are usually described as being *blazed* for the appropriate wavelength in the first order spectrum.

The spectral resolution of a particular grating is given by

$$R = \frac{\lambda}{\Delta\lambda} = mN \tag{1.42}$$

where N is the total number of grooves within the optical beam. However, in practice the resolution is usually set by either the pixel size of the detector or by the projected width of the spectrograph slit. For example, the red camera on the original Hale double spectrograph ($f_{cam} = 152\,mm$) has a beam size of $146\,mm$, so the formal spectral resolution with a 600 line/mm grating is therefore $R = 87,600$. This corresponds to a resolution of $0.069\,\text{Å}$ at $6000\,\text{Å}$, or, in linear units at the detector, $0.6\,\mu m$. In comparison, a 1-arcsec slit projects to a linear size of $25.6\,\mu m$, and the original TI CCDs used had a pixel size of $15\,\mu m$. Actual observations usually represent a trade-off between high signal-to-noise (as wide a slit-size as possible, to collect as much light as possible) and high resolution (as narrow a slit-size as feasible). A typical compromise is to match the slit width to two detector elements (Nyquist sampling); hence, observations with the Hale double spectrograph would be taken with a 1.2-arcsec slit, giving a resolution of ~3.3 Å with the 600 line/mm grating.

1.7 IMPROVING IMAGE QUALITY

The theoretical resolution of a telescope (the diffraction limit) is defined as

$$L = \frac{206265}{10^6} \left(\frac{\lambda}{D} \right) \tag{1.43}$$

where λ is the wavelength in microns, D the telescope diameter in metres, and L the resolution in arcseconds. Astronomical seeing, however, introduces distortions, with the result that the best optical images obtained from ground-based telescopes seldom have profiles with full-width half-maximum less than ~0.6 arcsec at optical wavelengths.

'Seeing' is the consequence of the passage of starlight through a variable-density medium – the Earth's atmosphere. Operationally, atmospheric seeing can be regarded as including two components: image motion, due to the light passing through severely turbulent regions; and a general broadening of the stellar profile, originating as the wavefront passes through regions of different densities (atmospheric cells). The latter have diameters of approximately 15–20 cm, so a large telescope is combining light from many cells at any point during an observation. Significant distortions can be introduced by poor air-flow within the dome itself. The final observed stellar point-spread function (PSF) is defined by both these refractive effects and by scattered light within the optics of the telescope and camera. The former effects generally dominate in the core of the image, while the latter are more important at larger radii (such as stellar diffraction spikes, due to light diffracted by the secondary mirror supports). Innovative designs have been proposed for eliminating many of these sources of scattered light. For example, in telescopes with off-axis foci, the secondary mirror lies outside the primary beam and therefore introduces no obstruction, with a consequent reduction in scattered light [K4]. The majority of telescopes, however, are still constructed along more conventional lines.

The most extensive analysis of the light profile of a star was carried out by King [K1], who used photographic techniques to determine the surface brightness at radii of more than 10,000 times the half-width at half peak intensity (HW) of the image – a dynamic range of almost 28 magnitudes (Figure 1.13). More recent observations using electronic detectors in the optical (B-band – matching the photographic data) and near-infrared (H-band) have confirmed the general form of the profile and its invariance, once scaled to the appropriate HW, with wavelength [R1]. Racine has shown that, once diffraction spikes are eliminated, the observations within 10 HW are an excellent match with the profile predicted by Kolmogorov turbulence theory, suggesting that atmospheric effects dominate in the inner part of the profile. However, the extended aureole at larger radius is almost certainly due to instru-

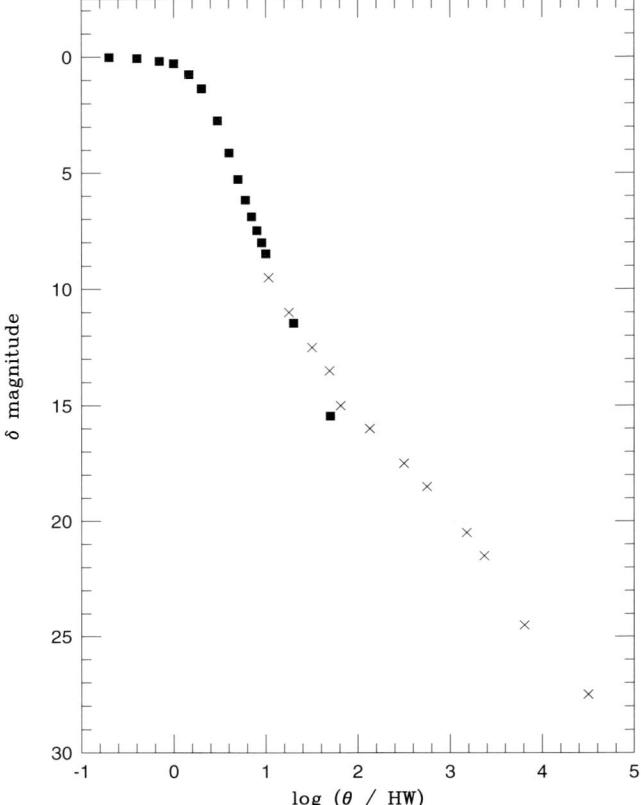

Figure 1.13. The stellar point-spread function, adapted from the measurements by King (crosses) and Racine (solid squares).

mental effects. In the case of King's analysis, Racine suggests that a significant component is due to scattered light from internal reflections in the 1-mm thick Schmidt plates used for the observations.

The improvement of seeing is currently a major priority at all large optical observatories, both through technological modifications (better temperature control, free air flow) and through the application of new techniques. Not only do such improvements permit higher spatial-resolution observations (resolving close binaries or structure in nearby galaxies), but concentrating the light from a point source also produces greater contrast with the sky background, and allows detection (or spectroscopy) at fainter magnitudes. Thus, a greater light-grasp can be achieved not only through the construction of new, larger mirrors, but also through improving the performance of existing telescopes. This present section provides an introduction to two techniques which have proven of considerable importance in studying low-mass dwarfs.

1.7.1   Speckle interferometry

Labeyrie [L1] showed that when radiation from a star passes through the Earth's atmosphere, the wavefront passes through regions of different densities (atmosphere cells) which introduce phase fluctuations. Integrating these fluctuations over time leads to the broad seeing disk illustrated in Figure 1.13. If the exposure time is limited to extremely short timescales (<100 msec) then some phase coherence remains. A single snapshot reveals a series of bright speckles which contain information on the image structure at a resolution matching the diffraction limit (Figure 1.14). Each individual observation includes only a few photons, and cannot provide sufficient signal to allow reconstruction of the image structure. However, many (hundreds) of such frames can be accumulated and combined, and can be analysed using Fourier techniques to reconstruct a diffraction-limited image.

Angular resolution decreases towards longer wavelengths – a circumstance that would seem to favour speckle observations at optical wavelengths. However, other factors favour observations in the near-infrared. First, the time-scale over which the atmosphere remains coherent, the correlation time, increases with increasing wavelength ($\propto \lambda^{6/5}$), permitting longer integration times, and a higher signal in each speckle frame in infrared observations. Second, the magnitude difference between any two main-sequence stars is smallest at infrared wavelengths. Even in Fourier space, binary companion searches are limited by the stellar profile, which sets the background level. As a result, the magnitude limit for detecting a secondary star depends on both the brightness of the primary star and the angular separation between the components. Typical observations can reach flux ratios of 60–100 (4–5 magnitudes) at separations of more than twice the diffraction limit. Working at near-infrared wavelengths therefore minimises the magnitude difference and maximises the chances for detection of very low-mass companions.

Infrared speckle observations can achieve diffraction-limited resolution – often a factor of 10 improvement over direct imaging – and are capable of detecting companions 100 times fainter than the primary at separations of $\sim 0\rlap{.}''25$. This technique has played a crucial role in allowing measurement of the orbits of

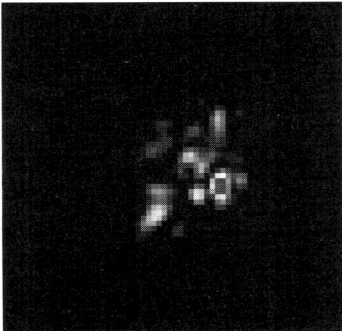

Figure 1.14. A speckle pattern.

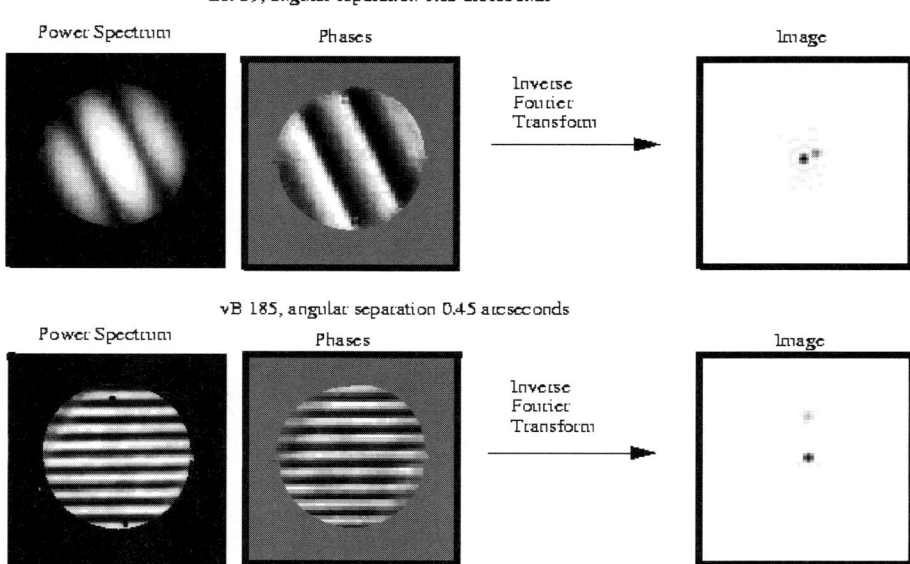

Figure 1.15. Speckle imaging analysis of two binaries in the Hyades. In each case, the left panels plot the Fourier amplitude and phase distribution, which are combined to give the diffraction-limited image on the right. Note that the orientation of the binary system is perpendicular to both amplitude and phase fringe patterns, while the spacing between the fringes increases, and the number of fringes decrease, as the separation decreases. (Illustration courtesy of J. Patience.)

binary stars and, (as described further in Chapter 8), the estimation of stellar masses. Figure 1.15 shows the results of Fourier analysis of speckle data for two representative binaries, two-dimensional 'images' of the derived phases and amplitudes. The cyclical fringe pattern is essentially an interferogram, with the period dependent on the separation of the two stars (wide separation, many fringes; narrow separation, few fringes) and the amplitude of the fringes dependent on the relative luminosity of the two sources. The orientation of the binary is perpendicular to the phase pattern. In principle, a binary of separation $\lambda/2D$ can be detected, since this produces one-half cycle of variation in the phase diagram – although this variation is detectable only for components of nearly-equal luminosity. This type of analysis has led to a more than threefold increase in the number of M dwarfs with masses determined to an accuracy of 20% or better.

1.7.2   Adaptive optics and interferometry

Speckle imaging 'freezes' the atmosphere in short exposures and reconstructs the image in Fourier space; adaptive optics techniques correct atmospheric wavefront distortions above the image plane of the telescope. In the simplest case, a tip–tilt mirror takes out image motion introduced by seeing and telescope vibrations; in more complex instrumentation, multi-element 'rubber' mirrors are used to eliminate higher-order distortions (see [B1] for an excellent introductory review). The efficiency of these systems is measured by the Strehl ratio – the ratio between the peak flux in the core of a point source and the value expected for diffraction-limited imaging.

The advantage of AO techniques is that they provide sharpened images of individual objects which can be fed directly into CCD cameras or onto the slit of a spectrograph. As a result, they reveal complex structures, which would defy identification in Fourier space. AO systems are driven by feedback loops, using relatively bright stars (12th magnitude for 4-metre class telescopes) as reference sources. The main limitation is that wavefront disturbances can be considered as uniform (and therefore correctable) over only a small angle, θ_0, the isoplanatic angle. This angle depends on the coherence length of the atmosphere, r_0, and inversely on the average distance of the atmospheric layer responsible for seeing distortions. The latter parameters are observatory dependent, but are typically 2–3 arcsec at optical wavelengths, rising to 10–20 arcsec at 2.2–3.3 μm ($\theta_0 \propto r_0 \propto \lambda^{6/5}$). Since a bright guide-star within the isoplanatic patch is required, only a small fraction of the sky is accessible ($\sim$1–2% at K-band, <0.01% at I) with natural guide-stars.

In the case of searches for low-mass companions, the potential primary star often serves as its own guide-star. For more general purposes, limitations in areal coverage will be addressed once laser guide-star technology becomes firmly established. A powerful (>5 kW) pulsed laser excites Na D-line emission in the menospheric layer of the atmosphere, at an altitude of $\sim$ 90 km. That bright patch serves as a guide-star. Systems are currently under development at most of the new generation of 8–10-m telescopes.

Interferometry has a long history of use in observations at radio wavelengths: data are combined from individual telescopes of modest aperture (5–10 m) separated by a much longer baseline ($\sim$100 m or more), giving a spatial resolution compatible with the latter. These techniques are being extended to near-infrared and optical wavelengths, both in ground-based observations (notably with the Keck telescopes) and from space.

1.8 SUMMARY

In the preceding section we outlined some of the more important techniques in observational astronomy, particularly those relevant to the studies of low-mass dwarfs described in the following chapters:

- the definition of co-ordinate systems and the measurement of proper motion and parallax in stellar astrometry;
- an outline of telescope and detector systems;
- a summary of the main aspects of broadband stellar photometry;
- an introduction to stellar spectroscopy and spectrographs;
- and a brief review of image-sharpening techniques.

The aim has been to provide a broad overview rather than a rigorous development, emphasising aspects which are of most practical use to the astronomical observer. More extensive discussions of each of these topics are included amongst the literature cited in the appended bibliography.

1.9   REFERENCES

B1   Beckers, J. M., 1993, *ARA&A*, **31**, 13.
B2   Bessell, M. S., 1979, *PASP*, **91**, 589.
B3   Bessell, M. S., 1990, *PASP*, **102**, 1181.
B4   Bessell, M. S., Brett, J. M., 1988, *PASP*, **100**, 1134.
B5   Boksenberg, A., 1972, in *Auxiliary Instrumentation for Large Telescopes: Proceedings of the ESO–CERN Conference*, eds. S. Lausten and A. Retz, p. 295.
B6   Bowen, I. S., 1962, in *Astronomical Techniques*, ed. W. A. Hiltner, University of Chicago Press.
B7   Buser, R., Kurucz, R. L., 1978, *A&A*, **70**, 555.
C1   Cousins, A. W. J., 1982, SAAO *Circulars*, No. 1.
C2   Culhane, J. L., Sanford, P. W., 1981, *X-ray Astronomy*, Charles Scribner's Sons, New York.
E1   Eccles, M. J., Sim, M. E., Tritton, K. P., 1983, *Low Light Level Detectors in Astronomy*, Cambridge University Press.
E2   ESA, 1997, The Hipparcos catalogue, ESA SP-1200.
F1   Filippenko, A., 1982, *PASP*, **94**, 715.
G1   Gunn, J. E., Carr, M., Rockosi, C., Sekiguchi, M. *et al.*, 1998, *AJ*, **116**, 3040.
H1   Hearnshaw, J. B., 1986, *The Analysis of Starlight*, Cambridge University Press.
H2   Henden, A., Kaitchuck, R. H., 1982, *Astronomical Photometry*, Van Nostrand Reinhold, New York.
J1   Jenkins, F.A., White, H. E., 1976, *Fundamentals of Optics*, McGraw-Hill Inc.
J2   Johnson, H. L., 1966, *ARA&A*, **4**, 193.
K1   King, I. R., 1971, *PASP*, **83**, 199.
K2   Kirchhoff, G., Part I: *Abhandl. d. Berliner. Akad.*, p. 63, 1861; p. 227, 1862. Part II: *ibid.*, p. 225, 1863.
K3   Kovalevsky, J., 1998, *ARA&A*, **36**, 99.
K4   Kuhn, J., Hawley, S. L., 1999, *PASP*, **111**, 601.
L1   Labeyrie, A., 1970, *A&A*, **6**, 86.
L2   Landolt, A. U., 1983, *AJ*, **88**, 439.
L3   Landolt, A. U., 1992, *AJ*, **104**, 340.
L4   Lane, A. P., 1979, *PASP*, **91**, 405.
L5   Lang, K. R., 1992, *Astrophysical data: Planets and stars*, Springer-Verlag, New York.
L6   Leinert, C., Richter, I., Planck, B., 1982, *A&A*, **110**, 111.
L7   Leggett, S. K., Allard, F., Berriman, G., Dahn, C. C., Hauschildt, P. H., 1996, *ApJS*, **104**, 117
L8   Levizzani, V. Prodi, F., 1988, *J. Geophys. Res.*, **93**, 5277.

L9   Longair, M. S., 1981, *High-Energy Astrophysics*, Cambridge University Press.
M1   Monet, D. G., 1988, *ARA&A*, **26**, 413.
O1   Ohlsson, J, 1932, *Ann. Lund. Obs.*, **3**.
O2   Oke, J. B., Gunn, J. E., 1982, *PASP*, **94**, 586.
O3   Osterbrock, D. E., 1997, *J. Hist. Astr.*, **25**, 1.
P1   Persson, S. E., Murphy, D. C., Krzeminski, W., Roth, M., Rieke, M. J., 1998, *AJ*, **116**, 2475.
P2   Pogson, N., 1854, *MNRAS*.
R1   Racine, R., 1996, *PASP*, **108**, 699.
R2   Roach, F. E., Gordon, J. L., 1973, *The Light of the Night Sky*, Geophysics and Astrophysics Monographs, V. 4.
S1   Sekiguchi, M., Iwashita, H., Doi, M., Kashikawa, N., Okamura, S., 1992, *PASP*, **104**, 744.
S2   Smart, W. M., 1971, *Spherical Astronomy*, 5th edition, Cambridge University Press.
T1   Tucker, W., Giacconi, R., 1985, *The X-Ray Universe*, Harvard University Press, Cambridge, Massachusetts.
V1   van Leeuwen, F., 1997, *Sp. Sci. Rev.*, **81**, 201.
V2   Vasilevskis, S., 1967, *AJ*, **72**, 583.
W1   Whittet, D. C. B., 1992, *Dust in the Galactic Environment*, the graduate series in astronomy, Institute of Physics Publishing, Bristol, Philadelphia, New York.
W2   Wilson, R., 1996, *Reflecting Telescope Optics I: Basic Design Theory and its Historical Development*, Springer-Verlag, Berlin.

2

Observational properties of low-mass dwarfs

2.1 INTRODUCTION

'It is a capital mistake to theorise before one has data.'
Arthur Conan Doyle

Astronomy is an observational science. In disciplines such as chemistry, biology and physics, carefully controlled laboratory experiments can be set up to test the effect of altering one specific variable, such as temperature, pressure or concentration. However, the physical conditions which pervade in most celestial objects are so extreme that there can be no question of their being matched in terrestrial laboratories, so direct experimentation is out of the question. Consequently, physical understanding of the conditions and processes which govern the formation and evolution of stars rests on our ability to interpret information carried by electromagnetic radiation. Observations of position as a function of time yield estimates of distance and of motion perpendicular to the line of sight; summing the energy output over all wavelengths gives the total (bolometric) luminosity; surface temperature can be estimated from the distribution of energy as a function of wavelength, and from the relative ionisation of different chemical species; individual spectral line profiles of various atoms allow us to estimate the density, gravity and gas pressure in the atmosphere; chemical composition influences the relative strengths of lines and bands in the spectrum; and the exact position of spectral lines tells us the radial velocity of the star.

We cannot vary the conditions prevalent on an individual star (as one astronomer puts it, 'we don't have a long enough stick to poke them with'), so observations of many individual stars are combined to build up a picture of how stars of different mass and composition evolve. This is a particularly powerful technique when the stars in question are drawn from the same star cluster and are therefore effectively coeval, but can also be illuminating when applied to stars in the immediate vicinity of the Sun (the solar neighbourhood). While the latter stars span a wide range of age and composition, the fact that distances can be measured with high precision makes this group a uniquely useful probe of M dwarf properties. Whether looking at individual stars or at ensembles, observations provide the yardstick for measuring the success of analytic and numerical models devised to explain the physical processes present.

With these issues in mind, this chapter reviews the observational characteristics which define M dwarfs and the recently-identified ultracool L dwarfs. An empirical and phenomenological approach is adopted, concentrating on qualitative inferences that can be drawn directly from observation. This lays the foundation for the more detailed consideration of the underlying physics discussed in succeeding chapters.

The next two sections provide an overview of the spectroscopic and photometric characteristics of low-mass dwarfs. It will become clear that there are substantial similarities between the latest K dwarfs and 'classical' M dwarfs. In particular, most of the molecular features that identify M dwarfs – TiO, H_2O, the metal-hydride bands – start to appear at spectral types earlier than M0. For that reason, in this book the term 'M dwarf' includes all stars with absolute visual magnitudes fainter than $M_V = 7.5$ – a definition which encompasses many stars formally classified as K5 or K7 dwarfs. Very low-mass (VLM) dwarfs are defined as having spectral types later than M6.5, therefore including members of the new spectral classes L and T.

2.2 SPECTRAL CLASSIFICATION: WHAT ARE M DWARFS AND L DWARFS?

2.2.1 Early observations

The advent of optical spectroscopy revolutionised astronomy in the mid-nineteenth century. For the first time, astronomers were placed in a position where they could probe the nature of stellar material. This ability was enhanced greatly when photography was used to record stellar spectra for relatively leisurely inspection, rather than relying on fleeting impressions made at dead of night under strained physiological conditions. Hearnshaw [H3] ably describes the early development of stellar spectroscopy and its transformation into a semi-exact science. From the outset it was clear that there is order in the way that stars behave: patterns amongst the various spectral features allowed certain stars to be grouped together. In 1864, Secchi, in the first serious attempt at spectral classification, placed stars under three headings: types I, II and III. Type III objects were 'coloured' stars with wide absorption bands in their spectrum, but it was not until 1904 that Fowler [F3] identified these spectral features as due to either the metal titanium or a compound containing titanium. Nonetheless, titanium oxide absorption bands – the defining feature of M stars – were clearly distinguished at the outset.

All of the early classification systems ordered stars from blue to red, reflecting the suspicion first voiced in 1874 by Vogel that this represented a scale of decreasing temperature, and perhaps an evolutionary sequence. A relatively small number of researchers – including Lockyer and (at least initially) Hale – argued that changes in the relative strengths of different absorption features (line-blanketing) was the primary cause of both spectral and colour variations. It was not until 1909 that Wilsing and Scheiner [W2] demonstrated conclusively that O-type and M-type stars did indeed lie at opposite extremes in the stellar temperature scale (see [H3] for a more extensive review).

The origination of the designation 'type M' lies with the scheme – devised at Harvard, by Mrs W. P. Fleming – for spectral classification outlined in the *Draper Memorial Catalogue* of 1890 (with the 'M' stemming from the fact that Secchi's class I and II stars were divided into eleven subclasses, A to K, omitting J). This catalogue included 10,351 of the brightest northern stars – a large, but not overwhelming number – and the Harvard system had some competitors, notably Lockyer and Vogel. However, as Hearnshaw explains, Pickering's championing of his local system, coupled with the irresistible force of Cannon's classification of the 225,300 stars in the Henry Draper catalogue (1918–1924), led to the Harvard spectral types becoming the *de facto* astronomical reference system by the 1920s.

Early spectral catalogues, based on visual or photographic data, were restricted to stars with bright apparent magnitudes. As a result, almost all M stars in these catalogues are giants rather than dwarfs – which became apparent only after Hertzsprung [H5] and Russell [R5] independently arrived at versions of the diagram that bears their names. Plotting luminosities against a temperature indicator (colour indices in Hertzsprung's case [H6], and spectral types by Russell [R6]) produced the wishbone-shaped diagram with 'giants' and 'dwarfs' (named by Russell) and a 'main sequence' (defined by Hertzsprung from his observations of Hyades and Pleiades cluster members). Figure 2.1 shows the modern incarnation of this diagram, based on stars with parallaxes measured by the ESA Hipparcos satellite. The initial H–R diagrams included few late-type (cool, low-mass) dwarfs, but nevertheless demonstrated that luminosity decreased with later spectral types further along the main-sequence.

The Harvard system provided a ready means of comparing the general properties of many stars, but only in an approximate fashion. Its utility was limited by its dependence on internal classification criteria: that is, the relative strengths of lines and bands in the programme star (Hβ against CN, for example) were measured and compared against a reference list of line ratios, and the spectral type assigned based on that comparison, rather than by matching against a set of standard stars. Moreover, the final calibration was rather coarse, with each spectral class spanning a wide range of physical properties, particularly with spectral type M. At Mount Wilson, Adams attempted to address the latter problem by adding a numerical qualifier, based on the absolute magnitude estimated from the relative strength of individual spectral lines. However, this approach had the unfortunate consequence that a change in the absolute magnitude calibration changed the entire spectral type scale.

Both problems were addressed by Morgan and Keenan in their definition of the MK spectral classification system (see [O2] for a thorough discussion of the genesis

The HR diagram

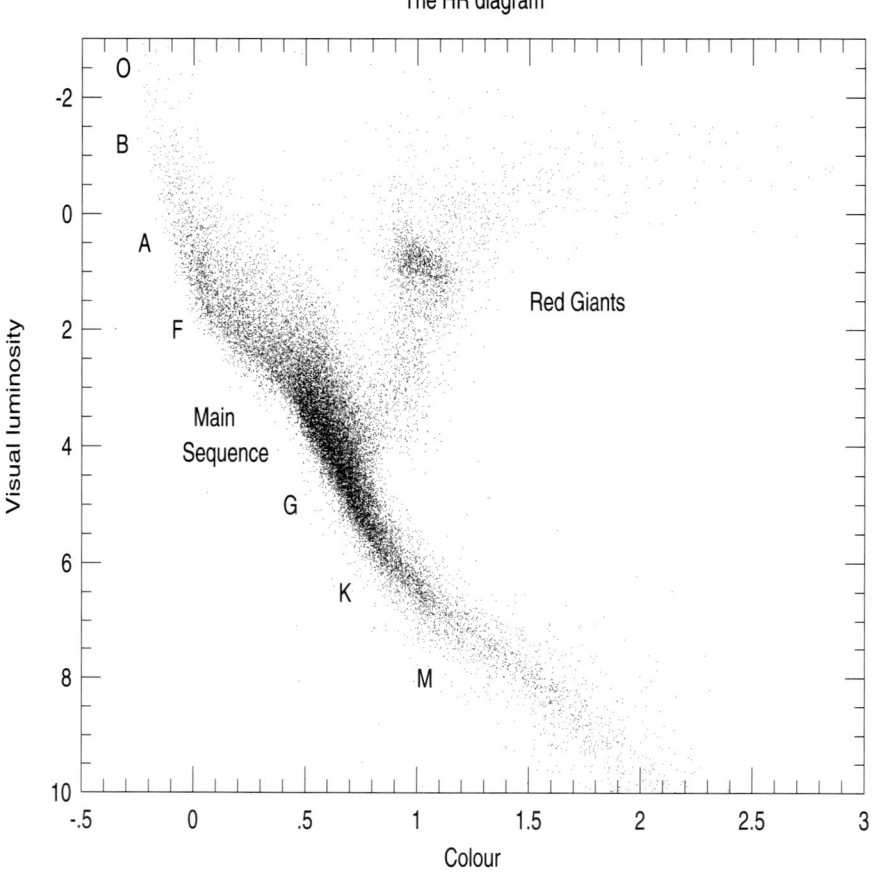

Figure 2.1. The Hertzsprung–Russell diagram, as defined by stars with parallaxes measured to an accuracy of better than 10% by the ESA *Hipparcos* satellite [E1].

of this system). The main achievement of their system was its empirical definition of spectral types with respect to a set of particular standard stars. Spectra of programme stars are classified by comparing the relative strength of specified features within a particular wavelength range (initially 3930–4860 Å) against observations, taken to the same signal-to-noise and at the same dispersion, of the standard stars. Moreover, the MK system added a formalised definition of luminosity class, with types I, III and V replacing the intermittently-used 'c' (narrow-lined), 'g' (giant) and 'd' (dwarf) qualifiers. The initial grid of standards was laid out in the MKK atlas of stellar spectra [M8]. The overall success of the system can be judged by its continued use (albeit in modified form) in current astronomical research. Many of those subsequent modifications and improvements have centred on the treatment of M dwarf stars.

2.2.2   An M dwarf classification system

Spectral classification is based purely on morphology – regularities in the appearance and disappearance of particular spectral features. The expectation is that a well-designed system, ranked by spectral variation, is also based on physical properties. Spectral type therefore provides a shorthand method of describing the overall physical characteristics of a given star, although it should be emphasised that the physical understanding (of why stars show spectral variations) post-dates, and is independent of, the definition of the classification system.

The traditional method of defining a spectral classification system is to take observations of an ensemble of stars – preferably of known absolute magnitude and luminosity class – and identify several key spectral features, prominent enough to be recognised easily, which show smooth behavioural trends when the individual spectra are suitably arranged. Thus, the hydrogen Balmer-series lines increase in strength from type O through B to type A, subsequently diminishing in strength through types F, G and K. Earlier (OB) and later (FG) stars can be distinguished by considering other features, such as the presence and strength of He I lines, the G-band (due to the CH molecule) or the H and K lines due to ionised Ca (Ca II).

The TiO bands which dominate optical spectra of M stars are an obvious choice as a primary indicator of spectral type. The next operational step in defining a workable classification system is to set up a grid of standards. Early observations were limited to stars with relatively bright apparent magnitudes, and, as a result, included few late-type M dwarfs. Moreover, most photographic observations before the 1940s were confined to the blue–green region of the spectrum, where spectral features saturate rapidly with decreasing temperature. As a result, the first MK system is of limited utility in M dwarf classification.

Recognising these limitations, the MK system was originally defined only for main-sequence stars of type M2 and earlier. With the development of larger telescopes and more efficient spectrographs and detectors, observations of later-type dwarfs accumulated, and it became necessary to extend the classification to these cooler dwarfs. However, with no generally accepted guidelines, several mutually incompatible systems arose. The two most widely used were the Yerkes system, developed by Morgan [M7] and Kuiper [K10], and tied primarily to the strength of TiO bands between 5800 and 6500 Å; and Joy's Mount Wilson system, which took TiO bandstrength in the blue spectral régime as the main indicator of spectral type. Based on different criteria at different wavelengths, these systems diverged in their classification of the later-type M dwarfs. The star Wolf 359 was classed as type M8 by Morgan, but as type dM6e by Joy. Thus it was necessary to determine what system had been used to classify a star before observations of different M dwarfs could be compared.

Part of the problem in classifying M dwarfs using visual or blue spectra is that those wavelengths lie far from the peak of the energy distribution, making it more difficult to obtain high signal-to-noise observations. While Keenan and MacNeil [K1] added to the wavelength coverage of the revised MK system, Boeshaar [B6] first included features as red as 6800 Å. Kirkpatrick and collaborators [K2] further extended the system to both longer wavelengths and later-type stars.[1] Their calibration – designated the KHM system – is based on both the relative strengths of spectral features in the range 6300–9000 Å and on the overall spectral slope across these wavelengths. Classification is through least-squares matching of flux-calibrated spectra against observations, made at the same resolution, of an extensive grid of standards. All spectral features contribute to some extent to the final type determination. However, most of the weight in the calibration rests with the stronger molecular bands (titanium oxide, vanadium oxide and calcium hydride), and the strength of individual features can be used to define a scale that approximately matches the KHM system. For example, measurements of the depth of the 7050 Å TiO bandhead provide spectral types for dwarfs earlier than type M6 (where the bandhead saturates), with the observations calibrated against data for KHM standard stars [R2].

Bessell [B5] has defined an alternative spectral-type calibration for M dwarfs. His system is related to Wing's [W3] giant-star spectral types, based on TiO bandstrength for earlier-type M dwarfs and on VO for the later-type stars where the TiO bands saturate. There is reasonable agreement ($\sim$0.5 spectral classes) with the KHM system in the former case, but Bessell assigns systematically earlier types to the later VLM dwarfs: for example, Gl 752 B (VB 10) is type M8 on the KHM system, but type M7 on Bessell's system. The more widely used KHM system is adopted in this book.

Figures 2.2 and 2.3 present optical spectra of representative K and M dwarfs, covering the blue–green and far-red wavelength regions, and identifying the principal spectral features. There are obvious correlations in the strength of individual lines and bands with changing spectral type. TiO is present, but weak, at type K7, and grows in strength until type M6, where most of the bandheads saturate. Bands due to several metal hydrides – MgH, FeH and, particularly, CaH – also first become detectable among K7 stars and grow in prominence with later spectral type. At approximately the same temperature as the redder TiO bands saturate, VO becomes evident at 7330–7530 Å. FeH bandheads also appear in the latest spectral types at 7786, 8692 and 9020 Å.

The strongest atomic lines at these wavelengths are Ca I at 4227 Å, the Na I doublets at 5890/5896 Å (the D lines, at the long wavelength limit of the blue–green spectra plotted in Figure 2.2) and 8183/8195 Å, the K I doublet at 7665/7699 Å and, to a lesser extent, the Ca II 'infrared' triplet at 8498, 8542 and 8662 Å. The last mentioned decline in strength towards later spectral types, besides being more difficult to detect as the 8432 Å TiO bandhead increases in strength. The remainder show a general increase in strength, although Ca I 4227 Å is swamped by TiO absorption at later spectral types. The KI doublet

---

[1] Those observations concentrate on the nearest stars, most of which are identified either by their number in Willem Luyten's catalogue of stars with annual proper motions exceeding 0.5 arcsec (the Luyten Half Second Catalogue, designated LHS, as in LHS 2924), or by their number from the nearby star catalogues compiled by Wilhelm Gliese and Hartmut Jahreiss (designation Gl or GJ; for example, Wolf 359 is Gl 406, the 406th star in Gliese's 1969 catalogue [G5]. See Section 7.5.1 for further details.

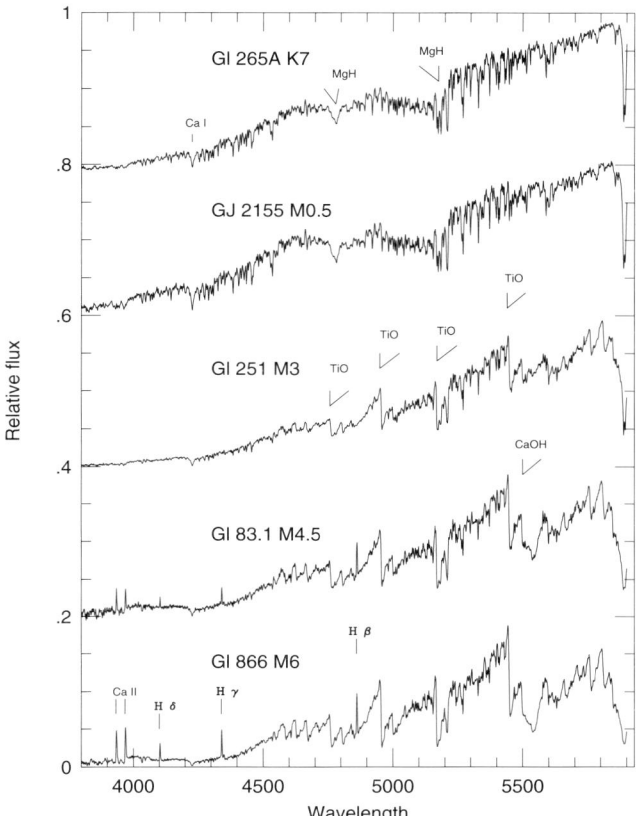

Figure 2.2. Blue–green optical spectra of M dwarfs, illustrating the main features used to calibrate spectral type. Prominent molecular bands and atomic features have been identified.

narrows in equivalent width between ~M7 and M9.5, before broadening dramatically at later types. Detailed M dwarf spectroscopic atlases are provided by Kirkpatrick *et al.* [K2], [K3] and, for the latest M spectral types, Tinney and Reid [T1].

2.2.3   Beyond M: spectral type L

M dwarfs were long regarded as defining the lowest extremities of the hydrogen-burning main sequence – the *Ultime Thule* of the H–R diagram. However, the marked improvement in sensitivity of wide-angle photometric surveys, particularly at near-infrared wavelengths, has resulted in the detection of increasingly fainter and cooler low-mass dwarfs. The most extreme of those objects have spectral characteristics which cannot be accommodated in class M, requiring the definition of a new spectral class – the first for almost half a century – class L.

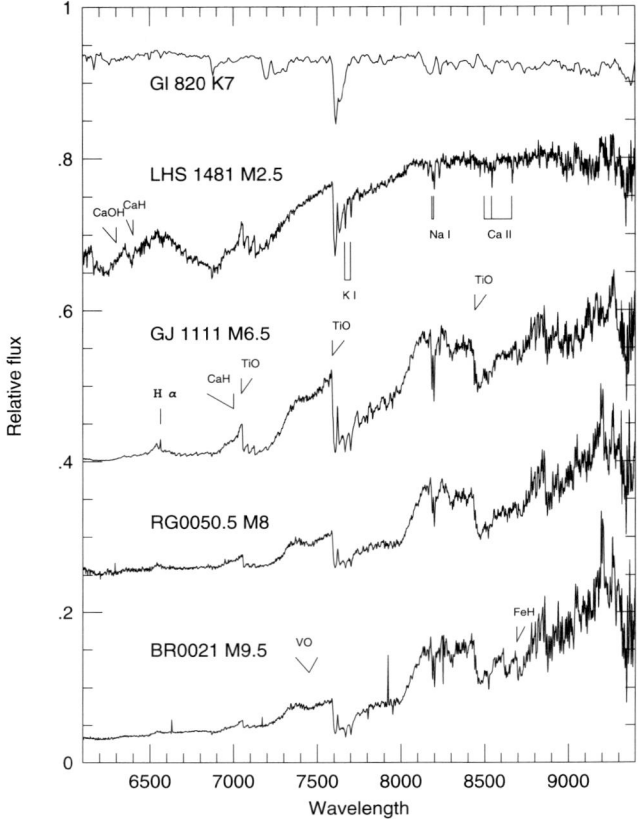

Figure 2.3. Red optical spectra of M dwarfs, illustrating the main features used to calibrate spectral types.

The first L dwarf, discovered in 1988, was GD 165B, an extremely red, low-luminosity companion of a DA white dwarf [B2]. Initial spectroscopic observations [K4] suggested puzzling dissimilarities between this dwarf and well studied late-type M dwarfs, such as VB 10, LHS 2924 and LHS 2065. However, with only a 2.3-m telescope, the signal-to-noise in the available spectrum was low, while the much brighter white dwarf companion limited coverage to longward of ~ 7500 Å. Finally, the (remote) possibility was raised of significant atmospheric contamination by metal-rich ejecta during the asymptotic giant branch or planetary nebula evolutionary phase of the degenerate companion. Thus, GD 165B was, by-and-large, considered a unique oddity.

Nonetheless, the quest for lower-temperature, lower-luminosity dwarfs continued, spearheaded by Kirkpatrick, Henry and collaborators, who compiled samples of 'ultracool' ($>$M7) dwarfs from proper motion catalogues [K5]. As described in more detail in Chapter 9, the breakthrough came in 1997. First, follow-up spectroscopy of candidate late-type dwarfs in the general field, selected using near-infrared photometry obtained with the prototype camera for the 2-Micron All-Sky Survey (2MASS, [S6]), resulted in the identification of one extreme ultracool dwarf, 2MASSP J0345432+254023, initially classed as type >M10 [K6]. Shortly thereafter, Ruiz *et al.* [R4] discovered a faint proper motion star, named Kelu 1 (Kelu is Mapuche for red), with an unusual optical spectrum, reminiscent of GD 165B.

The trickle turned into a flood with the first results from the wide-field near-infrared sky surveys, DENIS and 2MASS. Spectroscopy of ultracool candidates from the DENIS brown dwarf mini-survey [D1] revealed three isolated dwarfs with spectra similar to GD 165B. Similarly, follow-up observations of colour-selected candidates from the 2MASS survey resulted in the identification of more than 85 such objects between August 1997 and July 1999 [K8]. The latter observations had the advantage of the unparalleled light grasp of the Low Resolution Imaging Spectrograph [O1] on the Keck 10-m telescope, but even so, many L dwarfs are too faint to observe at $\lambda < 7000\,\text{Å}$. With more than 100 L dwarfs now known, GD 165B has been transformed from an oddity to a prototype.

The 2MASS dataset is sufficiently rich that it provides a well-sampled sequence for the new spectral class (Figure 2.4). As with the KHM M dwarf sequence, the primary classification is based on spectral behaviour over the wavelength range 6000–10000 Å. Specific features are identified in Figure 2.5. The main characteristic separating type L from type M is the diminishing strength of TiO, the primary criterion of spectral class M. VO remains prominent among the earliest L dwarfs, but is barely detectable in Kelu 1 (spectral type L2). The dominant molecular features are metal hydrides: CaH, FeH and CrH at far-red wavelengths, and MgH shortward of 6000 Å. The Na I doublet at 8183/8192 Å weakens with progressively later spectral type, but the resonance lines due to K I, Cs I, Rb I and the sodium D lines all become increasingly stronger. The behaviour of the K I 7665/7699 Å doublet is particularly interesting, broadening in a pronounced fashion at spectral type L4. The two components are essentially indistinguishable, with a combined equivalent width of more than 100 Å. The sodium D lines show similar behaviour in the few L dwarfs bright enough for observation at those wavelengths.

The physics underlying these variations is discussed in detail in Chapters 4 and 9. In qualitative terms, dust formation is believed to remove TiO and VO from the atmosphere, reducing the opacity. As a result the photosphere (the stellar 'surface') lies at relatively high depth, where pressure broadening leads to line widths of white dwarf proportions amongst the more abundant species.

Several of the spectra plotted in Figure 2.4 exhibit an absorption line at 6708 Å (for example, 2MASS J1146+2230 and 2MASS J0850+1057). This feature is due to lithium and, as explained in Chapter 3, identifies these objects as having masses below $0.06\,M_\odot$ – well within the brown dwarf régime. However, the theoretical models described in Chapter 3 suggest that stars with masses below 0.08–$0.085\,M_\odot$ can reach temperatures lower than the likely M dwarf/L dwarf boundary ($\sim 2{,}000\,\text{K}$). Thus, L dwarfs probably represent a mixture of very low-mass hydrogen-burning stars and degenerate brown dwarfs.

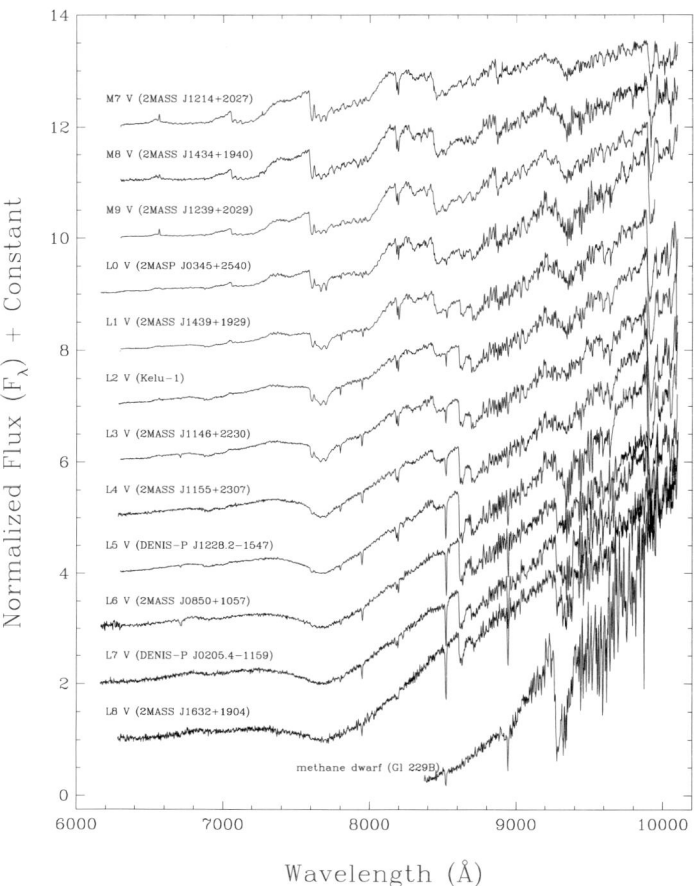

Figure 2.4. The L dwarf spectral sequence. (From [K8], courtesy of the *Astrophysical Journal*.)

Two final questions require discussion. First, why spectral type L? The original justification is given by Kirkpatrick [K7]. In brief, after eliminating possible confusion with white dwarfs, elliptical galaxies, supernovae and so on, only five letters – H, L, T, Y and Z – remain as unambiguous spectral type designations. Of those, L lies closest to M. Type T is reserved for Gl 229B-like methane dwarfs (Chapter 9). The second question concerns the devising of the most appropriate memnonic for the new spectral classification scheme OBAFGKMLT. This question is left as an exercise for the reader.

2.2.4   Infrared spectroscopy

M dwarfs emit most of their energy at near-infrared wavelengths between 1 and 2.5 μm. Until recently, the only observations that were possible were at low spectral

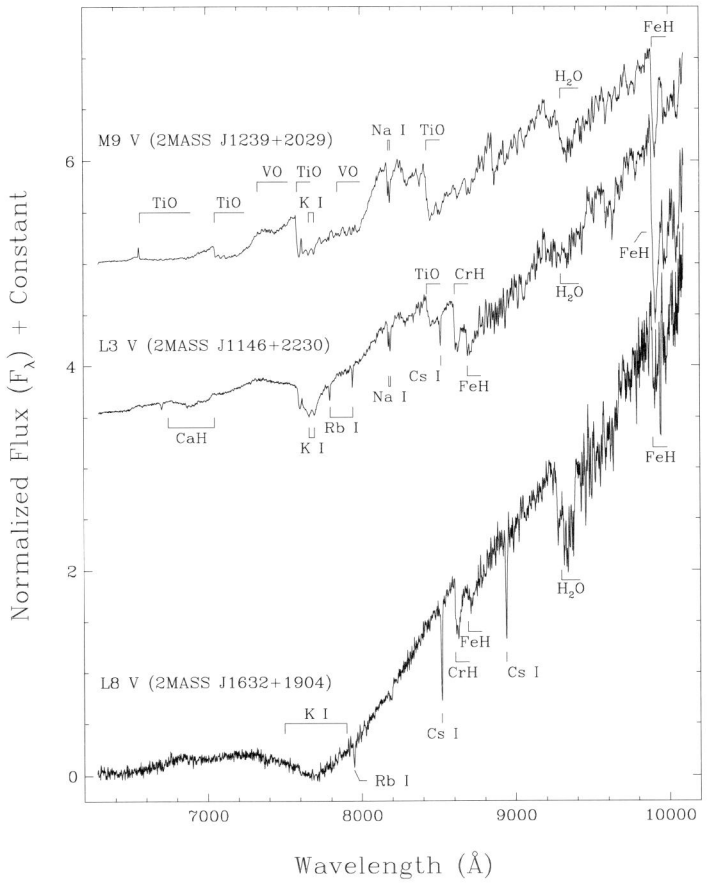

Figure 2.5. Key features defining spectral class L. (From [K8], courtesy of the *Astrophysical Journal.*)

resolution – effectively narrow-band photometry with a tuneable filter and a single-element detector [F2]. This state of affairs has changed radically with the availability of large format (up to 1,024-square) two-dimensional infrared detectors, and it is now possible to obtain near-infrared spectra that are comparable with optical data in both quality and resolution.

Figure 2.6 shows 1–3 μm spectra of early-, mid- and late-type M dwarfs. The most prominent features – striking even in the first low-resolution observations – are the broad absorption bands at 1.4, 1.8 and 2.6 μm due to water. These bands are also present in the Earth's atmosphere and, indeed, are effectively opaque at low altitudes above sea-level. Thus, only the wings of the stellar steam bands, broadened at the higher temperatures and pressures of the M dwarf atmospheres, are detectable from most astronomical observatories. However, there are a few dry, high-altitude sites – notably Mauna Kea, and parts of Antarctica – where it is possible to detect stellar

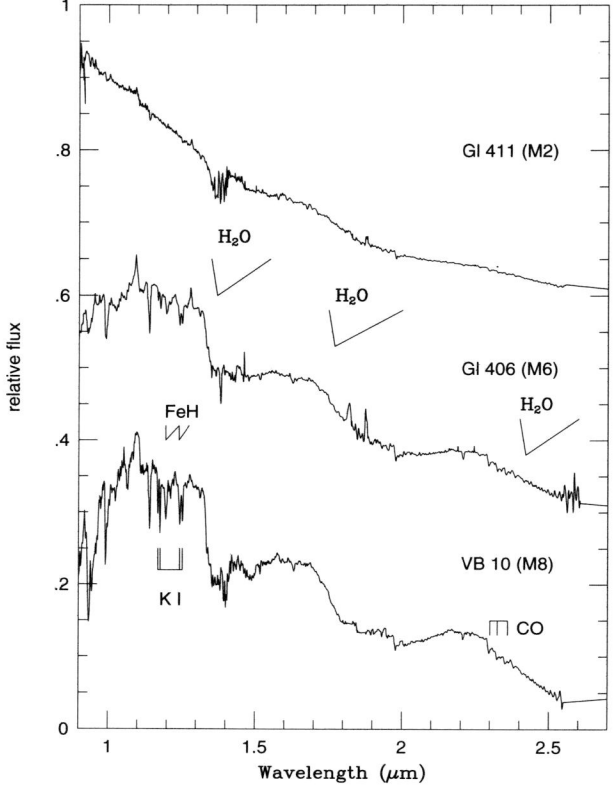

Figure 2.6. Near-infrared spectra of three late-type dwarfs: Gl 411 (M2), Gl 406 (M6) and VB 10 (M8). The most prominent features are the steam bands due to H_2O in the atmospheres of these cool stars, but CO and FeH bandheads as well as various atomic lines are also present. (Spectra courtesy of H. R. A. Jones.)

radiation across at least the two shorter wavelength bands. As with TiO at optical wavelengths, the steam bands are barely perceptible at spectral types K7 and M0, but grow in strength with decreasing temperature.

Other weaker molecular bands – notably CO at $2.15\,\mu m$ – are also present, while recent higher-resolution observations have identified FeH in the $1.1–1.2\,\mu m$ region [J2]. The latter authors also identified atomic lines due to Fe I and Mg I in the same region, while strong K I lines were detected previously by Kirkpatrick $et\ al.$ [K3], and Na I absorption may have been detected at $2.1\,\mu m$. These lines mirror the behaviour of the atomic lines at optical wavelengths, growing stronger in the later spectral types. Relatively few L dwarfs have been observed at these wavelengths, but the overall trend continues in these cooler objects. At some point, however, the temperature drops sufficiently to allow methane formation, and the resulting absorption bands remove over half the flux in the H ($1.6\,\mu m$) and K ($2.2\,\mu m$)

passbands. The low-mass brown dwarf companion Gl 229B [N1] was the first object of this class to be detected, but several similar objects are now known in the field (see Chapter 9). These dwarfs are designated spectral class T.

2.2.5 Mid-infrared and longer wavelengths

Observations of M dwarfs at mid-infrared ($\lambda \sim 5$–$100\,\mu$m) and sub-millimetre ($\sim 100\,\mu$m to 1 mm) wavelengths are extremely difficult, since there are no clean atmospheric windows, and faint targets must necessarily be detected against a bright, variable background. Despite those problems, 8–12 μm spectrophotometry has been obtained for a few stars (such as Gl 411 and Gl 406, by Aitken and Roche [B3]), while photometry exists for a number of (mainly) early-type M dwarfs. Spaceborne observations have higher sensitivity, since the overall transparency is higher and the background lower, and the IRAS satellite provided 12, 25, 60 and 100 μm data for a number of M dwarfs [W1], [M12], [M4]; but, with only a 20-cm telescope, was unable to obtain spectroscopy. The forthcoming SIRTF satellite should prove more effective.

Nonetheless, the available observations provide some indication of the behaviour of M dwarf atmospheres at long wavelengths. Figure 2.7 plots the 12 μm : 2.2 μm and 25 μm : 2.2 μm flux ratios for nearby M dwarfs detected by IRAS. The former ratio rises from $\sim 5\%$ at (I–K) ~ 1.5 (spectral type M1) to $\sim 10\%$ at (I–K) ~ 3 (spectral type M5). The longer wavelength flux is lower by approximately a factor of two.

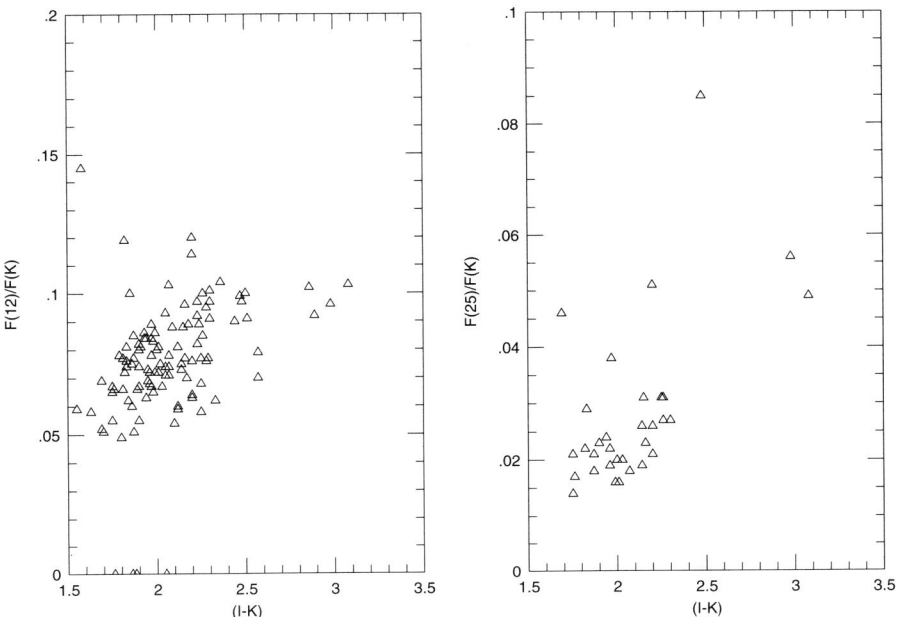

Figure 2.7. Infrared flux ratios for M dwarfs detected by IRAS.

In general, fluxes at wavelengths shortward of 20 µm are consistent with an extrapolation of a black-body, although observations of late-type dwarfs, where dust formation may be underway, are both scarce and imprecise. At longer wavelengths, however, M dwarfs are significantly brighter than expected. A number of stars (including Gl 65AB, Gl 285 and Gl 873) have fluxes at centimetre wavelengths which are four orders of magnitude above the black-body extrapolation [C2], while a few stars may also have been detected at sub-millimetre wavelengths [M4]. Active flare stars are more prone to detection (for example, see Lim *et al.* [L3], which describes the intermittent detection of Proxima Centauri at a wavelength of 20 cm). Mullan *et al.* [M13] note that the infrared/radio excess can be matched by a power-law, $F_\nu \propto \nu^{0.7}$, and suggest its origin in an ionised wind, although Lim and White [L4] argue against that hypothesis. Further observations of a larger number of targets are required to clarify the characteristic behaviour of M dwarfs at these long wavelengths.

2.2.6   Spectroscopic dwarf/giant indicators

M giants are typically 10^4 to 10^6 times more luminous than M dwarf stars of similar temperatures, implying (from equation (1.28)) that giants have radii 100–1,000 times larger than those of dwarfs. Both gravity and pressure are much lower in the more extended atmospheres, leading to spectroscopic differences with the more compact dwarfs. Atomic lines are significantly weaker in the giant stars. Indeed, the strength of Ca I 4226 Å is one of the principal luminosity discriminants for early-type M-stars in the revised MK system [K1], while the Na D lines (5890, 5896 Å), the 8184/8192 Å Na I doublet and the Ca II triplet (8498, 8542, 8662 Å) can also be used to discriminate between later-type dwarfs and giants. Balmer lines, in contrast, are stronger in giants earlier than type M3.

Molecular bands provide another luminosity-dependent spectral signature. Metal hydride bands are significantly stronger in dwarfs than in giants, and both MgH (in early M) and CaH (in later types) have been used to define luminosity classification. As Figure 2.8 shows, this classification can be extended to include young M-type T Tauri stars, which exhibit CaH bandstrengths intermediate between dwarfs and giants. The latter trait is consistent with the hypothesis that these stars are still contracting onto the main sequence (see Section 3.6). In contrast, CO and CN are significantly stronger in giants than in dwarfs; indeed CN is essentially absent from dwarfs.

Narrowband photometric systems can be designed to measure the strengths of these molecular features. Either specifically designed filters are used to measure the flux within an absorption band and at a nearby pseudo-continuum point, or the equivalent data are determined from flux-calibrated spectra. Matched against a suitable temperature indicator (broadband colours, spectral type), the on-band/pseudo-continuum flux ratio measures the bandstrength. Wing [W3] devised an eight-colour system which includes bands centred TiO (7120 Å) and the CN band (at 10395 Å), while Mould and McElroy [M10] used filters centred at 6880 and 7120 Å to compare the relative strength of the CaH 6880 Å and TiO $\gamma(0,0)$ bands. The latter

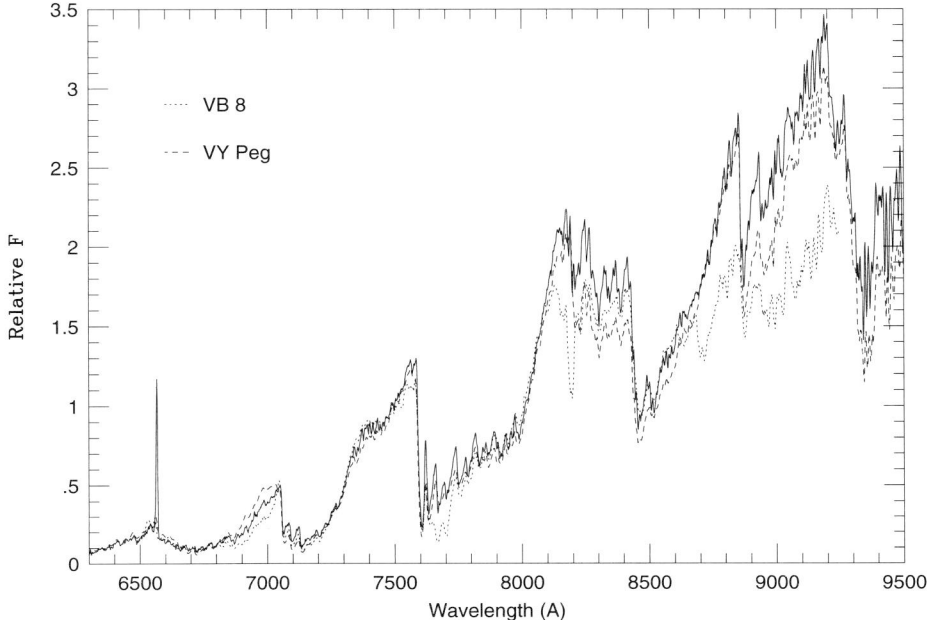

Figure 2.8. Luminosity-sensitive features of M-type stars. The solid line plots data for GG Tau Bb, a 10^6-year-old T Tauri star in a quadruple system. This spectrum is compared with the M7 giant VY Peg and the M7 dwarf VB 8. The main-sequence dwarf has the strongest atomic lines (notably K I at 7665/7699 Å) and CaH bands; GG Tau Bb has giant-like atomic features, but hydride bands intermediate in strength between VB 8 and VY Peg. (Courtesy of Russel White and the *Astrophysical Journal*.)

system has been developed to provide a spectral classification system for subdwarfs by Gizis ([G3], see also Chapter 11). In each case, the gravity-sensitive index is plotted against a gravity-insensitive index (such as spectral type or TiO band-strength), allowing separation of dwarfs and giants.

In the near-infrared, the 2.295 μm CO band is highly sensitive to luminosity class (Figure 2.9), and the CO, H_2O narrowband system [B1] was designed specifically to measure the strength of that band relative to the temperature-sensitive 1.9 μm steam band. Molecular luminosity-class indicators are more effective among later spectral types, where the bands are intrinsically stronger.

2.3 BROADBAND PHOTOMETRY OF M DWARFS

2.3.1 Spectral energy distributions, temperatures and radii

Spectral type can be used to estimate surface temperature. If a star lies on the main sequence, its intrinsic luminosity can be deduced by matching the temperature estimate against empirical or theoretical H–R diagrams. There are, however, limitations in using spectral type for this purpose. Figure 2.10 plots absolute

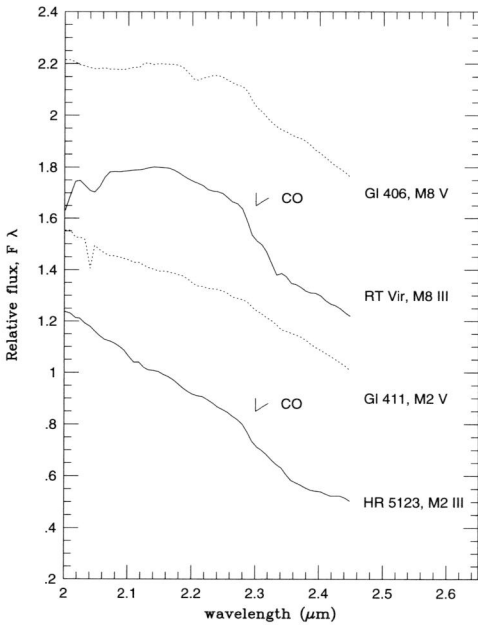

Figure 2.9. Infrared spectra of M dwarfs and giants. The giants have stronger CO ($\lambda \sim 2.295\,\mu$m) absorption than dwarfs of the same spectral type. (From [A3].)

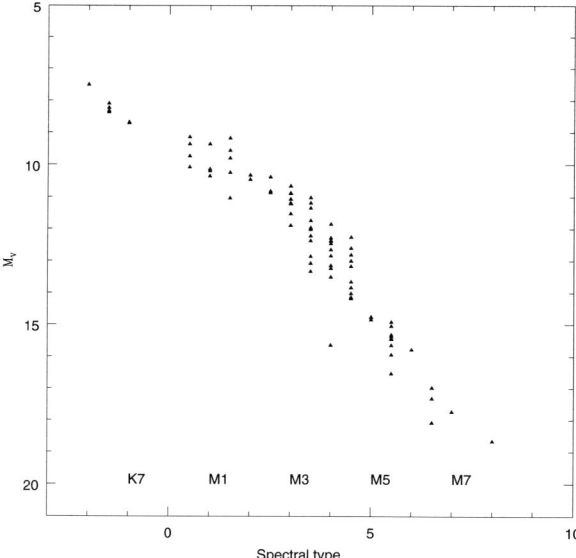

Figure 2.10. The relationship between absolute visual magnitude, M_V, and spectral type, defined by stars within 8 parsecs of the Sun.

visual magnitude against spectral type for stars currently known to lie within 8 parsecs of the Sun (see Appendix). Since types are quoted to only 0.5 subclasses and have an accuracy of at best ±0.5, the resultant H–R diagram is coarse-grained, with an rms dispersion of $\sigma(M_V) = \pm1.5$ magnitudes for early to mid-M dwarfs. Distances estimated from spectroscopic type alone are accurate to only a factor of two, and are therefore of limited value.

Broadband colours provide an alternative means of estimating stellar temperatures. Figure 2.11 illustrates how the Johnson/Cousins passbands sample the spectral energy distribution of late-type M dwarfs. Given the flux zero-points defined in Table 1.1, broadband photometry can be transformed to apparent flux densities, and approximate spectral-energy distributions determined. Figure 2.12 plots such data for four representative M dwarfs, scaling the distributions to match at the I-band. This demonstrates how the peak in the energy distribution shifts towards longer wavelengths with increasing spectral type, and how increased molecular absorption

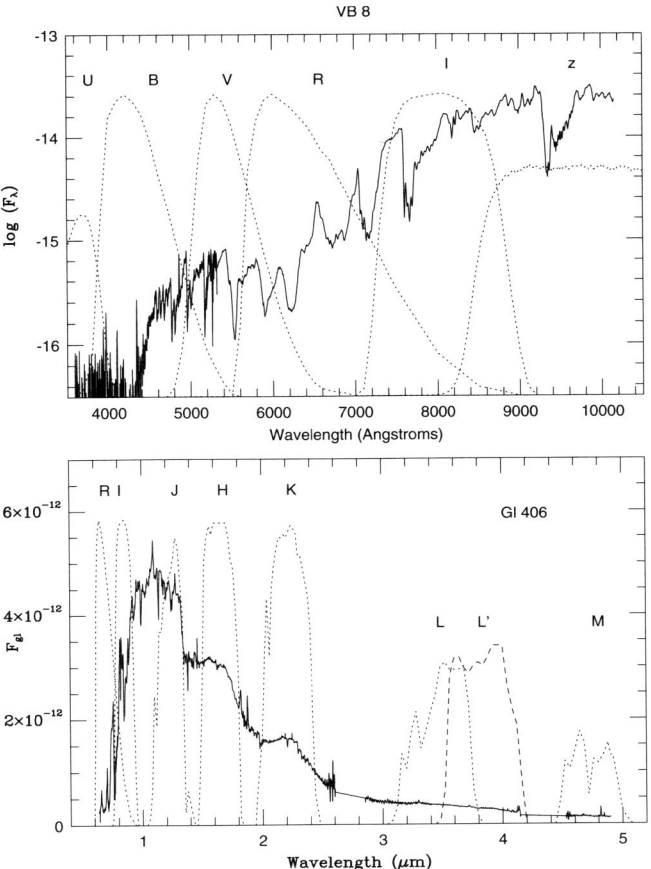

Figure 2.11. Broadband filter response curves.

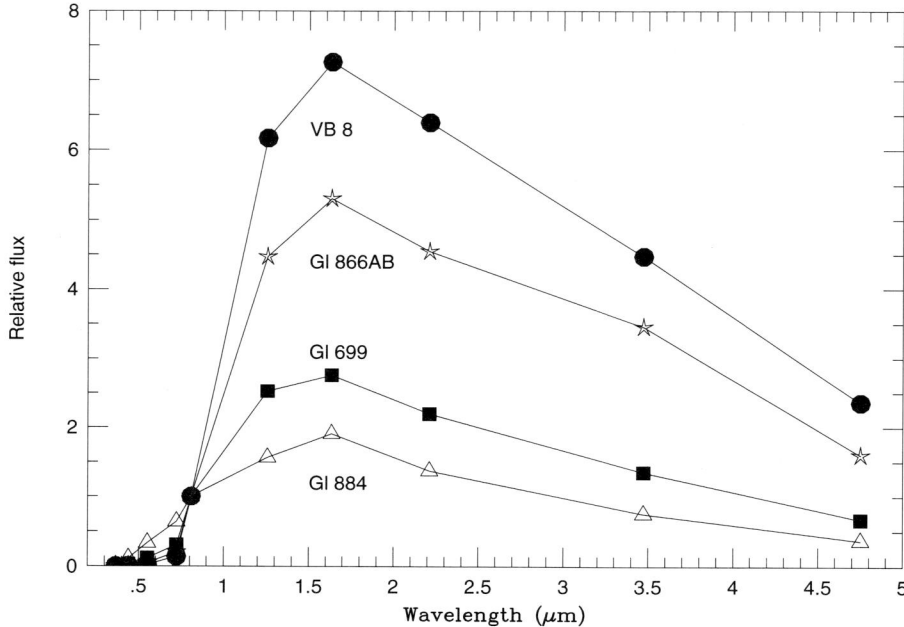

Figure 2.12. Broadband spectral energy distributions of M dwarfs: Gl 884, spectral type K5 (triangles); Barnard's Star, Gl 699, type M4 (squares); Gl 866AB, M5.5 (5-point stars); and VB 8, Gl 644C, type M7 (solid dots).

depresses the continuum at optical wavelengths. As a consequence, the fraction of the total flux emitted shortward of 7000 Å decreases from ~ 10% at spectral type K7 to <0.5% at M8.

Integrating the total energy beneath flux distributions such as those plotted in Figure 2.12 provides an estimate of bolometric magnitude, and hence luminosity. In undertaking these calculations, due allowance must be made for the fact that the near-infrared JHKLM passbands were chosen to avoid H_2O absorption in the terrestial atmosphere, and therefore also avoid the infrared steam bands in late-type M dwarfs (see Figure 2.6). Appropriate corrections (seldom exceeding 0.15 magnitudes) can be incorporated [B7], [L6]. M dwarfs have bolometric magnitudes ranging from $M_{bol} \sim 6.5$ at type K7 to $M_{bol} \sim 13$ for the M9/M9.5 dwarfs LHS 2924 and BR 0021-0216. These correspond to luminosities of $0.2 L_\odot$ to $4.5 \times 10^{-4} L_\odot$. L dwarfs reach substantially lower luminosities (as discussed further in Chapters 4 and 9).

In principle, luminosity determinations can be combined with measurements of stellar radii to determine effective temperatures using equation (1.28). In practice, direct measurements of radii of M dwarfs are rare. Infrared interferometers are capable of resolving the nearest M dwarfs, but these measurements are not yet of sufficient accuracy for useful radius determinations. At present, accurate data are limited to the M dwarf eclipsing binary systems YY Geminorum and CM Draconis.

The former, Gl 278 CD, is companion to the A-type binary, α Geminorum, and consists of a pair of M0 dwarfs, with $T_{eff}(1) = 3806$ K and $T_{eff}(2) = 3742$ K (± 200 K), masses of 0.62, 0.57 $M_\odot$, and radii of 0.66, 0.58 $R_\odot$ [L7]; the latter, Gl 630.1A, has a white dwarf common proper motion companion at a distance of ~ 380 AU, and consists of two M4 dwarfs, with effective temperatures of $\sim 3,150 \pm 100$ K, masses of 0.237, 0.207 $M_\odot$ and radii 0.252, 0.235 $R_\odot$ [L1], [M5]. CM Draconis has a high space motion relative to the Sun (163 km s^{-1}) and is probably one of the older members of the Galactic disk population.

The four binary components scarcely provide sufficient data to calibrate empirical radius and/or effective temperature relations. Determining temperatures using other techniques is a complex procedure, and is discussed in detail in Chapter 4. However, *representative* temperatures can be derived by matching black-body curves to broadband energy distributions, such as those illustrated in Figure 2.12. Pioneered by Greenstein *et al.* [G7], later studies have anchored black-body curves at either 2.2 μm [R1], [B3], [B4] or the L-band [T2]. The resultant temperature estimates are $T \sim 3,800 \pm 200$ K at M0 and $T \sim 2,100 \pm 200$ K at M9, implying radii of ~ 0.1 $R_\odot$ for VLM dwarfs such as VB8, VB10 and LHS 2924. From equation (1.29), this places the peak of the energy distribution in F_λ between 0.75 and 1.5 μm. The M dwarf and L dwarf temperature scale is discussed in more detail in Chapter 4.

2.3.2   Empirical colour–magnitude diagrams

Figures 2.2–2.5 show that the slope of the spectrum at optical wavelengths increases towards later types. Figure 2.12 shows that broadband colours provide a measurement of that slope. Plotting absolute magnitudes against those colours provides one of the fundamental tools of twentieth-century astrophysics – the colour–magnitude diagram, a proxy for the (L, T_{eff}) H–R diagram. Figure 2.13 plots the (M_V, (B–V)) diagram described by stars within 8 parsecs of the Sun (see Appendix). Stars with known unresolved binary companions are omitted from this figure and succeeding diagrams. Undetected companions may remain amongst the stars we have plotted, but as far as possible the measurements trace the behaviour of single stars.

The (M_V, B–V) diagram is the traditional stellar colour–magnitude diagram, a direct descendant of the 'photographic/photovisual' diagrams of the 1950s. However, this blue–green colour index is ill-suited for observations of M dwarfs, since the main-sequence steepens sharply at (B–V) magnitude ~ 1.4 (spectral type $\sim$ M1). There are two reasons for this. First, this temperature marks the onset of strong molecular bands at optical wavelengths; and second, the peak of the energy distribution moves longward of 8000 Å, placing the B and V bands on the high-frequency tail of the energy distribution. Less than 0.5% of the total flux is emitted in the V-band by spectral type M5, while the B-band contributes a mere 0.15%. The result is that a change in the (B–V) colour of only 0.15 magnitudes encompasses a decrease in M_V of ~ 4 magnitudes. Note, however, the shallower slope at $M_V > 12.5$.

One of the main applications of the colour–magnitude diagram is estimation of photometric parallaxes – using the observed colour to infer an absolute magnitude, which, coupled with the apparent magnitude, allows an estimate of distance. The

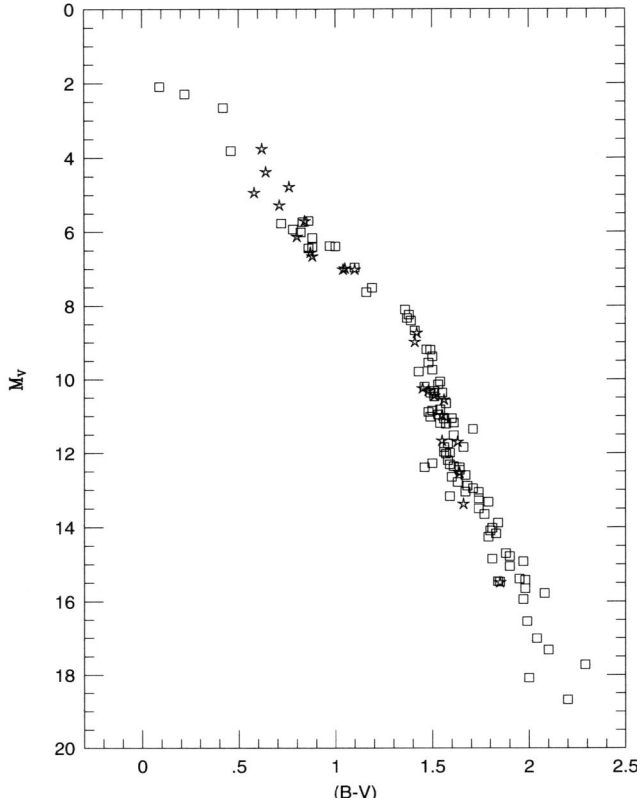

Figure 2.13. The (M_V, (B–V)) colour–magnitude diagram for nearby single stars within 8 parsecs of the Sun. Stars from the northern sample (appendix) are plotted as open squares, while southern stars are plotted as 5-point stars.

steep slope of the (M_V, (B–V)) relationship clearly mitigates against the effective use of this colour–magnitude diagram for late-type stars. The intrinsic dispersion on the main-sequence is ~ 1.5 magnitudes in M_V at a given (B–V), leading to distance uncertainties comparable to those derived from the spectral-type/M_V relation (Figure 2.10).

   Photometric parallax accuracy can be improved by using a colour with a larger dynamic range, defining an M dwarf main-sequence with a shallower slope than in (B–V). This can be achieved by choosing passbands providing a longer baseline in wavelength. Figure 2.12 shows that there is no advantage to choosing two passbands close to the peak in the spectral energy distribution, even though the stars are intrinsically more luminous at those wavelengths. The JHKL bands all lie near the flux maximum, so colours constructed from those passbands are insensitive to spectral type changes from M0 to M7. The most effective passband combination includes one filter close to the energy maximum in the near-infrared, with the second

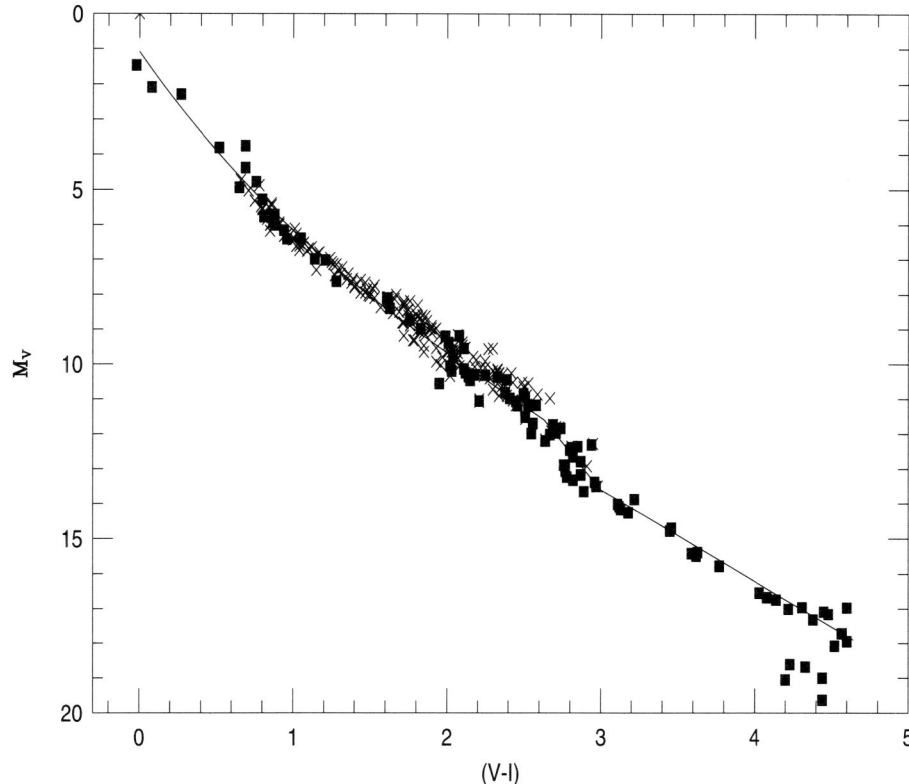

Figure 2.14. The $(M_V, (V–I))$ colour–magnitude diagram for nearby single stars. Crosses mark stars observed by the recent Hipparcos astrometric satellite; the remaining stars have accurate ground-based parallax measurements [M6]. Photometry taken from [B5] and [L5]. The solid line outlines the mean relationship listed in the text.

placed at sufficient distance in wavelength (usually in the optical) to provide the leverage necessary to measure small changes in the overall energy distribution.

The $(V–I)$ colour index offers many such advantages. Technologically, the I-band is the reddest well-calibrated passband measureable with photomultiplier tubes or CCD cameras, so both V and I can be measured with the same instrument (as opposed to combining optical and near-infrared observations, which require different instrumentation). The photometric system is also well defined, with an extensive grid of standards of all spectral types (see Chapter 1). Figure 2.14 demonstrates the advantages offered by $(V–I)$ for classifying early- and mid-type M dwarfs, although $(V–I)$ reaches a maximum for stars of $M_V \sim 18$ (spectral type M8) and becomes bluer for lower-luminosity stars. The latter behaviour probably reflects increased molecular absorption in the I-band by VO and FeH at these very low temperatures, as illustrated in Figure 2.3.

The main sequence changes slope at several points over the 20-magnitude range plotted in Figure 2.14 – at $M_V \sim 7$, ~ 8.5 and ~ 12. The last, at $(V–I) \sim 2.9$, is particularly abrupt. On either side of that point, the data are well matched by the following polynomials,

$$M_V = 3.98 + 1.437(V - I) + 1.073(V - I)^2 - 0.192(V - I)^3,$$
$$0.85 < (V - I) < 2.85 \quad (2.1)$$

and

$$M_I = 3.66 + 3.46(V - I) - 0.517(V - I)^2 + 0.0448(V - I)^3,$$
$$2.96 < (V - I) < 3.45 \quad (2.2)$$

The dispersions about these relationships are only $\sigma_{M_I} \sim 0.32$ magnitudes, but at $(V–I) \sim 2.9$ magnitudes (spectral type $\sim$ M4) stars span a range of 1.5 magnitudes in M_V, or a factor of four in luminosity. Possible physical reasons for each of these features are discussed in Chapter 3.

Colours based on optical and near-infrared wavelength passbands offer advantages in studying the lowest luminosity dwarfs. Figure 2.15 plots two such colour–magnitude diagrams, $(M_I, (I–K))$ and $(M_K, (V–K))$, together with the near-infrared $(M_K, (J–K))$ diagram. The photometry is taken from Leggett's [L5] compilation of data for M dwarfs. There is significantly more dispersion in colour at a given M_I/M_K among the early and mid-M spectral types $((I–K) < 3)$ than in Figure 2.14, perhaps partly reflecting problems in unifying all near-infrared data in a common system. The reddest points in $(I–K)$ are L dwarfs with known parallax and, at $M_I \sim 20.7$, the T dwarf Gl 229B. Methane absorption at 2.2 μm in the last object leads to the relatively blue $(I–K)$ and $(J–K)$ colours. The $(J–K)$ colour clearly provides no luminosity information for dwarfs with $M_K < 10$ (spectral types earlier than M8).

Passbands longward of 2.2 μm lie on the Rayleigh–Jeans tail of the energy distribution, and therefore show only limited variation with decreasing temperature and luminosity. $(K–L')$ colours range from ~ 0.15 mag. at spectral type M0 to ~ 0.65 mag. at M9 [L5]. Relatively few M dwards have 5 μm M-band data, with colours of ~ 0.2 mag. at spectral type M6 [B3].

Broadband colours can also be used to estimate bolometric magnitudes. The bolometric correction for a given passband is defined[2] as

$$BC_\lambda = M_{bol} - M_\lambda \quad (2.3)$$

To first order, BC_λ depends on the effective temperature of the star, and hence the broadband colour. The empirical relationship between BC_λ and a given colour can be calibrated through moderate-resolution spectrophotometry of a small number of representative standard stars. So far, the most thorough analysis has been undertaken by Leggett *et al.* [L6], who derive temperatures and bolometric magnitudes for 13 M dwarfs and three M-type halo subdwarfs by combining 0.35 to 5 μm spectrophotometry and broadband measurements. Figure 2.16 matches second-order polynomials to their derived corrections in the V-, I- and K-bands.

---

[2] Bolometric corrections are sometimes quoted in the opposite sense as $M_\lambda - M_{bol}$.

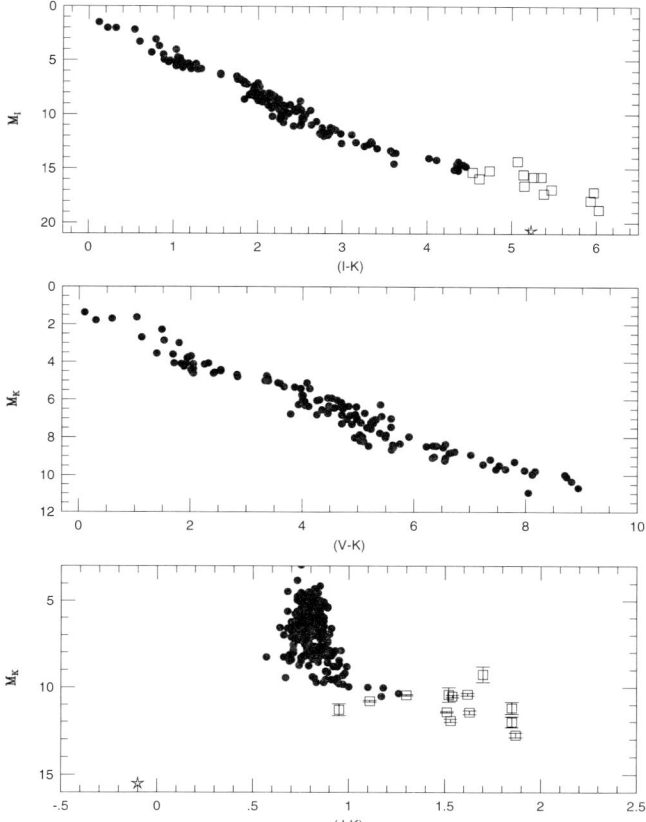

Figure 2.15. Optical/near-infrared colour–magnitude diagrams for nearby stars and brown dwarfs. The open squares in the (I–K) and (J–K) diagrams mark the locations of L dwarfs with parallaxes measured by the US Naval Observatory (see Chapter 9). The 5-point star is the methane dwarf, Gl 229B.

$$BC_V = 0.27 - 0.604(V - I) - 0.125(V - I)^2, \sigma = 0.059\,\text{mag}$$

$$BC_I = 0.02 + 0.575(V - I) - 0.155(V - I)^2, \sigma = 0.061\,\text{mag}$$

$$BC_K = 0.42 + 1.486(I - K) - 0.220(I - K)^2, \sigma = 0.048\,\text{mag} \qquad (2.4)$$

Unfortunately, the latest-type star in the [L6] sample is GJ 1111 (M6.5), and corrections become increasingly uncertain for later-type M dwarfs and L dwarfs. Nonetheless, the substantial corrections in BC_V and their rapid increase with redder colours (decreasing temperature) emphasise the distance of that band from the peak of the energy distribution. The K-band (and J-band) corrections, on the other hand, change by only 0.4 magnitudes between (I–K) ~ 2 (spectral type $\sim$ M2) and the reddest stars in the sample.

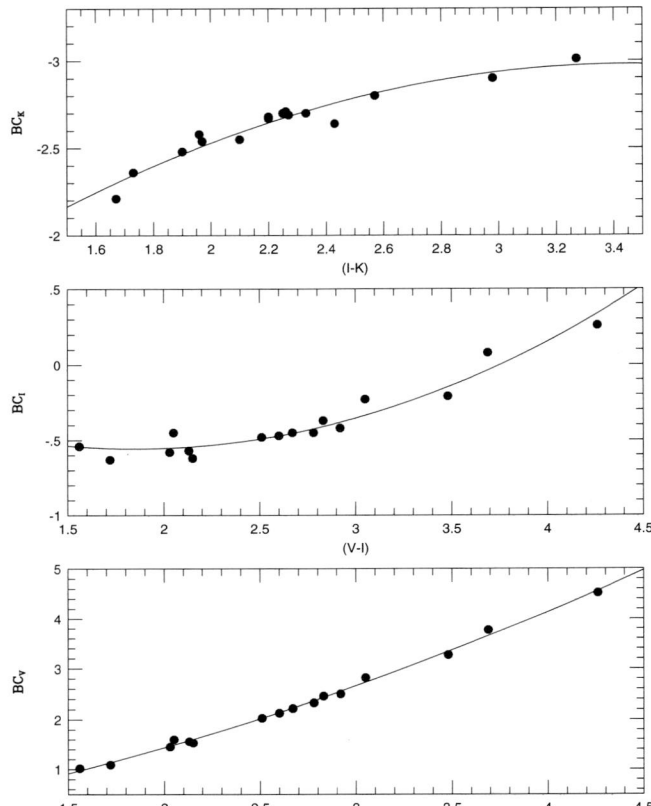

Figure 2.16. Bolometric corrections for the V, I and K broadband magnitudes as a function of (V–I) and (I–K) colours. The datapoints are from [L6], and details on the fitted relations are provided in the text.

2.4   ABUNDANCE EFFECTS ON THE H–R DIAGRAM

The Russell–Vogt theorem states that the position of a star on the H–R diagram is a function of its mass, abundance and age. The overwhelming majority of dwarfs in the vicinity of the Sun are members of the Galactic disk and have abundances within a factor of two of that of the Sun. The discussion in the previous two sections centred on their spectroscopic and photometric properties. This section briefly reviews the observational properties of lower metal-abundance subdwarfs in the Galactic halo (described further in Chapters 4 and 11).

2.4.1   Spectroscopic bandstrengths, absolute magnitudes and abundance

The term 'subdwarf' was devised by Kuiper [K9] to describe FGK-type stars which lie significantly below the sequence defined by the nearby stellar population in the

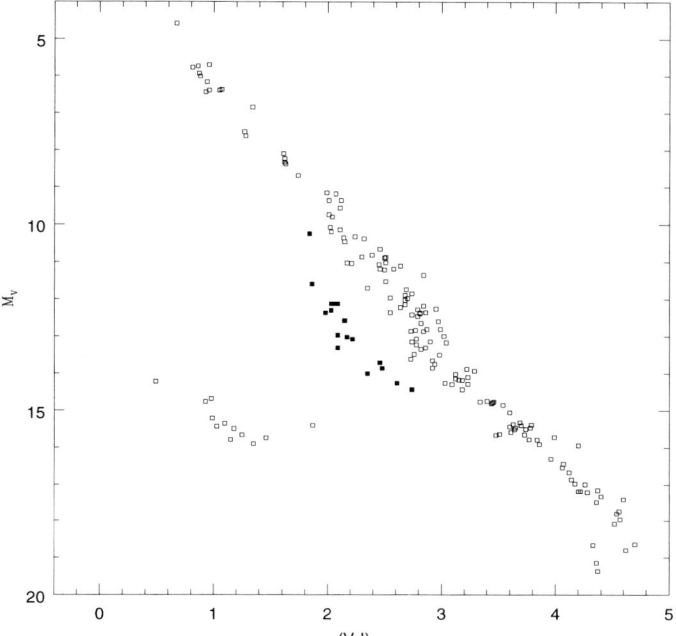

Figure 2.17. The subdwarf sequence in the $(M_V, (V–I))$ plane. The subdwarfs, plotted as solid points, are stars with π_{trig} determined by the US Naval Observatory [M6]. The degenerate white-dwarf sequence is also shown.

$(M_V, (m_{pg}–m_v))$ diagram. Most of these stars also have high space motions relative to the Sun, and several early studies (notably Chamberlain and Aller [C1]), recognised their low metal abundance relative to that of the Sun. Modern analyses show that the abundance distribution peaks at $[m/H] \sim -1.0$,[3] with a low abundance tail reaching $[m/H] = -4.0$. Sandage and Eggen [S4] first demonstrated that the subdwarf sequence lies below (or, rather, at higher temperatures) than the disk dwarf sequence in the $(\log(L), T_{eff})$ H–R diagram.

The subdwarf sequence(s) extends to the lower main sequence and spectral type M, with Kapteyn's Star (Gl 191; see Appendix) the classic example. Trigonometric parallax measurements of proper-motion stars show that subdwarfs can lie as much as two magnitudes below the disk main-sequence in the $(M_V, (V–I))$ plane, but apparently do not achieve absolute visual magnitudes or colours matching late M dwarfs (Figure 2.17). The reason for this behaviour lies in the lower metallicity, which leads to lower opacities and higher temperatures at a given mass (see Chapter 4). As a consequence, a larger proportion of the total luminosity is emitted at optical wavelengths (see Chapter 11).

---

[3] $[m/H]$ is the average metal abundance relative to the solar abundance $(Z = 2\%$ by mass) expressed on a logarithmic scale. Thus, $[m/H] = -1.0$ is equivalent to an average metal abundance one-tenth that of the Sun. See Section 4.3, equation (4.6)

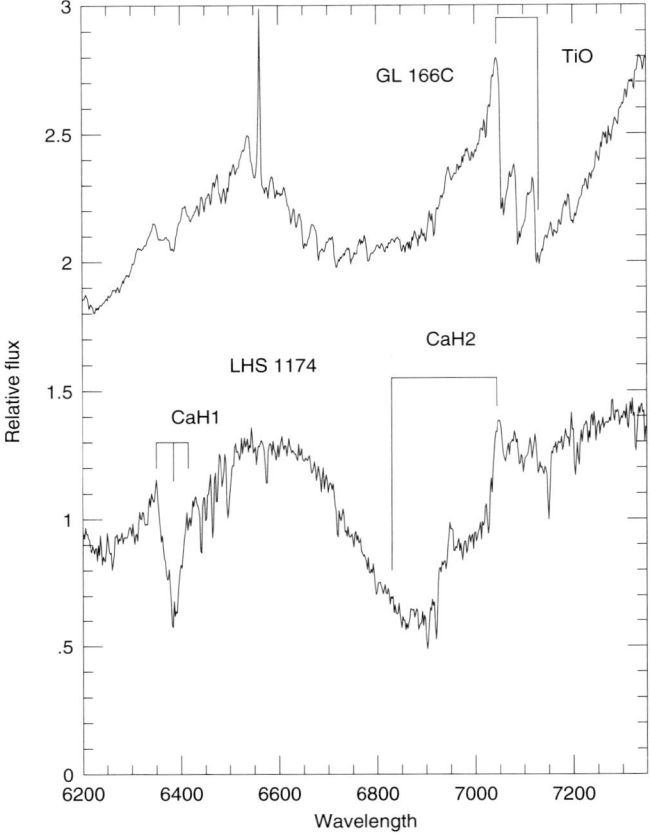

Figure 2.18. Red spectra of the near-solar abundance disk M dwarf Gl 166C, $M_V = 12.81$, and of the high-velocity halo subdwarf LHS 1174, $M_V = 12.97$. Despite the close similarity in absolute visual magnitude, there are striking spectroscopic differences, principally in the relative strength of the TiO and CaH bands.

Subdwarf stars are identified spectroscopically by comparing the relative strengths of the TiO and metal hydride bands – MgH in K subdwarfs and CaH in M subdwarfs. Figure 2.18 compares red spectra of the disk dwarf Gl 166C and the halo subdwarf LHS 1174. Both stars have absolute visual magnitudes of $M_V \sim 13$, but the subdwarf has only weak TiO absorption. The differences can be quantified using narrow-band spectroscopic indices, designed to measure the flux ratio at the base of a given band against a pseudo-continuum value; that is, by determining the depth of a given spectroscopic feature. This is the same technique employed in dwarf/giant discrimination (Section 2.2.6). Figure 2.18 illustrates the greatly increased strength in CaH absorption in an extremely metal-poor halo subdwarf as compared with a solar-abundance M dwarf of the same luminosity. Chapter 11 outlines how these measurements can be used to estimate approximate chemical abundances.

2.4.2   Effects on broadband photometry

Late-type halo subdwarfs have noticeably different broadband colours than the local disk stars. Plotting the ((B–V)–(V–I)) and ((V–R)–(R–I)) two-colour diagrams (Figure 2.19) shows that subdwarfs lie 0.1–0.3 magnitudes above disk dwarfs in both planes: that is, subdwarfs are redder in (B–V) at a given (V–I) and in (V–R) at a given (R–I). This mainly reflects reduced TiO absorption in the I-band.

There are also discernible differences in the behaviour at near-infrared wavelengths. Disk dwarfs and giants earlier than spectral type M0 outline identical sequences in the ((J–H), (H–K)) diagram, but diverge at later types, with (J–H) ~ 0.7 mag. for dwarfs. This probably reflects the change in the temperature gradient with the onset of convection, reducing the flux emitted in the H^- opacity minimum at 1.6 μm [M9], [M11]. (See also Sections 3.4 and 4.4.) Mould and Hyland [M9] first pointed out that K and M subdwarfs lie below main-sequence stars in this plane (Figure 2.20). Anticipating Chapters 4 and 11, the source of this behaviour lies

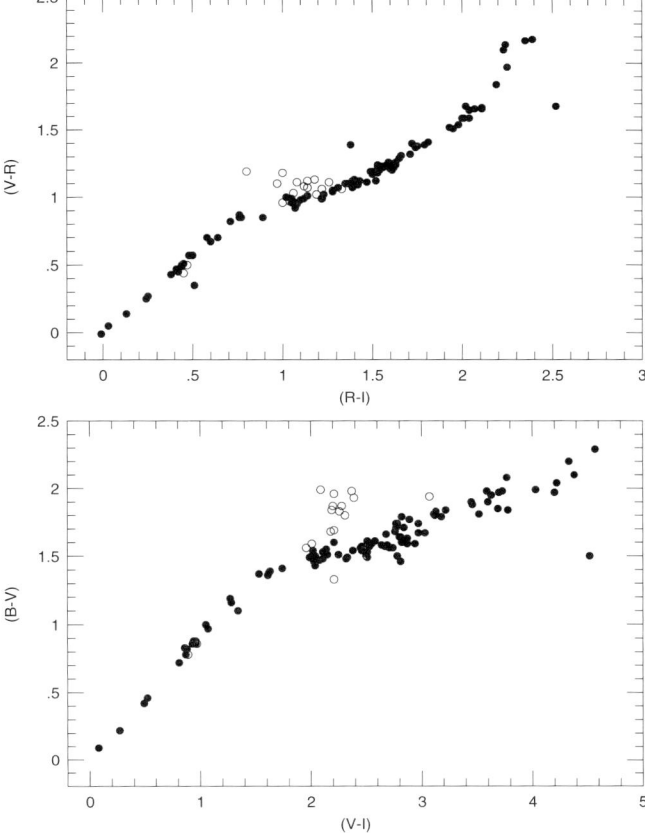

Figure 2.19. The ((B–V)/(V–I)) and ((V–R)/(R–I)) two-colour diagrams for main sequence stars. Disk dwarfs are plotted as solid points, while halo stars (photometry by Bessell) are plotted as open circles.

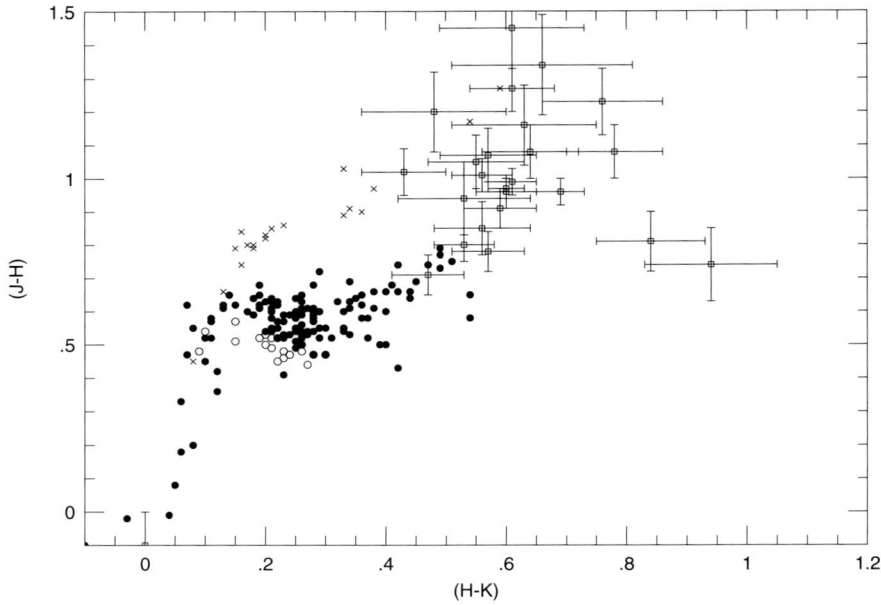

Figure 2.20. The near-infrared ((J–H)/(H–K)) diagram. Disk dwards are solid points, halo subdwarfs are open circles, and L dwarfs are open squares with error bars. Also plotted (as crosses) and data for red giants to illustrate the divergence of the sequences at (H–K) >0.12 magnitudes (spectral type ~K5).

with the increased role of pressure-induced H_2 opacity in the metal-poor atmo-spheres, with the increased opacity at 1.6 μm reducing the proportion of flux that is emitted in the H-band. Again, the offset is only ~0.1–0.15 magnitudes, providing limited dynamic range for quantifying deviations from the disk dwarf sequence. A number of stars with disk-like kinematics (moderate to high Galactic rotational velocities; see Chapter 6) lie close to subdwarfs in this JHK diagram, and are identified as 'old disk subdwarfs' by Mould and McElroy [M10].

A consequence of the reduced opacity in subdwarfs is that the main sequence terminates at a much brighter visual absolute magnitude than is the case for solar-abundance disk dwarfs. Luminosities at the end points, however, are more similar. LHS 1742a – the lowest luminosity subdwarf in the USNO parallax sample – has an absolute visual magnitude $M_V = 14.74$, and a (V–I) colour of 2.74 magnitudes. The bolometric magnitude is $M_{bol} = 12.5$, comparable to the luminosity of the disk dwarf VB 10 ($M_{bol} = 12.9$), which has (V–I) = 4.5 magnitudes and $M_V = 19.5$. LHS 1742a is close to the limit of the hydrogen-burning main sequence for halo stars (see Chapter 3).

2.5   ACTIVITY

Most M dwarfs in the Galactic disk appear to maintain nearly constant photometric properties. A subset, however, exhibit substantial, short-term luminosity variations –

stellar flares. As with most variable stars, the first flare star was discovered in the course of an unrelated survey project. Hertzsprung [H7] had taken a number of plates of the Carina region of the Milky Way in 1924. One star (subsequently designated DH Carinae) was brighter by approximately two magnitudes on one exposure, but had returned to its original magnitude in a subsequent exposure. Given that the plates for the project were taken over a relatively short period of time, it was clear that the variability was rapid (days rather than months) but sporadic, since only the one event was observed. With such sparse data and an apparently unique event, Hertzsprung even suggested that a possible cause might be a stellar/ planetary collision.

It was not until 1940 that two similar variables were identified by van Maanen [M1], [M2], Lalande 21258B (Gl 412B – WX UMa) and Ross 882 (Gl 285 – YZ CMi). Again these were serendipitous discoveries, made from plates taken for a separate project, and the observations suggested that the increase in brightness was a relatively short-lived phenomenon. However, the extremely short timescale of stellar flares did not become fully apparent until Carpenter's December 1947 observations of the high proper-motion pair L726-8 (Gl 65AB). Carpenter was working on obtaining a more accurate determination of the trigonometric parallax, taking multiple short exposures on a series of photographic plates. One of the plates showed the fainter component in the pair (UV Ceti) brightening by 2.7 magnitudes in the 20 minutes between consecutive exposures and fading by a factor of four over the succeeding 40 minutes [L8]. Later that same year, Joy and Humason's spectro-scopic observations [J5] of M dwarfs revealed particularly strong emission lines due to the Balmer series of hydrogen and to Ca II H and K during flares.

Since these initial discoveries, numerous other M dwarfs have been found to be similarly active (including VB 10, [H8]), while others, although not yet observed to flare, exhibit substantial Ca II and Balmer-line emission. Stars with Balmer emission are designated dMe stars. As on the Sun, magnetic fields are presumed to provide the mechanism for heating the outer atmosphere (chromosphere, transition region and corona) of the star. The magnetic field strengths for the handful of dwarfs with direct measurements are ~ 2–$3\,$kG [S1], [S2], [J1], comparable with that measured for large sunspots. The following sections present the general observational picture. A more thorough discussion of magnetic activity on M dwarfs is given in Chapter 5.

2.5.1   Chromospheric activity

The Sun is the only star of which detailed resolution of the density and temperature structure is possible. While the density drops monotonically, with increasing height above the photosphere, the temperature reverses its decline at 600 km above the visible surface and increases rapidly thereafter, reaching 10,000 K at a height of 2,000 km and 10^6 K at 5,000 km. The inner part of this structure is the chromosphere (named because of its strong reddish colour due to Hα emission), visible during eclipses. The outer regions, where the density falls to less than 10^{16} particles m^{-3}, form the solar corona. There is a thin transition region, with temperatures of 10^4–10^6 K, between the chromosphere and corona.

Table 2.1. The relative number of dM and dMe dwarfs of different
spectral types.

Spectral type	N (total)	N (dMe)	% dMe
M0	98	8	8
M0.5	88	9	10
M1	102	5	5
M1.5	105	6	6
M2	108	5	5
M2.5	170	13	8
M3	226	30	13
M3.5	292	59	20
M4	230	72	31
M4.5	134	55	41
M5	57	21	37
M5.5	30	18	60
M6	15	4	27
M6.5	10	5	50
M7	12	12	100
M8	6	3	50
>M9	3	0	0

Most M dwarfs exhibit core emission at the Ca II H and K lines, which are ground state resonance transitions with E.P. $= 4\,eV$. Hydrogen emission is less prevalent, since the hydrogen atoms must have a significant population in the $n = 2$ state (E.P. $= 10.2\,eV$) in order to produce Balmer emission lines. There is not a strong correlation between the strength of the Ca II and the Balmer emission lines, and many stars with substantial H and K emission exhibit Hα in absorption. As discussed in Chapter 5, this may point to a difference in the formation conditions. Ultraviolet and X-ray observations probe the high-temperature gas in the transition region and corona.

Joy and Abt [J4] originally divided M dwarfs into dM and dMe stars, based on the presence or absence of significant (>1 Å equivalent width)[4] Hα emission. Even the initial spectroscopic surveys demonstrated that dMe stars are more common among the later spectral types. Joy [J3], for example, found that 25 of 32 stars with spectral types later than M4 had detectable emission, while only 16 of the 78 M0–M4 stars were dMe dwarfs. Indeed, every star later than M5 in the [J4] survey is listed as an emission-line star, suggesting that this is a defining characteristic of all late-type dwarfs. This proves not to be the case, however. Giampapa and Liebert [G2] carried out the first systematic survey of Hα emission in VLM dwarfs and found chromo-spherically inactive stars even among stars as faint as $M_V = +19$: for example, LHS 2924, spectral type M9, has very weak Hα emission. Since then, spectroscopic observations of nearly all of the M dwarfs in the most recent version of the nearby-star catalogue ([G6], the CNS3) have been obtained, and Table 2.1 lists the division

---

[4] See Chapter 4 for a discussion of equivalent width.

of stars between dM and dMe as a function of spectral type [H1], [G4]. The proportion of dMe dwarfs appears to peak around spectral type M7, with emission becoming increasingly rare at later spectral types. There are, however, a few L dwarfs (such as Kelu 1, Figure 2.4) with detectable emission, so activity is not quenched completely amongst very low-mass dwarfs.

But is this apparent increase in activity amongst the lower luminosity stars real? As already discussed, increasing spectral type reflects decreasing temperature, decreasing total luminosity, a shift of the peak of the energy distribution towards longer wavelengths and a consequent reduction in the percentage of the total flux emitted at optical wavelengths. The net result is that the continuum flux in the vicinity of the Hα line is lower in later-type stars, so a weaker emission line is more easily detectable. In other words, an equivalent width of 1 Å at type M5 corresponds to a lower line-flux in Hα than in a line of the same equivalent width at type M0. The extent to which this dilution in the continuum flux affects the dMe statistics can be tested by calculating the ratio between the luminosity in the Hα line and the bolometric luminosity, L_α/L_{bol}. This ratio measures the strength of the active emission relative to the total energy budget of the star, and is thus a true measure of 'activity', irrespective of the continuum flux in a particular band.

Figure 2.21 plots the log of the L_α/L_{bol} ratio against bolometric magnitude.

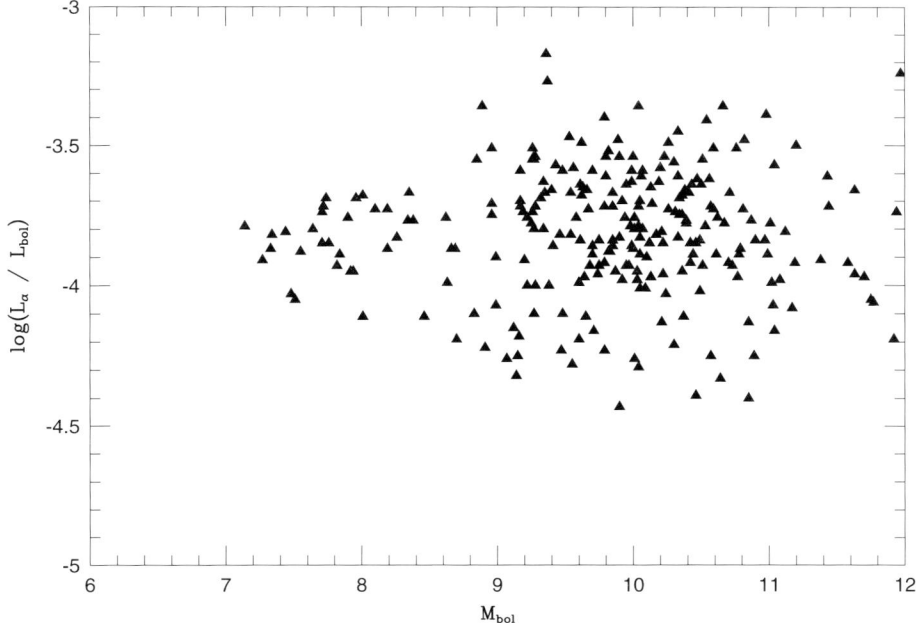

Figure 2.21. The log of the ratio between the luminosity emitted in the Hα line, L_α, and the total bolometric luminosity as a function of M_{bol} in dMe dwarfs within 25 parsecs of the Sun. The observations are from [H1].

Repeated observations of dMe stars show that most, if not all, exhibit night-to-night variability of up to a factor of two in the equivalent width (and hence the flux) in the Balmer lines. This intrinsic variability undoubtedly contributes to the scatter in the observed distribution. Nonetheless, the mean level of activity is effectively constant with luminosity to at least $M_{bol} \sim 12$. Late-type dMe dwarfs are not predominantly less active than early-type dMe dwarfs, implying that the higher dMe/dM ratio indicates a larger fraction of active stars. The reduced numbers of Hα detections amongst M8 and later dwarfs, however, suggests that the mean activity level may well decline among the latest spectral types.

2.5.2   The corona

The existence of the tenuous gas that forms the solar corona has been known (if not understood) since the first observations of a solar eclipse, and photographic studies of coronal structure have been a focal point of most eclipse expeditions since the 1850s. The invention of the coronagraph allowed optical observations of the brighter inner corona outside of eclipse, but detailed quantitative study of the high-temperature gas only became possible with the development of UV and X-ray instrumentation for sounding rockets and, subsequently, satellite observatories. Orbiting X-ray telescopes allowed the first searches for coronal gas around other stars, and one of the major surprises of the extensive survey of the X-ray sky by the Einstein satellite (launched in November 1978) was the discovery that M dwarfs are not only readily detectable at X-ray wavelengths, but that they are amongst the most luminous main-sequence stars at high energies. Figure 2.22(a) plots the X-ray luminosities derived from Einstein observations [V1]) of main-sequence stars with spectral types O to M, and shows that M dwarfs are only slightly less luminous in *absolute* terms than solar-type stars. Most are more luminous than Sirius. Comparing the energy output at X-ray wavelengths against that in the visual, M dwarfs can reach a fractional output L_X/L_{vis} of as much as 10% – a factor of 1,000 higher than the typical value for G-type stars.

Satellites with more sensitive, higher spatial – and spectral-resolution instrumen-tation – notably the Röntgensatellit (ROSAT) launched in 1990 – have provided direct observations of almost all of the K and M dwarfs within 7 parsecs of the Sun [S5]. While none of these stars are bright enough to allow even moderate-resolution spectroscopy, it is clear that the bulk of the emission is at relatively soft X-ray energies of < 1 keV. All but a handful of the K and M dwarfs within 7 parsecs were detected by ROSAT. As with Hα, coronal activity is measured by determining the fraction of the total luminosity emitted at X-ray wavelengths. Figure 2.22(b) plots those data for the ROSAT nearby-star observations, and shows that L_X/L_{bol} ranges from ~ 0.0001–0.1%, with lower-luminosity stars having relatively more coronal emission.

The chromosphere and the corona can be regarded as the inner and outer layers of a stellar atmosphere, and the mechanisms responsible for their activity are almost certainly related. Coronal activity is clearly a long-lived phenomenon, as evidenced by the ROSAT detection of significant X-ray flux from most of the local dM dwarfs,

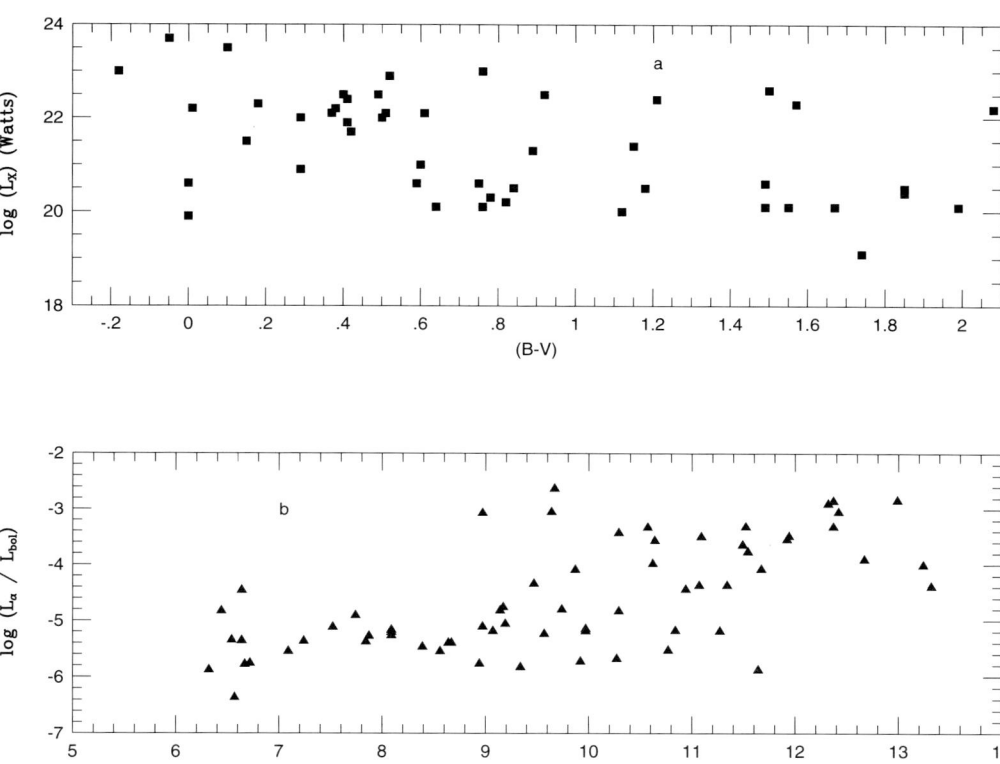

Figure 2.22. (a) The log of the X-ray luminosity of nearby main-sequence stars measured by the Einstein satellite. (b) The log of the ratio between the X-ray and bolometric luminosities for K and M dwarfs within 7 parsecs. (Data are from Schmitt *et al.* [S4].)

some of which are likely to have ages approaching the 10 Gyr of the Galactic disk. Indeed, several low-mass halo stars have been detected, albeit with extremely low X-ray luminosities.

2.5.3   Flares

M dwarf chromospheres, like the solar chromosphere, vary in their level of activity over a wide range of timescales. As already noted, Hα linestrengths can vary by a factor of two from one night to the next. Some stars, however, are prone to much more substantial outbursts, when the blue and ultraviolet flux can increase by a factor of 100 or more and emission lines strengthen by an order of magnitude in a matter of a few seconds or minutes, with the star returning to its quiescent state after

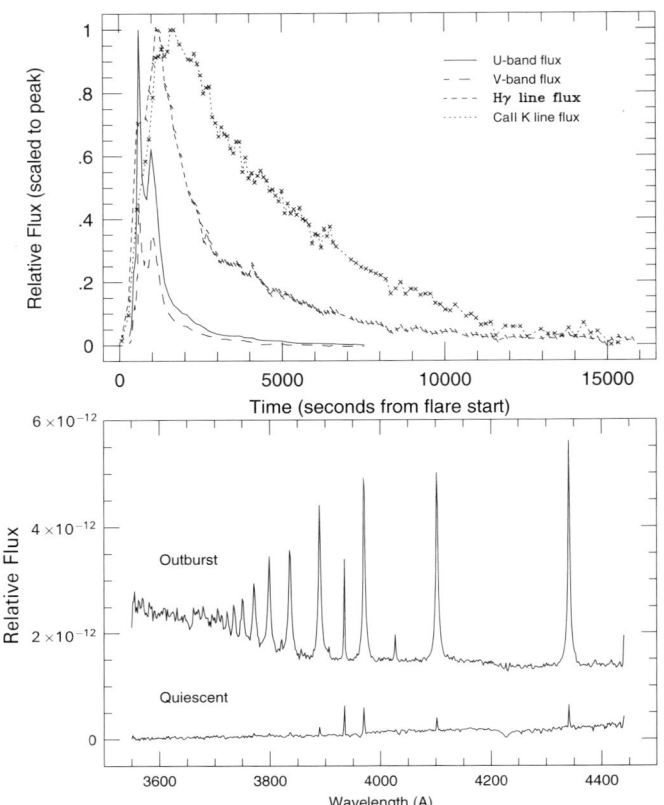

Figure 2.23. (a) Light curves of AD Leo during its outburst of 12 April 1985. The relative variations in the U-band and V-band are compared with the changing Ca II K and Hγ emission. (b) A comparison between the ultraviolet–blue spectrum of AD Leo in outburst and in quiescence. (Data are from [H2].)

minutes or hours. These are M dwarfs like UV Ceti, the prototype of a class of variable star known as flare stars.

Figure 2.23 plots line and continuum intensity variations during an outburst of the well-studied flare star Gl 388, also known as AD Leo [H2]. As pointed out originally by Bopp and Moffett [B7] in their pioneering high time-resolution observations, the continuum enhancement generally has a strong impulsive phase and a longer-lasting gradual phase, in accord with solar flare observations. Flares are energetic phenomena with an equivalent black-body temperature of some 9,000–10,000 K; hence the photometric U-band is preferred for observations. The emission lines rise somewhat more slowly and are enhanced for longer than the continuum. The dominant optical emission lines are those of the hydrogen Balmer series and Ca II H and K, which can attain line fluxes of 10–100 times the values seen during quiescence. Emission lines of He I and II, and atomic species such as Ca I, Fe I and

II are also commonly observed, while the ultraviolet reveals a rich spectrum of highly ionised emission lines such as C IV, Si IV and N V.

The statistics on flare frequency are sparse, with the seminal work being that of Lacy, Moffett and Evans [L2]. Using the excess flux in the U-band, E_U, as a measure of total flare energy, they found that intrinsically fainter (lower mass, later spectral type) flare stars had more frequent but less energetic flares. Brighter flare stars emit a higher fraction of their bolometric luminosity in flares. In the case of AD Leo, Pettersen *et al.* [P2] derive the relation

$$\log \frac{N}{T} = (15.0 \pm 2.1) - (0.62 \pm 0.09) \, E_U \tag{2.5}$$

for the cumulative distribution. This predicts, for example, an average time interval of $10^{6.08}$ seconds ($\sim$two weeks) between flares with energy $E_U = 10^{27}$ Joules. Observations of other flare stars suggest that the slope of this distribution changes with L_{bol}, with more frequent but less energetic flares on stars with smaller L_{bol}. There is also a statistical correlation between flare rise time and flare energy, such that faster flares are less energetic as a whole.

The most active flare stars emit $\sim$0.1% of their bolometric luminosity in the form of flares. Flares have also been observed on dM stars, where the fractional flare luminosity may be as low as 0.0001%. According to Pettersen [P1], it is the frequency of flaring, and not the average energy per flare, that has declined in dM dwarfs. Section 5.4.2 describes analysis and modelling of stellar flares.

2.6 MASSES OF M DWARFS

Mass is the fundamental parameter which determines the position that a star occupies on the main sequence. Unfortunately, the determinination of accurate masses is one of the most difficult observational problems in stellar astronomy, since Kepler's laws and Newtonian orbital dynamics provide the only method of directly measuring accurate stellar masses.[5] This limits observations to stars in binary systems with known heliocentric distances, permitting transformation of angular measurements to linear separations: where the angular separation between components is sufficient to allow measurement of both, but where the period is short enough to permit an orbit determination within a human lifetime. These criteria are met only by a small subset of the stars in the immediate solar neighbourhood, within 10–15 parsecs of the Sun.

Masses have been determined for about 30 M dwarfs, either in eclipsing binaries or wider systems with astrometrically determined orbits. Chapter 8 provides a description of the measuring techniques used. Figure 2.24 plots the resulting

---

[5] Gravitational microlensing also provides mass estimates, but the typical uncertainties involved in a single measurement are very substantial, while the actual lens is seldom observed directly. Thus, while this technique may provide a statistical determination of the stellar mass function, it is not yet useful for estimating masses of individual stars. See Section 8.3.3 for further discussion.

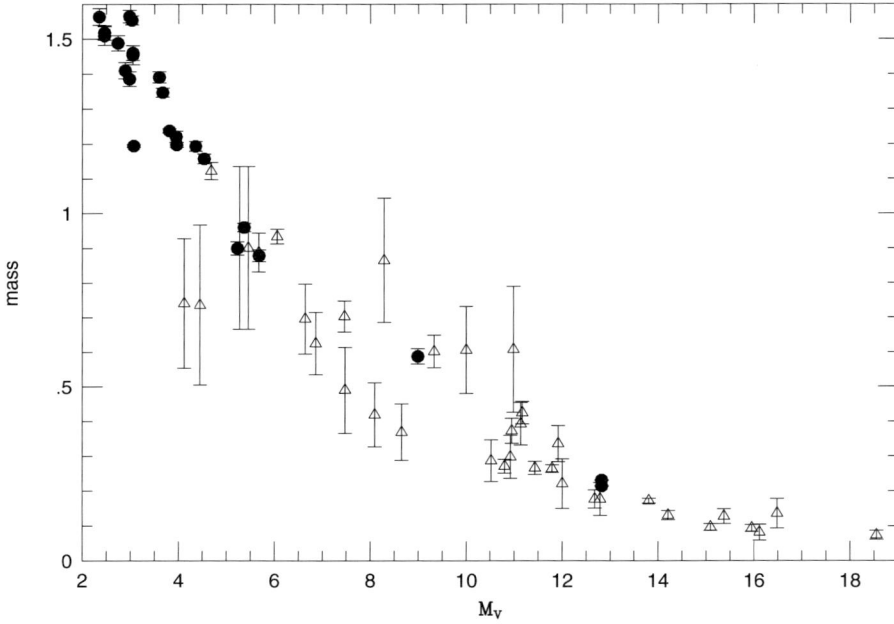

Figure 2.24. The (mass, M_V) relationship defined by stars in binaries with well-determined orbits. Solid points are eclipsing binaries (from [A2]) and open triangles are astrometric binaries (from [H4]).

mass–luminosity relationship at visual wavelengths. Defining M dwarfs as main-sequence stars with $M_V > 7.5$, the inferred masses range from $\sim 0.6\,M_\odot$ at spectral type M0 to less than $0.1\,M_\odot$. As discussed further in Chapter 3, the lower boundary lies close to the mass limit for hydrogen fusion, M_{HBL}. Some L dwarfs, with lower luminosities, fall below that limit. Objects with masses $M < M_{HBL}$ have no long-term energy source, and cool rapidly to extremely low luminosities. These are brown dwarfs.

2.7   STELLAR STATISTICS

Previous sections consider M dwarfs on a star-by-star basis. However, one of the most important scientific issues centres on the determination of their frequency as a function of mass – the mass function. This parameter is of fundamental importance for both star formation theory and Galactic structure. Low-mass stars and brown dwarfs are prime dark-matter candidates, and determining the number density of such objects has occupied the attention of many astronomers over the last 30 years. Chapters 7, 8 and 9 review the extensive studies of this issue. A general point worth emphasising is that statistical analyses are only as reliable as the initial definition of the parent sample. With that in mind, this introductory section concentrates on the

statistical properties of stars currently located within the immediate solar neighbourhood.

2.7.1   Binary and multiple star systems

An issue which must be addressed in statistical studies is the fact that stars (and
brown dwarfs) are not always found in isolation. Indeed, binary and multiple star
systems are relatively common. Their frequency can be quantified using two
parameters: the multiple star fraction,

$$m.s.f. = \frac{N_{bin} + N_{tri} + N_{quad} + \cdots}{N_{sys}} \qquad (2.6)$$

where N_{bin} is the number of stellar systems that are binary, N_{tri}, the number of
triples, and so on; and the companion star frequency,

$$c.s.f. = \frac{N_{bin} + 2 \times N_{tri} + 3 \times N_{quad} + \cdots}{N_{sys}} \qquad (2.7)$$

The m.s.f. gives the fraction of stellar systems which include more than one
component; the c.s.f. gives the probability that a star, picked at random, has a
more luminous companion.

   There are two main techniques used to search for binary and multiple systems:
direct imaging and radial velocity monitoring. The former is better suited to
detecting wide companions in long period orbits, with a detection efficiency that
decreases with distance. The latter searches for reflex motion in the primary star, and
is most sensitive to companions at small separations and high orbital inclinations. It
has an efficiency which depends on the apparent magnitude and spectral type of the
primary, rather than distance. Both techniques are discussed in more detail in
Section 8.3.

   Early surveys centred on radial-velocity monitoring of solar-type stars, and
suggested that the m.s.f. was close to 100% [A1]. Those analyses, however, were
based on magnitude-limited samples, and were therefore subject to several selection
effects – notably, a bias towards equal-mass spectroscopic binaries. Such binary stars
have twice the luminosity of a single star of the same spectral type and, as a result,
the effective distance limit for detecting these binaries is higher (by $\sqrt{2}$) in a
magnitude-limited sample; the sampling volume is therefore higher by $2^{3/2}$; and
the frequency of such systems is overestimated by a corresponding amount. More
recent surveys are based on volume-limited samples, with the requirement being a
fair sampling of the local stellar population. Of these analyses, the most influential is
the survey of solar-type stars undertaken by Duquennoy and Mayor [D2]. Their
dataset includes 164 primaries with spectral types between F7 and G9, and parallaxes
(from the CNS1) greater than 0″.045, although the sample is *not* complete within the
corresponding distance limit. Seventy-two (44%) of these stars are spectroscopic,
visual or common proper motion binaries, and, allowing for observational selection
effects (relatively few of the stars have high-resolution imaging, for example),

Duquennoy and Mayor deduce that the total binary fraction may be as high as 70% – which still implies that more than 80% of all stars have companions.

Lower-luminosity stars – in particular, M dwarfs – have received less attention for the usual reason: with fainter apparent magnitudes, observations are more difficult. Limiting a survey to M dwarfs alone should lead to an underestimate of the true binary frequency, since M dwarfs with main-sequence companions of earlier spectral type are excluded *a priori*. However, since M dwarfs are four times more common than all other stars combined, this selection effect is correspondingly reduced.

Initial investigations of M dwarf binarity suggest a significantly lower frequency than Duquennoy and Mayor's G-dwarf results. Fischer and Marcy [F1] combined results from a radial velocity survey of 70 M dwarfs [M3] with complementary imaging data, and deduce an overall binary fraction of only $\sim 35\%$. While these stars do not constitute a complete sample, Henry and McCarthy's survey [H4] of the northern ($\delta > -25°$) M dwarfs indicates a nearly identical multiplicity function -34.5%, albeit based on only 29 stellar systems, including eight binaries and two triples. Two of the multiple systems – 40 Eridani and Stein 2051 – have components which are not M dwarfs.

The Appendix to this book lists the current census of all systems known to be within 8 parsecs of the Sun. As discussed in the following section, this sample is likely to be substantially complete for declinations north of $-30°$ – a sample we shall refer to as the northern 8-parsec sample. While as-yet undiscovered low-luminosity companions cannot be ruled out completely, nearly all of the stars have been included in either radial-velocity monitoring programmes or high-resolution imaging surveys, and most have been scrutinised using both techniques. It is therefore unlikely that more than a few stellar companions remain hidden among this sample.

The 8-parsec sample includes a total of 138 main-sequence stars (including the Sun), nine white dwarf degenerates and two brown dwarfs (Gl 229B, Gl 570D) in 103 systems. These systems comprise 69 single stars (including four white dwarfs), 25 binaries (three of which have white dwarf components), seven triples and the quadruple systems Gl 570ABCD and Gl 644ABCD. The overall m.s.f. is thus only 33.0%, whilst the c.s.f. is 42.7%. Dividing the sample at $M_V = 7.5$, there are 118 M dwarfs in 86 systems, comprising 55 single stars; 26 binaries (three of which have non-M dwarf primaries); six triples (two with non-M dwarf primaries); and Gl 644. Combined with the G-dwarf results [D2], these data suggest that the stellar multiplicity fraction is mass dependent. Observations of young T Tauri stars suggest that there may also be environmental influences which determine the multiplicity fraction [G1].

Besides deriving the multiplicity fraction, the nearby-star data probe the distribution of mass ratios, $q = M_B/M_A$, and test whether they are dependent on the separation between components. This is a thorny issue, beset by observational selection effects. Early statistical studies suggested that there was a preference for mass ratios close to 1 – at least amongst close binary systems. However, a more thorough analysis suggested that this result stemmed from a bias towards double-lined spectroscopic binaries in the observational sample – binaries which, by

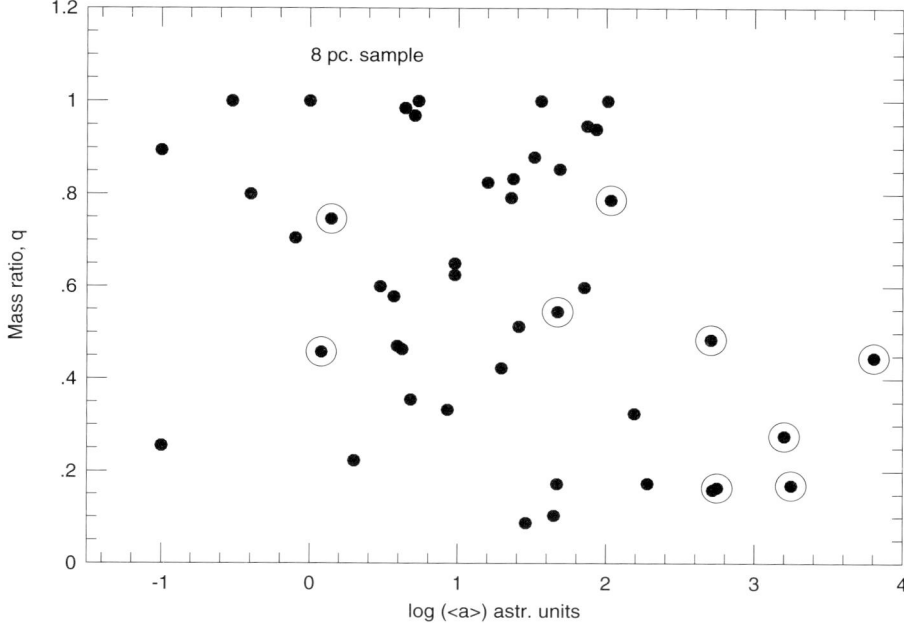

Figure 2.25. Mass ratio as a function of separation for binaries in the 8-parsec sample. Wide, third components of triple systems are plotted as encircled points.

definition, must include stars of near-equal mass. Trimble [T3] reviewed and re-analysed prior results, and concluded that on balance, the observational data favoured a distribution, $F(q) \propto q^{-1}$; that is, a preference for low mass-ratios. This is the form of the distribution expected if stars are selected at random from a mass function which rises with decreasing mass. However, it is important to bear in mind that most stars in the various surveys analysed in [T3] are at least as massive as the Sun.

Orbits have not yet been determined for all of the binaries in the 8-parsec sample, as systems with separations of tens or hundreds of AU have periods of centuries or more. However, if the orbital inclination is random, then the projected separation can be statistically related to the semi-major axis through the simple scaling

$$a = 0.80 \times \Delta \tag{2.8}$$

where Δ is the observed separation. Figure 2.25 shows the distribution of mass ratio, q, as a function of inferred semi-major axis for the northern 8-parsec binaries. In triple systems, mass ratios are measured for the two stars with minimum separation (i.e. $q_1 = M_B/M_A$) and for the ratio between the outermost component and the sum of the inner stars ($q_2 = M_C/M_A + M_B$). There is no clear correlation between q and a. However, if the mass ratio distribution is considered (Figure 2.26) then, contrary

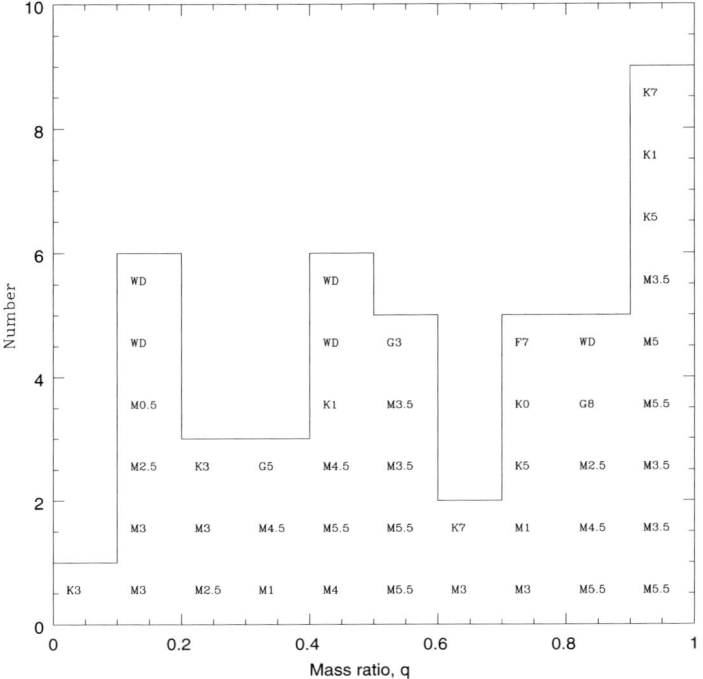

Figure 2.26. The distribution of mass ratios for binaries in the 8-parsec sample.

to results from analyses of higher-mass stars, there appears to be a tendency towards near equal-mass systems. Clearly a larger sample and better statistics are required before firm conclusions can be drawn.

2.7.2   The stellar luminosity and mass functions

The stellar luminosity function is defined as the number of stars per unit absolute magnitude per unit volume, $\Phi(M)$. Combined with an appropriate mass–luminosity relationship, this can be used to derive the mass function, $\Psi(M)\,dM$ or $\xi(\log M)\,d\log M$, which describes the relative frequency of formation of stars of different masses: that is, how a molecular cloud fragments.[6] Integrating the latter determines the stellar contribution to the local mass density – a parameter of significance in understanding Galactic dynamics and the nature of dark matter. The simplest method of determining $\Phi(M)$ is to conduct a stellar census within a specified volume. The crucial point lies in assessing the likely completeness of the resulting sample – an exercise which can be undertaken for the sample of stars within 8 parsecs of the Sun.

---

[6] M is used for magnitude, M is used for mass when discussing the luminosity and mass functions.

The most recent version of the *Catalogue of Nearby Stars*, supplemented by data
from the Hipparcos astrometric satellite [E1], lists 191 stars in 139 systems with
parallaxes exceeding 125 mas ($r < 8$ parsecs). Data for these stars are listed in the
Appendix, where the sample is divided into northern and southern subsets, setting
the boundary line at $\delta = -30°$. The rationale for this split lies in the distribution of
terrestrial observatories: more than 90% are in the northern hemisphere, reflecting
the distribution of land mass and population. As a result, northern skies have been
scrutinised in more detail, and for a longer period, than the southern skies. Most
nearby stars were discovered through their having high proper motion, and the
deepest such surveys are Luyten's Palomar proper-motion catalogues [L9]. Since
Palomar lies at a latitude of $+32°$, Luyten's observations became increasingly
difficult at southern declinations, and the declination $\delta = -30°$ provides a natural
division of the sky. This eliminates from consideration only the southernmost 25%
of the sky. In fact, comparison of Tables A.1 and A.2 show that the relative number
of *systems* north and south of this division is close to the expected ratio of $3:1$;
although there are almost 25% fewer *companions* known amongst the southern
sample.

But how complete is the current northern 8-parsec catalogue? Since most stars
were identified from proper-motion surveys, this introduces the possibility of a bias
against stars with low tangential velocities relative to the Sun. Luyten's main
catalogue, however, is the LHS (Luyten Half Second) catalogue, including stars
with $\mu > 0''.5 \, yr^{-1}$. These stars have extensive follow-up spectroscopy, and it is
unlikely that any LHS stars within 8 parsecs remain unidentified. The proper
motion limit corresponds to only $18 \, km \, s^{-1}$ at 8 parsecs, and fewer than 20% of
disk stars can be expected to have such low rates of motion. In fact, 14 of the 103
systems in Table A.1 have $\mu < 0.5 \, arcsec \, yr^{-1}$, suggesting that at most a handful of
such low-velocity systems remain to be discovered.

Sample completeness can also be tested by comparing the number distribution
with distance. Given the small number of systems in total, the northern 8-parsec
sample cannot be subdivided too finely. Table 2.2 lists the number of systems within
the distance limits 0–5, 5–8 and 8–10 parsecs for several luminosity intervals. Binary
and multiple star systems are binned in M_V based on the brightest star in the system.

Table 2.2. Distance distribution of nearby stars, $\delta > -30°$.

M_V	<5 parsec	$5 < r < 8$ parsec	$8 < r < 10$ parsec
≤ 4	3	6	10
$4 < M_V \leq 7$	4	16	29
$7 < M_V \leq 10$	2	13	22
$10 < M_V \leq 12$	8	26	55
$12 < M_V \leq 14$	5	26	47
>14	11	17	30
All	34	108	204

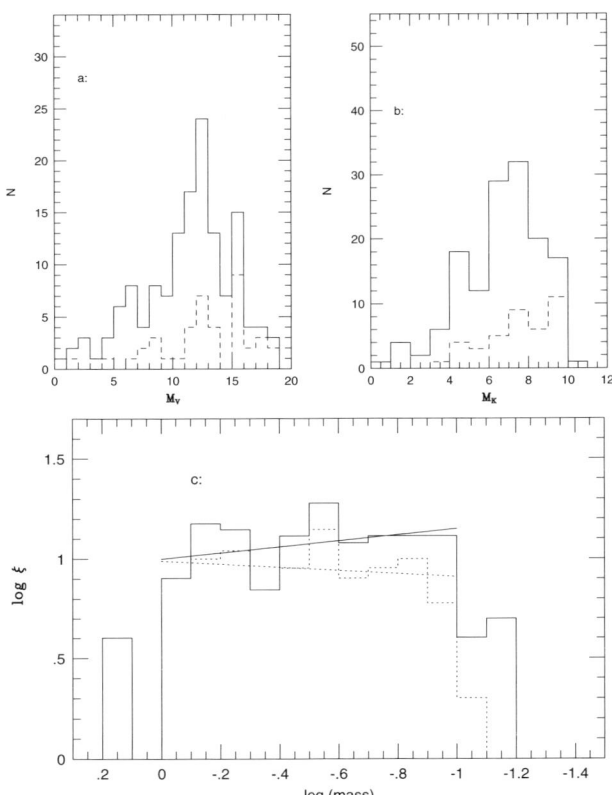

Figure 2.27. $\Phi(M_V)$ and $\Phi(M_K)$ for the northern 8-parsec sample and the corresponding mass function, $\xi(\log M)\, d\log M$. The dashed line in (a) and (b) shows the contribution from secondary components; the dotted line in (c) plots the mass function for stellar systems, excluding secondary components.

Volume sampling indicates that the observed star counts should increase by a factor of 3 between the first and second distance intervals, while there should be a factor of 2 difference between the numbers in the second and third columns. The expected ratios hold for $M_V < +14$. Taken at face value, the data for the faintest stars suggest a shortfall of a factor of 2 in numbers at $r > 5$ parsecs, although the deficit is less than 2σ given the sampling uncertainties in the 5-parsec sample. Overall, the northern 8-parsec sample is likely to be $\sim 90\%$ complete for H-burning stars.

Accepting these uncertainties, Figure 2.27 shows $\Phi(M_V)$ and $\Phi(M_K)$ derived from data for the main-sequence stars in the northern 8-parsec sample. These luminosity functions are constructed on a star-by-star basis, with the dashed lines indicating the contribution from stellar companions. The visual luminosity function reaches a maximum at $M_V \sim +12$, and declines toward fainter magnitudes; $\Phi(M_K)$ peaks at $M_K \sim 8$. The extended tail towards faint absolute magnitudes in $\Phi(M_V)$ reflects decreasing temperature and increasing bolometric corrections rather than a substantial change in luminosity; $\Phi(M_K)$ provides a better representation of the distribution with bolometric luminosity.

The stellar mass function is derived from the luminosity function data by applying the appropriate mass–luminosity (or mass–absolute magnitude) relationship. Figure 2.27(c) presents the results for the northern 8-parsec sample, and the systemic mass function (including only single stars and primary stars in multiple systems) is also plotted. Traditionally, the mass function is expressed as a power-law, $\Psi(M) = dN/dM \propto M^{\alpha}$ [S3]. Figure 2.27 shows that $\Psi(M) \propto M^{-1.15}$ provides a reasonable approximation to the star-by-star 8-parsec data for $1 > M/M_{\odot} > 0.1$; an exponent of $\alpha = -0.92$ matches the systemic function. The inferred total mass density due to main-sequence stars is $\sim 0.048\,M_{\odot}\,\mathrm{pc}^{-3}$. Chapter 8 provides a more extensive discussion of these results, and Chapter 9 extends analysis to the substellar régime.

2.8 SUMMARY

The main aim of this chapter has been to provide an empirical overview of the properties of M dwarfs, and to discuss how those properties can be used to illuminate broader issues such as star formation, nucleosynthesis, stellar magnetic fields, Galactic structure and galaxy formation. Defining M dwarfs as main-sequence stars with TiO absorption, their properties can be summarised as follows:

- Their spectra are characterised by molecular absorption bands, notably TiO, VO and metal hydrides at optical wavelengths; H_2O and CO in the near-infrared.
- They have absolute visual magnitudes between $M_V = 7.5$ and $M_V = 20$.
- Their luminosities range from $0.2\,L_{\odot}$ to less than $5 \times 10^{-4}\,L_{\odot}$.
- They have increasingly red optical and near-infrared colours with increasing spectral type: $1.5 < (V–I) < 5$ and $1 < (I–K) < 6$.
- Their effective temperatures lie between $\sim 3800\,\mathrm{K}$ (M0) and $\sim 2100\,\mathrm{K}$ (M9).
- Their radii lie between $0.5\,R_{\odot}$ and $0.1\,R_{\odot}$.
- They often have active chromospheres and coronae.
- Their masses lie between $0.6\,M_{\odot}$ and $0.1\,M_{\odot}$.

The newly recognised L dwarfs represent an extension of the M dwarf sequence to cooler temperatures, lower luminosities and lower masses, below the hydrogen-burning limit. TiO and VO become progressively less prominent in late L dwarfs, probably through their condensing as solid dust particles in the stellar atmosphere. Metal hydride bands and alkali lines dominate the optical spectrum.

Sub-stellar mass brown dwarfs can cool to temperatures below $\sim 1,200$ K, where methane forms in the atmosphere. This leads to substantial absorption in the H and K passbands, and blue near-infrared colours, as exemplified by the prototype, Gl 229B.

In the succeeding chapters, the properties of these cool dwarfs are discussed in more detail, giving more attention to the underlying physics required for theoretical modelling. The observations discussed in this chapter, however, represent the primary constraints on those models and the fundamental description of the nature of M dwarfs.

2.9   REFERENCES

A1   Abt, H. A., Levy, S. G., 1976, *ApJS*, **30**, 273.

A2   Andersen, J., 1991, *Astr. Ap. Rev.*, **3**, 91.

A3   Arnaud, K. A., Gilmore, G., Collier Cameron, A., 1989, *MNRAS*, **237**, 495.

B1   Baldwin, J. R., Frogel, J. A., Persson, S. E., 1973, *ApJ*, **184**, 427.

B2   Becklin, E. E., Zuckerman, B., 1988, *Nature*, **336**, 656.

B3   Berriman, G., Reid, I. N., 1987, *MNRAS*, **227**, 315.

B4   Berriman, G., Reid, I. N., Leggett, S. K., 1992, *ApJ*, **392**, L31.

B5   Bessell, M. S., 1991, *AJ*, **101**, 662.

B6   Boeshaar, P., 1976, PhD thesis, Ohio State University.

B7   Bopp, B. W., Moffett, T. J., 1973, *ApJ*, **185**, 239.

C1   Chamberlain, J. N., Aller, L. H., 1951, *ApJ*, **114**, 52.

C2   Cox, J. J., Gibson, D. M., 1985, in *Radio Stars*, ed. R. M. Hjellming, D. M. Gibson, Reidel, Dordrecht, p. 233.

D1   Delfosse, X., Tinney, C. G., Epchtein, N., Bertin, E., Borsenberger, J., Copet, E., de Batz, B., Fouquem P., Kineswenger, S., Le Bertre, T., Lacombe, F., Rouan, D., Tiphene, D., 1997, *A&A*, **327**, L25.

D2   Duquennoy, A., Mayor, M., 1991, *A&A*, **248**, 485.

E1   ESA, 1997, The Hipparcos catalogue, ESA SP-1200.

F1   Fischer, D. A., Marcy, G. W., 1992, *ApJ*, **396**, 178.

F2   Frogel, J. A. 1971, PhD thesis, California Inst. of Technology.

F3   Fowler, A., 1904, *MNRAS*, **64**, 16.

G1   Ghez, A. M., McCarthy, D. W., Patience, J. L., Beck, T. L., 1997, *ApJ*, **481**, 378.

G2   Giampapa, M. S., Liebert, J., 1986, *ApJ*, **305**, 784.

G3   Gizis, J. E., 1996, *AJ*, **113**, 806.

G4   Gizis, J. E., Monet, D., Reid, I. N., Williams, R., 1999, *AJ*, **118**.

G5   Gliese, W., 1969, *Catalogue of Nearby Stars*, Veröff. Astr. Rechen-Inst. Heidelberg, Nr. 22.

G6   Gliese, W., Jahreiss, H., 1991, *Third Catalogue of Nearby Stars*, preliminary version.

G7   Greenstein, J. L., Neugebauer, G., Becklin, E. E., 1970, *ApJ*, **161**, 519.

H1   Hawley, S. L., Gizis, J. E., Reid, I. N., 1996, *AJ*, **112**, 2799.

H2   Hawley, S. L., Pettersen, B. R., 1991, *ApJ*, **378**, 725.

H3   Hearnshaw, J. B., 1986, *The Analysis of Starlight*, Cambridge University Press.

H4   Henry, T. J., McCarthy, D. W., 1993., *AJ*, **106**, 773.

H5   Hertzsprung, E., 1905, *Zeitschrift für wissenschaftliche Photographie*, **3**, 429.

H6 Hertzsprung, E., 1907, *Zeitschrift für wissenschaftliche Photographie*, **5**, 86.

H7 Hertzsprung, E., 1924, *BAN*, **2**, 87.

H8 Herbig, G. H., 1956, *PASP*, **68**, 531.

J1 Johns-Krull, C. J., Valenti, J. A., 1996, *ApJ*, **459**, L95.

J2 Jones, H. R. A., Longmore, A. J., Allard, F., Hauschildt, P. H., 1996, *MNRAS*, **280**, 77.

J3 Joy, A. H. 1947, *ApJ*, **105**, 96.

J4 Joy, A. H., Abt, H. A., 1974, *ApJ*, **28**, 14.

J5 Joy, A. H., Humason, M., 1949, *PASP*, **61**, 133.

K1 Keenan, P., MacNeil, R. C., 1976, *An Atlas of Spectra of the Cooler Stars: Types G, K, M, S and C*, Columbus: Ohio State University Press.

K2 Kirkpatrick, J. D., Henry, T. J., McCarthy, D. W., 1991, *ApJS*, **77**, 417.

K3 Kirkpatrick, J. D., Kelly, D. M., Rieke, G. H., Liebert, J., Allard, F., Wehrse, R., 1993, *ApJ*, **402**, 643.

K4 Kirkpatrick, J. D., Henry, T. J., Liebert, J., 1993, *ApJ*, **406**, 701.

K5 Kirkpatrick, J. D., Henry, T. J., Simons, D. A., 1995, *AJ*, **109**, 797.

K6 Kirkpatrick, J. D., Beichman, C. A., Skrutskie, M. F., 1997, *ApJ*, **476**, 311.

K7 Kirkpatrick, J. D., 1998, in *Brown Dwarfs and Extrasolar Planets*, ed. R. Rebolo, E. Martín and M. R. Zapatero Osorio, ASP Conf. Ser. No. 134, p. 405.

K8 Kirkpatrick, J. D., Reid, I. N., Liebert, J., Cutri, R. M., Nelson, B., Beichman, C. A., Dahn, C. C., Monet, D. G., 1999, *ApJ*, **519**, 802.

K9 Kuiper, G. P., 1939, *ApJ*, **89**, 549.

K10 Kuiper, G. P., 1942, *ApJ*, **95**, 201.

L1 Lacy, C. H., 1977, *ApJ*, **218**, 444.

L2 Lacy, C. H., Moffett, T. J., Evans, D. S., 1976, *ApJS*, **30**, 85.

L3 Lim, J., White, S. M., Slee, O. B., 1996, *ApJ*, **460**, 976.

L4 Lim, J., White, S. M., 1996, *ApJ*, **462**, L91.

L5 Leggett, S. K., 1992, *ApJS*, **82**, 351.

L6 Leggett, S. K., Allard, F., Berriman, G., Dahn, C. C., Hauschildt, P., 1996, *ApJS*, **104**, 117.

L7 Leung, K.-C. Schneider, D. P., 1978, *AJ*, **83**, 618.

L8 Luyten, W. J., 1949, *ApJ*, **109**, 532.

L9 Luyten, W. J., 1976, *Proper Motion Survey with the 48-inch Palomar Schmidt Telescope*, University of Minnesota publications.

M1 van Maanen, A., 1940, *ApJ*, **91**, 503.

M2 van Maanen, A., 1940, *PASP*, **57**, 216.

M3 Marcy, G. W, Benitz, K. J., 1989, *ApJ*, **344**, 441.

M4 Mathioudakis, M., Doyle, J. G., 1993, *A&A*, **280**, 181.

M5 Metcalfe, T. S., Mathieu, R. D., Latham, D. W., Torres, G., 1996, *ApJ*, **456**, 356.

M6 Monet, D. G., Dahn, C. C., Vrba, F. J., Harris, H. C., Pier, J. R., Luginbuhl, C. B., Ables, H. D., 1992, *AJ*, **103**, 662.

M7 Morgan, W. W., 1938, *ApJ*, **87**, 589.

M8 Morgan, W. W., Keenan, P. C., Kellman, E., 1943, *An Atlas of Representative Stellar Spectra*, Chicago: University of Chicago Press.

M9 Mould, J. R., Hyland, A. R., 1976, *ApJ*, **208**, 399.

M10 Mould, J. R., McElroy, D. B., 1978, *ApJ*, **220**, 935.

M11 Mould, J. R., 1976, *A&A*, **48**, 443.

M12 Mullan, D. J., Stencel, R. E., Backman, D. E., 1989, *ApJ*, **343**, 400.

M13 Mullan, D. J., Doyle, J. G., Mathioudakis, M., Redman, R. O., 1992, in *The 7th Cambridge Workshop on Cool Stars*, ed. M. S. Giampapa and J. A. Bookbinder, ASP Conf. Ser. 26, 328.

N1 Nakajima, T., Durrance, S. T., Golimowski, D. A., Kulkarni, S. R., 1994, *ApJ*, **428**, 797.

O1 Oke, J. B., Cohen, J. G., Carr, M., Cromer, J., Dingizian, A., Harris, F. H., Labreque, S., Lucinio, R., Schaal, W., Epps, H., Miller, J., 1995, *PASP*, **107**, 375.

O2 Osterbrock, D. E., 1993, *50 Years of the MK Process*, Astr. Soc. Pacif. Conf., **60**, 199.

P1 Pettersen, B. R., 1987, *Vistas in Astronomy,* **30**, 41.

P2 Pettersen, B. R., Coleman, L. A., Evans, D. S., 1984, *ApJS*, **54**, 375.

R1 Reid, I. N., Gilmore, G. F., 1984, *MNRAS*, **206**, 19.

R2 Reid, I. N., Hawley, S. L., Gizis, J. E., 1995, *AJ*, **110**, 1838.

R3 Reid, I. N., Gizis, J. E., 1997, *AJ*, **113**, 2246.

R4 Ruiz, M. T., Leggett, S. K., Allard, F., 1997, *ApJ*, **491**, L107.

R5 Russell, H. N., 1911, Carnegie Inst. Washington Publ. No. 147.

R6 Russell, H. N., 1913, *The Observatory*, **36**, 324.

S1 Saar, S. H., Linsky, J. L., 1985, *ApJ*, **299**, L47.

S2 Saar, S. H., Linsky, J. L., Beckers, J. M., 1986, *ApJ*, **302**, 777.

S3 Salpeter, E. E., 1956, *ApJ*, **121**, 161.

S4 Sandage, A., Eggen, O. J., 1959, *MNRAS*, **119**, 278.

S5 Schmitt, J. H. M. M., Fleming, T. A., Giampapa, M. S., 1995, *ApJ*, **450**, 392.

S6 Skrutskie, M. F. *et al.*, 1997, in *The Impact of Large-Scale Near-IR Sky Survey*, ed. F. Garzon *et al.*, Kluwer, Dordrecht, p. 187.

T1 Tinney, C. G., Reid, I. N., 1999, *MNRAS*, **301**, 1031.

T2 Tinney, C. G., Mould, J. R., Reid, I. N., 1993, *AJ*, **105**, 1045.

T3 Trimble, V., 1990, *MNRAS*, **242**, 79.

V1 Vaiana, G. S. *et al.*, 1981, *ApJ*, **244**, 163.

V2 Vyssotsky, A. N., Janssen, E. M., 1951, *AJ*, **56**, 58.

W1 Waters, L. B. F. M., Coté, J., Aumann, H. H., 1987, *A&A*, **172**, 225.

W2 Wilsing, J., Scheiner, J., 1909, *Publ. Astrophys. Obs. Potsdam*, **19**, 1.

W3 Wing, R. F., 1973, *Spectral Classification and Multicolour Photometry*, IAU Symposium No. 50, ed. Ch. Fehrenbach and B. E. Westerlund, Reidel, Dordrecht, p. 209

3

The structure, formation and evolution of low-mass stars and brown dwarfs

3.1 INTRODUCTION

The development of a self-consistent theoretical description of the internal structure of stars, and the consequent construction of models which trace evolutionary behaviour, represents one of the major achievements of twentieth-century astrophysics. Most studies have centred on intermediate- and high-mass stars, with little consideration of M dwarfs with masses below $0.6\,M_\odot$. In part, this reflects the availability of more precise observational constraints, and also the greater analytic tractability in modelling higher-mass stars. Recently, however, the lower main sequence has attracted more attention, with a series of detailed models extending past the hydrogen-burning limit to the boundary between low-mass brown dwarfs and giant planets.

The aim of the present chapter is to concentrate on issues relevant to the structure of low-mass stars. The following section provides an historical overview of the development of stellar structure theory, allowing modern analyses to be placed in the proper context, while the succeeding section outlines the general principles. Thorough discussion of stellar structure theory can be found in textbooks devoted specifically to that subject, including Clayton [C4], Goldberg and Scadron [G3] and Hansen and Kawaler [H1], while the classic monographs by Chandrasekhar [C3] and Schwarzschild [S7] remain strikingly relevant in many areas.

3.2 A BRIEF HISTORY

The foundations of the modern theory of stellar structure were laid during the later stages of the nineteenth century in a series of analyses of the equilibrium configuration of gaseous spheres. The first such study – by the American meteorologist J. Homer Lane [L2], originated largely as a byproduct of an investigation of the surface temperature and density of the Sun. Deriving 54,000°F (30,000 K) for the former,[1] Lane hypothesised that at such temperatures the solar material was 'torn asunder' and could be treated as a perfect gas. Based on that assumption, he derived an analytical description of a self-gravitating sphere, and showed (implicitly) that a uniform contraction led to an increase in temperature: Lane's law – $rT(r) = constant$. Identical results were derived independently by Ritter [R1], who extended the mathematical treatment in a series of 18 papers published in *Weidemann Annalen* between 1878 and 1889.

Underlying this analysis are two important principles: the balance between the internal pressure and the self-gravity of the Sun, the principle of hydrostatic equilibrium; and the balance between the energy produced and that radiated from the surface, the principle of thermal equilibrium. Based on the principle of energy conservation, Helmholtz [H6] had previously identified gravitational contraction (more specifically, meteoric accretion) as a possible energy source for the Sun. Lane's analysis codified that process, while invoking Kelvin's [K1], [K3] mechanism of 'convective equilibrium' as a means of energy transport. The latter had been devised in modelling the terrestrial atmosphere. Kelvin [K2] later provided his own analysis of this issue, while the last word on the subject was provided by Emden's [E2] monograph *Gaskugeln*, which outlines the full series of equations describing gaseous spheres in hydrostatic equilibrium – the Lane–Emden equations.

An important inference which could be drawn from Lane's law was that stars might be expected to start their lives as diffuse, low-temperature objects which increase in temperature as they evolve and contract. Should the perfect gas law become invalid at some point (as was expected to happen at high densities, for example), then Lane's law would fail, and the temperature would decrease with time. This proposition ran contrary to the then-prevailing opinion that stars initially had high temperatures and cooled progressively with time (hence the reference to 'early' and 'late' spectral types in Pickering's Harvard classification system) and was largely ignored (except by J. Norman Lockyer). However, the discovery of the shape of the stellar distribution with temperature, as revealed by the Hertzsprung–Russell diagram (1911–1913) appeared to provide observational support for this hypothesis, with stars on the giant branch representing the 'ascending series' or contraction phase, while the main-sequence FGKM dwarfs were interpreted as the 'descending' cooling stage. It is interesting to note that this evolutionary model, although treated with some reservations, was still given serious consideration by Eddington [E1].

The gravitational contraction hypothesis also provided a means of estimating the solar lifetime by comparing the present-day luminosity with the available potential energy. This is the Kelvin–Helmholtz timescale:

$$t_{KH} = \frac{E_{PE}}{\bar{L}_\odot} \approx \frac{GM_\odot^2}{R_\odot \bar{L}_\odot} \sim 3 \times 10^7 \text{ yr} \tag{3.1}$$

[1] As Chandrasekhar ([C3], p. 176) has pointed out, Lane's analysis predates the derivation of Stefan's law (1879), and is based on a substantial extrapolation of the available observational results on emissivity as a function of temperature. Lane's method was not at fault.

where $\bar{L}_\odot$ is the average solar luminosity. The initial applications of this analysis by Helmholtz and Kelvin implied solar lifetimes of less than 100 Myr – at odds with the 'deep time' required by Hutton and Lyell's geological studies ('... no vestige of a beginning – no prospect of an end', [H9]). An alternative energy source was required, and Eddington suggested that such might be found in the 'knockabout comedy of atomic physics', where 'subatomic energy of some kind is liberated within the star, so as to replenish the store of radiant energy' [E1]. These suspicions were confirmed in the 1930s with the development of hydrogen fusion theory by Atkinson, von Weizsacker, Gamow, Bethe and others.

Eddington's own calculations, summarised in *The Internal Constitution of the Stars*, placed emphasis on radiative, rather than convective, energy transport. Originally described by Sampson [S1] and further refined by Schwarzschild [S6], radiative processes are of prime importance in stars with masses exceeding that of the Sun. Based on his models, Eddington derived a mass–luminosity relationship and a critical mass – the Eddington limit, where the energy generated in radiative pressure is sufficient to overcome self-gravity and disrupt the star.

Eddington's models followed the standard (at that time) assumption that iron was the most common element, leading to a mean molecular weight of $\mu \sim 2.2$, where μ is defined as the mean mass per particle measured in units of the mass of a hydrogen nucleus (proton), m_H. Part of Eddington's enthusiasm for this composition lay in the consequent balance between radiative pressure and gas pressure as a function of mass, allowing a 'cloud-bound physicist' to predict the properties of the stars ([E1], pp. 15–16). Payne [P1] had earlier demonstrated the uniformity of chemical composition from star to star, hinting at the dominance of H and He. By the end of the 1920s, Russell [R2] had demonstrated conclusively that hydrogen was the most abundant element in the Sun, with $\log(N_H/N_{Fe}) \sim 4.3$. Strömgren [S15], [S16] subsequently developed a set of models which showed that the H–R diagram could be represented as a series of stars of varying mass in hydrostatic and thermal equilibrium, subject to radiative energy transport.

The next major advance in understanding stellar structure came with the description of the properties of white dwarf stars in terms of degeneracy theory. Adams' [A4] spectroscopic observations of the companion of Sirius had produced the unexpected result that its spectral type was close to that of Sirius A, indicating a comparable temperature and a radius close to that of the Earth. Given a mass of $\sim 1\,M_\odot$ (confirmed by gravitational redshift measurements, [A5], but see [G4]), the implied average density is close to 1 ton per cubic inch. Fowler [F1] used Fermi–Dirac statistics to provide an explanation by applying the Pauli exclusion principle to an electron gas (that is, electron degeneracy). Chandrasekhar [C2], [C3] fully developed this area of astrophysics, extending the analysis to relativistic degeneracy, famously contested by Eddington. The latter aspect of the theory was applied observationally only with the discovery of pulsars in the late 1960s.

During the 1940s and 1950s, considerable progress was made in many areas, notably the development of nucleosynthesis theory [B9] and the calculation of evolutionary, rather than static, stellar models [T1], [S7], the latter calculated painstakingly step-by-step on hand-operated Brunswegger calculators.

Schwarzschild's work in particular led to what are amongst the first models of red giant stars, characterised as inhomogeneous gaseous spheres, in contrast to the homogeneous main-sequence models. Böhm-Vitense [B8] devised the mixing-length approximation as a means of simulating convection in cooler main-sequence stars and red giants. The continued improvement in accuracy of the input physical parameters (nuclear reaction rates, cross-sections, opacities) and the availability of increasingly more powerful computational power, particularly over the last decade, has led to the development of more complex and more realistic models (such as in [D2], [D3], [M1], [V2], [B2], [B10]).

3.3 GENERAL PRINCIPLES OF STELLAR STRUCTURE

3.3.1 The fundamental equations

The foundations of stellar structure theory rest on four fundamental equations [C3], [S7]. The first three equations are straightforward in character. The equation of continuity of mass describes the mass distribution for a spherical object:

$$\frac{dM(r)}{dr} = 4 \pi r^2 \rho(r) \tag{3.2}$$

where $M(r)$ is the mass distribution, and $\rho(r)$ is the density distribution. Second is the equation of hydrostatic equilibrium, which gives the balance between the attractive force of self-gravity and internal pressure support:

$$\frac{dP(r)}{dr} = - \frac{GM(r)}{r^2} \rho(r) = - g(r) \rho(r) \tag{3.3}$$

where $P(r)$ is the pressure distribution, and $g(r)$ is the gravitational acceleration. Third, the equation of thermal equilibrium, gives the balance between energy generation and energy loss through radiation:

$$\frac{dL(r)}{dr} = 4 \pi r^2 \rho(r) \epsilon(r) \tag{3.4}$$

where $L(r)$ is the luminosity, and $\epsilon(r)$ is the rate of energy generation.

The fourth equation describes the temperature distribution as a function of radius, and is therefore related to the mode of internal energy transport. There are three mechanisms: conduction, convection and radiation. Conduction plays no significant role in low-mass main-sequence stars and brown dwarfs, although it is important in white dwarfs and neutron stars. Radiation transport contributes significantly under some circumstances, and convection is by far the most important mechanism. The radiative and convective transport equations are:

$$\frac{dT(r)}{dr} = \frac{-3}{4ac} \frac{\kappa}{T^3} F(r) \qquad \text{radiative transport}$$

$$= \left(1 - \frac{1}{\gamma}\right) \frac{T}{P} \frac{dP}{dr} \qquad \text{convective transport} \tag{3.5}$$

where in the first equation, a is the Stefan–Boltzmann constant, c is the velocity of light, κ is the opacity and $F(r)$ is the flux as a function of radius, r: $F(r) = L(r)/4\pi r^2$; and in the second equation, γ is the ratio of specific heats C_p/C_v. Section 3.4 provides more details on energy transport.

In addition to these four fundamental equations, stellar modelling requires relations defining three parameters: pressure, energy generation and opacity. These are the three constitutive equations: the equation of state, which describes the pressure,

$$P = P(T, \rho, composition) \tag{3.6}$$

the equation of energy generation,

$$\epsilon = \epsilon(T, \rho, composition) \tag{3.7}$$

and the opacity equation,

$$\kappa = \kappa(T, \rho, composition) \tag{3.8}$$

The former two relationships are described in more detail below, and the various sources contributing to stellar opacities are outlined in Chapter 4.

Finally, these equations can only be solved given the appropriate boundary conditions. In the classical case, the four conditions are:

$$M(r) = 0 \text{ at } r = 0 \qquad P(r) = 0 \text{ at } r = R$$

$$L(r) = 0 \text{ at } r = 0 \qquad T(r) = 0 \text{ at } r = R$$

With the development of more accurate atmosphere models (see Chapter 4), the luminosity and temperature distributions predicted by those models can be used to set the exterior boundary conditions.

3.3.2   Gas polytropes and the Lane–Emden equations

Lane's original analysis of the surface temperature of the Sun was based on the analytical approximation that pressure could be described as a function of only density, $P = P(\rho)$. This approach – generalised by Emden and codified by Chandrasekhar – still offers an insight into the internal structure of low-mass stars and brown dwarfs during evolutionary phases where the pressure distribution takes the polytropic form,

$$P = K\rho^\gamma, \gamma = 1 + \frac{1}{n} \tag{3.9}$$

where n is the polytropic index and K is a constant. The resulting solutions are models of gas polytropes.

The Lane–Emden equations are dimensionless forms of the equation of hydrostatic equilibrium. Eliminating $M(r)$, equation (3.3) can be re-written as

$$\frac{1}{r^2} \frac{d}{dr} \left(\frac{r^2}{\rho(r)} \frac{dP(r)}{dr} \right) = -4\pi G \, \rho(r) \tag{3.10}$$

Substituting for $P(r)$ using equation (3.6) produces

$$\frac{1}{r^2}\frac{d}{dr}\left(\frac{r^2}{\rho}\gamma K\rho^{\gamma-1}\frac{d\rho}{dr}\right) = -4\pi G\rho \tag{3.11}$$

or, defining

$$\phi = \frac{\gamma}{\gamma-1}K\rho^{\gamma-1} \tag{3.12}$$

we have

$$\frac{1}{r^2}\frac{d}{dr}\left(r^2\frac{d\phi}{dr}\right) = -4\pi G\rho \tag{3.13}$$

ϕ is related to the gravitational potential, $\Phi(r)$:

$$\Phi(r) = \frac{GM(r)}{r^2} \tag{3.14}$$

as follows. Substituting for $M(r)$ in equation (3.11) produces Poisson's equation in spherically symmetric co-ordinates:

$$\nabla^2\Phi(r) = \frac{1}{r^2}\frac{d}{dr}\left(r^2\frac{d\Phi}{dr}\right) = -4\pi G\rho \tag{3.15}$$

Comparison of equations (3.10) and (3.12) shows that

$$\phi(r) = -[\Phi(r) - \Phi(R)] \tag{3.16}$$

since ρ, P and ϕ all vanish at the stellar surface, $r = R$.
Equation (3.9) can be rewritten as

$$\rho = \left(\frac{\gamma-1}{\gamma}\frac{\phi}{K}\right)^{\frac{1}{\gamma-1}} = \left[\frac{\phi}{(n+1)K}\right]^n \tag{3.17}$$

so substituting for ρ in equation (3.10) produces

$$\frac{1}{r^2}\frac{d}{dr}\left(r^2\frac{d\phi}{dr}\right) + 4\pi G\left[\frac{\phi}{(n+1)K}\right]^n = 0 \tag{3.18}$$

Defining ϕ_0 as ϕ at $r=0$, we introduce the scaled variable

$$\theta = \frac{\phi}{\phi_0} \tag{3.19}$$

and the scaled radius

$$\xi = \frac{r}{a} \text{ where } a = \left[\frac{4\pi G\phi_0^{n-1}}{((n+1)K)^n}\right]^{-0.5} \tag{3.20}$$

Rewriting equation (3.15) using these variables produces the Lane–Emden equations,

$$\frac{1}{\xi^2}\frac{d}{d\xi}\left(\xi^2\frac{d\xi}{d\theta}\right) + \theta^n(\xi) = 0 \tag{3.21}$$

where the boundary conditions are

$$\theta(0) = 1 \quad \text{and} \quad \left(\frac{d\theta}{d\xi}\right)_{\xi=0} = 0 \tag{3.22}$$

Cast in this manner, the density, pressure and temperature distributions all scale with θ. Thus, from equation (3.14)

$$\rho = \left[\frac{\phi_0}{(n+1)K}\right]^n \theta^n = \rho_0 \theta^n \tag{3.23}$$

where ρ_0 is the central density. Similarly, from equation (3.6)

$$P = K\rho^{\frac{n+1}{n}} = \frac{1}{K^n}\left(\frac{\phi_0}{(n+1)}\right)^{n+1} \theta^{n+1} = P_0\,\theta^{n+1} \tag{3.24}$$

Finally, given the perfect gas equation of state, $P_g = N_A k \rho T/\mu$, k is Boltzmann's constant, N_A is Avogadro's number, and μ is the mean molecular weight, then

$$T = \left(\frac{\beta\bar{\mu}}{N_A k}\right)\frac{P}{\rho} = \left(\frac{\beta\bar{\mu}}{N_A k}\right)\left(\frac{\phi_0}{n+1}\right)\theta = T_0\theta \tag{3.25}$$

where $\beta \equiv P_g/\rho$.

The location of the first zero in the Lane–Emden function, $\theta(\xi_1) = 0$, corresponds to the surface of the star, so ξ_1 is the scaled stellar radius

$$R = a\xi_1 \tag{3.26}$$

The total mass is given by

$$M = 4\pi a^3 \rho_0 \left[-\xi^2 \frac{d\theta}{d\xi}\right]_{\xi=\xi_1} = 4\pi a^3 \rho_0 m_1(n) \tag{3.27}$$

and the gravitational energy of a polytrope is

$$\Omega = -\frac{3}{5-n}\frac{GM^2}{R} \tag{3.28}$$

The polytropic approximation can be used to derive analytical models for low-mass dwarfs, as demonstrated by Burrows and Liebert [B11]. For an $n = 1.5$ polytrope (full convection or non-relativistic degeneracy), they derive

$$\rho_c \sim 5.99\left[\frac{3M}{4\pi r^2}\right] \propto M^2 \tag{3.29}$$

$$P_c \sim 0.77\frac{GM^2}{R} \propto M^{10/3} \tag{3.30}$$

$$R \sim 2.357\left[\frac{K'}{G}\right]M^{-1/3} \tag{3.31}$$

where ρ_c and P_c are the central density and pressure, and R is the radius. In the latter context, Zapolsky and Salpeter [Z1] provide the following analytical representation of the mass–radius relationship for very low-mass dwarfs:

$$R = 0.016 \left[\frac{M_\odot}{M} \right]^{\frac{1}{3}} \bigg/ \left[1 + \left(\frac{M}{0.0032\,M_\odot} \right)^{-0.5} \right]^{\frac{4}{3}} R_\odot \qquad (3.32)$$

Extending these calculations to include evolutionary effects, [B11] provides the following relationships:

$$T_{eff} \approx 27{,}700\tau^{-0.324} M^{0.827} \kappa_R^{0.088} \text{ Kelvin} \qquad (3.33)$$

$$L \approx 1.74 \times 10^{-6} \tau^{1.297} M^{2.04} \kappa_R^{0.35} L_\odot \qquad (3.34)$$

where τ is the age in Gyr, M is the mass in $M_\odot$, and κ_R is the Rosseland mean opacity (see Chapter 4). These relationships make no allowance for complications such as deuterium burning or the strong wavelength dependence of atmospheric opacities in low-temperature atmospheres, but provide order of magnitude estimates for these parameters in fully convective low-mass dwarfs ($M < 0.3\,M_\odot$).

3.3.3   Energy generation

Stars and brown dwarfs form through gravitationally-induced collapse of dense gas within molecular cloud cores – a process described in more detail towards the end of this chapter. During the formative phase and the initial stages of evolution, radiant energy is generated through the transformation of potential energy into heat. The Kelvin–Helmholtz contraction timescale (equation (3.1)) for an M dwarf is substantially longer than for the Sun: for example, CM Draconis A, the more massive component of the well-known eclipsing binary, has $M = 0.237\,M_\odot$, $R = 0.252\,R_\odot$ and $L = 0.005\,L_\odot$, giving $\tau_{KH} \sim 1.3\,\text{Gyr}$. However, these longer lifetimes are still only a small fraction of the Hubble time, the age of the Universe, estimated as 12–14 billion years (Chapter 6).

The primary energy source for M dwarfs, as with other main-sequence stars, is nuclear fusion. As the star contracts, the central temperature and density increase and, when the core temperature exceeds $T_c \sim 3 \times 10^6\,\text{K}$ (the exact value is density dependent), hydrogen fusion begins. Not all objects are sufficiently massive (have sufficient potential energy of collapse) to raise T_c above this threshold, and for those lower-mass objects – brown dwarfs – τ_{KH} is the appropriate luminous timescale. While this fact is mentioned in passing in early structural analyses of low-mass stars (see [L6], for example), the full consequences were emphasised in Kumar's [K7] explicit calculation of the hydrogen-burning limit (see Section 3.4.2).

The net result of hydrogen fusion is the conversion of four ^{1}H atoms to one ^{4}He atom. This conversion occurs through two series of reactions: the proton–proton chain (P–P cycle) and the CN (or CNO) cycle. The total energy, ϵ, generated by both processes is a function of core temperature (Figure 3.1). Interacting particles must overcome their respective Coulomb barriers, hence the rate of interaction increases with increasing kinetic energy (kT). In the case of the Sun, both reaction series are present, although the CN cycle contributes less than 10% of the total energy.

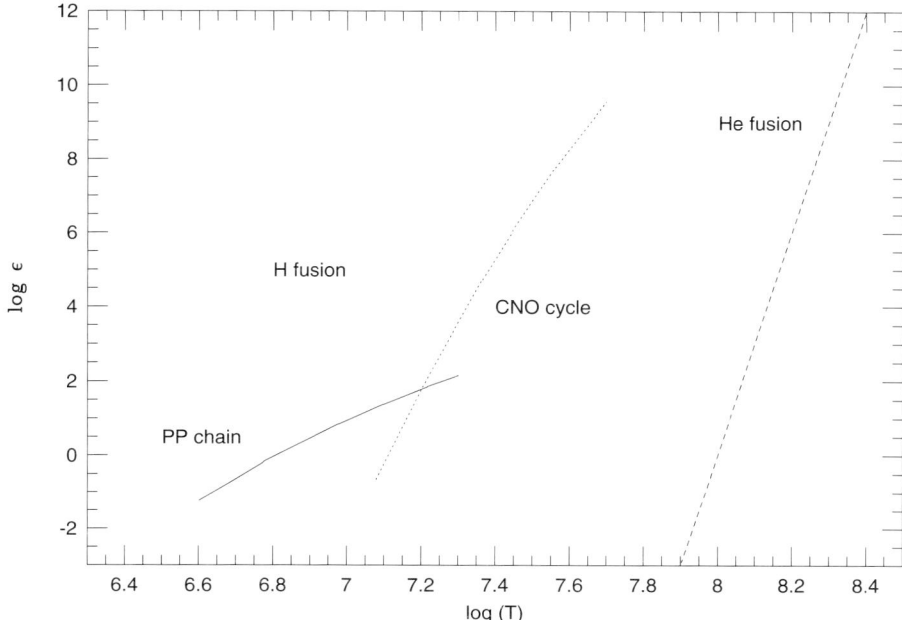

Figure 3.1. The relative energy-generation rates of the P–P chain, the CNO cycle and triple-α helium fusion. The calibrations are from [S7].

Thorough discussions of the physics of stellar nucleosynthesis are given by Clayton [C4] and Bahcall [B1]. Here we summarise only those factors relevant to low-mass dwarfs. Amongst those stars, the P–P cycle is the overwhelmingly dominant source of energy. The reactions in the basic chain are as follows:

$$p + p \rightarrow e^+ + \nu + d, \qquad Q = 1.442\,\text{MeV}, \quad 1^I$$

$$d + p \rightarrow {}^3\text{He} + \gamma, \qquad Q = 5.493\,\text{MeV}, \quad 2^I$$

$${}^3\text{He} + {}^3\text{He} \rightarrow {}^4\text{He} + p + p, \qquad Q = 12.859\,\text{Mev}, \quad 3^I \qquad (3.35)$$

where p denotes a proton (hydrogen nucleus); d, a deuterium nucleus; e^+, a positron; ν, a neutrino; and γ, a photon. Q is the energy released by each reaction (the neutrino in reaction 1^I carries $\approx 0.262\,\text{MeV}$). An alternative to the first step is the *pep* reaction

$$p + e^- + p \rightarrow d + \nu \qquad (3.36)$$

where e^- is an electron; but this reaction is of secondary importance.

The P–P cycle can be extended to reactions involving heavier elements if the temperature is sufficiently high (that is, in higher-mass M dwarfs). There are two subsequent reaction series: the **PP II** chain, starting with the products of reaction 2^I

$$^3\text{He} + {}^4\text{He} \rightarrow {}^7\text{Be} + \gamma, \qquad Q = 1.586 \text{ Mev}, \quad 3^{II}$$

$$^7\text{Be} + e^- \rightarrow {}^7\text{Li} + \nu, \qquad Q = 0.861 \text{ MeV}, \quad 4^{II}$$

$$^7\text{Li} + p \rightarrow {}^4\text{He} + {}^4\text{He}, \qquad Q = 17.347 \text{ MeV}, 5^{II} \tag{3.37}$$

and the PP III chain, which follows from reaction 3^{II}

$$^7\text{Be} + p \rightarrow {}^7\text{B} + \gamma, \qquad\qquad Q = 0.135 \text{ MeV}, \quad 4^{III}$$

$$^8\text{B} \rightarrow {}^4\text{He} + {}^4\text{He} + e^+ + \nu, \qquad Q = 18.074 \text{ MeV}, 5^{III} \tag{3.38}$$

For each of these reactions to occur, particles must overcome the Coulomb potential barrier surrounding each nucleus, given by ([C4], Chapter 1.5)

$$V = \frac{Z_1 Z_2 e^2}{R} = 1.44 \frac{Z_1 Z_2}{R} \text{ MeV} \tag{3.39}$$

where Z_1 and Z_2 are the atomic numbers of the two nuclei; e is the charge on an electron, and R is the separation in femtometres (10^{-15} m). Interactions occur only when $R \lesssim 2$–3 fm, requiring energies of $E > 0.5$ MeV. Particles in the stellar core follow a Maxwell–Boltzmann velocity distribution. The most probable energy of a particle is $kT/2$, and the average energy per particle is $\frac{3}{2}kT$, where

$$kT \sim 0.862 \times 10^{-7} T \text{ keV} \tag{3.40}$$

Since the central temperature of M dwarfs never exceeds 10^7 K, it is clear that quantum tunnelling effects are vital in overcoming Coulomb repulsion and initiating fusion. The efficiency of that process varies as $e^{-b\sqrt{E}}$, where e is the energy of the particle and b is a constant.[2] Nuclear reaction rates are determined by the convolution of the number of particles in the high energy tail of the Maxwell–Boltzmann distribution, $N \propto e^{-E/kT}$, and the tunnelling efficiency. The convolved function peaks sharply (the Gamow peak) at

$$E_0 = 1.22(Z_1^2 Z_2^2 A T_6^2)^{\frac{1}{3}} \text{ keV} \tag{3.41}$$

where T_6 is the temperature in units of 10^6 K.

Typical values for E_0 range from 10 to 30 keV for reactions in the P–P cycle. As T_c decreases with decreasing mass, the reaction rates for the PP II and PP III reactions also decrease and fail to achieve equilibrium. Indeed, for central temperatures below 8×10^6 K, reaction 3^I requires more than 10^9 years to reach equilibrium. At masses below ~ 0.25 M$_\odot$, temperatures are too low to allow significant production of ^{4}He, and the P–P cycle effectively terminates at reaction 2^I [B11].

When an M dwarf is on the main sequence, deuterium is produced through reaction 1^I and subsequently burned through reaction 2^I, while lithium is produced through reaction 4^{II} and burned through reaction 5^{II}. However, deuterium and lithium are both present at their interstellar abundances in a newly-formed star.

---

[2] $b = 31.28 Z_1 Z_2 \sqrt{A}$, where A is the reduced atomic weight, $A = A_1 A_2 / A_1 + A_2$, [C4].

Deuterium, with a primordial ratio $D/H \sim 2 \times 10^{-5}$ [T2] can be transformed to ^{3}He through reaction 2^I when the central temperature exceeds $\approx 6 \times 10^5$ K. This process, originally highlighted by Grossman [G5], leads to a pause in the luminosity evolution of dwarfs with masses exceeding $\sim 0.012\,M_\odot$, with the luminosity remaining approximately constant for between a few million years and 10^5 years, due to the additional energy source. This stage has been referred to as the deuterium-burning main sequence.

Primordial lithium is destroyed through reaction 5^{III}. However, the threshold temperature for initiating that reaction is relatively high: $\sim 2 \times 10^6$ K. As a result, objects with masses below $\sim 0.06\,M_\odot$ are predicted to retain an undepleted lithium fraction. This has become an important observational technique for verifying the nature of low-temperature brown dwarf candidates ([M2]; Chapter 9).

3.3.4 Energy transport

Energy generation occurs within the stellar core, and this energy must be transported to the surface to be released as radiation. As outlined in Section 3.1, convection is the dominant process in defining the internal structure of M dwarfs, with radiation playing a lesser, but still important, role. The criterion for stability against convection was stated originally by K. Schwarzschild: if a volume element is displaced in the presence of a temperature gradient with no change in the heat content (adiabatically), then the element expands or contracts to match the change in pressure. For an adiabatic expansion,

$$P = K\rho^\gamma \tag{3.42}$$

where $\gamma = C_P/C_V$, the ratio of specific heats at constant pressure and constant volume. The condition for stability (a restoring force on the displaced volume element) is

$$\left(\frac{P + dP}{K}\right)^{\frac{1}{\gamma}} > \rho + d\rho \tag{3.43}$$

or

$$\frac{1}{\gamma}\frac{1}{P}\frac{dP}{dr} > \frac{d\rho}{dr} \tag{3.44}$$

If we assume that the perfect gas law is valid,

$$P = NkT = \frac{N_A k \rho T}{\mu} \tag{3.45}$$

the stability criterion can then be rewritten as

$$\frac{1}{\gamma}\frac{1}{P}\frac{dP}{dr} > \frac{1}{P}\frac{dP}{dr} - \frac{1}{T}\frac{dT}{dr} \tag{3.46}$$

where dT/dr is the adiabatic temperature gradient. Hence we can define the superadiabatic temperature gradient:

$$\Delta \nabla T = \left(1 - \frac{1}{\gamma}\right) \frac{T}{P} \frac{dP}{dr} - \frac{dT}{dr} \tag{3.47}$$

The system is stable against convection for $\Delta \nabla T < 0$. Finally, since

$$C_p - C_v = \frac{Ak}{\mu} \qquad \text{and} \qquad \frac{dP}{dr} = -g(r)\rho \tag{3.48}$$

the convective inequality can be rewritten as

$$\left|\frac{dT}{dr}\right| \propto F\kappa < \frac{g}{C_P} \quad \text{for convection} \tag{3.49}$$

where F is the flux level and κ is the opacity. Thus, a system is unstable against convection if the flux level is high (as in the central regions of intermediate- and high-mass stars); if gravity is low (as in red giant envelopes); if the opacity if high; and if C_P is high, which is the case where abundant elements are undergoing ionisation. The last two circumstances hold to varying extent for M dwarfs and brown dwarfs.

Early-type M dwarfs have convective envelopes and radiative central regions, often referred to as radiative cores.[3] Approximately 90% of the total mass resides in the radiative core of a $0.55\,M_\odot$ M0 star, with the fraction dropping to $\sim 70\%$ by $0.4\,M_\odot$, or spectral type M2/M3 [D5]. At $\sim 0.3\,M_\odot$, spectral type $\approx$M4, the star becomes fully convective, and analyses suggest that the same circumstances prevail for brown dwarfs to temperatures of 1,500 K [B13].

Convection is generally modelled using the mixing-length approximation [B8], where each mass element participating in the convective motion is considered to rise or fall adiabatically over a distance l before achieving thermal equilibrium. The mixing length, l, is expressed in terms of the pressure scaleheight, H, with $\alpha = l/H$ typically taken as between 0.5 and 2. Describing this procedure as a theory of convection is something of an exaggeration, but the mixing-length approximation has been used almost exclusively in stellar modelling over the last 40 years. Canuto *et al.* [C1] have recently introduced a different approach, modelling convection using turbulent diffusion, but the technique has yet to be applied to M-dwarfs.

Finally, while convection is responsible for the bulk of energy transport in low-mass dwarfs, it is important to remember that every object has superimposed a thin radiative envelope – the atmosphere. The position of the boundary between convective and radiative transport is strongly dependent on wavelength-dependent effects tied to the complex, mainly molecular, opacities present in the atmosphere. The atmosphere determines how efficiently energy is released by the star, and therefore acts as a feedback mechanism, influencing the internal structure. Thus, uncertainties in atmospheric physics permeate inwards to produce corresponding uncertainties in the internal structure of these late-type dwarfs. These issues are described in more detail in Chapter 4.

---

[3] The radiative core should not be confused with the nuclear core, where fusion reactions are underway. The latter occupies only the central few percent of the radiative core.

3.3.5 The equation of state

The equation of state (EOS) describes the relationship between pressure, temperature and density within the stellar model, and therefore plays a pivotal role in
determining both the internal structure and the consequent predicted surface
properties. The Lane–Emden equations allow the full structure to be derived if the
system can be modelled as a gas polytrope. The polytropic form for the EOS is valid
under certain circumstances: in convection zones, where the temperature gradient is
adiabatic (n = 1.5); under conditions of full degeneracy (n = 1.0 or 1.5); and in an
ideal (perfect) gas (n = ∞).

Schwarzschild [S7] demonstrates that the (log T, log ρ) plane can be separated into
a number of régimes dependent on the dominant contributor to the EOS (Figure
3.2). At high temperatures and low densities, radiation makes a substantial

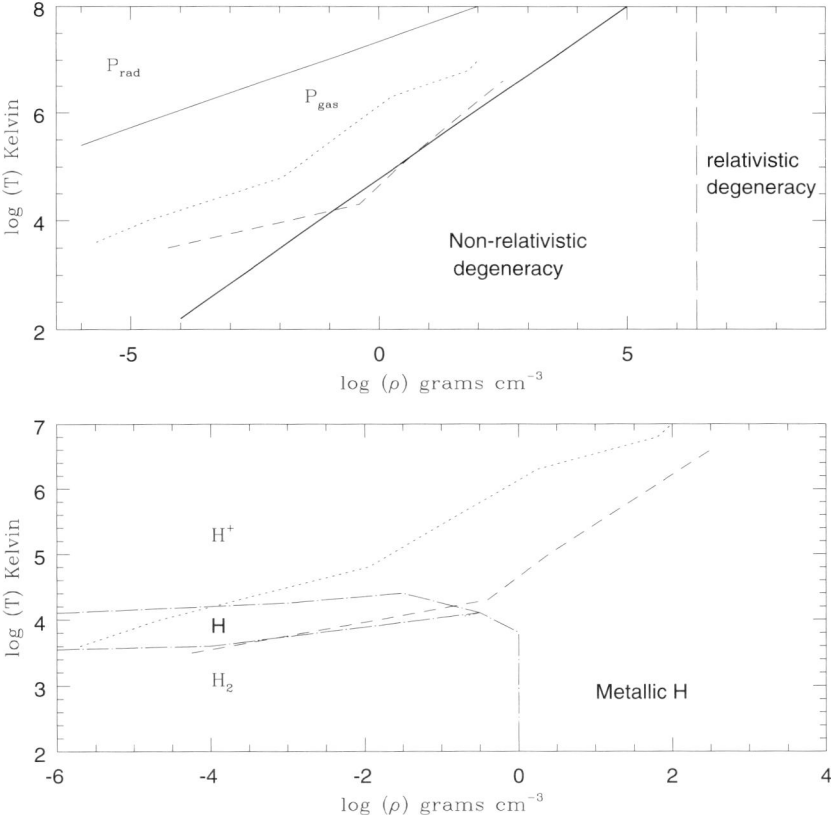

Figure 3.2. The (density, temperature) plane. The solid lines separate the régimes where radiation
pressure and degeneracy are dominant; the lower diagram identifies regions where hydrogen is ionised,
neutral, molecular or metallic. Both diagrams show the predicted radial density/temperature relationship
for 0.5 and 0.085 M☉ stars (dotted and dashed lines respectively, models from [G8]) and a 0.5-Gyr-old
0.05 M☉ brown dwarf (dash–dot line, from [B11]).

contribution to the total pressure. Under these circumstances, the gas pressure contribution can be modelled using the perfect gas law, so

$$P_{tot} = P_{rad} + P_{gas} = \tfrac{1}{3}aT^4 + NkT \qquad (3.50)$$

Electrons have half-integral spin, and are therefore governed by Fermi–Dirac statistics. As the density increases at a given temperature, a larger fraction of the available electron states are occupied, and the Fermi exclusion principle comes into play. Once all electronic states are occupied, the gas is fully degenerate. At non-relativistic velocities the electron pressure is defined by

$$P_E = \frac{8\pi}{15mh^3}p_0^5 \qquad (3.51)$$

and by

$$P_E = \frac{2\pi c}{3h^3}p_0^4 \qquad (3.52)$$

at relativistic velocities, where h is Planck's constant and p_0 is the maximum momentum. In the non-relativistic case, this corresponds to

$$P_E = K_1 \left(\frac{\rho}{\mu_E}\right)^{\frac{5}{3}} \qquad (3.53)$$

while the relativistic case gives

$$P_E = K_2 \left(\frac{\rho}{\mu_E}\right)^{\frac{4}{3}} \qquad (3.54)$$

where K_1 and K_2 are constants. Thus, the EOS for non-relativistic degeneracy is equivalent to an $n = 1.5$ polytrope, while a relativistically degenerate star can be modelled as an $n = 1.0$ polytrope.

The divisions in the ($\log T$, $\log \rho$) plane between regions where the dominant contribution to the EOS stems from radiation pressure, gas pressure and degeneracy, respectively, are outlined in Figure 3.2. Superimposed on the diagram are radial distributions predicted for 0.5, 0.085 and 0.05 $M_\odot$ dwarfs; that is M0 and M8/M9 stars, and a substellar-mass brown dwarf. These models show that neither radiation pressure nor relativistic degeneracy are important in determining the structure of either M dwarfs or brown dwarfs. However, non-relativistic electron degeneracy is significant in defining the equation of state in lower-mass objects, influencing in particular the location of the hydrogen-burning mass limit and the mass-radius relation (see Sections 3.4.2 and 3.5.3). Degeneracy becomes increasingly important with decreasing mass.

Figure 3.2 shows that fully-convective M dwarfs can be modelled with reasonable accuracy using simple polytropic approximations to the EOS and, as described in the following section, $n = 1.5$ polytropes have been applied to this purpose. However, over most of the mass range spanned by these objects, corrections must be introduced to allow for a variety of complicating factors. Partial degeneracy affects the innermost radii at low masses well before the star becomes fully degenerate; Coulomb interactions occur due to the presence of free charges (ions and electrons) throughout the star; there are perturbations due to the close approach of outer electron shells in ions, atoms and molecules; pressure ionisation occurs – a non-ideal gas ionises at temperatures lower than expected for an ideal gas; and the surface boundary conditions (notably the atmospheric opacities) affect the interior temperature and density distributions. All of these effects must be taken into account in computing the EOS in models of low-mass dwarfs. A brief history of the theoretical treatment of these issues is given by Saumon *et al.* [S5], who also present a thorough discussion of the physics underlying the most recent developments. The equations of state for hydrogen and helium derived in that paper are used in all of the recent models of low-mass stars, brown dwarfs and giant planets. The following section reviews the attributes of those models, concentrating on the predicted observable properties – notably, luminosities and temperatures.

3.4   MODELS OF LOW-MASS DWARFS

3.4.1   The development of low-mass models

The complex physical nature of M dwarfs coupled with their low intrinsic luminosities, and the consequent scarcity of high-quality observations, limited the extent of analysis as the underlying theory of stellar structure was being developed. Until the mid-1970s, observational constraints in the (M_{bol}, T_{eff}, mass) planes were set by a mere handful of binaries: YY Gem (Gl 278 C, or Castor C), Kruger 60AB (Gl 860AB), L726-8AB (Gl 65AB) and, to a lesser extent, VB 10 (Gl 752B). CM Draconis (Gl 630.1A) was added to the mix only in the mid-1970s [L1]. Nonetheless, there were a small number of far-reaching studies undertaken during the middle years of the twentieth century.

Early structural analyses, such as [O2], modelled low-mass stars as a radiative core surrounded by a convective envelope, but that treatment led to significant discrepancies between theoretical predictions and observations of late-type (>M4) dwarfs. Limber [L6] was the first to demonstrate that these discrepancies are resolved if the latter stars are fully convective: the shallower adiabatic temperature gradients lead to lower inferred T_c and M_{bol} for a given observed T_{eff}. In addition, Limber drew attention to two other important issues: the onset of electron degeneracy in the core, and the effects of H_2 formation and dissociation on the radial temperature gradient. As discussed further in the following section, the former process was first elaborated by Kumar [K7], [K8], who demonstrated that a minimum mass is required to achieve central temperatures sufficiently high for sustained hydrogen burning. Copeland *et al.* [C6] produced the first set of models which explicitly show the effects of the H_2 dissociation layer (Figure 3.3). The adiabatic gradient is reduced, leading to higher pressures and a reduced temperature gradient. Consequently, both luminosity and effective temperature are increased (although the radius is unchanged) as compared with models where H_2 is omitted. This results

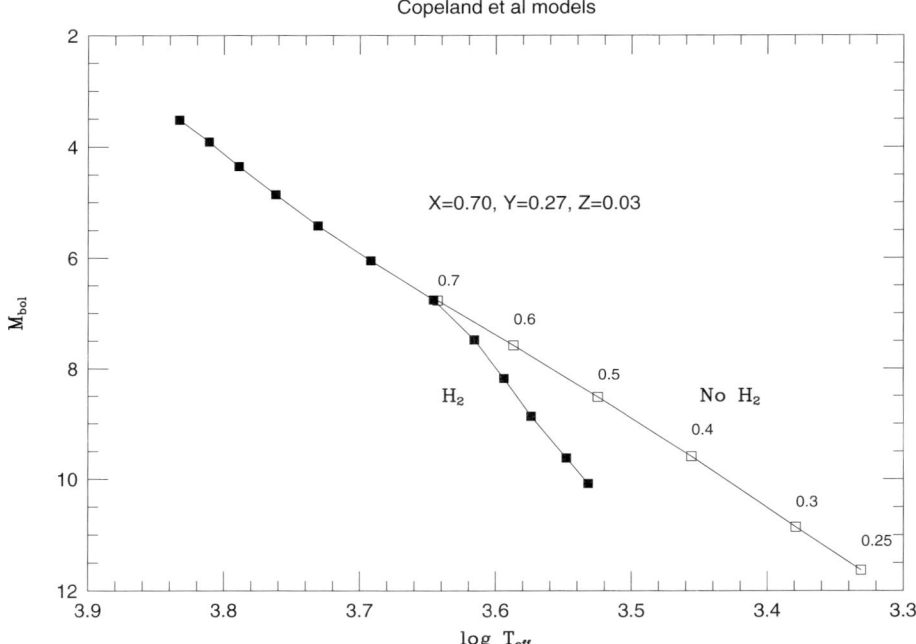

Figure 3.3. The effect of dissociation of the H_2 molecule on the form of the lower main sequence. The numbers 0.7–0.25 refer to the mass of the model. The models are from Copeland *et al.* [C6].

in a change of slope in the main sequence in the H–R diagram and in the mass–luminosity relationship. Both effects are observed – for example, see the steepening in slope at $M_V \sim 8.5$ in Figure 2.14.

The models of Copeland *et al.* were limited to masses exceeding $0.25\,M_\odot$. More extensive calculations were undertaken by Grossman, Graboske and collaborators, aimed explicitly at modelling the structure of VLM dwarfs above and below the hydrogen-burning limit [G5], [G6], [G7]. The final version of these models [G8] incorporated a number of significant innovations, including allowing for fusion of primordial deuterium as an energy source, using non-grey atmospheres to define the surface boundary conditions, extensive consideration of the effects of partial degeneracy in the stellar core, and incorporating non-ideal gas properties due to Coulomb interactions. Primordial deuterium burning occurs through reaction $2'$ of the P–P chain when T_c exceeds $\sim 7 \times 10^5$ K, and leads to a short-lived standstill in the rate of decline of L, T_{eff} for dwarfs with $M \geq 0.012\,M_\odot$. The initial calculations of Grossman *et al.* were based on a primordial abundance $N_D/N_H = 2 \times 10^{-4}$ and suggested that low-mass ($< 0.25\,M_\odot$) dwarfs would spend 10^6–10^8 years (τ_D increasing with decreasing mass) on a deuterium-burning main sequence, running almost parallel to the hydrogen-burning sequence, but at higher luminosities and temperatures. Current estimates place N_D/N_H closer to 2×10^{-5}, with a corresponding decrease in τ_D.

Adapting the boundary conditions to allow for non-grey opacities might seem to be a minor change, since the radiative layers constituting the stellar atmosphere include less than 1% of the total mass. However, these layers have a disproportionate effect in defining the internal structure in low-mass dwarfs, since they effectively define the entropy of the system. That entropy remains constant throughout the convective regions, the full interior in dwarfs with masses below $0.3\,M_\odot$. The only atmosphere calculations available for use by Grossman *et al.* (from [A8]) were limited in scope, but the data were sufficient to indicate the importance of allowing for non-grey opacities. Incorporating the latter led to both lower luminosities and lower temperatures for a given mass and age, reducing the mass-limit for hydrogen burning. The $(\log T, \log \rho)$ radial profiles of the [G8] 0.5 and $0.085\,M_\odot$ models are plotted in Figure 3.2. The former has a radiative core, and as a result $\rho \propto T^3$ (an $n=3$ polytrope) in the central regions, while $\rho \propto T^{1.5}$ (an $n=1.5$ polytrope) in the outer convective envelope. The $0.085\,M_\odot$ model is both fully convective and partially degenerate near the core, with $\rho \propto T^{1.5}$ at all radii except within the outermost radiative envelope.

Taken as a whole, the [G8] models reproduce qualitatively the change in slope in the main sequence ascribed by [C6] to H_2 dissociation near the lower boundary of the atmosphere, although [G8] place the inflection at $0.6\,M_\odot$ rather than $0.7\,M_\odot$. [G8] also predict a change in slope in the mass–luminosity relationship near that mass, partly due to non-equilibrium in the ^{3}He/H ratio as reaction $3'$ decreases in efficiency (with a consequent change in energy production), and partly due to increased molecular opacity sources. The net effect is a more rapid decrease in mass with decreasing luminosity for masses below $\sim 0.5\,M_\odot$ – a flattening of the mass–luminosity relationship, dM/dL, evident in Figure 3.4.

Subsequent model calculations by D'Antona and Mazzitelli [D1], [D2] revealed evidence for further structure in the mass–luminosity relationship at masses close to the hydrogen-burning limit. Besides the change in slope at $\sim 0.5\,M_\odot$, D'Antona and Mazzitelli identified a pronounced steepening in dM/dL at the lowest masses (Figure 3.4) – that is, a small change in mass corresponds to a substantial change in luminosity. This result, confirmed by Sienkiewicz [S8], Vandenberg *et al.* [V1] and more recent studies, has significant implications for the interpretation of the stellar luminosity function, $\Phi(M_V)$: specifically, a flat or declining luminosity function (number of stars per unit magnitude) may well be consistent with a rising mass function (number of stars per unit mass). This matter is discussed in Chapter 8.

Further developments made during the 1980s in modelling low-mass stars are well summarised in the reviews by Dorman *et al.* [D5] and Burrows and Liebert [B11]. Many of those models still placed heavy reliance on the polytropic approximation and most used grey atmospheres. The significant breakthroughs of recent years have been stimulated by, first, the development of the appropriate equation of state for metallic hydrogen [S5]; second, the inclusion of grain opacities in evolutionary models [B12], [B13]; and, finally, the marriage between interior models [B2], [B3] and the complex low-temperature atmosphere codes devised by Allard and Hauschildt [A6] and collaborators (see Chapter 4). The properties of these recent low-mass models are summarised in Section 3.4.3.

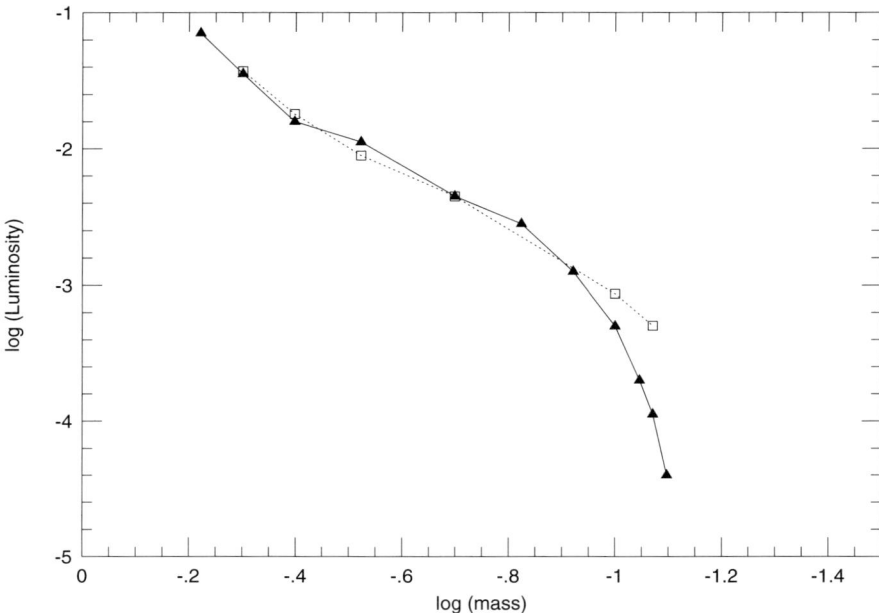

Figure 3.4. The mass–luminosity relationship derived by Grossman et al. ([G8], open squares) and D'Antona and Mazzitelli ([D2], solid triangles).

3.4.2   The minimum mass for hydrogen burning

The existence of a lower mass limit to the hydrogen-burning main-sequence for stars of a given abundance is tied to the onset of degeneracy in the stellar core during pre-main sequence evolution. As the protostar collapses, the central density rises and the core temperature, T_c, increases as potential energy is transformed to thermal energy. For solar-type stars, T_c is driven above $\sim 3 \times 10^6$ K and hydrogen fusion is initiated, providing a steady energy source which supports the star against further collapse.

Low-mass stars are required to collapse to higher densities than solar-type stars before achieving the critical T_c for fusion ignition. Since $GM^2/R \sim NkT$ in the perfect gas regime, and $M/R \sim$ constant for a given T_c, $\langle \rho \rangle \propto M^{-2}$. As the density increases, the core becomes partially degenerate, measured by the degeneracy parameter

$$\alpha_E = \frac{N_e h^3}{2(2\pi M - EkT)^{\frac{3}{2}}} \sim \frac{\text{electron chemical energy}}{kT} \tag{3.55}$$

If $\alpha_E < -4$, the material can be treated as a perfect gas, and Maxwell–Boltzmann statistics apply, while if $\alpha_E > 20$ the material is fully electron degenerate, and Fermi–Dirac statistics are appropriate. Figure 3.5 (adapted from [G8]) plots the predicted radial variation in α_E for a range of masses. Λ_F is the plasma parameter, the ratio between the Coulomb energy and kT. A plasma parameter $\Lambda_F < 0.1$ indicates weak

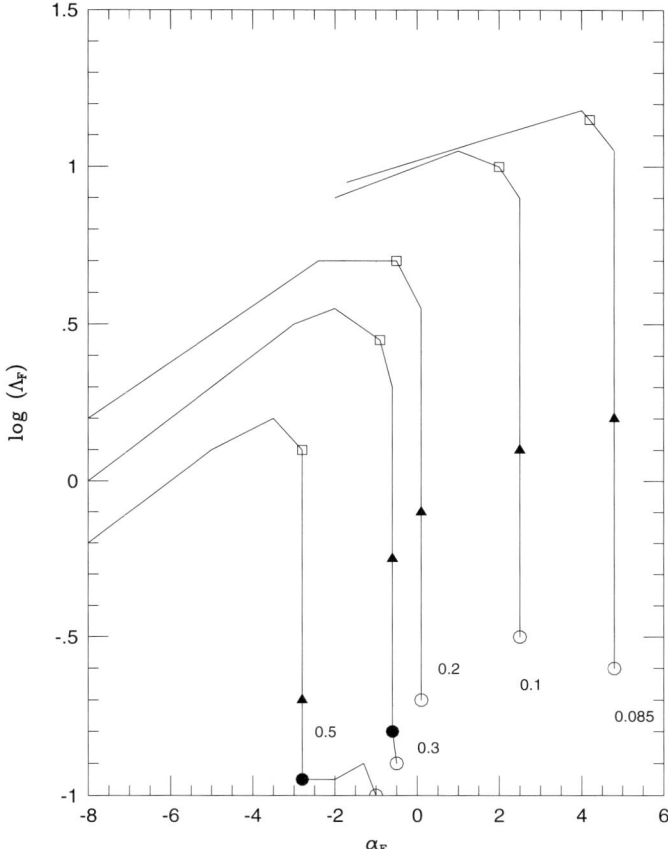

Figure 3.5. The Coulomb force, Λ_F, as a function of the radial variation in the degeneracy parameter, α_E for the 0.5, 0.3, 0.2, 0.1 and 0.085 $M_\odot$ models computed by [G8]. The open circle marks the centre of each model; the solid circle marks the lower boundary of the convective zone – the models are fully convective for $M < 0.3\,M_\odot$. The solid triangle is the core boundary (including 97% of the total mass), while the open square indicates where degeneracy becomes important.

screening of the nucleus, while $\Lambda_F > 10$ implies strong screening. The figure illustrates the growing degeneracy with decreasing mass, with α_E constant within the convective envelope.

As degeneracy becomes more significant in a contracting protostar, an increasing fraction of the liberated potential energy is absorbed in reducing the separation between degenerate electrons, rather than transformed to thermal energy. Degeneracy prevents the stellar radius from decreasing below $\sim 0.1\,R_\odot$. The consequence, originally pointed out by Kumar [K9], is that T_c reaches a maximum value, T_{max}. Subsequent evolution depends on the total mass of the dwarf, and on the degree to which nuclear fusion can continue to provide a stable energy source. There is not, in fact, a simple, clean division between 'stars' – where fusion makes a significant

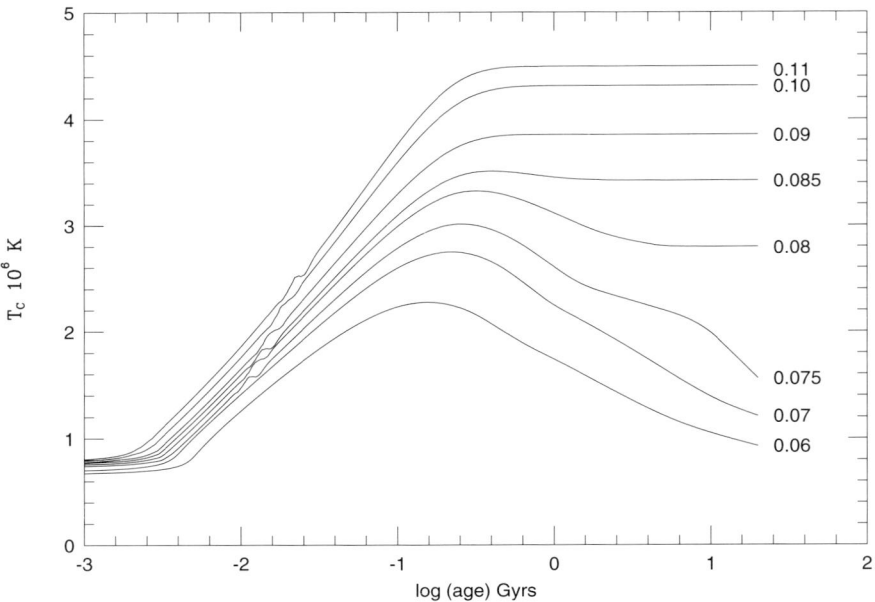

Figure 3.6. The evolution of core temperature with time predicted by Burrows *et al.* [B12], [B13] for dwarfs bridging the transition from main sequence, core-hydrogen-burning stars to brown dwarfs.

contribution to L_{tot} – and 'brown dwarfs' – which, to quote Burrows and Liebert, 'cool like a rock'.

A low-mass dwarf faces four possible futures after the onset of significant core degeneracy. These are illustrated in Figure 3.6 using models computed by Burrows *et al.* [B12], [B13]. In order of decreasing mass, they are:

• Hydrogen fusion continues at a sufficient rate to maintain $T_c \sim T_{max}$; the star remains on the main sequence, and the luminosity and effective temperature are constant for many Hubble times ($M \geq 0.09\,M_\odot$ in Figure 3.6).
• Degeneracy reduces T_c below T_{max}, but the temperature remains sufficiently high to permit continuing fusion, so the dwarf achieves a main-sequence configuration ($0.08-0.085\,M_\odot$ in Figure 3.6).
• Fusion has been initiated, but degeneracy (eventually) reduces T_c below the critical level for continuing hydrogen-burning, so the 'ex-star' cools as a brown dwarf ($0.075\,M_\odot$ in Figure 3.6).
• T_c never reaches sufficiently high levels to allow fusion to become a significant energy source; the object becomes a degenerate brown dwarf and cools to oblivion ($M \leq 0.07\,M_\odot$ in Figure 3.6).

The first and second scenarios result in stable low-mass stars; scenario (4) produces a brown dwarf, and the hydrogen-burning limit, M_{HBL}, lies somewhere

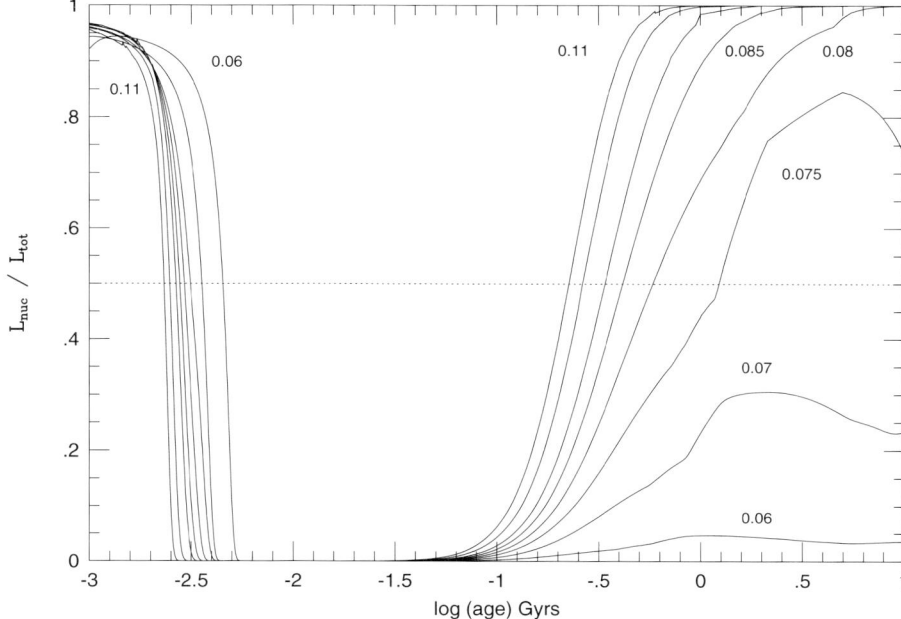

Figure 3.7. The fractional luminosity contributed by nuclear fusion reactions as a function of time for the [B12], [B13] models plotted in Figure 3.6. The initial burst of fusion marks deuterium burning which persists longest in the lowest-mass model.

between scenario (3) (which produces what has become known as a 'transition object') and scenario (4). The dividing line is usually placed at a mass where the fractional contribution to the total luminosity due to hydrogen fusion, L_{nuc}/L_{tot}, never exceeds 50% – an arbitrary definition.

Kumar's original estimate placed M_{HBL} at $\sim 0.07 \, M_\odot$. Figure 3.7 plots L_{nuc}/L_{tot} for the Burrows *et al.* models illustrated in Figure 3.6. These models predict that a $0.08 \, M_\odot$ solar-abundance dwarf achieves a stable luminosity and temperature after $\sim 2 \times 10^8$ years; that a $0.075 \, M_\odot$ dwarf is a transition object, able to sustain fusion for nearly 10^{10} years, but eventually fated to cool degenerately; and that a $0.07 \, M_\odot$ dwarf, while partially supported by fusion at ages of 1–2 Gyr, evolves as a brown dwarf. The formal hydrogen-burning limit is $M_{HBL} = 0.073 \, M_\odot$.

The calculations plotted in Figures 3.6 and 3.7 are for solar-abundance (disk) dwarfs. Decreasing the metal content leads to lower opacities and both higher effective temperatures and higher luminosities at a given mass. As discussed in Chapters 2 and 11, these effects lead to metal-poor stars defining separate main sequence(s), lying blueward of disk dwarfs in the H–R diagram. The higher surface temperatures in turn require higher core temperatures, with the result that M_{HBL} is driven to higher masses at the lower metallicities found in the Galactic halo. Both D'Antona and Mazzitelli [D4] and Baraffe *et al.* [B4] have computed low-mass models for a range of abundance. The former models predict that M_{HBL} rises from

$0.090\,M_\odot$ at a metallicity of $[Fe/H] = -0.7$ to $0.097\,M_\odot$ at $[Fe/H] = -2.3$. The latter value is almost $0.02\,M_\odot$ higher than the solar-abundance M_{HBL} computed by the same authors. Finally, lower opacities and higher transparency lead to more rapid cooling of substellar-mass brown dwarfs, and a correspondingly more rapid decrease in luminosity with time. As a result, direct observations of brown dwarf members of the Galactic halo are likely to task even the Next Generation Space Telescope.

3.4.3   Properties of current models of low-mass dwarfs

The 1990s saw substantial advances in modelling both the atmospheres and the internal structure of low-mass dwarfs. Similar advances are likely to occur over the succeeding decade or more. Nonetheless, consideration of the predictions made by the current generation of stellar models provides a snapshot of our present under-standing in the field. We consider the results derived from two sets of models: by the Tucson group – Burrows, Hubbard, Lunine, Saumon and collaborators; and by the Lyon group – Baraffe, Chabrier and collaborators.

For present purposes we concentrate on the Tucson calculations described by Burrows et al. ([B12] – the 'X' models) and the extension to lower masses in Burrows et al. [B13], while the Lyon models are from Baraffe et al. [B4], [B5]. The Tucson dataset is limited to low-mass dwarfs, spanning the mass range 0.009–$0.20\,M_\odot$, while Baraffe and Chabrier extend their calculations to $1\,M_\odot$, but do not sample the substellar régime at the same resolution in mass. Both sets of models adopt similar physics for the internal structure, with the equation of state based on Saumon and Chabrier's [S4] calculation. The main differences lie in the opacities and the boundary conditions: the [B12] Tucson models use grey atmospheres, although the later lower-mass calculations adopt non-grey boundary conditions, and both sets of models allow for dust formation and grain opacities; the Lyon calculations, on the other hand, use Allard and Hauschildt's model atmospheres (Chapter 4) to define the surface boundary conditions, but do not allow for extensive grain formation. This difference in formulation leads to a difference in the presentation of the model predictions: Burrows et al. limit their predicted properties to data in the theoretical plane (L, T_{eff}) and near-infrared (JHK) colours; Baraffe and Chabrier, with the advantage of more realistic atmospheres, provide colours and magnitudes (BVRIJHK) in addition to (L, T_{eff}). Taken together, these two sets of models provide a measure of the reliability of our current understanding of the structure of low-mass dwarfs.

The basic characteristics of low-mass stars and brown dwarfs are best illustrated by considering their evolution in luminosity and effective temperature as a function of mass. Figures 3.8 and 3.9 plot those predictions for the Tucson models. Both diagrams show a clear bifurcation in the behaviour at $\tau > 2$–$3\,$Gyr, with hydrogen-burning stars maintaining nearly constant properties after settling into their main-sequence configurations, while brown dwarfs show a continued decline in both L and T_{eff}. The $\sim 0.075\,M_\odot$ transition objects exhibit a standstill in that descent for up to $\sim 10\,$Gyr (close to the age of the Galactic disk), before fusion is extinguished and

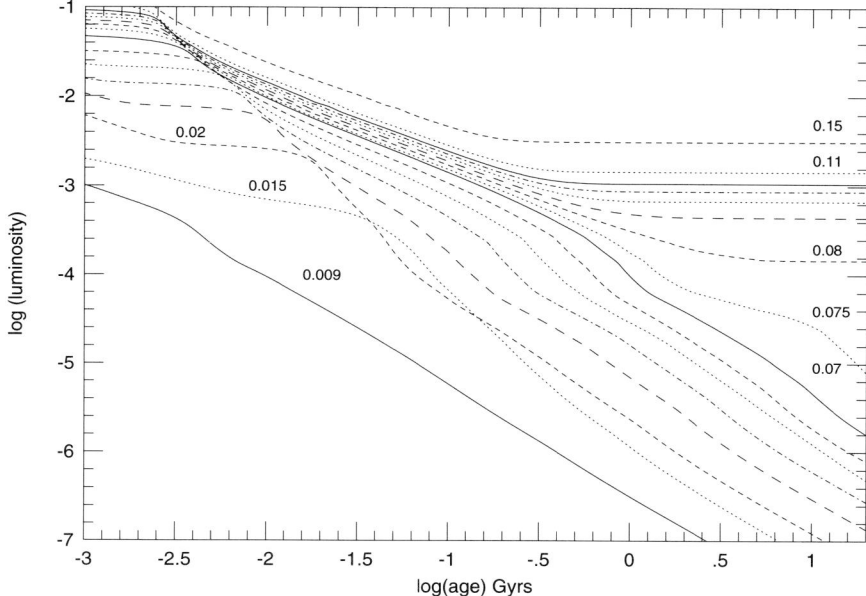

Figure 3.8. Luminosity evolution of low-mass dwarfs as computed by the Tucson group. The models range in mass from 0.009 to 0.15 $M_\odot$ (the tracks are coded as in Figure 3.9). The divergence of brown dwarfs from main-sequence stars, and the equivocal behaviour of transition-mass objects, is evident at ages exceeding ∼1 Gyr.

cooling takes over. The transition between M dwarfs and L dwarfs is generally placed at ∼2,000 K. Thus, these models predict that early-type L dwarfs encompass a mix of both *bona fide* old, hydrogen-burning stars and younger (<2 Gyr-old) brown dwarfs.

The effective temperature of almost all objects – save those of the lowest mass – is predicted to remain nearly constant at ages of less than $\tau \sim 10^7$ years. At the same time, the luminosity decreases by more than an order of magnitude. This obviously reflects the rapidly-decreasing radius as the dwarf contracts either onto the main sequence or to a degenerate configuration. Figure 3.10 shows the behaviour predicted by the Tucson models. Fusion of primordial deuterium provides an additional energy source during this same time period for dwarfs with masses exceeding ∼ 0.012 $M_\odot$, as evidenced by a plateau in the evolution of the luminosity, effective temperature and radii. This behaviour must be taken into account when determining statistical parameters, such as the mass function, for young star clusters.

The same dichotomous behaviour is evident in the evolution of the mass–luminosity relationship with time (Figure 3.11). For ages up to $\tau \sim 3 \times 10^8$ years, the luminosity decreases over the full mass range with increasing age, with deuterium burning slowing the rate of decline at $\tau \leq 10^7$ years, $M \gtrsim 0.012 M_\odot$. At ages exceeding 3×10^8 years, the lowest mass stars have achieved thermal and hydrostatic

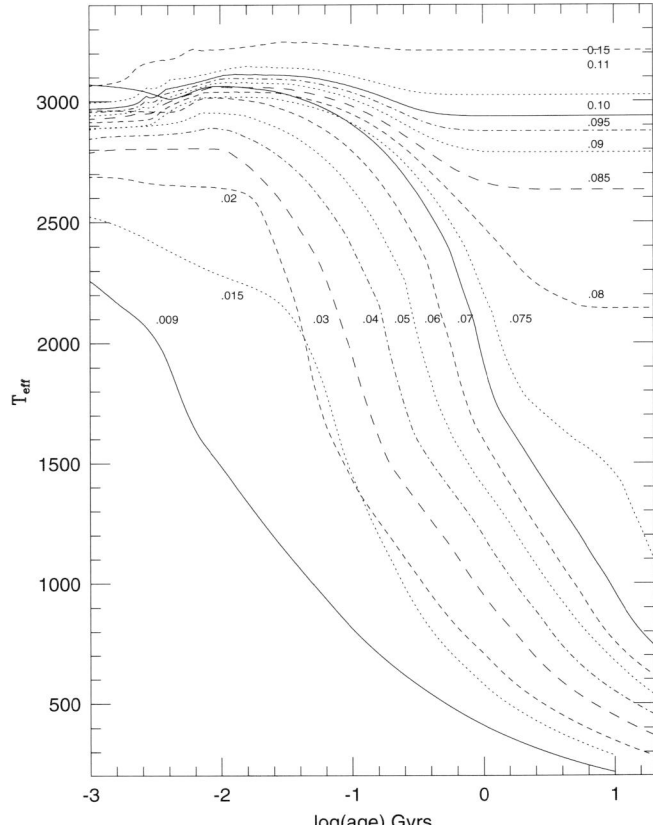

Figure 3.9. The evolution of effective (surface) temperature with time, for the same models illustrated in Figure 3.8.

equilibrium, but L continues to decline at substellar masses. This divergence was first emphasised by D'Antona and Mazzitelli, and clearly eliminates the possibility of determining masses for brown dwarfs unless age is a known quantity. The latter requirement underlines the utility of open star clusters in mass function analyses.

Finally, Figure 3.12 shows evolutionary tracks in the $(\log L, T_{eff})$ H–R diagram. The Tucson models predict $T_{eff} \sim 1{,}900\,\mathrm{K}$ and $\log(L) \sim 10^{-4}\,L_{\odot}$ as the lowest value attainable by a hydrogen-burning star; the Lyon models place the limit at somewhat higher temperatures and luminosities: $\sim 2{,}000\,\mathrm{K}$ and $\sim 10^{-3.9}\,L_{\odot}$. Lower-mass objects evolve through the full temperature range between $\sim 3{,}000\,\mathrm{K}$ and $< 500\,\mathrm{K}$, but at different rates depending on the mass. Note that lower-mass brown dwarfs are predicted to have higher luminosities for a given T_{eff}, reflecting their larger radii. The main consequence of this temperature evolution is that the observed spectral type of substellar-mass objects changes with time, complicating the interpretation of observations of L dwarfs.

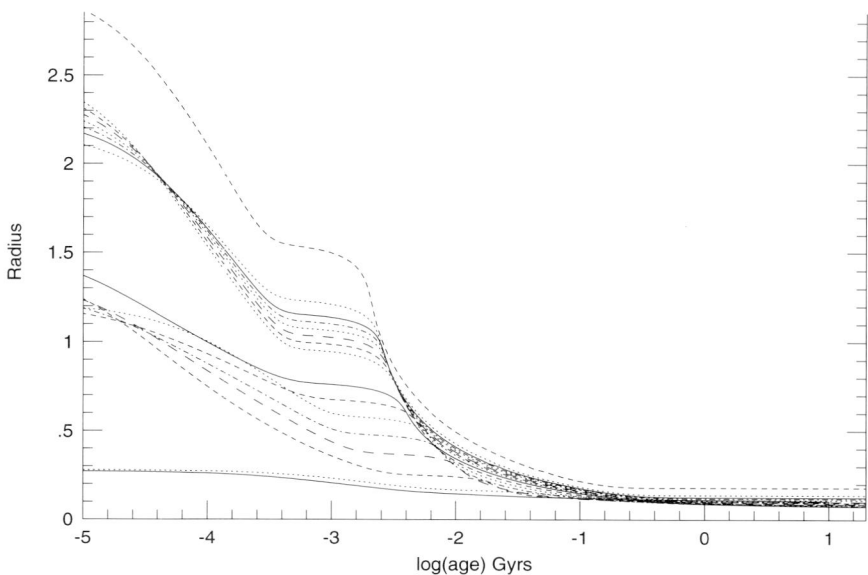

Figure 3.10. The variation with time of radii of low-mass dwarfs, as predicted by the Tucson models. As in the $L(\tau)$ and $T_{eff}(\tau)$ diagrams, the initial plateau reflects energy generated through deuterium burning.

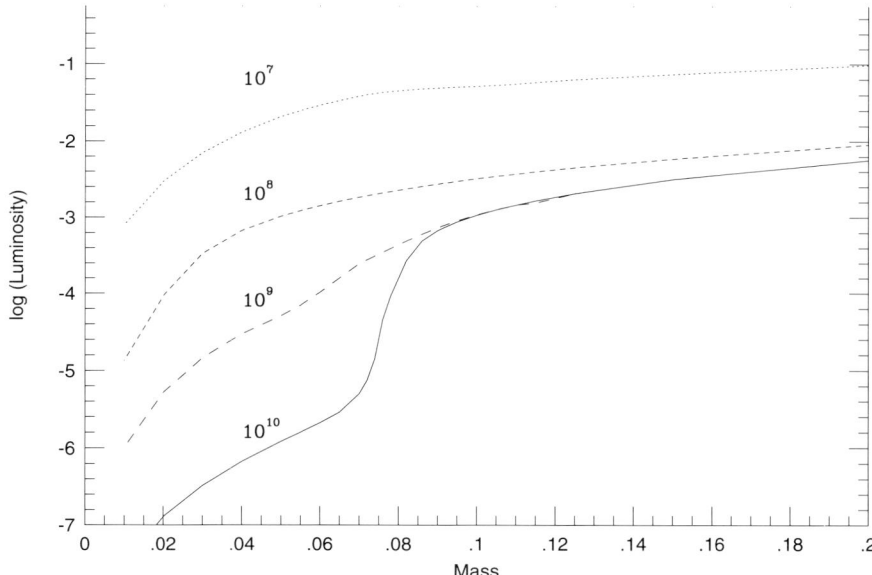

Figure 3.11. The mass–luminosity relationship at various ages, as predicted by the Tucson low-mass dwarf models. The substantial evolution in luminosity at low masses originally highlighted by D'Antona and Mazzitelli is clearly evident.

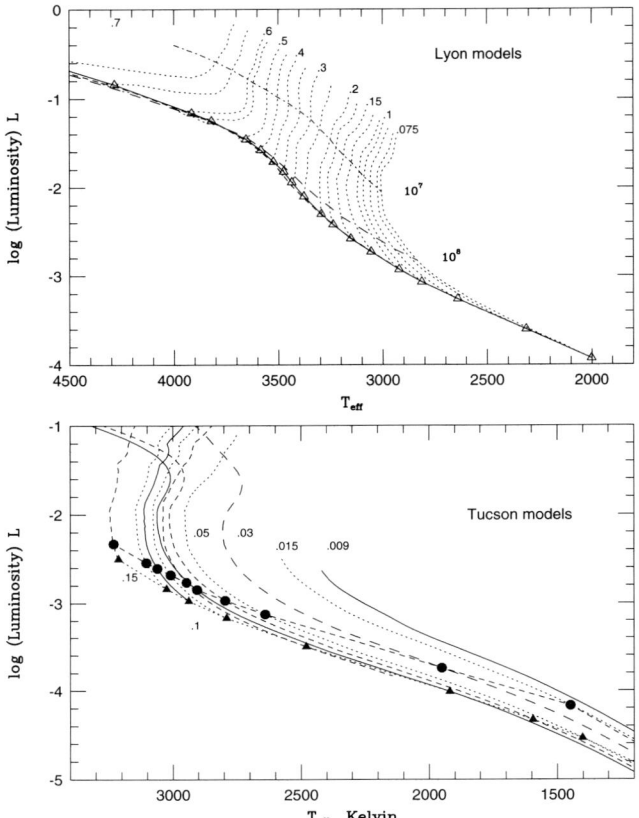

Figure 3.12. Evolutionary tracks of low-mass dwarfs. The lower diagram plots results from the Tucson group for masses between 0.15 and 0.009 $M_\odot$ (coded as in Figure 3.9). Solid dots mark ages of 10^8 years, and solid triangles 10^9 years. The upper diagram plots predictions from the [B5] Lyon models. Representative tracks are identified by their mass and isochrones for ages 10^7 (dash–dot line), 10^8 (dashed line), 10^9 (short-dashed line) and 10^{10} years (solid line), with the open triangles marking the main-sequence location of each model.

3.5 MATCHING MODELS AND OBSERVATIONS

The crucial test of these theoretical models is the comparison between their predicted properties and the observed characteristics of low-mass dwarfs. These comparisons can be undertaken in either the observational or theoretical planes: for example, the H–R diagram comparison can be made in either the $(\log L, T_{eff})$ plane or the (absolute magnitude, colour) plane. In both cases, uncertainties in the transformation between observed and theoretical quantities contribute significantly to the uncertainties in the final comparison. Theoretical models of low-mass stars and brown dwarfs showed marked improvement throughout the 1990s, and this improvement is likely to continue. For present purposes, the Tucson and Lyon datasets are taken as representative of current models of disk dwarfs, supplemented by D'Antona and Mazzitelli's [D4] models of metal-poor halo stars. (The reader should check contemporary literature for the latest improvements.)

3.5.1   The Hertzsprung–Russell diagram

Figure 3.13(a) compares the observed and predicted locations of the lower main-sequence on the H–R diagram. The empirical temperatures and luminosities are those of the mean M-dwarf scale described in more detail in Chapter 4 (see Table 4.1), supplemented by data for the few L dwarfs with measured trigonometric parallaxes. The theoretical relationships plotted are for 0.6 and 10 Gyr-age Tucson

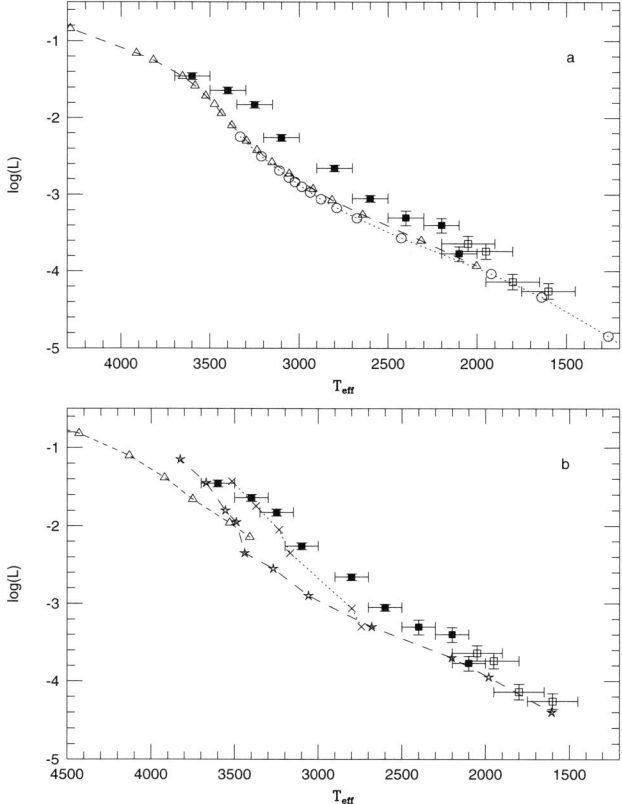

Figure 3.13. Comparison of the theoretical and observational H–R diagrams. (a) compares the Tucson (open circles) and Lyon (open triangles) solar abundance models for age 10 Gyr against the empirical M-dwarf sequence (solid squares) and data for L dwarfs with measured trigonometric parallaxes (open squares). (b) matches the same observational data against the theoretical main sequences from [C6] (triangles), [G8] (crosses), and [D2] (stars). All of the models produce significantly higher temperatures than the observed sequence.

models and 1 and 10 Gyr-age models from [B5]. Locations of individual models on the 10-Gyr isochrones of both datasets are indicated. The models are identical with those plotted in Figure 3.12.

The two sets of theoretical calculations are in good agreement. The Lyon models are ~ 50–100 K cooler and ~ 0.05–0.1 dex less luminous in $\log(L/L_\odot)$ for masses above $0.08\,M_\odot$, although differences near M_{HBL} lead to the $0.075\,M_\odot$ Lyon model being ~ 400 K hotter and ~ 0.5 dex brighter than the Tucson counterpart. Neither set of models, however, provides a perfect match to the empirical sequence. As Figure 3.13(b) shows, this is not a new problem: theoretical models have been consistently 200–300 K hotter than the observationally-defined M dwarf sequence for more than a quarter of a century. The [G8] models provide a partial exception, matching the observed location of early M dwarfs, but overestimating both luminosity and temperature to an increasing extent with decreasing mass.

Since the [B5] models incorporate the Allard and Hauschildt model atmosphere calculations, the Lyon models also predict colours and absolute magnitudes in the observational plane. Figure 3.14 compares those predictions against the observed $(M_V, (V–I))$ and $(M_I, (I–K))$ distributions described by single disk dwarfs with accurate $(\sigma_\pi/\pi < 10\%)$ trigonometric parallax determinations. Besides the solar-abundance models, the [B5] predictions for $[Fe/H] = -0.25$ dex (a closer match to the mean abundance of the Galactic disk) are also shown. Both models are for ages of 10 Gyr, and are limited to $M > 0.075\,M_\odot$; that is, hydrogen-burning dwarfs in their main-sequence configurations.

These models represent a significant advance over all previous analyses, but Figure 3.14 shows that there is still room for improvement. The solar abundance models are significantly bluer than the observations for $M_V > 10$, as might be expected given the higher temperatures evident in Figure 3.13. The inflection in the main sequence at $M_V \sim 9.5$ $(M \sim 0.5\,M_\odot, (V–I) \sim 2.1)$ in the $[Fe/H] = 0$ models is due to H_2 formation and dissociation near the photosphere, while the subsequent flattening at $M_V \sim 13$ marks the increasing importance of degeneracy. Neither the 'step' in the empirical main sequence at $M_V \sim 12.5$, $(V–I) \sim 3$ nor the 'hook' at $M_V > 17.5$ are reproduced by these (or any other) models. Moreover, the Lyon $[Fe/H] = -0.25$ models are predicted to be redder than the solar abundance models at $M_V > 15$. All of these effects may well reflect incomplete treatment of opacities in the model atmospheres.

Theory and observations are in better agreement in the $(M_I, (I–K))$ plane, with the solar abundance isochrone providing a good match to the observed sequence. This may be due partly to the lesser importance of molecular opacities at 0.8 and 2.2 µm – an hypothesis supported by the theoretical/empirical mass–luminosity comparisons discussed in the following section. However, the good agreement between theory and observation at near-infrared wavelengths does not extend to the cooler L dwarfs: both the Tucson and Lyon models predict a maximum (J–K) colour of ~ 1.0, while the reddest L dwarfs have (J–K) colours exceeding two magnitudes. As discussed further in Chapters 4 and 9, the redder colours are probably due to dust formation within these cool atmospheres, requiring even more complex theoretical models.

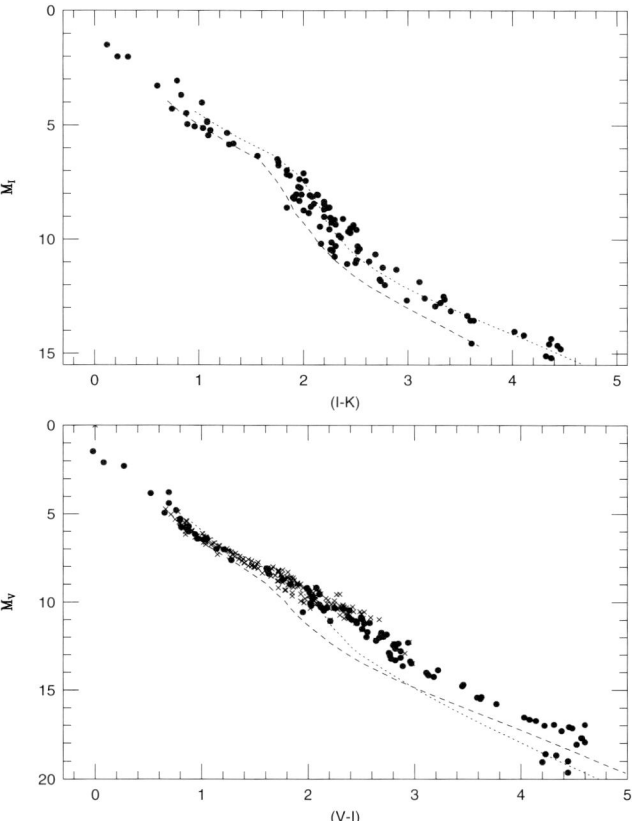

Figure 3.14. The [B5] (M_V, (V–I)) and (M_I, (I–K)) main sequences compared with observations of nearby stars. Solid points mark stars drawn from the 8-parsec sample; crosses are more distant stars with Hipparcos parallax measurements. The dotted line is the theoretical solar-abundance sequence; the dashed line marks the [Fe/H] = −0.25 dex predictions.

As discussed in the previous section, low-mass subdwarfs outline different colour–magnitude relationships than disk dwarfs. Both D'Antona and Mazzitelli and the Lyon group have predicted (M_V, (V–I)) relationships for a range of abundances. It should be noted that these models adopt different helium abundances: [D4] assume that the helium mass fraction increases from 23% at [Fe/H] = −2.3 to 24% at [Fe/H] = −0.7, as would be expected given nucleosynthesis in massive stars and recycling; [B4] adopt Y = 0.25 for all models. Both sets of isochrones are predicted to have a characteristic S-shape due to H_2 dissociation at ∼0.5 $M_\odot$, and the dominance of degeneracy at masses below ∼0.12 $M_\odot$. The qualitative accuracy of these predictions has been confirmed by deep colour–magnitude data for globular clusters (Section 11.5).

Figure 3.15 compares both sets of predictions against data for nearby subdwarfs

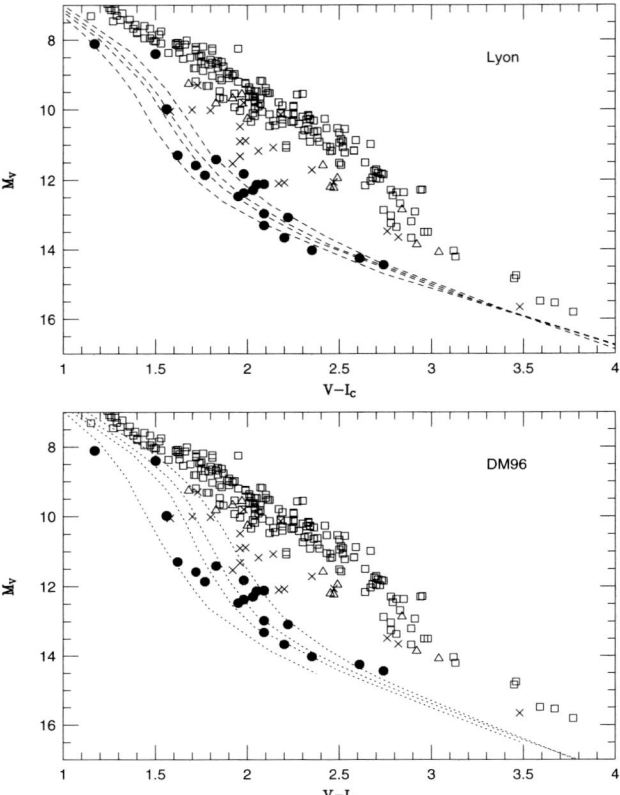

Figure 3.15. Theoretical main sequence relationships for subdwarfs, as predicted by D'Antona and Mazzitelli [D4] (DM96) and Baraffe *et al.* [B4] (Lyon). The former models are for [Fe/H]= −2.3, −1.3, −1.0 and −0.7; the latter for abundances of −2.0, −1.5, −1.3 and −1.0. The disk main sequence (open squares) is defined by the nearby stars plotted in Figure 3.14; data for the halo subdwarfs are taken from [M5] and [G2]. Solid points mark extreme subdwarfs; crosses mark intermediate subdwarfs (see Chapter 11).

with accurately determined trigonometric parallaxes. The observational sample has been divided into intermediate and extreme subdwarfs (a classification discussed further in Chapters 4 and 11). The former stars are likely to have abundances [M/H] > −1.5, while the latter are more metal-poor. The models match the extreme subdwarfs, but not the intermediate abundance stars. Uncertainties in the upper boundary conditions (the atmospheres) are likely to make a significant contribution to the discrepancy.

3.5.2   The mass–luminosity relationship

Mass is the fundamental stellar parameter, but the determination of stellar masses is extremely difficult observationally. (Chapter 8 includes extensive discussion of the

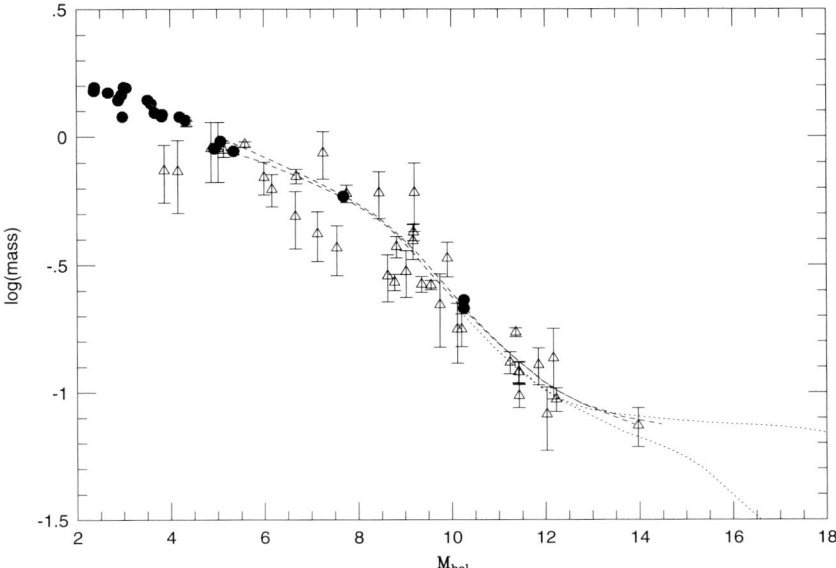

Figure 3.16. The mass–luminosity relationship for low-mass stars. Solid points denote data for eclipsing binaries (from [A7]); triangles mark astrometric binaries (from [H7], [H8]). The dashed lines are 1 and 10 Gyr isochrones from the Lyon models; the dotted lines (limited to masses below $0.2\,M_\odot$) are the 0.6 and 10 Gyr Tucson predictions.

methods currently available for this purpose.) At present, masses have been measured for only a small number of disk stars in binary systems, and no empirical data are available for metal-poor subdwarfs. Thus, theoretical/empirical comparisons are limited to near-solar abundances. The mass–luminosity relationships predicted by both the Tucson and Lyon solar-abundance models prove to be in reasonable agreement with the available observational calibrators. Figure 3.16 compares these relationships (for ages of 0.6 and 10 Gyr and 1 and 10 Gyr respectively) against empirical measurements of binary star components (see Table 8.1). The agreement is well within the observational uncertainties. In similar fashion, the Lyon models provide a reasonable match to the observed distribution in the (M_K, mass), (M_V, mass) and (M_I, mass) planes. On the other hand, the scatter amongst the empirical data is sufficiently high that it is not possible to determine whether the mismatch between theory and observation in the colour–magnitude plane (Figure 3.14) stems from calibration problems in one or several passbands. This is not surprising given the shallow slope of the mass–absolute magnitude relationships below $\sim 0.2\,M_\odot$, and lack of data at very low masses.

3.5.3 The mass–radius relationship

Since theoretical models disagree with the empirically-defined temperature scale but are in reasonable agreement with observed luminosities, disagreement between theoretically and observationally inferred stellar radii might be expected – and this proves to be the case. Figure 3.17(a) presents the expected mass–radius relationship for low-mass main-sequence stars and 0.6 and 10-Gyr-old brown dwarfs; the Lyon and Tucson models are in good agreement in the range of overlap in mass. As outlined above, the radius is expected to decrease almost linearly with decreasing mass until close to $0.1\,M_\odot$, where degeneracy becomes the dominant source of pressure support, preventing the radius decreasing below $\sim 0.08\,R_\odot$. The radius increases with decreasing mass as degeneracy becomes less important: Jupiter, at $0.009\,M_\odot$, is essentially non-degenerate.

Directly determined radii are available only for the components of YY Geminorum and CM Draconis. However, given a temperature scale, it is possible

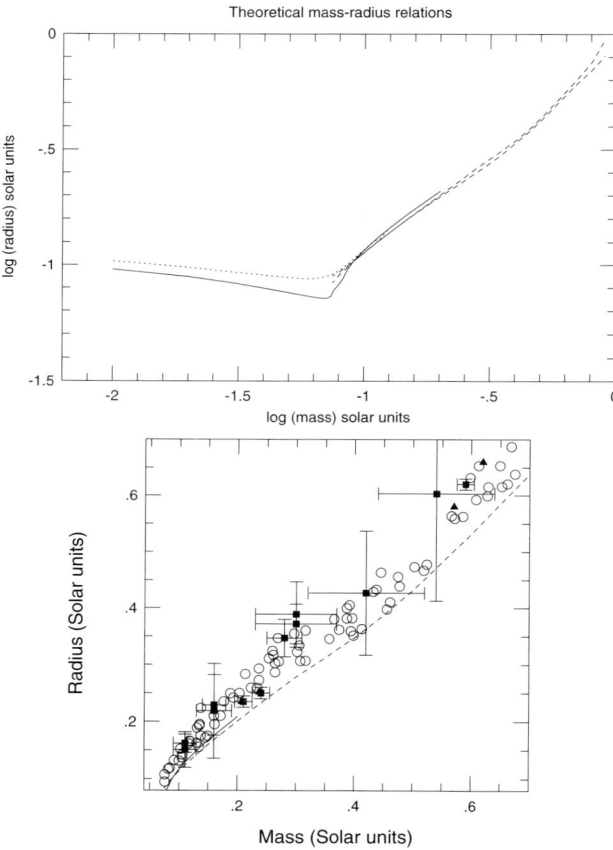

Figure 3.17. The mass–radius relationship. (a) plots the predicted behaviour of the 0.6 (dotted line) and 10 Gyr (solid line) Tucson models and the 1 and 10 Gyr Lyon models (dashed lines). (b) compares the observed and predicted relationship within the mass range spanned by M dwarfs: open circles are photometrically calibrated radii; solid squares are data from Popper's [P2] analysis of astrometric binaries; solid triangles mark the locations of the CM Dra and YY Gem components. (From [C5], courtesy of the *Astrophysical Journal*.)

to calibrate a photometric index as a temperature indicator, and (as for Table 4.1) derive photometric radii from

$$R = \left(\frac{L}{4\pi\sigma}\right)^{0.5} T^{-2} \qquad\qquad (3.56)$$

Luminosity calibration is straightforward given an accurate parallax and multi-wavelength photometry.

Clemens *et al.* [C5] have derived temperature calibrations for both the (V–I) and (I–K) colour indices, calibrating these relationships using well-studied stars from the photometric data compiled by Leggett *et al.* [L5]. They are consistent with the mean spectral-type calibration given in Chapter 4 – unsurprisingly, since the latter relation is also based on the [L5] calibrators. Masses are estimated using the empirical (M_K, mass) relation derived by Henry and McCarthy [H7] (see Figure 3.16). Figure 3.17(b) plots the mass–radius relationship outlined by single stars in the 8-parsec sample with the requisite photometry, and compares it with the theoretical calibrations. Data for astrometric binaries are from Popper [P1], in which photometric techniques were employed to estimate radii. The directly measured radii of the M dwarf eclipsing binaries, YY Gem and CM Dra, are also plotted. There is an obvious offset between theory and observation in the expected sense: the theoretical temperatures are higher, requiring smaller theoretical radii if the observed and predicted luminosities are to agree.

As yet, there is no unambiguous means of choosing between theoretical and empirical temperature (and hence radius) calibrations. However, since the theoretical relation is predicted to vary smoothly with mass, similar behaviour can be expected in the colour–magnitude diagram, modified by opacity-induced effects. In fact, the (M_V, (V–I)) main sequence shows an extremely abrupt feature at (V–I) ~ 3, which is reproduced in other colour–magnitude or spectral type/absolute magnitude diagrams. The fact that this feature is present over a wide wavelength range suggests that the effect is global, rather than due to absorption affecting one particular passband. An obvious possibility is a sharp change in radius over a relatively small range in temperature, and the proximity to $0.3\,M_\odot$ suggests a correlation with the onset of full convection. Alternatively, Clemens *et al.* [C5] hypothesise that the effect might be tied to changes in the interior structure – perhaps separation of iron and hydrogen in the stellar core. Whatever the cause, this 'step' in the main sequence remains unexplained by current theoretical models.

3.6 THE FORMATION OF LOW-MASS STARS

Star formation is a fundamental (perhaps *the* fundamental) astrophysical process, underlying heavy element production and the overall evolution of galactic systems. As yet, we have only a partial understanding of the various mechanisms which govern the morphological rearrangement of clouds of diffuse gas into self-gravitating spheres in hydrostatic equilibrium. These uncertainties reflect the complexity of the process, particularly dealing with gas magnetohydrodynamics over scales of up to several parsecs. A full consideration of star formation theory and observations lies beyond the scope of this book. In this section we sketch the basic outline and refer the reader to the review of Shu *et al.* [S9], and Hartmann's monograph [H2], for more detailed treatment.

3.6.1   Molecular clouds: the initial phase

Stellar birth takes place within giant molecular clouds (GMCs). These systems have masses of 10^5 to $\sim 5 \times 10^6 \, M_\odot$, sizes of 10–60 parsecs, densities exceeding 10^9 particles cm^{-3}, and temperatures of 10–30 K. Relatively complex molecules can form at these densities and temperatures, including HC_3N, CH_3OH and $HC_{11}N$, in addition to SO_2, CO and CS. GMCs lie close to the mid-plane of the Galactic disk, and are evident in the optical merely as dark, absorbing patches, silhouetted against the background stellar population. With typical internal extinctions of $A_V > 25$ magnitudes, these complexes are impervious to study at even near-infrared (1–2 μm) wavelengths. However, thermally-induced vibrational and rotational transitions in the constituent molecules produce emission at millimetre wavelengths, and these emissions can be observed, and the clouds mapped, by millimetre and sub-millimetre radio telescopes [S3].

As might be expected, molecular hydrogen is the most abundant species present in GMCs. However, since H_2 lacks any strong emission lines, most observations centre on CO, adopting a standard CO/H_2 conversion factor to transform the observed line flux to an estimate of the total molecular mass.[4] Given a CO flux measurement in units of (Kelvin km s^{-1}), the Galactic conversion factor is $\alpha_{Gal} \sim 2.3 \pm 0.3 \times 10^{20}$ molecules cm^{-2} (K km s^{-1})$^{-1}$. The factor is derived either from observations of individual clouds, where the velocity dispersion of the gas can be used to estimate a dynamical mass, which can be compared to the mass in CO inferred from the measured total flux, and by matching the observed production rate of high-energy cosmic rays (see [S17]).

Initial millimetre observations were made using single-dish radio telescopes with diameters of 10–25 m, achieving typical angular resolution of 40–60 arcsec. Over the last decade, interferometric arrays, combining several individual dishes (usually each of diameter 6–10 m) with baselines of up to 200–300 m, have improved the resolution to 1–2 arcsec. The latter is comparable to optical and near-infrared observations and provides a linear resolution of a few hundred AU in the nearby Taurus (140 parsecs) and Orion (500 parsecs) star-forming clouds. This is close to the scale expected for the pre-planetary solar nebula (Chapter 10), and has led to insight concerning the early stages of planetary formation.

The original single-dish observations revealed that GMCs are extremely inhomogeneous, with high-density regions only a few parsecs in size [S2], [B7]. These clumps include smaller regions of even higher density – known as cloud cores – which mark

---

[4] CS and ^{13}CO are often used to trace the density distribution in high-density regions where emission from the more common isotope ^{12}CO is saturated.

the sites where individual protostars are forming. The driving mechanism behind the formation is gravitational collapse, suitably modified by the magnetic fields known to be present within the GMCs [S9]. The initial collapse phase is generally believed to be simple spherical accretion onto an incipient protostellar core. However, most clumps are expected to have some degree of ordered rotation due to gravitational interactions with other clumps. As the collapse progresses, conservation of angular momentum leads to the formation of an accretion disk around the central protostar [P3], [S9]. In principle, millimetre interferometry permits observations at sufficient velocity resolution to detect both the overall inflow of collapsing material, and rotation within the disk, although the complex velocity fields have not yet allowed an unambiguous identification of that process.

As the protostar becomes more massive, it also becomes hotter and more luminous, with

$$L \approx \frac{GM\dot{M}}{R_*} \tag{3.57}$$

where M is the accretion rate, and R_* is the protostellar radius [S9]. At some point, the global infall decreases substantially, either due to radiation pressure or to the effect of winds generated by the star. The exact mechanism remains to be elucidated, but what is clear is that a bipolar outflow develops along the axis perpendicular to the disk, leading to gas velocities of 50–$100\,\mathrm{km\,s^{-1}}$ and the spectacular structures evident in Herbig–Haro (HH) objects (Figure 3.18, colour-plate section).

The timescales involved in these early evolutionary phases are somewhat uncertain. For a protostar to be visible as an HH object, the parent GMC must be disrupted at least partially, either by shock fronts encountered while crossing spiral arms, or by internal winds generated by massive protostars. GMCs are generally estimated to have lifetimes of $\sim 10^7$ years. In contrast, the typical age estimated for a protostar once it becomes accessible to optical/infrared observation lies in the range of a few $\times 10^5$ to $\sim 10^6$ years. These age estimates may be compatible, since a stellar core may well spend considerable time continuously accreting material before it is 'born' as a protostar. Given these uncertainties, the duration of star formation within a given GMC also remains unclear, but probably does not exceed a few million years.

3.6.2   Young stellar objects in the field

Giant molecular cloud complexes span a wide range of environments, with the extremes being typified by the high star density in the massive Orion star formation region centred on the Trapezium cluster, and the more dispersed, low-mass GMCs in the Taurus complex. The former includes more than 3,000 stars, and will probably evolve to a relatively high-density open cluster, perhaps comparable to the Pleiades, while the individual clouds in the latter system are likely to produce no more than a few dozen stars in total. As discussed further in Chapter 8, there are suggestions that the stellar mass function may vary depending on the environment, with a bias towards low-mass star formation in the lower-mass Taurus-like systems.

Once free of the parent GMCs, low-mass protostars are identifiable as T Tauri stars – pre-main sequence stars with late-type spectra (K or M for masses less than $\sim 1\,M_\odot$), and strong emission lines, particularly due to the hydrogen Balmer series (for example, see the spectrum of GG Tau Bb, plotted in Figure 2.8). The systems are usually divided into classical T Tauri stars (cTTs) with strong emission lines due to both chromospheric and accretion disk emission and 'weak-line' T Tauri stars (wTTs) which exhibit only chromospheric emission. At this point in their evolution, protostars are fully convective and evolve towards the main sequence along near-vertical (constant temperature) tracks in the H–R diagram (as illustrated by the Lyon models plotted in Fig 3.12). The latter are known as Hayashi tracks, following the first extensive analysis of pre-main sequence stars by Hayashi [H3], [H4]. Since protostars have radii which are significantly larger than main-sequence stars of the same temperature (Figure 3.10), the luminosities are correspondingly higher.

While these protostars are no longer embedded in the high-density gas of the parent GMC, many (all?) T Tauri stars retain circumstellar gaseous disks until ages of several Myr. These are evident through the detection of excess radiation at infrared wavelengths (JHKL) over that expected from extrapolating a black-body with a temperature matching the optical observations. Initially, those excesses were interpreted as circumstellar shells of dust grains [M4]. However, the addition of longer-wavelength data showed that the energy distribution required emission from a range of temperatures, and was more consistent with a disk configuration [R3]. In particular, the mass inferred for the molecular gas in many cTTs is sufficient to obscure the central star at visible wavelengths if the gas is distributed in a circumstellar sphere[5] [A1]. The presence of bipolar outflows in a variety of systems [S12] also pointed to an axisymmetric distribution of gas and dust. HL Tauri was the first system to be resolved, with millimetre observations revealing a 2,000-AU diameter disk-like structure [B6], while Keplerian motion was first identified conclusively in the 600-AU diameter disk surrounding GM Aurigae [K5]. More recently, HST observations have provided additional evidence for disks, both set against the background of the Orion Nebula (so-called protoplanetary disks or 'proplyds' – Figure 3.19, colour-plate section) and from direct near-infrared images ([S13], [S14], Figure 3.20, colour-plate section).

Improvements in the spatial resolution and sensitivity of millimetre interferometric arrays has led to a clearer picture of the evolution of circumstellar disks in T Tauri systems [K4]. The disk is believed to form during the initial stages of protostellar collapse, as gas with sufficient angular momentum to avoid falling directly onto the central mass settles into a centrifugally-supported disk, which grows outward with time. Velocity maps of the disk in HL Tau reveal infall over most of the outer regions, with ordered rotation present only within the central 100–200 AU [H5], [K6]. GM Aur, on the other hand, has Keplerian motion throughout,

---

[5] The central star *is* obscured in systems where our view lies along the equatorial plane of the disk.

and the lower energies suggest that this is an older system. Bipolar outflows, both collimated jets (Herbig–Haro objects) and molecular flows, are common during these 'earlier' stages. The driving mechanism is believed to arise through coupling between the magnetic field (which entrains the outflow) and viscous accretion within the circumstellar disk. The details of that mechanism – notably, the source of viscosity – remain under investigation (see, for example, [A3]).

Approximately 50% of cTTs retain optically-thick circumstellar disks at ages of $\tau \sim 3$Myr, and a few systems are known with ages of 10 Myr. Mid-infrared observations indicate that some cTTs with optically thick disks have central holes [S10], suggesting that the disk dissipates outwards from the inside, either through more efficient grain agglomeration at high densities or dynamical effects cleaning out the interior orbits. In most cases the disk becomes optically thin by $\tau = 5$–10 Myr. This sets an upper limit for the formation time of gas-giant planets (as discussed further in Chapter 10). Particulate material survives for substantially longer periods as 'debris disks', identified first through IRAS detection of far-infrared excess radiation from a number of nearby stars, notably Vega [A9]. In a few systems, including β Pic [S11], the disk has been imaged directly through reflected light at optical and infrared wavelengths (see Chapter 10).

Most 'low-mass' T Tauri stars studied to date actually have masses which are close to that of the Sun. A high fraction of these systems also prove to be binaries [G1] leading to limitations on the sizes of the circumstellar disks in these systems, and sometimes to the formation of circumbinary disks [J1]. It is only within the last few years that observations have extended to sufficiently faint magnitudes to identify a large number of pre-main sequence stars which will evolve to become either M dwarfs or substellar-mass brown dwarfs. The latter are discussed in more detail in Chapter 9.

3.7 POST-MAIN-SEQUENCE EVOLUTION AND THE DEATH OF THE MILKY WAY

Low-mass stars have main-sequence lifetimes which are orders of magnitude longer than the present age of the Universe (currently estimated as 12–14 Gyr). Little consideration has been given to the later phases of evolution of these objects, perhaps because the prospects of observing an evolved M dwarf are slight. Eventually, however, even $0.1\,M_\odot$ dwarfs exhaust their central energy cores, cease hydrogen core-burning and evolve off the main sequence. The final stages of their existence – perhaps the ultimate luminous phase of the Universe – have been examined by Laughlin *et al.* [L4], the only detailed study of its kind to date.

The starting point for the [L4] analysis is the set of low-mass main-sequence models calculated by Laughlin and Bodenheimer [L3]. The latter have properties close to those of the Tucson [B12] dataset outlined in the earlier sections of this chapter. Figure 3.21 shows the predicted evolution in luminosity and effective temperature of a $0.1\,M_\odot$ M dwarf. After arriving on the main sequence via the Hayashi track, the luminosity and temperature increase, reflecting the gradual

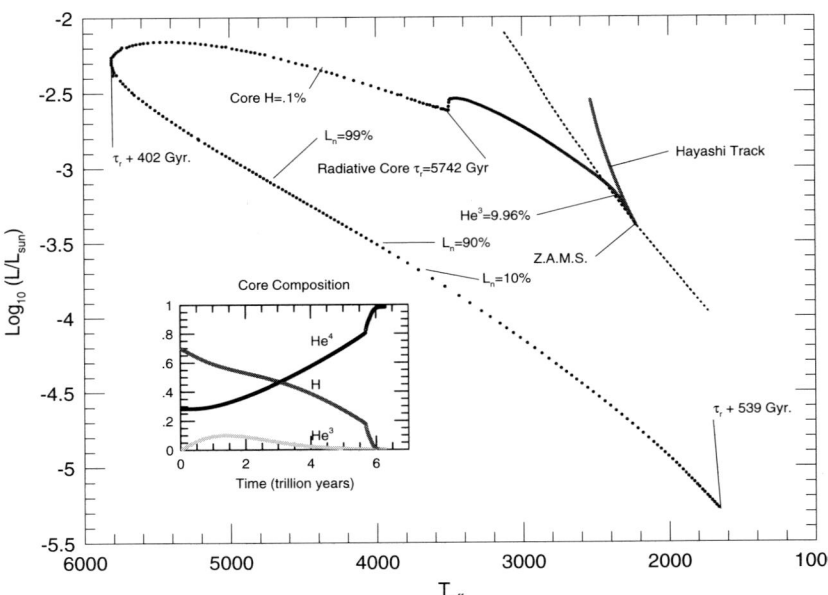

Figure 3.21. The evolution of a 0.1 $M_\odot$ M dwarf as predicted by Laughlin et al. [L4]. The inset diagram shows the relative mass fractions of hydrogen, ^{3}He and ^{4}He as a function of time. (Courtesy of Greg Laughlin and the *Astrophysical Journal*.)

increase in the ^{3}He fraction and the consequent increased mean molecular weight. At an age of 1.4×10^{12} years (1.4 trillion years), the core temperature rises sufficiently to permit reaction $3'$ in the PP chain, ^{3}He is converted to ^{4}He. and the ^{3}He mass fraction decreases from its maximum value of 9.96% as equilibrium PPI burning is established (see Figure 3.21, inset).

During the initial 1.4 trillion years, the stellar core expands and reduces ρ_c, reacting to the higher-energy generation rate. However, the hydrogen content dwindles as ^{4}He becomes the majority constituent, and the core is forced to contract and becomes hotter to maintain $\epsilon(R)$ at a sufficiently high rate to preserve hydrostatic equilibrium. The surface temperature increases, reflecting the increase in T_c. At an age of 5.74 trillion years, the hydrogen fraction has been reduced to only 16%, the opacity is low enough to invalidate the convection criterion, and the star develops a radiative core. The stellar radius, which has increased by only $\sim 10\%$ over the main-sequence value, contracts, leading to a sharp dip in luminosity.

Once the radiative core is established, the 0.1 $M_\odot$ post-MS dwarf mimics solar-type stars in evolving through hydrogen core-burning and shell-burning phases, although in this case the initial hydrogen mass-fraction is only 16%, and timescales are measured in tens of Gyr. The surface temperature and luminosity continue to increase as the star is forced to contract even further to maintain $\epsilon(R)$ despite the

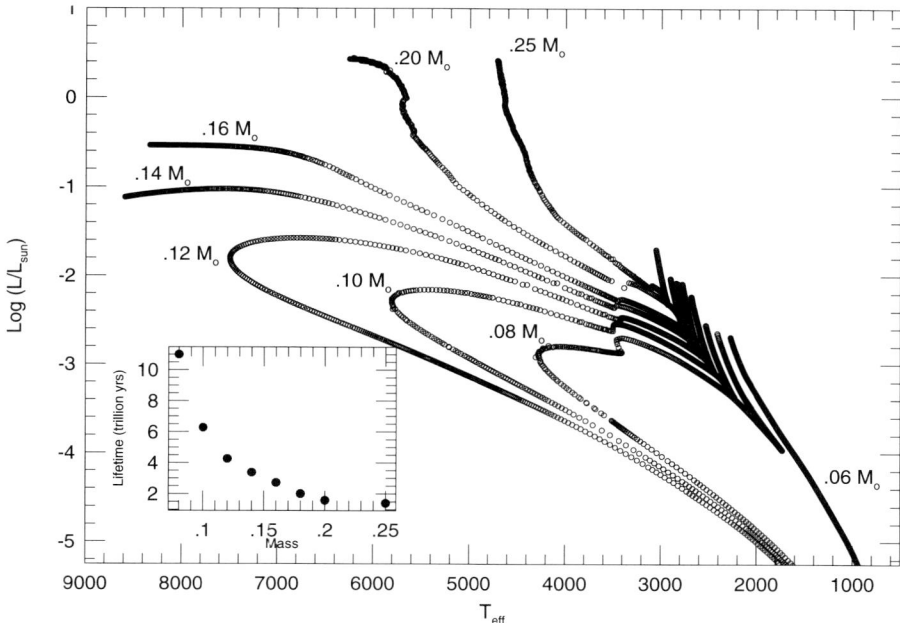

Figure 3.22. Evolutionary tracks for low-mass dwarfs. The inset diagram indicates the main-sequence lifetime as a function of mass. (Courtesy of Greg Laughlin and the *Astrophysical Journal*.)

depleted fuel supply. T_c never comes close to the temperatures required for helium burning. Degeneracy becomes the dominant source of pressure support, conduction becomes the main form of energy transport, and the luminosity eventually starts to decline.

The $0.1\,M_\odot$ star reaches a maximum surface temperature of $\sim 5{,}800$ K (comparable with the present-day Sun) at an age of nearly 6 trillion years. The hydrogen-burning shell is still active at that point, and continues to supply most of the luminosity over the succeeding ~ 80 Gyr. Figure 3.21 indicates the locations where the fractional contribution to the luminosity due to nuclear burning, L_N, reaches 99%, 90% and 10%. $\epsilon(R)$ continues to decrease after passing the temperature maximum, as does the radius and hence the luminosity. All nuclear fusion eventually ceases when the dwarf cools to temperatures below 2,000 K as a helium white dwarf. Laughlin *et al.* note that the predicted final composition is 99% helium and 1% hydrogen (by mass) – an extremely efficient use of the initial fuel source.

Figure 3.22 shows comparable evolutionary tracks for a range of masses from a $0.06\,M_\odot$ brown dwarf to a $0.25\,M_\odot$ star with a main-sequence lifetime of merely ~ 1 trillion years. The brown dwarf shows the continuous decline in (L, T_{eff}) that is expected given the absence of any hydrogen fusion; stars with masses $\le 0.16\,M_\odot$ evolve in a manner similar to that already described for the $0.1\,M_\odot$ dwarf, although the radiative core develops at a progressively earlier stage with increasing mass, leading to a higher mass fraction of hydrogen in the envelope of the final degenerate.

The 0.2 and 0.25 $M_\odot$ dwarfs, however, follow a somewhat different evolutionary path towards the end of their post-main sequence lifetimes. In lower-mass dwarfs, T_c decreases after core degeneracy sets in; in these higher-mass objects, T_c continues to increase, fuelled by more active hydrogen shell-burning. This added energy drives the radius to values exceeding $1\,R_\odot$ – more than five times the main-sequence radius. These stars ascend a giant branch, albeit as yellow(?) giants, paradoxically hotter than their main-sequence progenitors.

The timescales involved in all of these calculations are extremely long: the full evolution of a $0.1\,M_\odot$ from protostar to white dwarf exceeds 400 Hubble times. Several interesting consequences are highlighted by [L4]. First, the evolved 0.14 and $0.16\,M_\odot$ dwarfs spend several Gyr at luminosities of ~0.1–$0.3\,L_\odot$ and effective temperatures higher than that of the Sun. These periods of time are similar to the present age of the Solar System. On that basis, [L4] suggest that this might provide an opportunity for Earth-like life to develop in previously frigid terrestrial planetary companions at the appropriate distance from the post-MS 'M dwarfs'. As discussed in Chapter 10, a jovian-mass planet has already been discovered around the nearby $\sim0.25\,M_\odot$ M-dwarf, Gl 876.

The galaxy inhabited by these hypothetical lifeforms would be very different from our own. Setting aside the possibility that the mass density in the Universe might exceed the critical value and provoke a re-collapse, most of the available baryonic matter would be locked up in low-mass stars or stellar remnants. As a result, little interstellar gas and dust would be expected, resulting in insignificant continuing star-formation. This, in turn, would lead to a dearth of intermediate- and high-mass stars. The highest-temperature stellar objects visible might well be F-type post-MS low-mass dwarfs, although the sheer number of such stars should serve to maintain the total luminosity of the Milky Way for ~ 10 trillion years. Eventually, however, the Galaxy will consist of only degenerate stellar remnants, with temperatures of $\sim60\,\mathrm{K}$ maintained by dark matter interactions [A2], producing a total luminosity (summed over the entire Milky Way) of only $1\,L_\odot$.

3.8 SUMMARY

The basic principles of stellar structure were established during the middle years of the twentieth century. It is only within the last decade, however, that it has become possible to compute reliable models for very low-mass stars and brown dwarfs. We have reviewed the development of those models in this chapter. Starting with the four fundamental equations of stellar structure, the Lane–Emden equations for gas polytropes, index $n = 1.5$, allow an approximate description of the properties of fully-convective low-mass stars and brown dwarfs. Accurate analyses, however, demand more detailed calculations – notably, an equation of state which takes into account non-linear effects due to pressure ionisation, Coulomb interactions and, in particular, increasing degeneracy.

   Degeneracy forces star-like objects with masses below $\sim 0.1\,M_\odot$ to maintain a nearly constant radius of $\sim 0.1\,R_\odot$, with a corresponding decrease in the central temperature with decreasing mass. Once that temperature drops below the critical value for initiating hydrogen fusion, the only energy sources are short-lived deuterium fusion and contraction, and the luminosity and surface temperature decrease rapidly with time. Current models place the boundary between low-mass stars and brown dwarfs at a mass of $\sim 0.075\,M_\odot$. A comparison between the latest models and observations of low-mass stars and brown dwarfs reveals that, while there is good agreement in the predicted luminosities, discrepancies in the effective temperature scale remain. The models are typically 200 K hotter than the observed sequence for spectral types between M1 and M8, with corresponding offsets in the theoretical and empirical mass–radius relationships. Finally, we have provided an overview of current theories of the formation and evolution of low-mass dwarfs.

3.9   REFERENCES

A1   Adams, F. C., Emerson, J. P., Fuller, G. A., 1990, *ApJ*, **357**, 606.
A2   Adams, F. C., Laughlin, G., 1997, *Rev. Mod. Phys.,* **69**, 337.
A3   Adams, F. C., Lin, D. N. C., 1993, in *Protostars and Planets III*, ed. E. H. Levy and J. I. Lunine, Tucson, University of Arizona Press, p. 721.
A4   Adams, W. S., 1914, *PASP*, **27**, 236.
A5   Adams, W. S., 1925, *Proc. Nat. Acad. Sci.*, **11**, 382.
A6   Allard, F., Hauschildt, P. H., 1995, *ApJ*, **445**, 433.
A7   Andersen, J., 1991, *Astr. Ap. Rev.*, **3**, 91.
A8   Auman, J. J., 1967, *APJS*, **14**, 171.
A9   Aumann, H. H., Gillett, F. C., Beichman, C. A., de John, T., Houck, J. R., Low, F. J., Neugebauer, C., Walker, R. G., Wesselius, P. R., 1984, *ApJ*, **278**, L23.
B1   Bahcall, J. N., 1989, *Neutrino Astrophysics*, Cambridge University Press.
B2   Baraffe, I., Chabrier, G., Allard, F., Hauschildt, P. H., 1995, *ApJ*, **446**, L35.
B3   Baraffe, I., Chabrier, G., Plez, B., 1996, *ApJ*, **459**, L91.
B4   Baraffe, I., Chabrier, G., Allard, F., Hauschildt, P. H., 1997, *A&A*, **327**, 1054.
B5   Baraffe, I., Chabrier, G., Allard, F., Hauschildt, P. H., 1998, *A&A*, **337**, 403.
B6   Beckwith, S., Sargent, A. I., Scoville, N. Z., Masson, C. R., Zuckerman, B., Phillips, T. G., 1986, *ApJ*, **309**, 755.
B7   Blitz, L., Thaddeus, P., 1980, *ApJ*, **241**, 676.
B8   Böhm-Vitense, E., 1958, *Zs. f. Ap.*, **46**, 108.
B9   Burbidge, E. M., Burbidge, G. R., Fowler, W. A., Hoyle, F., 1957, *Rev. Mod. Phys.*, **29**, 547.
B10 Burrows, A., Hubbard, W. B., Lunine, J. I., 1989, *ApJ*, **345**, 939.
B11 Burrows, A., Liebert, J., 1993, *Rev. Mod. Phys.* **65**, 301.
B12 Burrows, A., Hubbard, W. B., Saumon, D., Lunine, J. I., 1993 *ApJ*, **406**, 158.
B13 Burrows, A., Marley, M., Hubbard, W. B., Lunine, J. I., Guillot, T, Saumon, D., 1997, *ApJ*, **491**, 856.

C1   Canuto, V. M., Goldman, I., Mazzitelli, I., 1996, *ApJ*, **473**, 550.
C2 Chandrasekhar, S. 1931, *ApJ*, **74**, 81
C3   Chandrasekhar, S., 1939, *An Introduction to the Study of Stellar Structure*, Dover Publications.
C4   Clayton, D. D., 1983, *Principles of Stellar Evolution and Nucleosynthesis*, University of Chicago Press.
C5   Clemens, J. C., Reid, I. N., Gizis, J. E., O'Brien, M. S., 1998, *ApJ*, **496**, 352.
C6   Copeland, H., Jensen, J. O., Jörgensen, H. E., 1970, *A&A*, **5**, 12.
D1   D'Antona, F., Mazzitelli, I., 1983, *A&A*, **127**, 149.
D2   D'Antona, F., Mazzitelli, I., 1985, *ApJ*, **296**, 502.
D3   D'Antona, F., Mazzitelli, I., 1994, *ApJS*, **90**, 467.
D4   D'Antona, F., Mazzitelli, I., 1996, *ApJ*, **456**, 329.
D5   Dorman, B., Nelson, L. A., Chau, W. Y., 1989, *ApJ*, **342**, 1003.
E1   Eddington, A. S., 1926, *The Internal Constitution of the Stars*, Cambridge University Press.
E2   Emden, V. R., 1907, *Gaskugeln*, Teubner.
F1   Fowler, R. H., 1927, *MNRAS*, **87**, 114.
G1   Ghez, A. M., Neugebauer, G., Matthews, K., 1993, *AJ*, **106**, 2005.
G2   Gizis, J. E., 1997, *AJ*, **113**, 806.
G3   Goldberg, H. S., Scadron, M. D., 1981, *Physics of Stellar Evolution and Cosmology*, Gordon & Breach Publishers.
G4   Greenstein, J. L., Oke, J. B., Shipman, H., 1985, *Quart. J. R. astr. Soc.*, **26**, 279.
G5   Grossman, A. S., 1970, *ApJ*, **161**, 619.
G6   Grossman, A.S., Graboske, H.C. 1971, *ApJ*, **164**, 475.
G7   Grossman, A.S., Graboske, H.C. 1973, *ApJ*, **180**, 194.
G8   Grossman, A.S., Hays, D., Graboske, H.C. 1974, *A&A*, **30**, 95.
H1   Hansen, C.J., Kawaler, S.D. 1994, *Stellar Interiors*, Springer-Verlag, New York, Berlin.
H2   Hartmann, L., 1998, *Accretion Processes in Star Formation*, Cambridge Astrophysics Series, No. 32, Cambridge University Press.
H3   Hayashi, C., 1961, *PASJ*, **13**, 450.
H4   Hayashi, C., 1966, *ARA&A*, **4**, 171.
H5   Hayashi, M., Ohashi, N., Miyama, S. M., 1993, *ApJ*, **418**, L71.
H6   von Helmholtz, H., 1854, Kant commemoration lecture, Königsberg.
H7   Henry, T. J., McCarthy, D. W., 1993, *AJ*, **106**, 773.
H8   Henry, T. J., Franz, O. G., Wasserman, L. H., Benedict, G. F., Shelus, P. J., Ianna, P. A., Kirkpatrick, J. D., McCarthy, D. W., 1999, *ApJ*, **512**, 864.
H9   Hutton, J., 1788, *Theory of the Earth*, Edinburgh.
J1   Jensen, E. L. N., Koerner, D. W., Mathieu, R. D., 1996, *AJ*, **111**, 2431.
K1   Lord Kelvin (W. Thomson), 1862, *Collected Papers*, **3**, 255.
K2   Lord Kelvin (W. Thomson), 1887, *Phil. Mag., ser. 5*, **23**, 287.
K3   Lord Kelvin (W. Thomson), 1897, *Collected Papers*, **5**, 205.
K4   Koerner, D. W., 1997, *Origins Life Ev. Bio.*, **27**, 157.
K5   Koerner, D. W., Sargent, A. I., Beckwith, S. V. W., 1993, *Icarus*, **106**, 2.
K6   Koerner, D. W., Chandler, C., Sargent, A. I., 1995, *ApJ*, **452**, L69.

K7   Kumar, S. S. 1963a, *ApJ*, **137**, 1121.

K8   Kumar, S. S. 1963b, *ApJ*, **137**, 1126.

L1   Lacy, C., 1977, *ApJ*, **218**, 444.

L2   Lane, J. H., 1869 *Amer. J. Sci.*, 2nd ser., **50**, 57.

L3   Laughlin, G., Bodenheimer, P., 1993, *ApJ*, **403**, 303.

L4   Laughlin, G., Bodenheimer, P., Adams, F. C., 1997, *ApJ*, **482**, 420.

L5   Leggett, S. K., Allard, F., Berriman, G., Dahn, C. C., Hauschildt, P. H., 1996, *ApJS*, **104**, 117.

L6   Limber, D., 1958, *ApJ*, **127**, 387.

M1   Maeder, A., Meynet, G., 1989, *A&A*, **210**, 155.

M2   Magazzu, A., Martìn, E. L., Rebolo, R., 1992, *ApJ*, **392**, 159.

M3   Mazzitelli, I., Moretti, M., 1980, *ApJ*, **235**, 955.

M4   Mendoza, E. E., 1966, *ApJ*, **143**, 1080.

M5   Monet, D. G., Dahn, C. C., Vrba, F. J., Harris, H. C., Pier, J. A, Luginbuhl, C. B., Ables, H. D., 1992, *AJ*, **103**, 638.

O1   O'Dell, C. R., Wen, Z., Hu, X. H., 1993, *ApJ*, **410**, 696.

O2   Osterbrock, D. E., 1953 *ApJ*, **118**, 520.

P1   Payne, C. H., 1925, Stellar Atmospheres, *Harvard Coll. Obs. Monographs*, No. 1.

P2   Popper, D. M., 1980, *ARA&A*, **18**, 115.

P3   Pringle, J. E., 1981, *ARA&A*, **19**, 137.

R1   Ritter, A., 1878, *Weidemann Annalen*, **5**, 543.

R2   Russell, H. N., Adams, W. S., Moore, C. E., 1928, *ApJ*, **68**, 1.

R3   Russell, H. N., 1929, *ApJ*, **70**, 11.

R4   Rydgren, A. E., Cohen, M., 1985, in *Protostars and Planets II*, ed. D. C. Black and M. S. Matthews, Tucson, University of Arizona Press, p. 371.

S1   Sampson, W. 1894, *Mem. RAS*, **51**, 123.

S2   Sargent, A. I., 1977, *ApJ*, **218**, 736.

S3   Sargent, A. I., Welch, W. J., 1993, *ARA&A*, **31**, 297.

S4   Saumon, D., Chabrier, G., 1991, *Phys. Rev. A.*, **44**, 5122.

S5   Saumon, D., Chabrier, G., Van Horn, H. M., 1995, *ApJS*, **99**, 713.

S6   Schwarzschild, K., 1906, *Gott. Nacht.*, **41**.

S7   Schwarzschild, M., 1958, *Structure and Evolution of the Stars*, Dover Publications.

S8   Sienkiewicz, R., 1982, *Acta Astr.*, **32**, 275.

S9   Shu, F. H., Adams, F. C., Lizano, S., 1987, *ARAA*, **25**, 23.

S10 Skrutskie, M. F., Dutkevitch, D., Strom, S. E., Edwards, S., Strom, K. M., Shure, M. A., 1990, *AJ*, **99**, 1187.

S11 Smith, B. A., Terrile, R. J., 1984, *Science*, **226**, 1421.

S12 Snell, R. L., Loren, R. B., Plamback, R. L., 1980, *ApJ*, **239**, L17.

S13 Stapelfeldt, K. R. *et al.*, 1998a, *ApJ*, **502**, L65.

S14 Stapelfeldt, K. R. *et al.*, 1998b, *ApJ*, **508**, 736.

S15 Strömgren, B. 1932, *Zs. f. Ap.*, **4**, 118.

S16 Strömgren, B. 1933, *Zs. f. Ap.*, **7**, 222.

S17 Strong, A. W. *et al.*, 1988, *A&A*, **207**, 1.

T1   Tayler, R. J., 1954, *ApJ*, **120**, 332.

T2   Tytler, D., Fan, X. M., Burles, S., 1996, *Nature*, **381**, 207.
V1   Vandenberg, D. A., Hartwick, F. D. A., Dawson, P., Alexander, D. R., 1983, *ApJ*, **266**, 747.
V2   Vandenberg, D. A., 1992, *ApJ*, **391**, 685.
Z1   Zapolsky, H. S., Salpeter, E. E.. 1969, *ApJ*, **158**, 809.

4

The photosphere

4.1 INTRODUCTION

The photosphere is the surface layer of a star which emits the radiation that we observe photometrically and spectroscopically. Due to the high gravity and pressure in the M dwarf atmosphere, the photosphere is remarkably thin, extending a mere 100–200 km into the star. Nevertheless, it performs the crucial tasks of moderating the energy flow from the nuclear reactions within, regulating the surface temperature and luminosity, and producing the line and continuum radiation that we observe. Many M dwarfs show evidence of an additional, tenuous, outer atmosphere (chromosphere, transition region and corona) located above the photosphere, probably produced by heating in regions of strong magnetic fields. Its presence may well modify the photospheric behaviour in some subtle, or even critical, fashion; however, we defer discussion of this region to Chapter 5, and concentrate here on the photosphere as a separate entity.

Analysis of the photosphere is accomplished using numerical models together with observations of the emergent spectrum. The observations often consist only of broadband colours, although a detailed, high-resolution spectrum provides much more information and consequently more insight into the atmospheric structure. With the twin tools of models and observations, it is possible to determine the temperature, gravity and metallicity of the star, and to infer such quantities as the mass, radius and luminosity.

The history of stellar atmospheres is rich with the insight of many of the great astronomers of the early twentieth century. Hearnshaw [H3] provides a detailed historical account from which we summarise the principal methods and results as background for the understanding of the modern analysis. It is our intention to provide the reader with the tools necessary to understand and evaluate current photospheric modeling efforts, and to then present the results together with observations and interpretation. We make no attempt to give a full explanation of the process by which modern numerical models are produced. Several excellent treatises are available which treat this difficult and complex problem in the detail it deserves. See, for example, the standard text by Mihalas [M3].

4.2   HISTORICAL PERSPECTIVE

In the early 1800s, Wollaston observed dark lines superposed on the continuous spectrum of colours formed by passing sunlight through a narrow slit, and then through a prism. Fraunhofer subsequently mapped the lines, named the strongest by letter designations still in use today (for example the D lines, now known to be sodium resonance lines, and the H and K lines, now identified as resonance lines of ionised calcium), and measured the wavelengths of several hundred lines. Understanding the formation of those dark lines, and using them to interpret the physical conditions at the surface of the star, comprises the field of stellar atmospheres. This work occupied astronomers for much of the first half of the twentieth century; among the triumphs – well before the age of modern computers – were the identification of temperature as the fundamental parameter determining the spectral type of a star; the description of gravity differences between the 'bright, narrow-lined' (giant) and 'faint, broad-lined' (dwarf) stars of the same spectral type; and the realisation that the solar composition of elements was shared by almost all stars, with hydrogen being the most abundant element by several orders of magnitude.

Advances in theoretical physics in the early part of the twentieth century, including the discovery of the photoelectric effect and the development of quantum mechanics, led to the concept of quantised energy levels that were available for an electron to occupy within an atom, and paved the way for the interpretation of the dark lines as the absorption of a photon with the particular energy necessary to excite an electron to a higher energy level. Figure 4.1 illustrates a schematic energy-level diagram for the first three bound levels plus continuum of the hydrogen atom. The excitation potential, ϕ_{ij}, is the energy required to excite an electron from level i to level j, and is related to wavelength by $\phi_{ij} = hc/\lambda$, where h is the Planck constant and c is the speed of light.

Boltzmann showed that, in thermodynamic equilibrium, the relative populations in the levels i, j are given by:

$$\frac{n_i}{n_j} = \frac{g_j}{g_i} e^{-\phi_{ij}/kT} \tag{4.1}$$

where k is the Boltzmann constant. Equation (4.1) allows the calculation of the number of atoms in each state if the temperature T, the statistical weights of the levels, g_i and g_j, and the excitation potential, ϕ_{ij}, are known. Alternatively, if the number of atoms in each state can be estimated from observations of an absorption line, the temperature can be determined. Saha extended the study of level populations to include ionisation (that is, transitions from a bound level to the continuum, which had first been discussed in detail by Eddington several years earlier), and

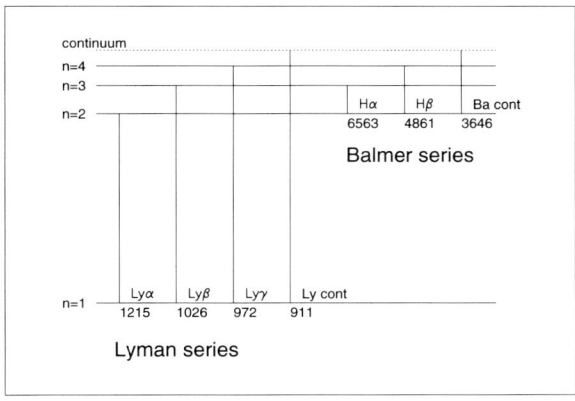

Figure 4.1. Schematic of energy levels of the hydrogen atom, illustrating bound–bound (absorption) and bound–free (ionisation) transitions in the Lyman and Balmer series. The wavelengths in Ångstroms of the transitions are indicated.

deduced the Saha equation:

$$\frac{N^+ n_e}{N} = \frac{u^+}{u} \frac{2(2\pi m)^{3/2}}{h^3} (kT)^{3/2} e^{-\chi_{ion}/kT} \tag{4.2}$$

where N^+ and N are the number densities of ions and neutrals, respectively, n_e is the electron density, the us are partition functions (similar to the statistical weights, g, in the Boltzmann equation) and χ_{ion} is the ionisation potential.

The Saha equation was instrumental in understanding the differences between the 'bright, narrow-lined and redder' stars of the same spectral type as the more common solar-like stars. Pannekoek reasoned that stars of the same spectral type must have nearly the same temperature, and hence the brighter stars must be much larger (giants). (He implicitly assumed that the masses were similar, which was later confirmed by binary star measurements.) A larger size implies a lower atmospheric density in the absorbing layer, and hence a higher degree of ionisation at a given temperature, since $N^+/N \propto n_e^{-1}$. Spectroscopic analysis thus gave a lower temperature for a giant of the same spectral type as a dwarf, in agreement with the observations which showed that the giants were redder. The narrower lines in the giants were a result of reduced pressure (collisional) broadening, which affects the line profile (see below).

Armed with these theoretical tools, and a 'stellar atmosphere' which consisted of a continuum source overlaid by an absorbing layer of constant temperature and pressure, Payne used the extensive spectroscopic data available at Harvard to show that the changes in the strength and location of stellar absorption lines for stars of different spectral type could be attributed to the changing excitation and ionisation of different atomic species as a function of temperature in the absorbing layer. The most striking example is the appearance of the hydrogen Balmer lines, which comprise transitions from the n = 2 excited level of hydrogen (see Figure 4.1), and which are visible in the optical spectrum as the Hα line at 6563 Å, the Hβ line at 4861 Å, and the Hγ line at 4340 Å. It was well known that these lines reached their maximum strength in stars of spectral type A (this is, in fact, the basis of the 'A' designation), becoming weaker both towards the hotter O stars and the cooler M stars. Payne explained this as the trade-off between having a large neutral hydrogen population in the ground state (in the cool stars), and having hydrogen being mostly ionised (in the hot stars), leading to a maximum in the number of hydrogen atoms having electrons in the first excited state at an intermediate temperature of about 10,000 K. The application of this balancing act between excitation and ionisation to a variety of atomic species that were sensitive to different temperatures, allowed her to quantitatively define a temperature scale for all spectral types (though the number of M dwarfs was small, and confined to the earliest spectral types due to the faintness of the objects).

In addition, Payne compared the line strengths of various atomic species in numerous stars of the same temperature, and found that most stars had similar values. Thus compositional differences were not common, with the majority of stars sharing the solar distribution of metallicity. Further, she found that the strength of the hydrogen lines relative to the metal lines indicated a hydrogen abundance that was several orders of magnitude larger than the metals – a result which would later prove to be accurate, though it was mistrusted at the time.

A final important piece of background physics was the description of the line profile, which reflects the probability for photon absorption as a function of wavelength. The Lorentz damping profile, shown in Figure 4.2, arises from the Heisenberg uncertainty principle. The energy levels in an atom are not infinitely

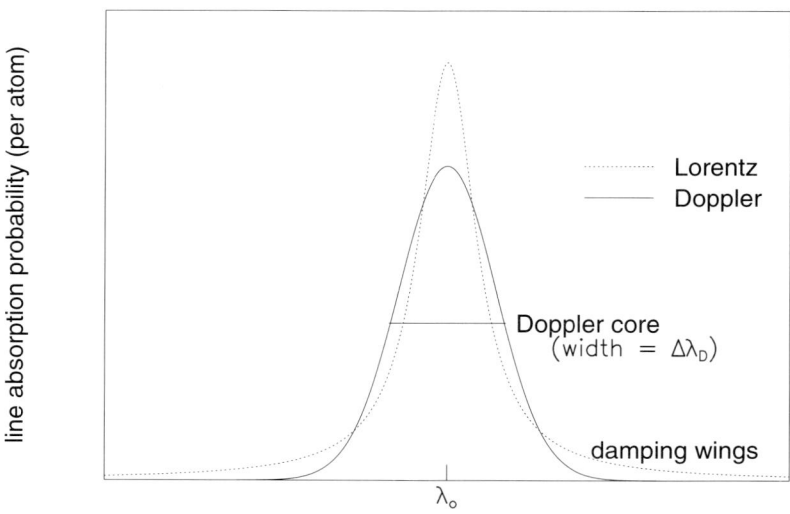

Figure 4.2. The line absorption profiles due to random thermal motions (Doppler) and finite energy width (Lorentz). Pressure broadening also has a Lorentz profile. The resulting Voigt profile is not a simple sum of the component profiles, although the properties of the Doppler core and damping wings are preserved.

sharp; each has a finite energy width ΔE and lifetime Δt such that $\Delta E\,\Delta t = \hbar$. Transitions with shorter lifetimes, such as resonance transitions, have large transition probabilities and large energy widths, leading to broader profiles, which are characterised by the damping constant γ ($\gamma \propto$ transition probability). The width of the Lorentz profile is typically only $10^{-4}\,\mathring{A}$, but the broad *damping wings* of the Lorentz profile can make an important contribution to the opacity at wavelengths far from the line centre. (For M dwarfs, we shall see that the Lorentz profile resulting from van der Waals interactions between neutral atoms and molecules will dominate the line profile except at the line centre.)

A second important effect on the line profile, which usually dominates at wavelengths near the centre, or *Doppler core*, of the line, is a result of random motion of the atoms in the atmosphere leading to a Doppler effect on the wavelength at which the atom can absorb a photon. The wavelength is modified by an amount $\Delta\lambda_D$, such that $\Delta\lambda_D/\lambda = v/c$, where the random velocity, v, depends on the mass of the atom, m, and the temperature, T; $v = \sqrt{2kT/m}$. An M dwarf with a temperature of 3,000 K has a Doppler width (for hydrogen) $\Delta\lambda_D \sim 7\,\mathrm{km\,s^{-1}}$, corresponding to $0.15\,\mathring{A}$ at $H\alpha$.

The Lorentz and Doppler profiles are shown schematically in Figure 4.2, which plots the absorption coefficient, κ, (the probability for photon absorption) as a function of wavelength. The combined (Voigt) profile consists of a Doppler core and Lorentz damping wings, and may be calculated analytically for any transition with known atomic properties, at a given gas temperature (see Böhm-Vitense [B4]; Mihalas [M3]).

In the late 1920s, Minnaert introduced the concept of the line equivalent width to quantitatively measure the total absorption in the line. The equivalent width is the

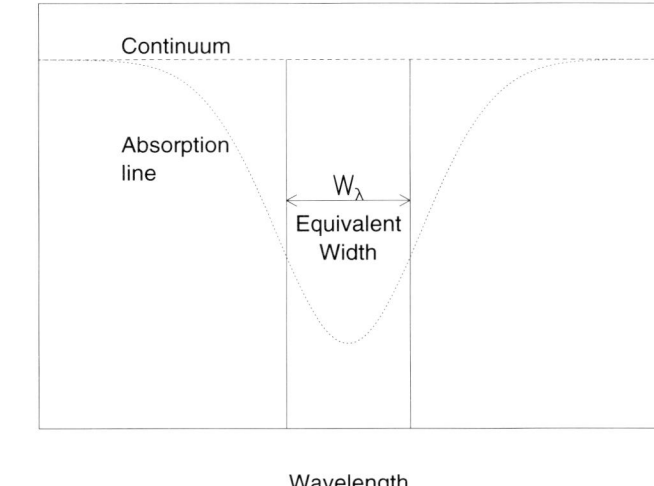

Figure 4.3. The area under the continuum of the absorption line (dotted) is the same as the area within the rectangle of width W_λ (the equivalent width) and height equal to the continuum flux.

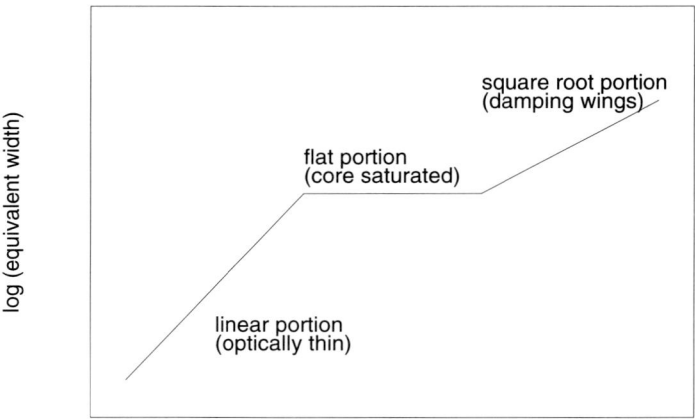

log (number of absorbers)

Figure 4.4. A schematic curve of growth, showing the progression in equivalent width that an absorption line would experience as the number of atoms increases. In practice, one observes many lines of the same species with varying strength to map the curve of growth (it being impossible to vary the number of atoms on a star many parsecs away!).

width in Å of a perfectly dark line that subtends the same area under the stellar continuum as the actual line (see Figure 4.3). He then showed that a plot of equivalent width (EW) against the number (N) of absorbing atoms had a character-istic shape that was nearly universal. This famous diagram is known as the 'curve of growth', and is shown schematically in Figure 4.4. The curve of growth divides into three parts, which Schutz explained in terms of the line profile. When the line is 'optically thin' ($\tau < 1$; see Section 4.3), each atom added to the layer will absorb radiation, hence the EW rises linearly with the number of atoms for weak lines. In the flat part of the curve of growth, the core of the line, where the probability of absorption is high, has saturated; every available photon of that wavelength is already being absorbed, so the addition of more atoms does not change the amount of absorption. Thus the EW barely changes as N increases. Finally, when there are enough atoms present, the probability of being absorbed in the damping wings becomes significant and the EW begins to rise as the square root of N (the square root dependence comes from the explicit expression for the damping profile, see [B4], [M3]). Since the number of atoms in the absorbing layer of the star cannot be varied, the curve-of-growth method is applied by observing the equivalent widths of a distribution of weak and strong lines (corresponding to low and high transition probabilities, respectively), produced by a given atomic species such as Fe I. The curve of growth can then be calibrated in absolute units using solar measurements. The curve-of-growth method remained the principal tool for abundance analysis until well into the latter half of the twentieth century.

In the late 1930s a final important puzzle was solved by Wildt, who showed that the principal source of continuous opacity in the Sun and other cool stars (including the M dwarfs) is the H$^-$ ion. It had been known for some time that stars did not radiate as perfect black-bodies, since the Balmer jump was easily observable in hot stars. Thus, colour temperatures determined by matching observed flux gradients to a black-body spectrum were in conflict with the ionisation temperatures found from analysis of spectral lines, with the colour temperatures being generally hotter. A source of continuous opacity was required in the optical wavelength region to bring the colour temperatures into agreement. It was thought by many that this opacity was due to a large number of metal transitions providing a 'line blanketing' effect. The first crude model atmospheres were being developed in the 1930s; Biermann and Unsöld had independently found that an atmosphere comprised of 2/3 metals and 1/3 hydrogen could provide the needed opacity. This result was sharply at odds with that of Payne, and later Russell, who had found hydrogen to be thousands of times more abundant than the metals.

The opacity due to the H$^-$ ion, which consists of an electron loosely bound to a neutral hydrogen atom, resolved both the temperature and abundance discrepancies. The second electron can be ionised by photons with E > 0.7 eV, or $\lambda < 1.7$ μm. Thus any optical or near-infrared photon can ionise an H$^-$ ion, providing a source of continuous opacity which depends on the hydrogen abundance (rather than requiring a large metal abundance).

The period from 1940 to the mid-1960s was marked by increasing sophistication of the atmospheric models for solar and hotter stars, including particularly the first approximate description of convection and its application to hot stars by Bohm and Böhm-Vitense. Little work was done on the M dwarfs, both for lack of observational material and because of the daunting task of describing the atomic and molecular opacity sources. We shall pick up the modelling discussion again in Section 4.5, after first reviewing the methods by which models are produced, and taking a closer look at the opacity sources. The complexity and sheer volume of the modern model results is rather daunting. They should be perused while keeping in mind the cautionary words of Armin Deutsch, who in 1966 wrote in a paper entitled *Even Simpler Methods of Abundance Determination from Stellar Spectra* [D1]: 'Of course, we must recognise that these methods produce results that are rough and not fully reliable. But this is preferable to the delusion that we can improve our results by adducing an inapplicable model, however sophisticated; or by processing irrelevant data, in quantities however vast.'

4.3   THEORETICAL CONSIDERATIONS

To understand the nature of the photosphere and how its structure is determined, we begin with some basic theoretical concepts. Additional basic material may be found in the introductory text of Böhm-Vitense [B4], while a more detailed and rigorous treatment is contained in Mihalas [M3]. Note that some of the equations in the following discussion have been presented in previous chapters; they are gathered here for clarity.

The photosphere may be characterised in its simplest form by three quantities: the effective temperature, T_{eff}, the surface gravity, g, and the mean metallicity, Z. The effective temperature is defined by:

$$\sigma T_{eff}^4 / \pi = F \tag{4.3}$$

where F is the total flux emerging from the photosphere, integrated over all wavelengths, and σ is Stefan's constant. The effective temperature thus represents the temperature of the star if it emitted as a perfect black-body. Ideally, the temperature determines the ionisation states and level populations of the atoms and molecules in the atmosphere through the Boltzmann and Saha equations (4.1) and (4.2), and thus the frequencies at which absorption and emission will primarily occur. The effective temperatures of M dwarfs are $\sim 2100 < T_{eff} < 3{,}800$ K, while the lower mass L dwarfs reach temperatures below 1,500 K.

The surface gravity is given by:

$$g = GM/R^2 \tag{4.4}$$

where M and R are the mass and radius of the star, respectively. The gravity primarily determines the pressure and density of the atmosphere. A higher-density atmosphere has a larger number of particles which can interact with the radiation field. Surface gravity is reported as $\log g$; the surface gravity of the Sun is $\log g \sim 4.44$, and for the L and M dwarfs it ranges from $\log g \sim 4.6$–5.3.

The mean metallicity is defined as:

$$Z = M_{metals}/M \tag{4.5}$$

where M_{metals} is the total mass contained in all elements except for hydrogen and helium, and M is again the mass of the star. Another common expression used to characterise the metallicity is the metal abundance of the star:

$$[m/H] = \log 10(n_m/n_H) - \log 10(n_m/n_H)_\odot. \tag{4.6}$$

Thus, for example, a star with a metal abundance of one tenth the solar value has $Z = 0.1$ and $[m/H] = -1$. The use of m to represent any metal in the star presumes that the distribution of metals in the star is the same as for the Sun.[1] As we shall see, the metallicity greatly affects the ability of the gas in the photosphere to absorb and emit radiation, since elements more massive than H and He have many more energy levels available for intercepting and producing photons.

A typical photospheric model is computed by assuming that the gas is arranged in homogeneous, plane-parallel layers of height dz, and is in a steady state. We can then apply the equation of hydrostatic equilibrium

$$dP/dz = -\rho g \tag{4.7}$$

which says that the pressure gradient is balanced by the force of gravity. An equation of state connects the pressure, density and temperature. For example, a perfect gas has the usual equation of state

$$P = \rho k T / \mu_M \tag{4.8}$$

[1] This may not always be the case; there is evidence that metals such as the α elements C, N, and O do not always occur in the solar ratio to the iron peak elements (see Chapter 6).

where μ_M is the mean molecular weight of the gas. The equation of state used for M and L dwarf atmospheres is much more complicated, requiring the calculation of partial pressures for many different atomic and molecular species (and, for the lowest mass objects, dust grains) at each height in the atmosphere. The equation of state changes with height, since molecules will form at different temperatures, and hence contribute varying partial pressures at different heights.

A second equilibrium condition is the requirement of energy balance in each layer, usually expressed as flux conservation,

$$dF/dz = 0 \qquad (4.9)$$

which says that all energy absorbed in a photospheric layer is exactly balanced by energy leaving the layer, so that the net flux change is zero. In contrast to earlier-type stars, the energy is carried by both convection and radiation in an M dwarf photosphere; the flux in equation (4.9) has both convective and radiative components. The treatment of convection is therefore an important aspect of the models. Modern models use a mixing length formulation for the convective flux [M3],

$$F_c = \frac{1}{2}\, \rho\, C_p \bar{v}_{conv}\, T\, \frac{l}{H} (\nabla_{atmos} - \nabla_{blob}) \qquad (4.10)$$

with the mixing length parameter (the ratio of the mixing length, l, to the pressure scale height, H) set to values between 1 and 2. C_p is the specific heat at constant pressure, $\bar{v}_{conv}$ is the average velocity of a convective blob, and ∇_{atmos} and ∇_{blob} are the temperature gradients in the atmosphere and in the blob, respectively. (See Section 3.3.4 for a discussion of the adiabatic temperature gradient which determines when convection will occur.) The convective flux depends strongly on C_p, which in turn depends on metallicity. C_p is particularly large in the atmospheres of M dwarfs, where molecules contain considerable amounts of internal energy; hence the importance of treating convection explicitly at all depths in the photosphere.

The radiative flux depends on the local density and metallicity through the optical depth, τ_ν at frequency ν,

$$\tau_\nu = \int_z \rho \kappa_\nu\, dz \qquad (4.11)$$

and the equation of radiative transfer,

$$\mu dI_\nu/d\tau_\nu = I_\nu - S_\nu. \qquad (4.12)$$

We will refer to the local quantity $\rho\kappa_\nu$ as the 'opacity' of a layer, and the integrated quantity given in equation (4.11) as the optical depth, τ_ν, although many authors use the terms interchangeably. The quantity I_ν is the intensity along a particular direction θ ($\mu = \cos\theta$, not to be confused with the mean molecular weight). The radiative flux is the integral of the intensity over all directions and frequencies. Equation (4.12) says that the change in intensity, dI_ν of a ray travelling through a gas with optical depth $d\tau_\nu$, is given by the initial intensity I_ν modified by the source function S_ν, which describes the balance between absorption and emission within the layer, at frequency ν. Complete descriptions of the derivation and usage of equation (4.12) are given in [M3].

The optical depth, τ_ν, accounts for the interaction between the photospheric matter and the radiation field through the important parameter κ_ν, the mass extinction coefficient. κ_ν is also referred to as the absorption coefficient, as in Section 4.2, but it actually contains both an absorption and a scattering component. κ_ν has units of cm^2/gm, and is essentially a cross-section per unit mass, describing how likely it is that a photon will undergo an interaction with a particle in the gas. The distance that a photon travels before it undergoes an interaction is called the photon mean free path. The opacity, $\rho\kappa_\nu$, has the units of inverse length and can be thought of as the inverse of the photon mean free path at a particular frequency. The optical depth – the integral of the opacity over a distance z – then represents the number of mean free paths a photon must travel to escape from a depth z in the atmosphere. A general rule of thumb is that radiation emitted at optical depth $\tau_\nu < 1$ (a height less than one photon mean free path) will escape the atmosphere and be observed, while radiation from larger optical depth will be absorbed or scattered before it can escape. Stars with more metals (larger Z) will in general have more absorption and scattering, and thus a larger κ and greater optical depth over smaller distances.

Observations of the photospheric spectrum at different frequencies provide a map of the atmosphere, reflecting the opacity (and hence temperature, density and metallicity) of different layers, depending on the optical depths at the frequencies that are sampled. For example, the optical depth at the line centre frequency ν_o of a strong absorption line will be very large even at a small physical distance z into the photosphere, since κ_{ν_o} is large (the atom or molecule has a high cross-section for photon interaction). Radiation from $\tau_{\nu_o} = 1$ will come from a very shallow distance in the atmosphere. Conversely, the continuum radiation emanating at a frequency ν between strong lines will experience less opacity and will reach $\tau_\nu = 1$ much deeper into the atmosphere, reflecting the conditions at those layers.

From the above equations it is clear that specifying the effective temperature, surface gravity, and metallicity of a star provides boundary conditions on the flux, pressure and density, and constraints on the interaction between radiation and matter, which lead to a self-consistent determination of the optical depth at each frequency, and finally a prediction of the radiation emitted at each frequency. (This is from the point of view of the modeller. The observer, in contrast, has data showing the radiation emitted at each frequency, and would like to use the observations to infer the temperature, gravity and metallicity of the star.)

In practice, the procedure is to postulate an initial temperature and density structure, solve the above equations in each layer, perform numerical integrations over the equilibrium equations, and determine the corrections to the initial structure that must be applied to satisfy the equilibrium and boundary conditions. The solution is then iterated until convergence, and the quantity of interest for comparison with observations (the emergent spectrum) is computed. Although this qualitative description appears straightforward, in practice it is often difficult or impossible to carry out. Many workers have spent countless hours devising clever and innovative ways to solve the model atmospheres problem [C1].

In general, determining the interaction between the radiation field and the matter, which is hinted at by equations (4.11) and (4.12), is the most complicated and difficult part of the procedure. In particular, specifying the exact form of κ_ν at all frequencies is often an intractable problem, particularly when considering the vast number of transitions available for excitation in a typical metal atom or simple molecule.

A common assumption which greatly simplifies the problem is to assume that the atmosphere is in local thermodynamic equilibrium (LTE) – the condition that the radiation is locally in equilibrium with the matter, and that both may be characterised by a single temperature. In this case, the velocity distribution of the particles in the gas is described by a Maxwell–Boltzmann distribution of a given temperature, and the radiation field is described by a Planck function of the same temperature. The level populations of all states of the matter are then completely determined by the temperature (via the Boltzmann and Saha equations) allowing immediate calculation of κ_ν for all transitions being considered. The remaining problem, even for LTE models, is to specify sufficient transitions in the dominant atoms and molecules so that the computed opacity encompasses all of the important radiative processes.

In M and L dwarf photospheres, the effective temperature is low while the surface gravity is large, leading to high photospheric densities. Both low temperature and high density lead to increased opacity, since 1) essentially all atoms will be neutral with many transitions having low excitation energy; 2) many molecular species can form, contributing even more low-energy transitions; and 3) the number density of atoms and molecules will be large, increasing the probability of interaction with the radiation. These factors tend to move the atmosphere toward LTE, and all current photospheric models for cool dwarfs do assume LTE in the computations.

4.4   TREATMENT OF OPACITY

The description of the opacity is the single most important factor in producing an accurate photospheric model for a low-mass star. The reader may have encountered references to 'grey' atmospheres (computed assuming the opacity is frequency-independent), or atmospheres calculated using the Rosseland mean opacity (also frequency-independent, but weighted at each depth such that the integrated flux is correct at the surface; see [M3]). While adequate for some earlier-type stars, and particularly for calculations at deep layers in the stellar interior, these approximations are not valid for M and L dwarf photospheres. To illustrate the complexity of their spectra, Figure 4.5 shows a spectrum of the famous M8 dwarf VB 10, together with a black-body of the same effective temperature, the continuum expected from H^- and a model fit. The important molecular bands of TiO, VO, FeH, H_2O, and CO are indicated. It is clear that the simple black-body and H^- approximations to the continuum flux distribution have been greatly distorted by the (mostly molecular and

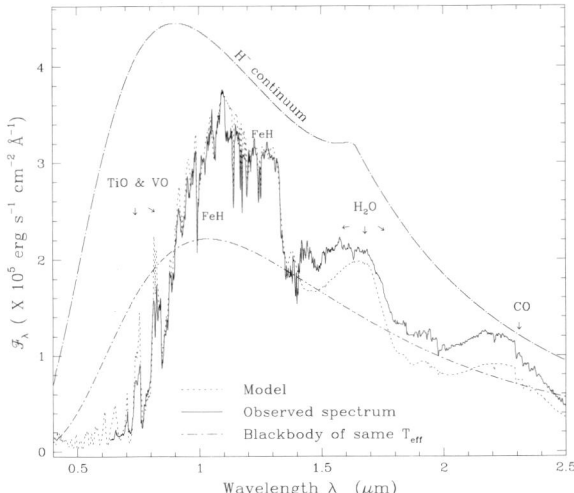

Figure 4.5. The observed spectrum of the M8 dwarf VB 10 is shown together with a model fit. The idealised black-body and H⁻ spectra are clearly inadequate to describe the complicated structure stemming from the significant molecular line opacity. (Courtesy of F. Allard and P. Hauschildt, and the *Astrophysical Journal* [A3].)

strongly frequency dependent) opacity, and detailed models are required to understand the spectra of these cool dwarfs.

The important opacity sources in a stellar atmosphere involve bound–bound (line), bound–free (continuum), and free–free (continuum) processes due to both absorption and scattering. In the bound–bound case, absorption occurs when a photon from the radiation field is absorbed by an atom (or molecule) in the gas and the excited state is subsequently de-excited by a collision between the atom and another particle (typically an electron, or another atom). The photon is thus destroyed, and its energy returned to the gas. Absorption therefore couples the radiation field to the matter through collisions. The collisions occur with a rate that depends on the velocity distribution of the particles, and thus reflects the local temperature of the gas. If absorption dominates the opacity, the condition of LTE will be approached, and the assumption of LTE is acceptable.

Scattering, in contrast, occurs when an excitation of an atom (or molecule) by a photon is followed by immediate radiative de-excitation (which emits a photon). No collisional processes are involved, so scattering is only weakly dependent on the local matter temperature. An important attribute of bound–bound scattering is that transitions have a finite energy width, such that a slight change in frequency may occur between the absorbed and re-emitted photons. This allows, for example, photons which are absorbed at the line centre frequency with high probability to be re-emitted at a frequency in the line wing. The probability of absorption is much smaller for a line wing photon (see Figure 4.2), so the emitted photon may travel a large distance through the atmosphere before being re-absorbed. In this way the radiation field can be quite decoupled from the local matter temperature. When scattering dominates the opacity, the assumption of LTE is no longer valid.

The bound–bound (line) opacities are used to describe transitions between bound levels, including the extensive, closely-spaced, rotational and vibrational bands produced in molecules. In practice TiO, CaH and other oxides and hydrides in the visible spectrum, and H_2O and CO in the infrared spectrum, dominate the line opacity under most conditions. Current references for line lists for these molecules may be found in [A2], [A4], [H1]. The huge number of transitions (literally hundreds of millions of lines for H_2O alone) makes the detailed calculation of κ_ν at individual frequencies a daunting task. As an alternative, various approximations to the opacity have been used (carbon [C1] contains a description of many of these methods). Examples include the straight mean (SM), harmonic mean (HM), just overlapping line approximation (JOLA), and opacity distribution function (ODF) methods which approximate the opacity in a given frequency interval by some average value, smoothly varying function, or histogram of values based on the frequency of occurrence of high and low opacities in the interval. A more rigorous but computationally expensive method is the opacity sampling (OS) procedure where the true opacity is calculated on a grid of pre-specified frequency points, including all transitions that contribute opacity above a given threshold at that frequency. The OS method represents the current state-of-the-art, but the models still fail to adequately represent the observations in some important frequency intervals. There is much work yet to be done both in the compilation of better transition data for important molecules, and in the treatment of the line opacity in the models.

Bound–free and free–free continuum processes, in contrast, are straightforward to compute analytically for atomic and molecular ionisation (dissociation) and recombination (formation), and for Thomson and Rayleigh scattering. Unfortunately none of these processes is very important in the cool, high-pressure environment of the M dwarf photosphere. Figure 4.6 shows the absorption coefficients for the most important contributors to the opacity in solar metallicity and low ($[m/H] = -2.5$) metallicity models. The solar metallicity model is dominated by molecular line absorption due to H_2O in the infrared and TiO in the optical, while the low metallicity model has far less TiO opacity in the optical and is controlled by collision-induced H_2 absorption in the infrared.

Allowed transitions in the H_2 molecule occur only at electric quadrupole and higher-order moments. However, H_2–H_2 opacity may be produced as a result of photon interactions with the temporary electric dipole induced during a collision between two H_2 molecules. This process is also known as pressure-induced absorption, since the number of H_2 collisions depends strongly on the pressure. In low metallicity atmospheres, H_2 is by far the most abundant molecule and the H_2–H_2 opacity dominates at wavelengths longer than about 1 μm. Linsky [L5] carried out an initial semi-empirical investigation of the H_2–H_2 opacity, but detailed quantum mechanical calculations are required to properly model this and other collision-induced absorption (CIA) opacities; for example, those induced during H_2–H and H_2–He interactions [L4]. Borysow [B5] provides an excellent review of current modelling efforts of the CIA opacities in cool stars.

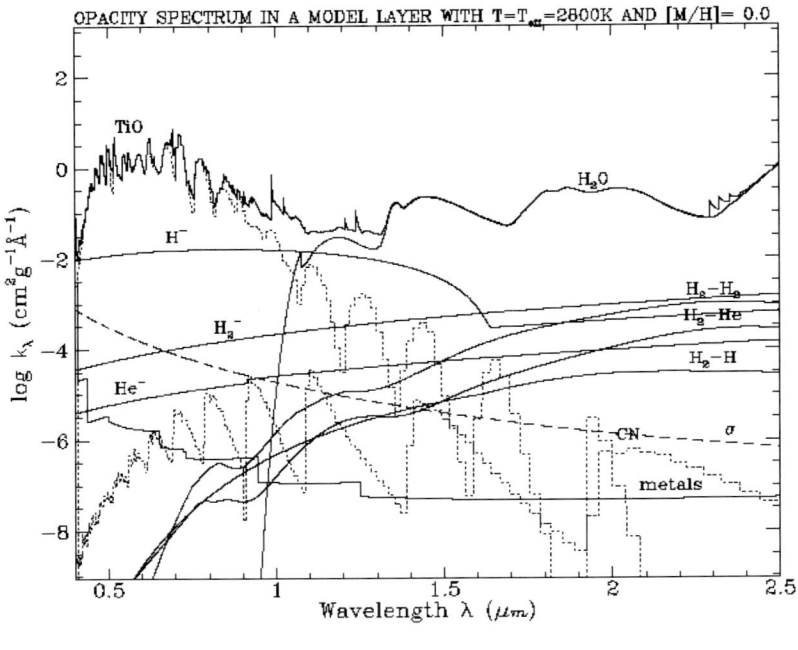

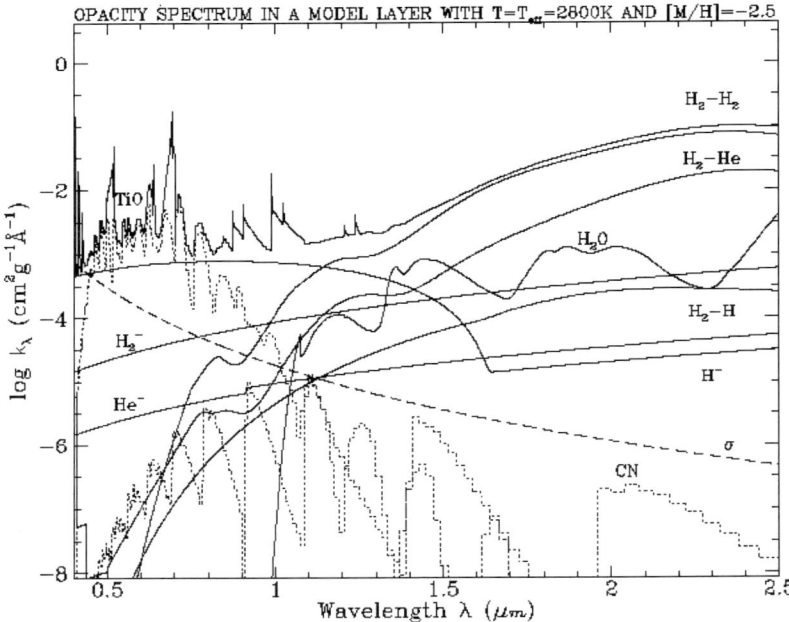

Figure 4.6. Wavelength dependence of opacity sources for solar and $[m/H] = -2.5$ models with $T = 2{,}800\,\mathrm{K}$ and $\log g = 5.0$. (Courtesy of F. Allard and P. Hauschildt, and the *Astrophysical Journal* [A3].)

Another important effect in the high-pressure M and L dwarf atmospheres is the presence of significant line-broadening due to van der Waals forces acting on the energy levels of the atomic and molecular species. This effect is known as 'pressure' or 'collisional' broadening, and it contributes to the Lorentz damping profile of the line with a (classical) damping constant [A4],

$$\gamma_{vdW} = 17C_6^{2/5} v^{3/5} N_p \qquad (4.13)$$

where v is the relative velocity between the interacting particles, N_p is the number density of the perturbing particles, and C_6 is an interaction constant. C_6 can be determined exactly for perturbations by atomic hydrogen, and has been given various approximate forms for molecular hydrogen interactions, which are more common in M and L dwarf atmospheres [S1]. Typical widths for strong absorption lines in M dwarfs are a few Å. As discussed above, increasing the energy width of the bound–bound transitions increases the importance of scattering (as opposed to true absorption) in the calculation of the opacity. This tends to move the atmosphere away from LTE.

There are several other broadening mechanisms that have a smaller effect on the line profile. The microturbulent velocity, which has a Gaussian profile, is a statistical necessity in higher-mass stars, but is generally less important for low-mass dwarfs, amounting to only 1–2 km s^{-1}. This is much less than the pressure-broadening or the width of the Doppler core due to random thermal motions. Stark broadening, which has a Lorentz profile, results from energy level perturbations due to an ambient electric field. It is generally not important in these mostly neutral photospheres, although it can play an important role in shaping the line profiles in the chromosphere. Rotation, and the effects of magnetic fields, can make important contributions to the line profiles for some stars (see Chapter 5).

4.5 PHOTOSPHERIC MODELS

There are several steps that must be undertaken in producing (and understanding) photospheric models. First, the underlying assumptions in the computation of the model must be known (for example: which atoms and molecules are included in the opacity? – what types of opacity approximations are made? – is the atmosphere in LTE? – is convection included as an energy transport mechanism?). Next, the results of the model computation must be examined, particularly noting how the structure of the model atmosphere varies with different assumptions. The latter provides insight into the sensitivity of the model calculations to the input physics, and will be discussed in this section. Finally, the atmosphere models are used with a spectrum synthesis program to produce an emergent spectrum that can be compared to observations, as described primarily in Section 4.6, although comparison with observations for some early models is briefly mentioned here. (For a comprehensive discussion of M dwarf models through 1996, see the excellent review in Allard *et al.* [A4]; a summary of newer models may be found in [A2].)

The models described in this section were computed under the assumptions of plane–parallel geometry, homogeneous layers, LTE, microturbulent velocities between 1–$2\,\mathrm{km\,s^{-1}}$, and with convection treated using a mixing length theory with mixing length parameter between 1 and 2, unless otherwise noted. Model results are typically shown as the run of temperature and density with optical depth, height, column density (mass), or gas pressure. The optical depth is reported at a standard continuum wavelength, τ_{std} ($\lambda_{std} = 1.2\,\mu\mathrm{m}$ in the NextGen models described below), or the Rosseland mean optical depth, τ_R, is used. The height, z, is measured outward from the photosphere boundary, taken to be the location where $\tau_{std} = 1$. The column density, N, and column mass, $\bar{m}$, are the integrals through the atmosphere of the number and mass densities, respectively: $N = \int_r n\,dr$, and $\bar{m} = \int_r \rho\,dr$. By convention, the outermost layer considered in the model has $r = 0$, and r increases inward, thus the column density and column mass also increase inward. Note that r and z are both height variables, but are defined in the opposite sense, and have different zero-points. The gas pressure, which we use here, is related to the column mass by the hydrostatic equilibrium requirement (equation (4.7)). At a given depth, $P_g = \bar{m}g$, so if the gravity is constant (as is generally assumed), the gas pressure is proportional to the column mass.

The first attempts at modelling main-sequence dwarfs with low effective temperatures were made in the mid-1960s by Gingerich and Kumar [G1], who used only continuum opacities and assumed radiative equilibrium (convection was neglected); Tsuji [T2], who included a few molecules (CaH, H_2O and CO) and attempted some models with convection; and Auman [A6], who included convection and a more detailed description of the water opacity employing an HM opacity treatment. Auman noted several results which remain true today: that convection must be included, as it is important to very small optical depth ($\tau < 0.01$) in the photosphere; that some broadband colours are not monotonic functions of T_{eff} and hence care must be taken in defining colour-temperature relationships that connect the models to observations; and that the strengths of the water bands depend critically on the detailed temperature structure and the treatment of convection. He mentioned the inadequacy of the water-band line list, but the lack of near-infrared spectra did not allow a comparison with observations.

The next improvements in the models came with Mould [M4], [M5], [M6], who used a version of the Kurucz ATLAS code [K3], including convection and treating atomic and molecular line opacities with an ODF description. The line lists were much more extensive than those of the previous decade, and included the important TiO molecule which dominates the optical opacity in the M dwarfs, along with H_2O and CaH, but not CO. He found that TiO is a good temperature indicator in most models (bands become stronger with decreasing temperature), but the TiO band-strength weakens at the lowest temperatures. Also, CaH is particularly sensitive to gravity, especially in the cooler models. Comparison of the model results with the few (quite uncertain) infrared observations in the water bands that were available at that time provided the first indication that the water opacities were being over-estimated in the models – a condition that still exists today.

The Mould models remained in common use for almost 20 years, as the

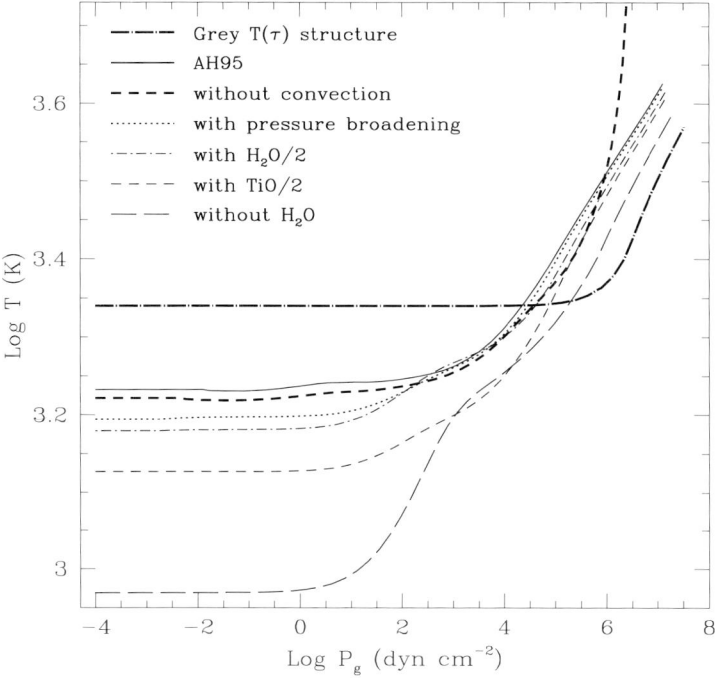

Figure 4.7. The effect of changes in model computations. The model has $T = 2,800\,K$, $\log g = 5.0$ and solar metallicity. (Figure courtesy of F. Allard and P. Hauschildt, and the *Annual Review of Astronomy and Astrophysics* [A4].)

measurement and compilation of atomic, and particularly molecular, line data slowly progressed. The next important sets of M dwarf models were produced in the early 1990s, in the PhD theses of Allard [A1] and Ruan [R2]. In 1995, Allard and Hauschildt published improvements on her models using the PHOENIX code [A3], and Brett extended the Ruan work using the MARCS code [G3], [B1], [B7]. Both groups obtained reasonable fits to the optical spectrum, and found that the water bands from the computed models were much too strong compared to the (now much improved) infrared observations [A3], [B6].

Both Allard and Brett employed extensive water and TiO line lists: Allard used SM and JOLA opacities for these molecules and an OS method for the atomic and molecular lines, while Brett used an OS method for all the molecular opacities.[2] Figure 4.7 shows the effects of varying the TiO and water opacities on one of the [A3] models. The exact values of the opacities are more important for TiO than for water, although ignoring water altogether results in a dramatic cooling of the outer atmosphere by CO.

---

[2] Brett also showed that current versions of the Kurucz ATLAS models – previously used by Mould and widely used for earlier type stars – are inadequate for M dwarfs, due to the lack of many molecular (especially water) opacities [B6].

The mid–late 1990s saw the computation of several grids of models by Allard, Hauschildt and co-workers. The models published in 1995 are known as the 'base' grid [A3]. Subsequent grids (an upgraded version of the 'base' grid, and the 'extended' grid) were also made available to the community but were not formally published. Some results are discussed in [A4]. One must beware (as we shall see in the next section) that the rapidly changing nature of the models during this time rendered the comparison with the observations quite fluid. The exact generation of models used could produce different results, for example, in the temperature scale (Figure 4.13). The situation improved in 1999 with the publication of the NextGen models, incorporating more extensive line lists and an opacity sampling treatment for all molecules, for atmospheres with effective temperatures greater than 3,000 K [H1]. These will most likely remain the standard for some years to come, and we shall concentrate on these models in our discussion of the spectra.[3]

Figure 4.8 is a comparison of the models produced during the years 1969–1999. There is reasonable agreement, although the Auman model is cooler, and the early Allard models ('base' and 'extended') are hotter than the others; the NextGen models are in closer agreement with the Brett and Mould models at large depth than the previous generations of Allard models. Though the models appear to be nearly the same, the spectra that are produced can be quite different in detail. This serves as a warning that the results of the computations are sensitive to the adopted physics, especially in the opacity treatment. The formation depths ($\tau_{line} \sim$ unity) for several strong molecules are indicated on the figure; these are generally at $\tau_{std} \ll 1$ or $\log \tau_{std} \ll 0$. Some models (Auman, Mould, Brett) do not extend far enough into the optically thin regimes of the important molecular species to include all of the molecular opacity.

Figure 4.9 shows the run of temperature (left panels) and standard optical depth (right panels) with gas pressure, for a range of NextGen models. The top two panels indicate the effect of varying the effective temperature while holding the gravity and metallicity fixed. The outer region of the lower-temperature atmosphere has greater optical depth at a given pressure, since lower-temperature conditions favour more molecule formation and hence greater opacity at a given physical height. At larger depth, the situation reverses as the H^- continuum opacity becomes dominant in the higher temperature stars. Variations in gravity and metallicity (middle and bottom panels respectively) both indicate that the pressure–opacity relationship controls the temperature structure. Plots of temperature as a function of $\log \tau_{std}$ show essentially no differences between these atmospheres. In the case of varying gravity, the higher-gravity atmosphere reaches a given optical depth at a higher pressure than a lower-gravity atmosphere, due to the higher density and consequent greater abundance of molecules. The effect is even more pronounced in the low-metallicity atmospheres, where the difference in pressure at $\log \tau_{std} = 0$ is nearly two orders of magnitude between the solar metallicity and $[m/H] = -2$ atmospheres. In this case, the explanation is the lack of metals and hence lower opacity in the low-metallicity

---

[3] At temperatures below 3,000 K, many species of dust grains can condense out of the gas phase in the atmosphere, and the effects of dust on the model structures become increasingly important. Models that include dust are discussed later in this chapter.

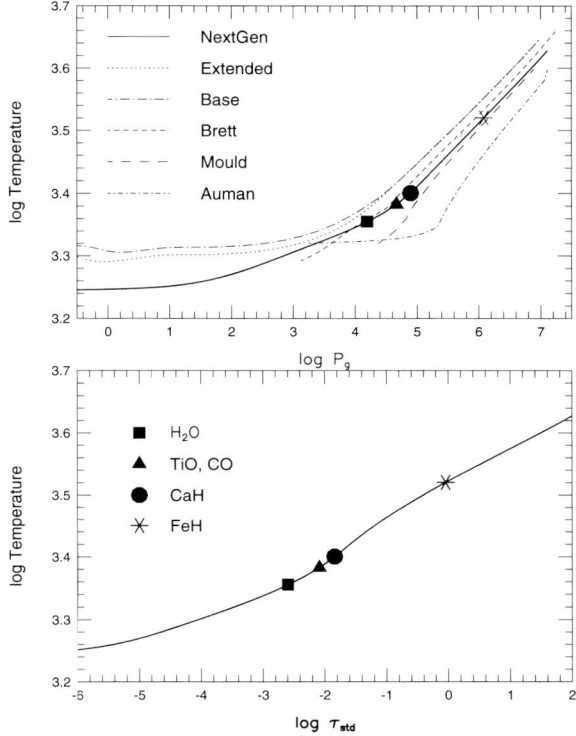

Figure 4.8. Comparison of a standard model computed by several authors, as indicated. The model has T = 3,000 K, $\log g$ = 5.0, and solar metallicity, except for the Mould model which has $\log g$ = 4.75. The formation depths of the molecular species H_2O, TiO, CO, CaH and FeH are given on both panels; the lower panel shows that the molecules generally form at small continuum optical depth.

atmospheres, which leads to diminished optical depth at a given pressure. It is therefore possible to look deeper into the atmosphere, to conditions of much higher pressure before attaining significant continuum optical depth. However, greater pressure and density do not always result in a higher molecular abundance for every species. When we examine the emergent spectra from these atmospheres we shall see that the complicated interplay between the metallicity and the pressure can influence molecule formation in a subtle fashion.

4.6   PHOTOSPHERIC DIAGNOSTICS

4.6.1   The comparison with observations

With the models in hand, it is necessary to produce diagnostics that can be compared directly to the observations. In other words, the modeller has an atmosphere giving temperature as a function of optical depth (for example), while the observer has a

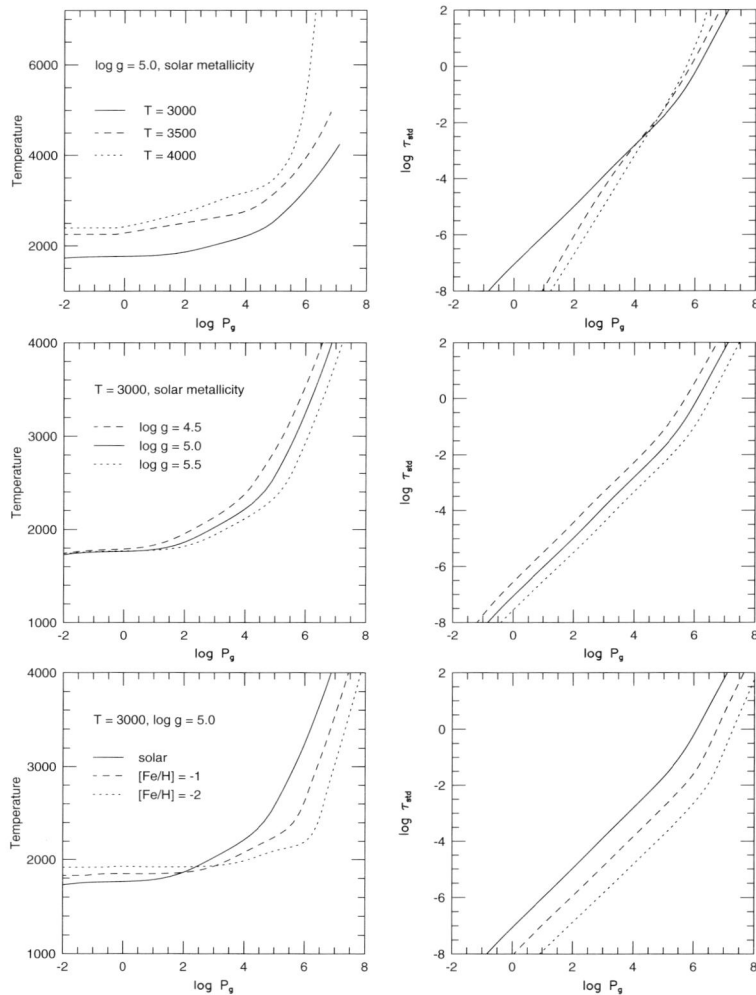

Figure 4.9. The NextGen models [H1], illustrating structural changes with temperature, gravity and metallicity. The left panels show the run of temperature with gas pressure, the right panels the change in continuum optical depth (τ_{std}). The line types apply to both panels. (Models courtesy of P. Hauschildt.)

spectrum showing various absorption lines superposed on a background continuum source. In order to determine whether the model accurately predicts the strengths and profiles of the absorption lines, and hence whether the model is an acceptable representation of the physical conditions on the star, a synthetic spectrum must be produced from the model. Most model atmosphere codes produce a spectrum as part of the model output. However, researchers who have access to the published models but not the atmosphere code, may just as easily use their own spectrum synthesis programs to produce an emergent spectrum (although care must, of course, be taken that the assumptions used in the code to produce the models are compatible with those in the spectrum synthesis code). We shall see in Chapter 5 that modellers of the outer atmosphere adopt these photospheric models as boundary conditions, and use other spectrum synthesis codes to investigate the consequences of adding a chromosphere or flare energy [H2], or to search for evidence of photospheric magnetic fields [J1].

Note that the opacities used in the spectrum synthesis code may be much more detailed, and computed on a much finer wavelength grid, since the underlying atmosphere is already given and is assumed not to change. Variables such as temperature and density are therefore known at every atmospheric level, and, although the opacity computation is still time consuming, it does not feed back into the atmospheric structure as it does in the model computation. Thus, for example, line profiles may be produced by using the full absorption profile (Figure 4.2) in the computation, rather than approximating the opacity in the line as a single value at the central frequency, which is often the expedient method when producing the model.

Previously, we described some of the classical methods of stellar atmospheres theory and the comparison with observations, including the concept of the line equivalent width and the curve of growth. These methods are still applicable, and indeed useful, in obtaining a qualitative feel for the model output and the match to the observations. With the current models, it is now also possible to obtain a quantitative fit to the spectral data. The output of the models is a synthetic spectrum, with line profiles computed on scales of an Ångstrom or less. The synthetic spectrum is convolved with filter profiles to produce photometric colours (see Figure 2.9), or rebinned to match the instrumental resolution and compared directly to an observed spectrum. Narrow-band indices centred on important molecular features such as the optical bands of CaH and TiO also provide a useful measure by which to compare models and observations.

Figure 4.10 shows the emergent spectra over the optical and near-infrared wavelength regime corresponding to the models in Figure 4.9. These are PHOENIX spectra which have been convolved with a Gaussian profile ($\sigma = 5\,\text{Å}$) to smooth out most of the spurious effects introduced by the opacity sampling method, while retaining the overall shapes of the important molecular features. Clearly the temperature has the strongest effect on the output spectra, while varying the gravity produces a barely noticeable effect in the optical region (a slight strengthening of the CaH bands, as first noted by Mould). Varying the metallicity is seen both in the strengthening of the hydride bands (CaH, FeH) and weakening of the oxide bands (H_2O, TiO). These effects are described in more detail below.

4.6.2   Temperature determinations

Most M dwarfs have gravity $\log g \sim 5.0$, and near-solar metallicity, making temperature the most important unknown parameter. Finding the temperature from a comparison of models with observations results in a spectral type: temperature calibration (using spectral data), or a colour–temperature calibration

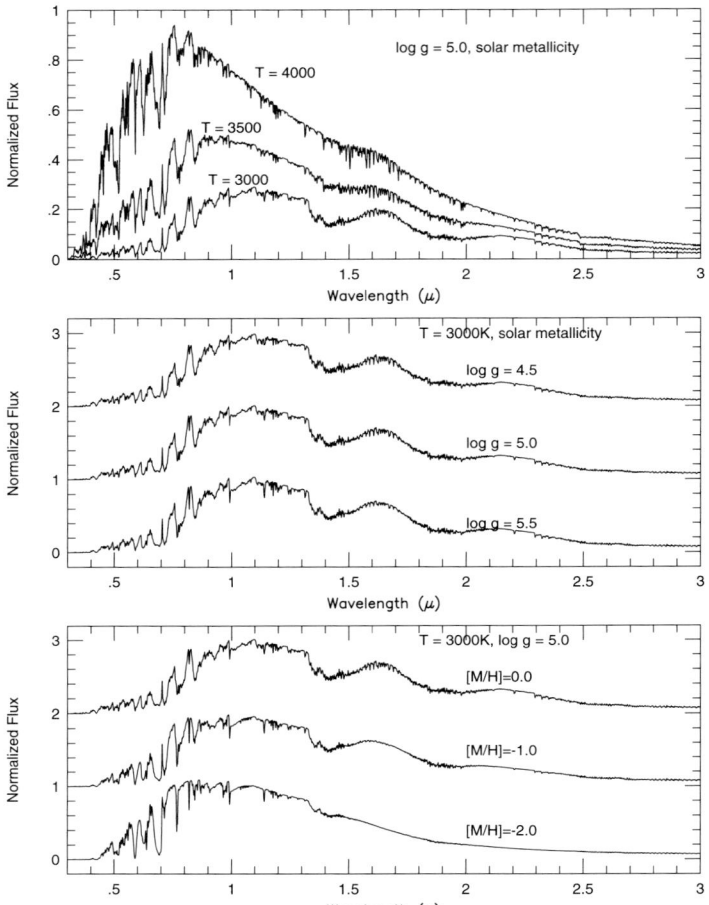

Figure 4.10. Spectra from the NextGen models shown in Figure 4.9, illustrating the changes that occur when the temperature, gravity and metallicity are varied. The temperature causes a marked change in the spectrum, shifting the wavelength peak to the blue and reducing the strength of the molecular bands in the hotter models. The changes with gravity are quite subtle, while the metallicity changes are most noticeable in their effect on the infrared water bands, which nearly disappear at low metallicity. (Model spectra courtesy of P. Hauschildt.)

(using photometric data). Both calibrations are in common use, and provide methods of obtaining temperatures for large samples based on the results of a few well-studied stars.

The Kirkpatrick spectral types (see Chapter 2) are usually adopted as the standard for M and L dwarfs. Temperature scales have been tied to these spectral types using optical and infrared spectra together with several versions of the Allard models (base grid [K1], upgraded version of base grid [L2], extended grid [G2]). No comprehensive temperature calibration has yet been carried out for the NextGen grid.

Photometric colour–temperature relationships work fairly well in the infrared as shown by Berriman [B1], [B2], who determined bolometric luminosities and 'black-body equivalent temperatures' (found by anchoring the black-body at a continuum point in the K-band) using a range of colours from U through M. Tinney [T1] used a similar procedure with blackbodies anchored in the L-band and obtained consistent results, while Leggett [L3] used infrared photometry to infer temperature, metallicity and mass for a number of late M and L dwarfs; her temperatures are given in 500 K bins, and agree within those boundaries with the other color–temperature estimates.

The spectroscopic and photometric temperature estimates are combined in Table 4.1, to define a spectral type–(V–I) colour–temperature relationship for solar-metallicity M dwarfs. Typical uncertainties in the temperature are ± 100 K; the spectral types are known to within 0.5 of a spectral class and the colors are measured to ~0.05 magnitudes. Figure 4.11 shows optical spectra for M0–M6 dwarfs, together with NextGen model spectra. The general features of the observed spectra are reproduced by the models, but there are clearly differences in fine detail. For example, the TiO bands between 6600 and 6800 Å, and between 7050 and 7250 Å, are generally too strong in the models compared to the observations, as is the CaH band between 6950–7050 Å. The M2 and M4 spectra are better fit than the hotter or cooler models.

Table 4.1. Fundamental properties of M dwarfs.

Spectral type[1]	V–I[2] Kelvin	Tempera-ture[3] $(R/R_\odot)$	Radius[4] $(M/M_\odot)$	Mass[5] $(10^{-2} L/L_\odot)$	Lumi-nosity[6] $\mathrm{g\,cm^{-2}\,s^{-1}}$	Log gravity[7]	Prototype
M0	1.92	3,800	0.62	0.60	7.2	4.65	Gl 278C[8]
M1	2.01	3,600	0.49	0.49	3.5	4.75	Gl 229A
M2	2.15	3,400	0.44	0.44	2.3	4.8	Gl 411
M3	2.46	3,250	0.39	0.36	1.5	4.8	Gl 725A
M4	2.78	3,100	0.36	0.20	0.55	4.9	Gl 699
M5	3.70	2,800	0.20	0.14	0.22	5.0	Gl 866AB
M6	4.06	2,600	0.15	0.10	0.09	5.1	Gl 406
M7	4.56	2,400	0.13	~0.09	0.05	5.2	Gl 644C (VB 8)
M8	4.33	2,200	0.12	~0.08	0.03	5.2	Gl 752B (VB 10)
M9	4.37	2,100	0.09	~0.075	0.015	5.4	LHS 2924[9]

[1] Spectral types on the Kirkpatrick scale.
[2] V–I colours on the Cousins system, from the Appendix, except for YY Gem [M1] and LHS 2924 [L1].
[3] Temperatures compiled from several photometric and spectroscopic calibrations (see text for references).
[4] Radii calculated from $L = 4\pi R^2 \sigma T^4$.
[5] Mass estimates from 8-parsec sample for each spectral type, except for M0 and M9 (see notes).
[6] L calculated from M_{bol} using $M_{bol}(\mathrm{Sun}) = 4.62$, $L_\odot = 3.9 \times 10^{33}$ ergs s^{-1}.
[7] $\log g$ calculated from $g = GM/R^2$.
[8] Gl 278C (YY Gem) is a well-observed eclipsing binary consisting of two dM0e stars; we use it for the prototype, although it is not in the 8-parsec sample. The values given are the average of the two stars. The (V–I) colour comes from [M1].
[9] LHS 2924 is the closest M9 dwarf, at a distance of 10.5 parsecs.

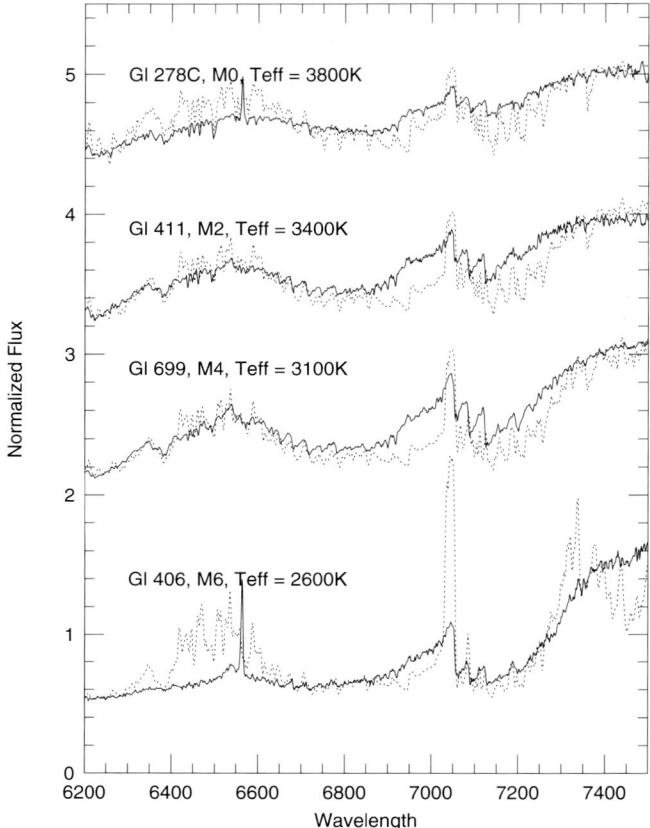

Figure 4.11. Optical spectra for the prototype stars with spectral types M0, M2, M4 and M6 in Table 4.1, together with NextGen model spectra. Models have temperatures from Table 4.1, $\log g = 5.0$, and solar metallicity, and have been convolved with a Gaussian of $\sigma = 5\,\text{Å}$. The data and models are normalised at $\lambda = 7400\,\text{Å}$. (Model spectra courtesy of P. Hauschildt.)

Figure 4.12 shows the infrared spectra of the M2 and M6 dwarfs, highlighting the persistent problem with the water opacities first discussed by Auman. As explained by Allard [A4], the infrared water opacities are systematically overestimated by opacity sampling techniques due to the large overlap of the H_2O lines forming the principal bands. In addition, many of the hot (steam) bands are not included in current lists. These deficiencies tend to offset one another. Fortuitously, the use of the SM opacity approximation for water in the Allard base models [A3], which underestimates the opacity, actually gave better fits to the observations than the more rigorous OS method of Brett [B6]. It appears now that the inclusion of dust opacity in the cooler M dwarf, and L dwarf, atmospheres is probably vital to the models (as discussed in Section 4.7). Preliminary models including dust [A2] provide a better fit to the spectral data as shown in the figure.

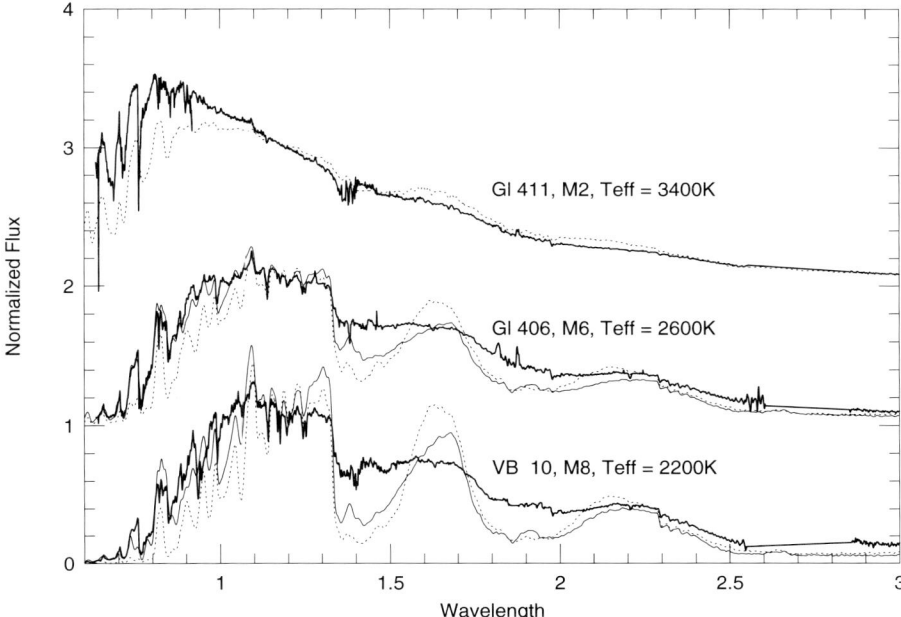

Figure 4.12. Gliese 411 (M2), Gliese 406 (M6) and VB 10 (M8) spectra from 0.6–3.0 μm (dark solid lines) are shown together with NextGen model spectra of 3,400 K, 2,600 K and 2,200 K respectively (dotted lines). The spectra are normalised at the standard model wavelength of 1.2 μm. Preliminary models including dust are shown as the thin solid lines for the 2,600 K and 2,200 K models. The preliminary dust models provide a much better fit in the optical, and show some improvement in the infrared water bands. (Model spectra courtesy of F. Allard and P. Hauschildt.)

The (V–I)–temperature relationship from Table 4.1 is shown in Figure 4.13 (solid points), and agrees well with the one given in [A2] (solid line). It is somewhat cooler than earlier relations obtained from the Allard base and extended model grids (dotted line). Note that the V–I colour saturates at spectral type M7, and becomes bluer at later types (an effect first noted by Auman; see above). This effect is also seen as the 'hook' in the colour–magnitude diagrams of Chapter 3; for example, Figure 3.14.

Although the temperature estimates from spectra and photometry using the latest models now agree fairly well, they are in disagreement with the structural model calculations. As described in Chapter 3, the structure models tend to produce higher temperatures at a specified mass or luminosity. In other words, an observed M dwarf spectrum, when compared to an atmospheric model grid, matches a model with temperature T_o. However, the observed mass and luminosity of that M dwarf, when compared to a structural model grid, would suggest that a different model, with $T \sim T_o + 200$ K should have provided the best fit. It is not clear whether the problem lies in the structural models – producing too high a temperature for a given mass star – or in the atmospheric models and spectral synthesis, producing atomic line and molecular band strengths and continuum radiation that suggest too low a

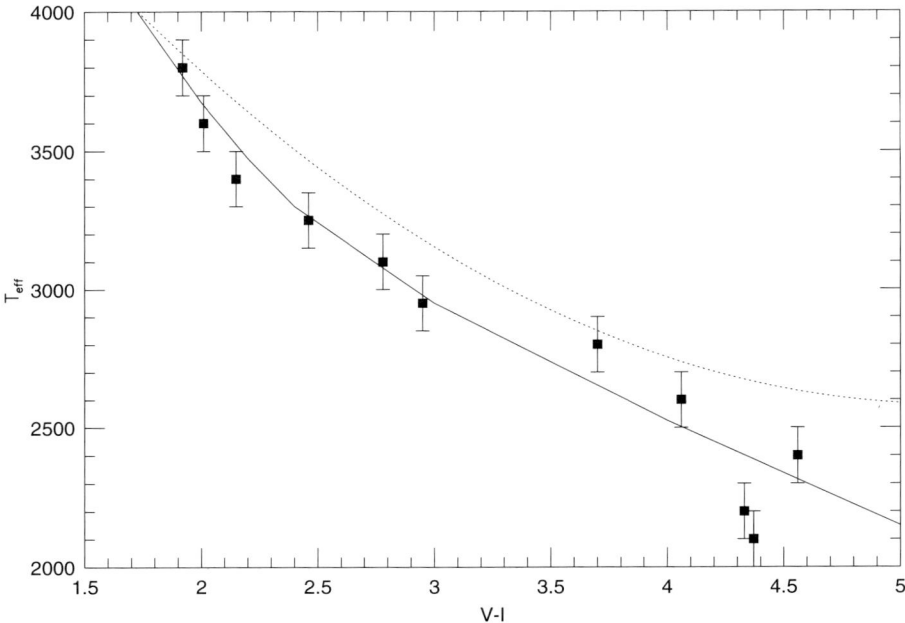

Figure 4.13. The (V–I) colour–temperature relationship from Table 4.1 (points and error-bars). The solid line is the relationship for the NextGen models from [A2]; the dotted line is the relationship from [C2] which they derived from the data and model fits (to Allard upgraded base models) given in [L2]. The difference in temperature amounts to nearly 500 K for the lowest-temperature M dwarfs. The lower-temperature scale is currently favoured.

temperature for the same star. As pointed out in Chapter 3, the effective temperature of the star depends critically on the atmospheric boundary conditions that are adopted in the structural calculation, so the problem is clearly coupled. It is ultimately a result of inadequate opacity data, and the approximate treatment of the existing opacity data, and hopefully will be rectified in the near future.

4.6.3   Gravity and radius determinations

Figure 4.14 shows high-resolution (2 Å) model spectra in the optical wavelength region, illustrating in finer detail the spectral changes due to varying gravity and metallicity. As described in Section 4.5, the higher gravity models reach optical-depth unity at higher pressure, resulting in increased molecular abundances and somewhat stronger bands. Figure 4.14 indicates that the CaH band at 6900 Å is deeper in the higher-gravity model, but the TiO band at 7050 Å is barely affected. Mould attributed the TiO behaviour to the competition for oxygen between CO, H_2O, TiO and several other oxides. At higher pressure, H_2O formation is favoured, and the abundance of the TiO molecule is essentially unchanged (at solar metallicity). In general, the dependence of the spectrum on gravity is more subtle

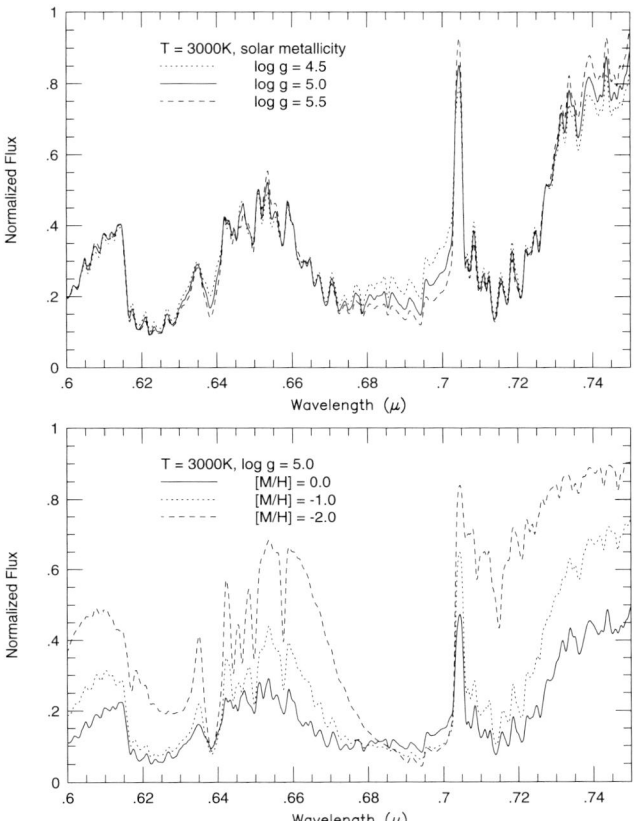

Figure 4.14. Higher-resolution spectra from the NextGen models of Figure 4.9, showing the detailed changes in the optical (red) CaH and TiO bands with varying gravity and metallicity. CaH becomes stronger with higher gravity, while TiO remains about the same strength. Alternatively, with lower metallicity CaH appears stronger, while TiO becomes weaker. The continuum is also noticeably bluer in the lower-metallicity models. (Model spectra courtesy of P. Hauschildt.)

than the dependence on metallicity, and requires high-resolution observations to resolve.

To date, the only analysis of such high-resolution observations is that of Valenti *et al.* [V1], who obtained R = 120,000 spectra of the dM3.5 star Gliese 725B. They used NextGen model atmospheres with their own spectrum synthesis program for TiO, and found T = 3202 ± 70 K, in agreement with the temperature scale in Table 4.1. They also found gravity, $\log g = 4.82 \pm 0.14$, and metallicity, [m/H] = -0.91 ± 0.07, which seems somewhat metal-poor for the temperature they obtained. This may be a result of not using the hydride bands in their analysis (see next section). Fitting high-resolution spectra with the NextGen models to obtain these parameters for a large sample of stars is clearly a profitable avenue for future work.

Using only photometric data, Leggett *et al.* [L2] employed an alternative approach to finding the gravity. They first used the observed position of a star in the H–R diagram together with an empirical mass–luminosity relationship [H4], to estimate the mass (see Fig. 2.23, and discussion in Chapter 8). They compared the masses obtained in this way to theoretical predictions from the structure models of the Lyon group (see Figure 3.16) and found good agreement. The structure models give an estimate of the radius which can be combined with the mass via equation (4.4) to determine the surface gravity.

Alternatively, the luminosity of the star may be determined from the (observed) bolometric magnitude, and together with a temperature estimate the radius can be found from equation (4.3), with $F = L/4\pi R^2$. The mass is found from the same empirical mass–luminosity relationship [H4]; the mass and radius determine the gravity, as before.

The latter method for determining mass and luminosity has been applied to the 8-parsec sample [R1], [C2], and the results are given in the Appendix. Using these data we have computed radii and gravities for each star in the Appendix. Table 4.1 gives the average mass, luminosity, radius and gravity together with a 'prototype' star for each spectral type; the prototypes are members of the 8-parsec sample except for the M0 prototype (YY Gem, a well-observed eclipsing binary with independent mass and radius determinations for each component) and the M9 prototype (LHS 2924, the closest M9 dwarf at 10.5 parsecs).

4.6.4   Metallicity

Known or suspected low-metallicity stars – which are referred to as subdwarfs due to their position below the solar metallicity main sequence in an H–R diagram (see Chapters 2 and 11) – must have their spectral types, metallicities and temperatures determined self-consistently; simply applying the relationships from Table 4.1 will lead to an underestimate of the temperature by several hundred degrees. In general, metallicity affects the atmosphere by reducing the opacity at a given height, hence the low metallicity atmospheres are expected to reach $\tau = 1$ in the continuum at a greater physical depth in the atmosphere, where the gas pressure and temperature are higher (as illustrated in Figure 4.9). This explains why a low-metallicity star appears hotter, and has a stronger blue continuum, at the same mass and luminosity as a solar-metallicity star. The lack of metals also renders the hydrogen opacity sources more important – particularly the pressure-induced H_2 dipole opacity in the infrared. This leads to a further redistribution of flux from the infrared into the optical in metal-poor stars.

Initial attempts at the identification of metal-poor low-mass stars [M4], [M5] used the change in the infrared colours due to the increased H_2 dipole opacity in the H- and K-bands (hence bluer J–H colour) to separate subdwarfs from disk dwarfs in the J–H versus H–K two-colour diagram (see Figure 2.18). However, the change in the colours is rather small, and is difficult to measure observationally. In addition, models of varying metallicity are not yet able to reproduce the observed colours in this diagram (see Figure 3 of [L3]), making it difficult to assign quantitative metallicities to the subdwarfs.

Analysis of the spectral changes in metal-poor stars is a more promising avenue for abundance determination. Figure 4.14 shows that changes in metallicity have a larger effect on the spectrum than changes in gravity, but that there was some overlap; for example, the CaH bands become stronger both in higher gravity and in lower-metallicity stars. To disentangle these effects requires observations of several spectral diagnostics that respond differently to changes in these parameters. In particular, the key to metallicity determination in M subdwarfs is to compare the strength of hydride (single-metal) features to species composed of two metals (for example, TiO). The hydride features actually appear to increase in strength with decreasing abundance, while the double-metal species rapidly decrease in strength. The explanation for this apparently paradoxical behaviour lies in the variation of the partial pressures of the species contributing to the line and continuum opacities. The strength of a spectral feature depends on the ratio between these opacities, and hence on the relative number densities of the relevant ions, atoms and molecules. In the optical wavelength regime, the H^- ion is the dominant source of continuum opacity (see Figure 4.6). The number density, $n(H^-)$, decreases slowly with decreasing metal abundance (since the electrons are contributed primarily by metals). The number density of TiO is reduced more rapidly – partly because TiO is a double-metal molecule, and partly because the high pressure in metal-poor atmospheres leads to most of the available oxygen forming H_2O. As a result, $n(TiO)/n(H^-)$ decreases sharply, and TiO absorption weakens at lower metallicity. In contrast, the number density of neutral hydrogen increases as the metallicity decreases, and this compensates partially for the lower absolute abundance of metals in the formation of the single-metal hydride molecules [C3], [B3]. As a result, the ratios n(hydride)/$n(H^-)$ do not change substantially amongst metal-poor stars.

[M6] estimated that the iron hydride (FeH) number densities, and hence bandstrengths, vary with gravity and metal abundance as

$$n(FeH)/n(H^-) \propto g^{0.5} Z^{0.25} \qquad (4.14)$$

The gravity dependence in this relationship is another manifestation of the need for a high-pressure atmosphere to form molecules, and demonstrates why the hydride bands are so much weaker in red giant stars. Other hydrides (CaH, NaH, AlH, CrH) show similar behaviour which, combined with the reduced TiO absorption, leads to hydride absorption bands dominating the spectra in M subdwarfs.

Gizis [G2] and Leggett [L2], [L3] pioneered much of the study of metallicity effects in low-mass stars. Gizis empirically defined several narrowband indices centred on optical (red) CaH and TiO bands in the $\lambda\lambda$ 6200–7500 Å region, in order to take advantage of the differing behaviour of the two molecules with metallicity variation. Since the TiO bands are abundance-sensitive, CaH bandstrength is a better measure of spectral type (temperature) in M subdwarfs. Figure 11.4 illustrates the separation of the disk, subdwarf M (sdM) and extreme subdwarf M (esdM) stars using these indices, and spectral-type sequences for the sdM and esdM stars are shown in Figures 11.5 and 11.6.

To quantify the abundances for each group, representative data can be compared to atmospheric models. Figure 4.15 shows a spectrum of the sdM1.5 dwarf LHS 482 together with model fits covering a range of metallicity, temperature and gravity. The best-fitting model has T = 3,600 K, $\log g$ = 5.0 and [m/H] = −1.0. An extensive set of observations and least squares fits (to the Allard extended models) define the spectral type–temperature calibration for the sdM and esdM stars given in Table 4.2 [G2]. The metallicities are typically [Fe/H] ∼ −1 for the sdM stars, and [Fe/H] ∼ −2 for the esdM stars.

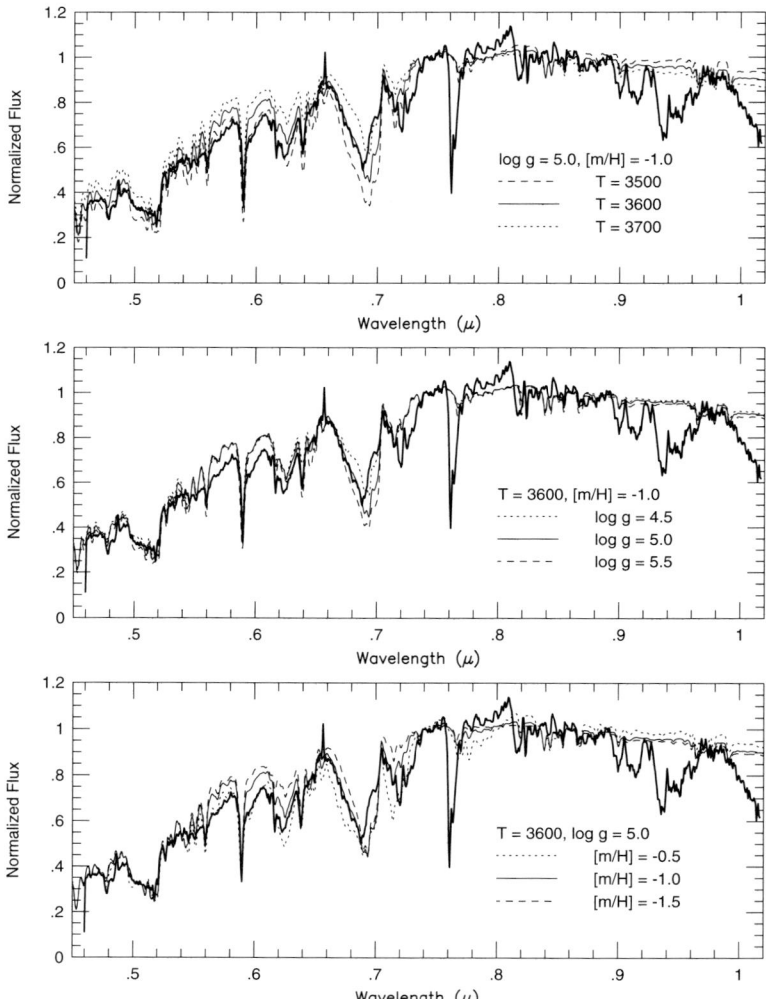

Figure 4.15. The sdM1.5 subdwarf LHS 482 spectrum (dark solid line) compared with model fits with differing temperature (*top*), gravity (*middle*) and metallicity (*bottom*). The model with T = 3,600 K, $\log g$ = 5.0 and [m/H] = −1.0 provides the best fit. (Model spectra courtesy of P. Hauschildt.)

Table 4.2. Temperatures of M subdwarfs and extreme
subdwarfs

Spectral type	T_{eff}	Prototype[1]
sdM0.5	3,700	LHS 307
sdM1.5	3,600	LHS 482
sdM2.5	3,500	LHS 20
sdM3.5	3,500	LHS 2497
sdM5.0	3,400	LHS 407
sdM7.0	3,200	LHS 377
esdM0.5	3,700	LHS 3259
esdM2.0	3,600	LHS 161
esdM3.0	3,500	LHS 1174
esdM4.0	3,400	LHS 375
esdM5.5	3,300	LHS 1742a

[1] Stars are identified by their numbers in the Luyten Half-Second
(LHS) catalogue.

4.7 DUSTY ATMOSPHERES IN THE LOWEST-MASS STARS AND BROWN DWARFS

Tsuji and collaborators [T3] were the first to suggest that dust formation might be important in M dwarfs, although dust had already been identified for many years in M giants. Dust formation depletes many important metals, including (especially) titanium. The most abundant dust grains include corundum ($Al_2 O_3$), perovskite ($CaTiO_3$), iron (Fe), enstatite ($Mg SiO_3$) and forsterite ($Mg_2 SiO_4$). As in the case of low metallicity, the hydrides increase in strength while the double-metal molecules weaken with the advent of dust formation. Thus the 'dusty' atmospheres are dominated by hydride bands, while TiO and VO become weak and, in the lowest-temperature stars, disappear altogether. A second important effect of adding dust to the models is the accompanying greenhouse heating which occurs when the extra dust-grain opacity is included. This enhanced continuum opacity at infrared wavelengths traps outgoing radiation, heating the atmosphere and redistributing the flux into different wavelength regions. The atmospheric heating results in increased dissociation of H_2O, so that spectra from models that include dust have weaker water bands. The dust will eventually settle gravitationally into deeper layers, leaving the atmosphere with depleted metal abundances but without the dust opacity. The time-scale over which this occurs is still open to question, and current modelling efforts include models both with and without dust opacity. Tsuji and collaborators [T3], [T4] have computed models including dust and TiO depletion, while Allard [A2] presented preliminary work on extending the NextGen models to include dust and settling. Approximately 1,000 dust species are included in the computation of the 'NG-dusty' models. Figure 4.12 illustrates the improved fit of the NG-dusty models, particularly in the optical spectra of dwarfs cooler than 3,000 K.

Besides the hydride bands, the most prominent features in the optical L dwarf spectra are alkali lines, including the relatively rare metals caesium, rubidium and (if $M < 0.06\,M_\odot$) lithium, as shown in Figure 4.16 (from [K2]). The strength of the alkali lines likely stems from the substantially reduced opacity at optical wavelengths in these objects. Not only have TiO and VO virtually disappeared, but dust formation also removes the main electron donors, and the H$^-$ and H$_2^-$ opacities (see Figure 4.6) are correspondingly reduced. As a result, we see deep into the L dwarf atmosphere at optical wavelengths, and the increased column density of neutral alkali atoms leads to very strong lines.

The effect is particularly pronounced for the resonance lines of K I at 7665 and 7699 Å, which broaden to equivalent widths of several hundred Å at spectral type L4. The Na I resonance doublet at 5890/5896 Å (the D lines) shows a similar

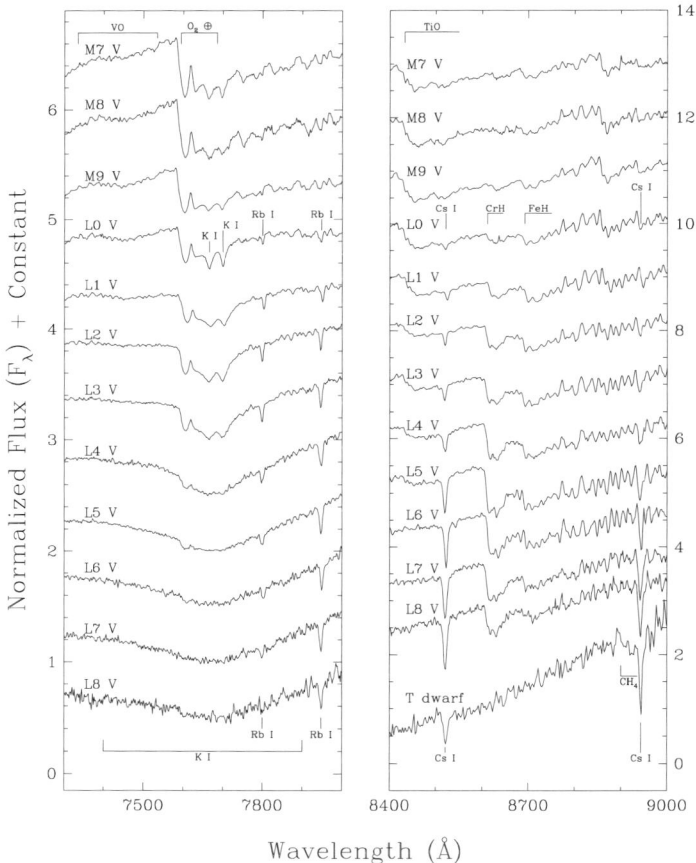

Figure 4.16. A series of late M and L dwarf spectra showing the increasingly strong alkali metal resonance lines with decreasing temperature. Hydride bands of CrH and FeH are also visible. (Figure courtesy of J. D. Kirkpatrick and the *Astrophysical Journal* [K2].)

behaviour (the $8192/8196$ Å doublet is not a resonance pair). The fact that these elements have stronger lines than the other alkali metals reflects the relative abundances of the species: on a scale where the number of hydrogen atoms is defined as $[H] = \log_{10} N(H) = 12.0$, $[Na] = 6.31$; $[K] = 5.13$; $[Li] = 3.31$; $[Rb] = 2.41$; and $[Cs] = 1.12$. The transparency of the atmosphere is such that the column densities of sodium and potassium are sufficiently high to place the resonance lines on the damping (square root) part of the curve of growth (see Section 4.2 and Figure 4.4). Figure 4.16 illustrates the extremely broad damping wings of KI at mid–late L spectral types.

Additional work on dust formation in very low temperature atmospheres has been carried out by Fegley and Lodders [F1] Burrows and Sharp [B8], who computed atmospheres including some 120 condensates and extending over a temperature range from 2,200 K down to 100 K. Those calculations allow a first estimate of the L dwarf temperature scale [K2]. TiO is predicted to condense onto grains of perovskite ($CaTiO_3$) at temperatures between 2,300 and 2,000 K; in comparison, the TiO bands are observed to reach maximum strength at a spectral type of M7–8 and have disappeared (except for the 8432 Å band) by type L2. Vanadium oxide is next to disappear, forming solid VO at temperatures between 1,900 and 1,700 K; that species is strongest at type M9.5 and disappears at type L4. CrH is expected to be entirely in the metallic phase at $\sim 1{,}400$ K, and Figure 4.16 shows that the 8611 Å band decreases in strength beyond type L5. All of the alkali lines are predicted to form chlorides at temperatures below 1,200 K, although sodium may condense onto high albite ($NaAlSi_3 O_8$) at somewhat higher temperatures; the observations show the $8192/8196$ Å doublet as barely visible by spectral type L8. Finally, carbon is predominantly found in CO among the L dwarfs, but preferentially forms methane at a temperature between 1,500 and 1,200 K. This point marks the transition between spectral type L and type T (the methane dwarfs, see Chapter 9). Combining these diagnostics with the observed change in the overall spectral energy distribution leads to the tentative temperature scale outlined in Table 4.3. The available evidence suggests that L dwarfs have radii near $0.1\,R_\odot$, and luminosities ranging from $-3.7 < \log L/L_\odot < -4.7$; the luminosity depends critically on the age at young ages. Masses are less than about $0.08\,M_\odot$, but no direct measurements are yet available.

Table 4.3. Temperatures of L dwarfs.

Spectral type	T_{eff}	Prototype[1]
L0	2,000	2MASP J0345+2540
L2	1,900	Kelu 1
L4	1,750	2MASS J1155+2307
L6	1,600	2MASS J0850+1057
L8	1,400	2MASS J1632+1904

[1] The names of the prototypes from the 2MASS survey are defined in [K2]. Kelu 1 is discussed in [R3].

As described in Chapter 9, work is proceeding to better define these quantities. In general, theoretical models for cool, methane-dominated (T-type) brown dwarfs appear on a firmer footing than studies of the hotter L dwarfs, although this may reflect the fact that very few are known. In addition to the Burrows models, Marley *et al.* [M2] and Allard *et al.* [A5] describe models of very low-temperature atmospheres. These are compared against the observations and discussed in more detail in Chapter 9. The modelling of L dwarf and T dwarf (and extrasolar giant planet) atmospheres is a very young and active field, and the reader will undoubtedly find many new references not mentioned here.

4.8   SUMMARY

Atmospheric parameters of M and L dwarfs may be found by matching synthetic spectra from model atmosphere calculations to observed data. High resolution spectral data allow quite accurate determinations, but even photometry (particularly in the infrared) can be used with some success. The models are now good enough that fine analysis of very high-resolution spectra allows metallicity, gravity and temperature to be determined with good precision. Lacking this data, gravity may be found by indirect methods using luminosity and temperature information to infer mass and radius. Metallicity may be determined by comparing the molecular bandstrengths of hydrides (CaH) to double metal species (TiO). Thus, the temperature, gravity and metallicity of a low-mass star are now accessible using optical and infrared photometry and spectroscopy together with the current generation of atmospheric models. Improvements in the next few years will come with the acquisition of higher-resolution optical and infrared spectra, and from continuing work on the molecular and dust opacity data and the treatment of the opacities in the models.

4.9   REFERENCES

A1   Allard, F. 1990, PhD thesis, University of Heidelberg.
A2   Allard, F. 1999, in *Very Low Mass Stars and Brown Dwarfs in Stellar Clusters and Associations*, ed. R. Rebolo, (Cambridge University Press), in press.
A3   Allard, F., Hauschildt, P. H., 1995, *ApJ*, **445**, 433.
A4   Allard, F., Hauschildt, P. H., Alexander, D. R. A., Starrfield, S., 1997, *ARA&A*, **35**, 137.
A5   Allard, F., Hauschildt, P. H., Baraffe, I., Chabrier, G., 1996, *ApJ*, **465**, L123.
A6   Auman, J. 1969, *ApJ*, *157*, 799.
B1   Berriman, G. B., Reid, I. N., 1987, *MNRAS*, **227**, 315.
B2   Berriman, G. B., Reid, I. N., Leggett, S. K., 1992, *ApJ*, **392**, L31.
B3   Bessell, M. S., 1982, *Pub. Ast. Soc. Australia*, **4**, 417.
B4   Bohm-Vitense, E., 1989, *Introduction to Stellar Astrophysics*, Volume 2, *Stellar Atmospheres*, Cambridge University Press.

B5   Borysow, A., 1993, in *Molecules in the Stellar Environment*, ed. U.G. Jørgensen (Lecture Notes in Physics, 428), Springer, p. 209.

B6   Brett, J. M., 1995a, *A&P*, **295**, 736.

B7   Brett, J. M., 1995b, *A&AS*, **109**, 263.

B8   Burrows, A., Sharp, C. M., 1999, *ApJ*, **512**, 843.

C1   Carbon, D. F., 1979, *ARA&A*, **17**, 513.

C2   Clemens, J. C., Reid, I. N., Gizis, J. E., O'Brien, M. S., 1998, *ApJ*, **496**, 352.

C3   Cottrell, P.L., 1978, *ApJ*, **223**, 544.

D1   Deutsch, A. J., 1966, in *Abundance Determinations in Stellar Spectra*, ed. H. Hubenet (IAU Symposium No. 26), 112.

F1   Fegley, B., Lodders, K., 1996, *ApJ*, **472**, L37.

G1   Gingerich, O., Kumar, S. S., 1964, *AJ*, **69**, 139.

G2   Gizis, J. E., 1997, *AJ*, **113**, 806.

G3   Gustafsson, B., Bell, R. A., Eriksson, K., Nordlund, A., 1975, *A&A*, **42**, 407.

H1   Hauschildt, P. H., Allard, F., Baron, E., 1999, *ApJ*, **512**, 377.

H2   Hawley, S. L., Fisher, G. H., 1992, *ApJS*, **78**, 565.

H3   Hearnshaw, J. B., 1986, *The Analysis of Starlight*, Cambridge University Press.

H4   Henry, T. J., McCarthy, D. W., 1993, *AJ*, **106**, 773.

J1   Johns-Krull, C. M., Valenti, J. A., 1996, *ApJ*, **459**, L95.

J2   Jones, H. R. A., Tsuji, T. 1997, *ApJ*, **480**, L39.

K1   Kirkpatrick, J. D., Kelly, D. M., Rieke, G. H., Liebert, J., Allard, F., Wehrse, R., 1993, *ApJ*, **402**, 643.

K2   Kirkpatrick, J. D., Reid, I. N., Liebert, J., Cutri, R. M., Nelson, B., Beichman, C. A., Dahn, C. C., Monet, D. G., Gizis, J. E., Skrutskie, M. F., 1999, *ApJ*, **519**, 802.

K3   Kurucz, R. L., 1970, *ATLAS: A Computer Program for Calculating Model Stellar Atmospheres*, Smithsonian Astrophysical Observatory Special Report No. 309, Cambridge.

L1   Leggett, S. K., 1992, *ApJS*, **82**, 351.

L2   Leggett, S. K., Allard, F., Berriman, G., Cahn, C. C., Hauschildt, P. H., 1996, *ApJS*, **104**, 117.

L3   Leggett, S. K., Allard, F., Hauschildt, P. H., 1998, *ApJ*, **509**, 836.

L4   Lenzuni, P., Chernoff, D. F., Salpeter, E. E., 1991, *ApJS*, **76**, 759.

L5   Linsky, J. L., 1969, *ApJ*, **156**, 989.

M1   Maceroni, C., Rucinski, S. M., 1997, *PASP*, **109**, 782.

M2   Marley, M. S., Saumon, D., Guillot, T., Freedman, R. S., Hubbard, W. B., Burrows, A., Luninie, J. I., 1996, *Science*, **272**, 1919.

M3   Mihalas, D., 1978, *Stellar Atmospheres*, 2nd edition, W. H. Freeman and Company.

M4   Mould, J. R., 1976, *A&A*, **48**, 443.

M5   Mould, J. R., Hyland, A.R., 1976, *ApJ*, **208**, 399.

M6   Mould, J. R., Wyckoff, S. 1978, *MNRAS*, **182**, 63.

R1   Reid, I. N., Gizis, J. E. 1997, *AJ*, **113**, 2246.

R2   Ruan, K. 1991, Ph.D. thesis, National University of Australia (also reported as Kui (1991)).

R3   Ruiz, M. T., Leggett, S. K., Allard, F. 1997, *ApJ*, **491**, L107.
S1   Schweitzer, A., Hauschildt, P. H., Allard, F., Basri, G. 1996, *MNRAS*, **283**, 821.
T1   Tinney, C. G., Mould, J. R., Reid, I. N. 1993, *AJ*, **105**, 1045.
T2   Tsuji, T. 1966, *PASJ*, **18**, 127.
T3   Tsuji, T., Ohnaka, K., Aoki, W. 1996a, *A&A*, **305**, L1.
T4   Tsuji, T., Ohnaka, K., Aoki, W., Nakajima, T. 1996b, *A&A*, **308**, L29.
V1   Valenti, J. A., Piskunov, N., Johns-Krull, C. M. 1998, *ApJ*, **498**, 851.

5

Stellar activity

5.1   INTRODUCTION

The presence of strong magnetic fields on the stellar surface manifests itself in many ways, referred to collectively as 'stellar activity'. The magnetic field is probably formed by the action of an interior dynamo, although its workings are not well understood. On the surface, the field may be organised in active regions, as on the Sun, leading to starspots, plages, prominences, and other solar-type phenomena. The magnetic field heats the tenuous atmospheric layers above the photosphere by non-radiative processes, and these heated layers are seen as the chromosphere, transition region and corona, giving rise to readily recognisable emission lines in the optical, ultraviolet and X-ray spectral regions. Radiation from the outer atmosphere may also feed back into the stellar photosphere, causing secondary effects. Perhaps the most spectacular manifestations of stellar activity are the occasional flares, during which the star emits an enormous amount of energy in a very short time. Some flares are accompanied by mass ejections which affect both the star and its immediate neighbourhood. Clearly, stellar activity has a profound effect on the surface layers which we observe. The magnetic field may also play an important role in the formation, interior structure and evolution of the star, although our understanding of these processes is in its infancy.

In this chapter we first provide a brief overview of current dynamo theories to provide context for understanding the observations. We then examine the evidence for magnetic fields on M dwarfs from direct observation of the photosphere and from the presence and structure of the outer atmosphere. Observations of solar-like activity, including spots and flares, are discussed and contrasted with the solar case. An extensive section on global activity properties observed in large samples of M dwarfs – both in the field and in clusters – describes the current observational situation. We finish with a brief discussion of new observational results for the lowest mass stars and brown dwarfs, and a summary attempting to construct a coherent picture of stellar activity in low-mass dwarfs.

5.2   THE MAGNETIC FIELD

5.2.1   Dynamo theory

Galileo first observed sunspots in the early 1600s, but it was George Ellery Hale –
using the spectroheliograph at Mount Wilson Observatory in the early 1900s – who
identified them as regions of strong magnetic field [H1]. The modern theory for the
production of the field began with the work of Parker [P1], who proposed a dynamo
mechanism for generating and sustaining magnetic fields in the convection zone of
the Sun. This model and its many subsequent revisions have come to be known as
the $\alpha\Omega$ dynamo. The nomenclature refers to the two forces operating on the plasma.
An initial poloidal field is subject to shearing forces by the interior radial differential
rotation in the Sun, resulting in the generation of an internal toroidal field (the Ω
effect).[1] The toroidal field suffers small-scale cyclonic motions with non-zero helicity,
generating a new poloidal field by the α effect:

$$\delta B_{pol}/\delta t = \alpha B_{tor} \tag{5.1}$$

where B_{pol} and B_{tor} refer to the poloidal and toroidal magnetic fields respectively,
and α is a term that depends on the kinetic helicity of the plasma (the rotational
velocity field and its curl). (DeLuca and Gilman [D1] and Durney et al. [D8] provide
detailed descriptions and derivations of the magnetohydrodynamic equations
describing the α effect.) Note that both the Ω and α effects are a consequence of
stellar rotation.

The toroidal field is twisted by the turbulent convection zone creating bundles of
field which break off from the overall structure. Magnetic buoyancy causes these
'flux tubes' to rise to the surface and appear as loop-like structures, with opposite
polarity at each of their footpoints. Individual flux tubes occur over much of the
solar surface (although they cover only a small area in total). Sunspots are organised
areas where many flux tubes have grouped together to make a large and noticeable
magnetic region (or 'active region'). The latitude dependence of sunspots – which are
observed to appear at mid-latitudes early in the solar cycle and at low latitudes late
in the cycle – is explained in $\alpha\Omega$ models by the propagation of the interior dynamo
wave towards the equator as the α effect works to cancel the toroidal field and
produce the new poloidal field. The observed tilt of sunspots relative to the solar
equator can be reproduced with detailed modelling of the forces on flux tubes as they
rise through the convection zone. Fisher et al. [F5] found that the Coriolis force was
the major contributor to the tilt.

As described by Parker [P2], the global toroidal field must be located at the base
of the convection zone, in a region of weak magnetic buoyancy. This allows the
toroidal field to be maintained for a timescale of several years, corresponding to the
period of the solar cycle. It may also be necessary to allow the flux tubes time to

---

[1] Figure 5.1 shows the Babcock model which depends on the surface *latitudinal* differential rotation, and
should not be confused with the interior *radial* differential rotation required in the $\alpha\Omega$ model.

attain the large field amplitude ($\sim 10^4$–10^5 Gauss) needed in the interior for them to appear with the $\sim 1,000$–$2,000$ Gauss fields that are observed at the surface [D7].

The regeneration of the poloidal field with opposite polarity marks the end of the 11-year solar cycle which is observed. Obtaining the opposite polarity at the end of the cycle places constraints on both the magnitude and sign of α. Leighton [L3] showed that the Sun must have differential rotation which increases inward at a fairly high rate to meet these constraints. Recent measurements indicate that the solar convection zone has nearly constant radial differential rotation at active region latitudes [S4]. These observations have led theorists to locate the regeneration region, where the α-effect operates, at the radiative–convective boundary (the base of the convection zone) where a large differential rotation is expected but cannot yet be directly measured. Thus, the current $\alpha\Omega$ dynamo model has both the toroidal and poloidal fields being generated deep inside the Sun. The surface phenomena which we observe are merely tracers of the interior, where all the action is taking place.

An alternative, phenomenological model of the magnetic field evolution during the solar cycle was proposed by Babcock [B1], using the observed behaviour of sunspots on the solar surface. The Sun rotates faster at the equator than at the poles; this latitudinal differential rotation produces an equatorial toroidal field from an initially simple dipole (poloidal) field. The toroidal field attains maximum shear first at mid-latitudes, so flux tubes (sunspots) appear there at the beginning of the cycle. Later in the cycle, as the toroidal field becomes more concentrated toward the equator by the continued action of surface differential rotation, sunspots appear at lower latitudes, as observed. Because the toroidal field is not strictly azimuthal, but retains some poloidal component, the sunspots rise to the surface slightly tilted (as observed), and with opposite polarity in each hemisphere. As their location migrates toward the equator during the solar cycle, the toroidal components cancel and the poloidal components re-establish a poloidal field with the opposite polarity. These ideas are illustrated in Figure 5.1.

In the Babcock model, the toroidal field is produced by an Ω effect (in this case from *latitudinal* differential rotation), and must be somehow transported to the base of the convection zone where flux tube generation occurs, for the same reasons given above. On the other hand, the regeneration of the poloidal field is accomplished without an α effect, simply as a natural consequence of the surface behaviour. A combination of the Babcock and Leighton models has been described by Durney [D7]; that model also includes the observed effect of poleward meridional circulation [S7] which enhances the re-establishment of the poloidal field.

Two common features of these dynamo models are the presence of differential rotation (hence rotation), and the location of the toroidal field at the interface between the convective and radiative zones in the solar interior. For this reason, the models are often referred to as 'shell' dynamos. Although many other details differ, these features are apparently required in order to produce the observed global, cyclic magnetic field on the Sun.

The cyclic behaviour is most easily observed in the sunspots, but also includes effects on the solar luminosity and the distribution of luminosity with wavelength. In particular, the Sun has a bluer colour temperature and emits more total luminosity at

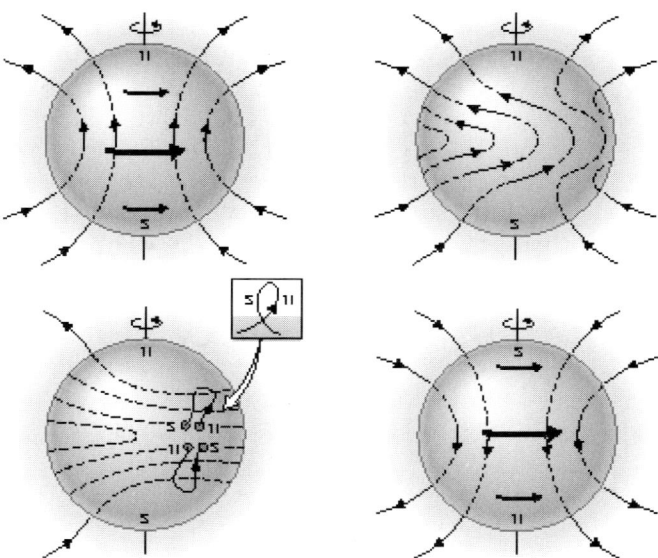

Figure 5.1. The Babcock model of the magnetic dynamo. (a) The solar magnetic field is initially a poloidal field. (b) Differential rotation drags the 'frozen-in' magnetic field lines around the Sun, converting the poloidal field into a toroidal field. (c) Turbulent convection twists the field lines into magnetic ropes, causing them to rise to the surface as sunspots, the polarity of the lead spots corresponding to the original polarity of the poloidal field. (d) As the cycle progresses, successive sunspot groups migrate toward the equator where magnetic field reconnection re-establishes the poloidal field, but with the original polarity reversed. (From Carroll and Ostlie, *Introduction to Modern Astrophysics*, Addison Wesley Longman, 1996.)

solar maximum. (The maximum of the solar cycle is defined as the time of strongest magnetic activity – largest spot coverage, most energetic flares, and so on). However the Sun also possesses other, less obvious, magnetic features, including the magnetic network and the intranetwork field. The magnetic network is thought to be produced by the break-up of active regions, and is thus related to the global cyclic field. The intranetwork field, in contrast, does not appear to vary much during the solar cycle, and Durney has suggested that it is generated by a turbulent dynamo, unrelated to the large-scale global dynamo which is responsible for the cyclic activity [D8]. Rosner had earlier proposed a 'distributed dynamo' model that incorporated some of these ideas to explain X-ray activity in late type stars [R6].

A turbulent dynamo produces magnetic fields by random convective motions in the convection zone, and does not require rotation (or differential rotation) or a radiative–convective boundary layer for its operation. The field is not stored for long periods, or organised in either time or space. Field is generated, and quickly forms flux tubes which rise to the surface, appear for a short time, merge with other regions of opposite polarity and are destroyed. No cycles are expected, and the coverage of the active regions should be rather uniform over the surface, rather than concentrated at mid–low latitudes. These are all features which are observed in the solar intranetwork field. The attractive feature of the turbulent dynamo model for low-mass stars is that it can operate under conditions where the interior is completely convective (that is, in dwarfs below $\sim 0.25\,M_\odot$; see Chapter 3). In these objects, no radiative–convective boundary layer exists for storing the global toroidal field required by the shell dynamo models. The observational predictions of the turbulent dynamo are: a weaker (or no) dependence on rotation, no evidence for cyclic behaviour; and uniform coverage of active regions. As we shall see, these are observed features of the magnetic activity on M dwarfs. There remains some question as to whether the large field strengths which have been observed can be generated by a turbulent dynamo. There is much theoretical work yet to be done to place the model on a firm analytical and numerical foundation, but it does appear to be a promising candidate for generating magnetic fields in low-mass objects, where solar-type shell dynamo models are not applicable.

5.2.2   Magnetic field observations

Direct measurements of the magnetic field are accomplished by utilising the Zeeman effect – which is the splitting of degenerate atomic levels into separate components in the presence of a magnetic field. The normal Zeeman effect results from splitting the $(2l + 1)$ degenerate m_l levels of the electron orbital with quantum number l. The selection rule $\Delta m_l = 0, \pm 1$ means there are three spectral lines formed in transitions to this l level, as shown in Figure 5.2.

The $\Delta m_l = 0$ transition is known as the π component of the line, and retains the original wavelength (in this simple case; more complicated levels can have several π components with small wavelength shifts). The wavelength offsets of the $\Delta m_l = \pm 1$ levels, known as the σ components, are given by Johns-Krull and Valenti [J2].

$$\Delta\lambda = \pm \frac{e}{4\pi mc}\lambda^2 g_{eff}|B| \tag{5.2}$$

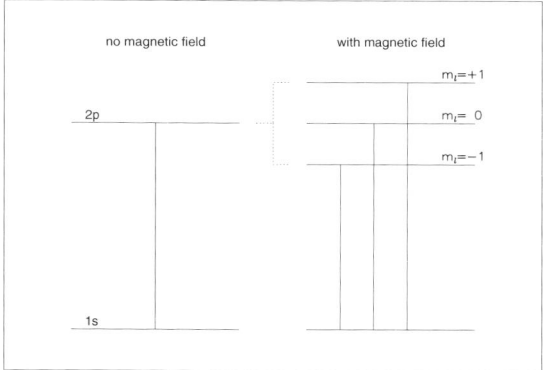

Figure 5.2. The splitting of a degenerate atomic level into its magnetic sub-levels in the presence of a magnetic field. The original transition formed only one spectral line; in the presence of a magnetic field three spectral lines are formed. The magnitude of the splitting depends on the Lande g_{eff} factor for the level, and on the strength of the magnetic field, as shown in equation (5.2).

Table 5.1. Active dMe dwarfs.

Gliese number (or other)	Name	Spectral type
65B	UV Cet	M6e
HD 289114	PZ Mon	K2e
278C	YY Gem	M0e
285	YZ CMi	M4.5e
GJ 2079	DK Leo	K7e
388	AD Leo	M3e
473B	FL Vir	M5.5e
494	DT Vir	M0.5e
517	EQ Vir	K5e
551	Prox Cen	M5.5e
630.1A	CM Dra	M4.5e
644C	VB 8	M7e
719	BY Dra	K7e
752B	VB 10	M8e
799A	AT Mic	M4.5e
803	AU Mic	M0e
820AB	61 Cyg AB	K5e, K7e
873	EV Lac	M3.5e
896A	EQ Peg	M3.5e

To facilitate observation, the splitting should be as large as possible, and lines are chosen with large Lande g_{eff} factors (magnetically sensitive levels), and long wavelength. However, infrared ($\lambda \sim 1$–$5\,\mu m$) spectrographs have typically been less sensitive than optical spectrographs (a situation which is gradually changing), so there is a trade-off in choosing the lines to observe.

Additional measurement uncertainty is introduced by the presence of thermal, turbulent, collisional and rotational broadening of the lines, which tend to mask the magnetic Zeeman signature. Careful analysis of the line profiles using model atmospheres and spectral synthesis (see Chapter 4) is required to accurately recover the magnetic-field information from the observed spectra. Methods of analysis fall into two categories: 1) comparison of magnetically sensitive and insensitive lines from the same star; for example, by observing two lines near to one another in wavelength, with very different g_{eff} values [R4]; or 2) comparison of a magnetically sensitive line in two different stars, one active and the other inactive [J2]. The comparison star must be chosen to be a close match in spectral type, evolutionary state and other physical properties to the star of interest, so that differences in their spectra can be attributed to the effects of the magnetic field. Magnetically insensitive lines can be compared between the two stars to test the validity of this assumption. The magnetic field strength B and the filling factor f (percentage of the stellar surface covered by field) are separately determined by assuming that $f\%$ of the star is covered by field with strength B and the other

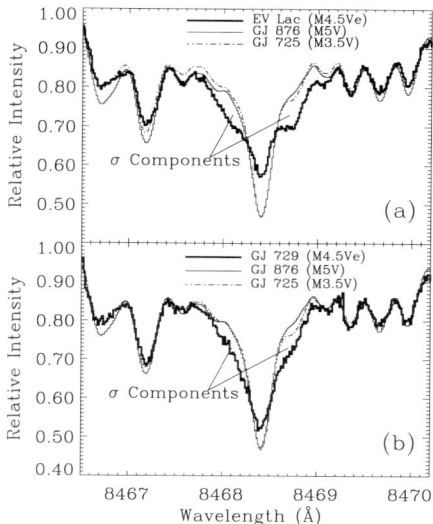

Figure 5.3. The Zeeman-broadened Fe I 8468 Å line observed in the active stars EV Lac and Gl 729. The σ components are clearly visible when the line profile is compared with inactive stars of similar spectral type. (Figure courtesy of C. Johns-Krull and the *Astrophysical Journal.*.)

$(1 - f)\%$ has no field. Synthetic model spectra are fit to the data to determine the two parameters B and f.

Field strengths in excess of ~ 600 G have been measured on a number of solar-type G and K stars, using the Robinson method [M1]. M dwarfs, being much fainter, present a significant observational challenge. At present, magnetic fields have been observed on only a few of the brightest dMe dwarfs (recall from Chapter 2, that dMe dwarfs are those with Hα in emission, indicating the presence of a chromosphere, and hence magnetic activity). Initial efforts by Saar and collaborators in the mid-1980s centred on Fourier transform spectroscopy, using the Robinson method to observe Ti I lines in the infrared [S2], [S3]. These resulted in detections for the dMe dwarf AD Leo and the K5e dwarf EQ Vir.[2] A more recent application using infrared echelle spectroscopy resulted in detections of DT Vir and YZ CMi [S1].

The second method has been employed with success by Johns-Krull, Valenti and collaborators [J2], [V1]. The magnetically sensitive Fe I line at 8468.4 Å gives an observed Zeeman splitting of the σ components of ± 83.7 mÅ kG^{-1}. Figure 5.3 illustrates the pronounced σ components of the Fe I line in the active stars compared to the inactive stars. The high-resolution coudé echelle observations shown in the figure were used to measure magnetic fields on Gl 729 and EV Lac.

[2] Active stars are commonly referred to by their variable-star names. Table 5.1 shows the Gliese numbers and variable-star names for a number of well-known dKe and dMe stars.

Table 5.2. Magnetic field measurements of dMe stars.

Name	Spectral type	B	$f(\%)$
Gl 171.2a	K2e	2,800	60
EQ Vir	K5e	2,500	80
DT Vir	M0.5e	3,000	50
AD Leo	M3e	3,800	73
		4,000	60
Gl 729	M3.5e	2,600	50
EV Lac	M3.5e	3,800	50
YZ CMi	M4.5e	4,200	67

Current results are summarized in Table 5.2, indicating that active M dwarfs generally have field strengths of a few thousand Gauss covering approximately half of the stellar surface. In the near future, new data should allow a comparison of field strength and filling factor with mass, spectral type, and atmospheric diagnostics such as Hα emission and X-ray emission.

The sensitive observations obtained in the late 1990s indicate that one-component models – where a single field strength and covering factor are assumed – do not fit the data particularly well (see Figure 5.4). A range of field strengths and filling factors are required to fit the excessively broad σ components that are observed. Evidently the surface magnetic field has a complicated spatial structure. Another, perhaps concurrent, possibility is that the field strength changes with height over the line formation region.

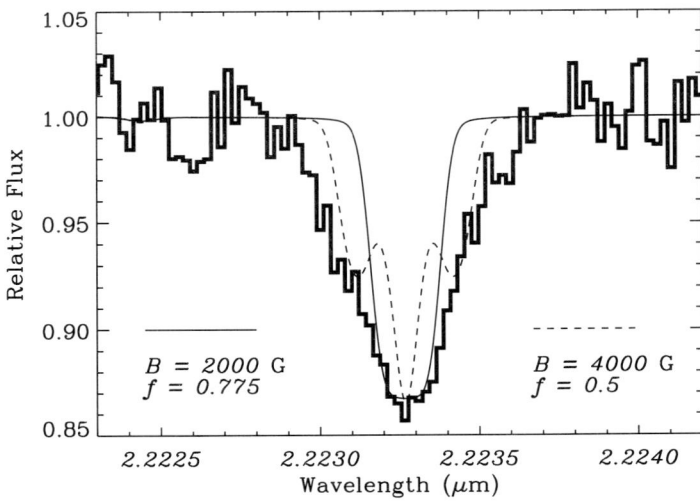

Figure 5.4. A single field strength and filling factor are insufficient to characterise the observed, Zeeman-broadened, line profile in AD Leo. (Figure courtesy of C. Johns-Krull.)

   Additional information on the spatial (surface) structure can be obtained by using the polarisation signature of the Zeeman effect. The π component of the magnetically sensitive line is linearly polarised, while the σ components are circularly polarised; one component is left circularly polarised in regions of positive polarity field while the other is right circularly polarised. In regions of negative polarity field the opposite is true, so if the star is equally covered with regions of positive and negative polarity field (by uniform coverage of spots which represent both polarities), the net circular polarisation signature is zero. Thus, circular polarisation measurements are useful primarily to measure globally organized fields, such as the overall dipole fields seen in Am stars and in individual sunspots. Such measurements for M dwarfs indicate that there is no evidence for an organised global field on Gl 729 at the 1% level on daily or yearly time-scales [V1]. If the field is organised into large active regions, as on the Sun, the circular polarisation signature should vary slightly as these regions rotate in and out of view. The fact that this variation is not observed at such a stringent level argues that the field may be composed of many small active regions, so that statistical fluctuations are not important.

   The time-dependent studies in [V1] show that the fields on Gl 729 and AD Leo do not vary on time-scales of days (similar to the rotation period) or years (similar to a solar cycle period). These magnetic field measurements are in agreement with other chromospheric and coronal indicators which show that solar-like magnetic cycles have not yet been observed on dMe stars. However, the Hα emission varies considerably during the course of the observations. Apparently, the chromospheric heating is not closely coupled to changes in the surface magnetic field – at least on these time-scales.

5.3   THE OUTER ATMOSPHERE

The outer atmosphere of the Sun begins at the temperature minimum region (TMR) and extends outward through the chromosphere, transition region and corona. Figure 5.5 shows the distribution of temperature with height and column mass for the solar VAL3C chromospheric model [V4]. Travelling outwards from the photosphere (height $= 0$ marks $\tau_{std} = 1$ in the continuum, the photospheric 'boundary'), the temperature first decreases through the TMR where the energy balance still reflects radiative equilibrium. At the height where magnetic (non-radiative) heating becomes important in the overall energy balance, the temperature reverses its decline and begins to rise, reaching a plateau at $\sim 7{,}000$ K (the chromosphere) before rising abruptly through the transition region and levelling out at $\sim 10^6$ K in the corona (not shown in the figure). VAL3C is a semi-empirical model, found by postulating an atmospheric structure, computing the emergent line profiles, comparing these with observations, and changing the structure until a good match is obtained between the model profiles and the data. The locations of formation for the principal diagnostic lines in the solar chromosphere – Hα, Ca II K, and Mg II k – are indicated on the figure. Chromospheric models for AD Leo are superposed on the solar model,

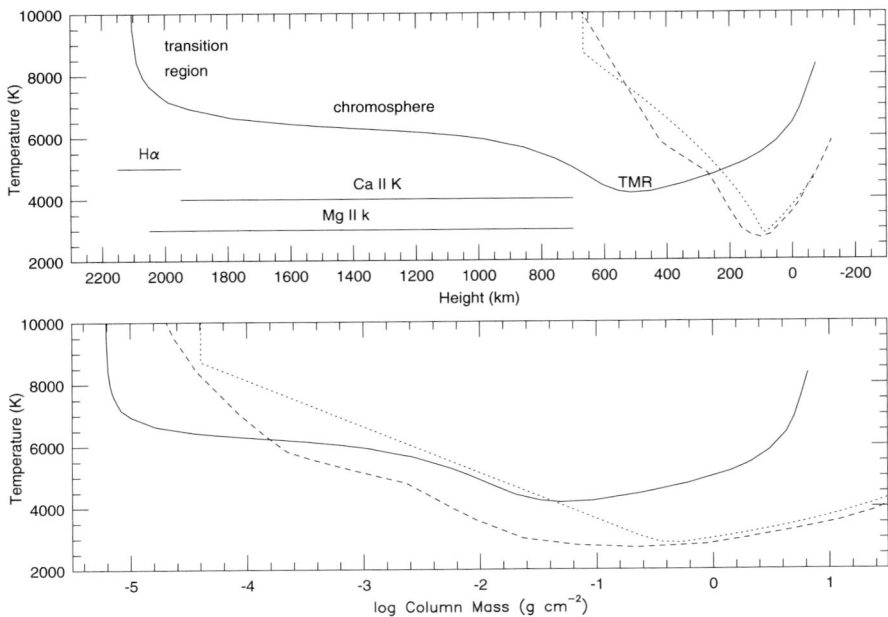

Figure 5.5. The VAL3C semi-empirical model of the solar chromosphere is the solid line. Two chromo-spheric models for AD Leo are shown: a semi-empirical model from Mauas and Falchi [M6] (dashed line), and a schematic model from Hawley and Fisher [H5] (dotted line).

illustrating the compressed nature of the chromosphere in the lower mass, higher gravity M dwarfs.

If the solar atmosphere were to be strictly in radiative equilibrium with the energy diffusing outward from the hot core, there would be no temperature rise outside of the photosphere, and the temperature would slowly approach interstellar values as the atmospheric density and radiative flux decreased. The thermal bifurcation model for the solar atmosphere proposed by Ayres and collaborators asserts that radiative equilibrium does occur in the quiet regions of the photosphere, and that only in the active regions (within magnetic flux tubes) is there a chromospheric temperature rise [A6], [A7]. The solar model shown in Figure 5.5 is then misleading, representing a global average over disparate regions and matching the physical properties of neither. Unfortunately, we are not yet able to resolve surface features on M dwarfs, and must use globally averaged values in our analysis[3]. In the following section, we refer to the outer atmosphere as though it is uniform and ubiquitous over the surface, keeping in mind that lateral inhomogeneity of considerable complexity is almost certainly present.

---

[3] Our inability to resolve surface features also means that the voluminous solar literature on spatially resolved structures such as prominences, spicules and bright points (see [Z2]) is not yet directly applicable to the study of low-mass stars.

5.3.1   The chromosphere

A distinctive feature of the chromosphere is the extended temperature plateau where the density falls by several orders of magnitude while the temperature remains nearly constant. The plateau is the result of a balance between the non-radiative (magnetic) heating, and radiative cooling, principally from collisionally excited emission lines of hydrogen, ionised calcium and ionised magnesium. The reason for the plateau was succinctly described by Ayres [A5], and is summarised here.

The collisional excitation which drives the cooling results primarily from collisions with electrons, and hence depends linearly on the electron density. Models with constant heating per unit mass (the simplest assumption) thus require nearly constant electron density to maintain the balance between heating and cooling that is required to keep the temperature constant. At chromospheric temperatures, the electrons are supplied by hydrogen. Constant electron density is achieved by a nearly perfect match between the increasing partial ionisation of hydrogen (freeing more electrons into the atmosphere) and the decrease in the total number of hydrogen atoms available imposed by the requirement of hydrostatic equilibrium. The chromospheric temperature plateau thus spans the hydrogen partial-ionisation region, which is quite extended (in height) in low-gravity giants, somewhat smaller in solar-type dwarfs, and very compressed in high-gravity M dwarfs. This provides a natural explanation for the observed correlation between the width of the Ca II K line and luminosity class in G and K stars, known as the Wilson–Bappu effect [W2].

The radiation that characterises the solar chromosphere comes from the same lines that provide the cooling – namely, the Lyman and Balmer series lines of hydrogen and the resonance lines of Ca II and Mg II. In low-mass stars the density is higher, the heating and cooling are larger, and the hydrogen lines are particularly strong. The entire Balmer series is usually in emission in dMe stars; the primary source of chromospheric cooling switches from the Ca II and Mg II lines in the Sun to the hydrogen Balmer lines in M dwarfs [L7]. In addition, the Ca II resonance lines near 3900 Å are difficult to observe in these cool stars, as the continuum flux is greatly reduced due to the low effective temperatures. The combination of strong hydrogen lines and ease of detection makes the Hα line at 6563 Å the principal diagnostic of an M dwarf chromosphere, in contrast to the study of chromospheres in solar-type stars where the Ca II K line is the diagnostic of choice.

The current generation of semi-empirical chromospheric models of M dwarfs do not provide a good fit to the data. There is a serious problem in fitting several different lines (Ca II, Mg II and H Balmer and Lyman series) with the same model, which has been attributed to inhomogeneous surface structures on the star. A continuing question is the relative coverage of plage and spot regions. Some recent references are the models shown in Figure 5.5, which fit the hydrogen Balmer lines but predict Ca II as too weak [M6] and too strong [H5] respectively; Giampapa and collaborators [G4], who found that the Mg II emission is too strong when the models predict the correct Ca II flux; and a series of papers by Houdebine and collaborators, showing that a very high coronal pressure (log column mass ~ -2.9 at the beginning of the transition region temperature rise, compared to ~ -4.5 in the models shown in Figure 5.5) is required to match both the hydrogen Balmer and Lyman line fluxes. The largest set of models by the latter authors [H16] had moderate success in reproducing Hα and Ca II K observations in early M dwarfs, but they noted that significant plage areas were necessary to describe the continuum variations. There is much work to be done in refining the models, which may lead to a better understanding of the distribution of surface activity on these stars.

The question of what heats the solar chromosphere is still under considerable debate, and very little work has been carried out on low-mass stars. An overview of the solar discussion is given by Kalkofen [K2].) There is agreement that Alfven waves traveling along the magnetic field lines (in the flux tubes) must play a role, but the details of the models have yet to converge. Acoustic heating (shocks) may play some role on early type stars and the Sun (the so-called basal chromosphere, [S5], [C2]), but is unlikely to be important in the relatively denser and cooler M dwarf atmospheres. The possibility that flaring activity in the corona provides enough downward-directed energy to heat the chromosphere has been explored briefly [C6], although this begs the question of how the flaring is initially produced in the corona.

5.3.2   The transition region

When hydrogen becomes nearly completely ionised, the chromospheric energy balance is disrupted by the sudden lack of collisionally induced cooling, since there is no longer a continuing supply of electrons to drive it. The atmosphere begins to heat up, and soon a situation of thermal runaway occurs. Between about 10^4 K and 10^6 K there are many resonance transitions in abundant elements that contribute strongly to the cooling; in fact, the radiative cooling function from an optically thin, collisionally excited gas peaks at temperatures near 10^5 K (see Figure 5.6; also Figure 5 in [C8]). The heating is not enough to maintain these temperatures, and the temperature must continue to rise (counter-intuitively) until it reaches a million degrees and the cooling is reduced, in the corona. The very thin layer between chromospheric and coronal temperatures is known as the transition region. In this region with an extremely steep temperature gradient, the energy balance is primarily between the heating due to thermal conduction and the radiative cooling:

$$\mathrm{d}F_c/\mathrm{d}z = -n_e n_H P(T) \tag{5.3}$$

where the Spitzer conductivity $F_c = -\kappa_o T^{5/2}\, \mathrm{d}T/\mathrm{d}z$ and $P(T)$ is the radiative loss function shown in Figure 5.6. Fisher and collaborators [F4] described how the temperature structure of the transition region could be obtained by numerically evaluating equation (5.3) for the Sun, and later have applied this formalism to M dwarfs [H5].

The principal emission lines that are produced in this region come from upper ionisation states of elements such as C, N, O, S and Si. Commonly observed lines are in the ultraviolet; for example, C II 1335 Å, He II 1640 Å, C III 1174 Å, C IV 1550 Å, S IV 1392 and 1401 Å, and N V 1240 Å. In the early 1980s, NASA's International Ultraviolet Explorer (IUE) satellite allowed access to the ultraviolet wavelength region for stellar exploration, and produced a wealth of information for solar-type

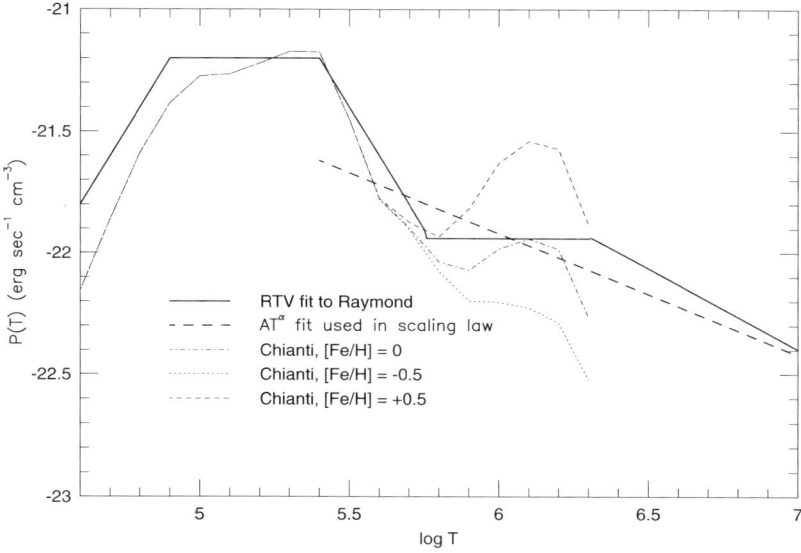

Figure 5.6. Radiative losses are shown as a function of temperature for a hot, optically thin plasma. The RTV analytical piecewise fit to the Raymond and Smith [R2] radiative loss function (solid line) and the AT^α approximation used in the RTV scaling law (long-dashed line) are shown. The other relationships are recent results from the Chianti database [D5], for several metallicities.

stars. Since M dwarfs are intrinsically faint, only the nearest and brightest were detected [L7], and even those were faint except during flares [B9], [H4]. Tight correlations between the fluxes emitted in chromospheric and transition region emission lines are found for solar-type stars [A8], but the M dwarfs differ systematically, showing less Mg II flux at a given transition region flux [O1]. Oranje attributed this behaviour to the shifting of chromospheric emission from Mg II into the H Balmer lines in the lowest-mass stars.

Recently, the NASA Extreme Ultraviolet Explorer (EUVE) satellite provided short wavelength ($70\,\text{Å} < \lambda < 760\,\text{Å}$) data which probe this temperature regime. Observations of F–M dwarfs are summarised by Mathioudakis *et al.* [M5]. They find good correlation between the strength of the EUV emission lines and other activity parameters such as Mg II and X-ray emission, with the M dwarfs again showing departures in Mg II. However, they propose that the Mg II flux is reduced because the chromosphere is 'saturated'. As we shall see, 'saturation' is currently a common theme in interpreting activity observations; sections 5.4.1 and 5.5.3 contain more discussion.

5.3.3   The corona

The corona is the outermost part of the stellar atmosphere, extending as much as a stellar radius above the photosphere, with a typical temperature of a few million degrees. Coronal emission is primarily at soft X-ray wavelengths, from collisionally-excited emission lines of high ionisation states of iron and other heavy elements. Magnetic fields confine the (ionised) coronal gas, as evidenced by solar images showing that the X-ray emitting regions are large loop-like structures whose footpoints are magnetic active regions in the photosphere. Emission from solar coronal plasma is also observed along open magnetic field lines; this directed outflow forms the solar wind.

The energy balance within a coronal loop was independently described in the classic papers of Rosner, Tucker and Vaiana (RTV) [R7], and Craig, McClymont and Underwood [C5]. In the simplest case, the heating is constant per unit mass and cooling is due to conduction and radiation. RTV provided a piecewise analytical fit to the radiative losses as a function of temperature, P(T) (from [R2]). Their fit, together with a power-law approximation in temperature with index α,

$$P(T) = A\,T^{\alpha} \tag{5.4}$$

is shown in Figure 5.6. The power-law approximation is valid from $\log T \sim 5.5$–7. More recent radiative loss calculations using the Chianti database [D5] are also shown, for solar, metal-rich and metal-poor atmospheres. The current solar metallicity relationship is still quite close to the RTV fit.

With these approximations, RTV found a relationship between the loop length L, pressure P, and apex temperature T_A which has no free parameters. The 'coronal loop scaling law' is

$$T_A = C(PL)^{1/3} \tag{5.5}$$

where the constant, C, depends on the values of A and α in the cooling function. With $A = -18.9$ and $\alpha = -1/2$ as shown in Figure 5.6, the value of $C = 1.4 \times 10^3$ given in RTV is obtained. Solar observations indicate that this loop model reproduces the empirical values for L, P and T quite well. Similar loop models have been used to describe the coronae of M dwarfs, particularly during flares [F3], [H5], [H7], [G3], [C9].

Description of the emitting material in the transition region and corona is accomplished using the formalism of the differential emission measure (DEM). The flux emitted in an optically thin emission line is given by $F = G(T)EM$, where the emission measure EM is an integral over the electron density squared:

$$EM = \int n_e^2\,dz \tag{5.6}$$

appropriate for two-body collisional excitation processes. $G(T)$ is a function incorporating the atomic physics, including the elemental abundance, the ionisation equilibrium that determines the abundance of the particular ion, and the collisional excitation rate. The latter two quantities depend sensitively on the temperature. Exact forms for $G(T)$ may be found in [H5]; calculations for a given elemental transition and temperature are straightforward, if tedious. The emission measure is thus an experimentally determined quantity describing how much plasma must be at

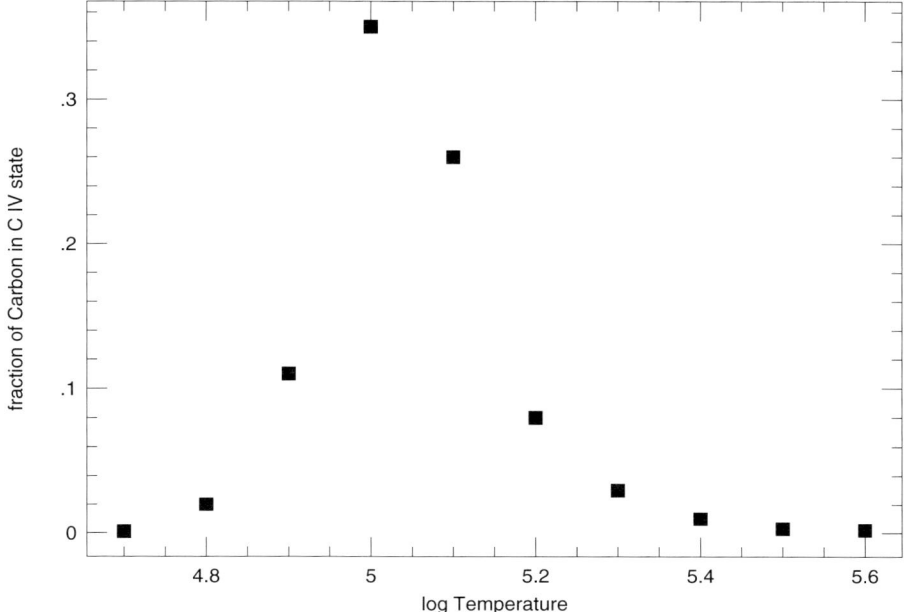

Figure 5.7. The C IV ionisation equilibrium. (From Arnaud and Rothenflug [A4].)

a given temperature to produce the line fluxes that are observed. Emission measures for individual lines are typically shown as distributions versus temperature, with a characteristic U-shape reflecting the inverse dependence on the ion abundance, which is a sharply peaked function of temperature, as shown in Figure 5.7 for C IV (from [A4]). The maximum fraction of carbon that occurs in the C IV state is ~35%, at a temperature of 10^5 K. At lower and higher temperatures, very little carbon is in the C IV state, and hence a very large amount of material would have to be at that temperature to produce the observed line flux. This leads to large inferred emission measures at temperatures far from the peak, and a minimum in the emission measure distribution at the peak formation temperature of the ion.

Observing several ions with different formation temperatures allows the mapping of the amount of emitting material as a function of temperature. Figure 5.8 shows the emission measure distributions for the C II, Si III, Si IV, C IV and N V ions observed with IUE during the strong 1985 flare on AD Leo, and for the C II, C IV and N V observations during a quiescent (non-flaring) time-period on the star. The DEM, $\xi(T)$, is defined from the models as

$$\xi(T) = \frac{n_e^2 T}{|dT/dz|} \tag{5.7}$$

a quantity which has the same units as the emission measure. $\xi(T)$ is a differential quantity, representing the change in the amount of emitting material with the

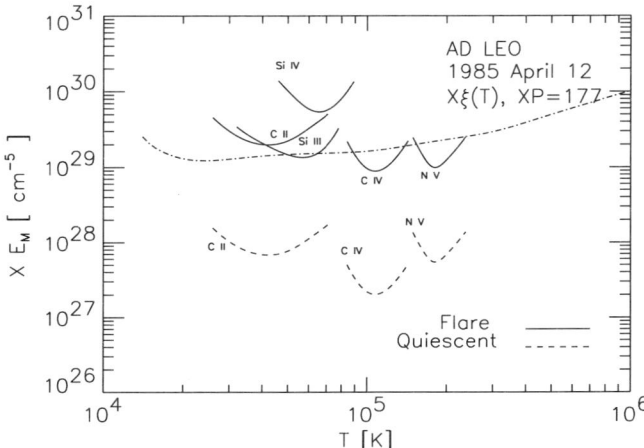

Figure 5.8. Emission measure distributions for several ions observed during a flare on AD Leo, together with a theoretical DEM curve. (Data from [H5], [H6].)

temperature gradient.[4] The DEM may be easily computed if a model structure (temperature and density as a function of column mass) is available for the transition region and corona. In the transition region, the energy equation (5.3), together with the equation of state and a form for the radiative losses (such as in Figure 5.6) produces a structure which leads to the $\xi(T)$ shown in Figure 5.8. The figure demonstrates that this simple energy balance structure does not agree well with the AD Leo flare data. In the corona, coronal loop scaling laws (as in equation (5.5), but see also [F3] for a description of loop scaling laws applicable during flare conditions) determine the structure allowing $\xi(T)$ to be computed [C9].

The Einstein, EXOSAT and ROSAT satellites have provided extensive information on the X-ray emission from low mass stars. Surprisingly (in the 1980s), M dwarfs were found to be relatively bright X-ray sources. Nearly complete surveys out to several parsecs, together with larger surveys at greater distances, have been carried out. Survey results for early and mid-M dwarfs were described in Chapter 2 (see Fig. 2.22); new results for the late M and L dwarfs are given in Section 5.6.

ROSAT observations include minimal spectral information, generally the ratio of emission in a 'hard' band (1–2.4 KeV) to a 'soft' band (0.1–1 KeV). The M dwarf ROSAT data are not well fit by the theoretical emission from coronal loops with a single apex temperature. A common practice is to fit the data with two temperature components, one with high $T_A(> 10^7 \, \text{K})$ which contributes the harder emission, and one with low $T_A(\sim 10^6 \, \text{K})$ which dominates the emission in the softer bandpass. These temperature components have been interpreted as representing two different populations of coronal loops ([G3], but see also [G8] for a cautionary viewpoint).

---

[4] Volume forms of the emission measure and differential emission measure are also in common use, see [C9].

The whole procedure of two-temperature fits is questionable, since very high ISM column densities to the source (indicating significant absorption of the softer emission by interstellar material) and low abundances (thus the emission measures must be higher in order to obtain the observed emission) are also commonly obtained from the fits, and do not agree with other observations [C9]. The difficulty probably lies in over-interpreting the available data; further progress in this field awaits higher spectral resolution measurements which should be provided by the Chandra X-Ray Observatory.

5.4   SPOTS AND FLARES

Obvious manifestations of magnetic activity on the Sun are the dark sunspots (active regions) which move across the solar disk, and the energetic flares which are typically associated with them. In this section we examine the evidence for starspots and stellar flares, and compare their properties with the solar analogues.

5.4.1   Spots and spot cycles

The solar dynamo gives rise to large active regions where many flux tubes are collected together to form pairs of sunspots, with opposite polarity. The evolution in the number and location of sunspots provides strong observational evidence for an 11-year cycle in the dynamo production of magnetic fields. Spots are a common feature in young, active solar-type stars; observations in open clusters show that the G and K dwarfs often exhibit strong *rotational modulation* in their light curves. Photometric rotational modulation is attributed to the presence of starspots rotating in and out of view, and is distinguished by its periodicity (the rotation period of the star) and smoothly varying, often sinusoidal, shape. Figure 5.9 shows two examples of photometric light curves for solar-type dwarfs from a survey of the Pleiades [K4].

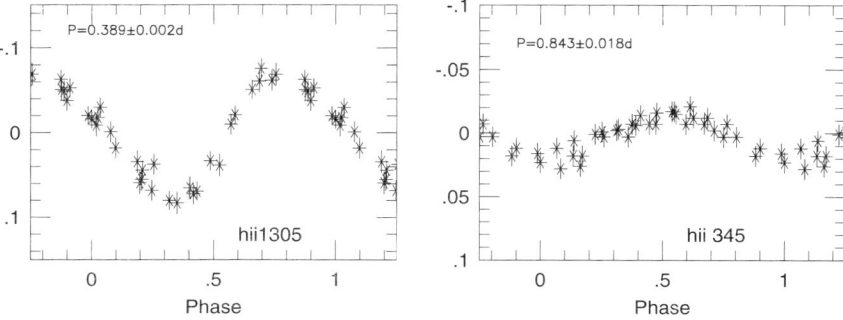

Figure 5.9. Light curves of solar-type Pleiades dwarfs showing photometric modulation typical of starspots rotating across the stellar surface. The derived periods are given on the figures. (Figure courtesy of A. Krishnamurthi [K4].)

The canonical picture says that as a solar-type star ages, its rotation slows due to angular momentum loss, and its magnetic activity therefore weakens because of the dynamo dependence on rotation. The photometric variation then becomes more difficult to observe, though there are examples of cyclic photometric variability in solar-type field stars [L9]. Magnetic activity in older solar-type dwarfs is more easily monitored in the Ca II K line, produced in plage areas surrounding the starspots. An extensive programme to observe Ca II K line variations in solar type dwarfs was begun in 1966 by Olin Wilson at Mount Wilson Observatory, and has continued through the efforts of Vaughan, Baliunas and collaborators, culminating in the large compilation in [B3]. They found that young stars generally had rapid rotation and high levels of chromospheric activity, but rarely exhibited cyclic behaviour. In contrast, older stars had slower rotation, lower levels of activity, and often exhibited smooth cycles similar to the solar cycle. The Sun falls into the latter category. Evidently the dynamo in these solar-type stars begins to produce cyclic behaviour at some stage in the star's evolution depending on its age, rotation, and perhaps other (as yet unknown) factors. They also found a small subset of stars with very low levels of activity and evidently no cyclic behaviour. Figure 5.10 plots three of their light curves, illustrating these phenomena.

Close binary stars can be extremely active, as exemplified by the RS CVn (consisting of ≈ solar-mass subgiant and dwarf stars) and BY Dra systems (consisting of K-type dwarfs). These tidally-locked systems have fast rotation, which promotes shell-dynamo magnetic activity. They exhibit photometric and chromospheric rotational modulation indicative of large starspots. RS CVn systems have been subjected to extensive observation and analysis using Doppler imaging techniques to dissect the line profiles and model the spot parameters and their distribution on the surface [R3]. (These techniques have not yet been applied to the fainter M dwarfs, so are not discussed further here.) There is as yet no conclusive evidence for cyclic behaviour of the magnetic activity in these systems [L2], [P7].

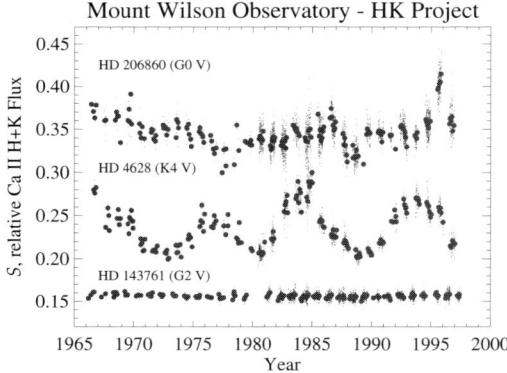

Figure 5.10. Light curves of the Ca II K index in G and K stars, from the Mount Wilson survey. Stars with less activity are more likely to show cycles (compare HD 206860 with HD 4628); a small subset of stars show very low activity and no cycles (HD 143761).. (Figure courtesy of S. Baliunas.)

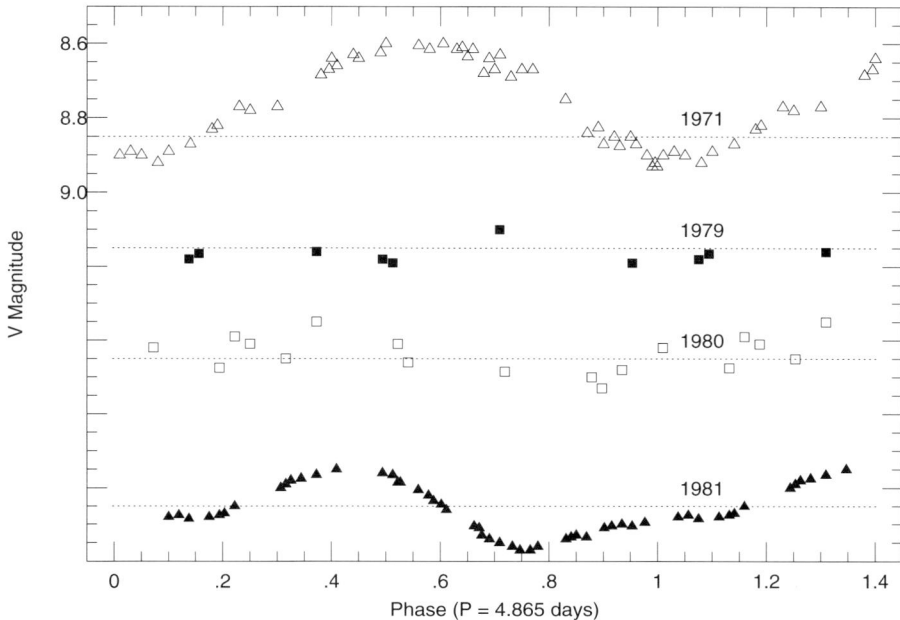

Figure 5.11. Light curves of AU Mic, spanning a period of 10 years. Clear rotational modulation with a period of 4.865 days is seen in 1971 and 1981; the intervening years show variability but no modulation. The data are plotted on the same magnitude scale as indicated in the 1971 light curve. An approximate mean magnitude V = 8.85 is shown by the dotted lines for each year. (Data are from [R5].)

Based on the observations of these other active stars, M dwarfs were expected to exhibit starspots and possibly spot cycles. However, considerable effort in the 1970s and 1980s failed to produce many convincing detections. Although most active stars showed variable light levels, few had the periodic light curves typical of a spot rotating in and out of view. A summary of the sparse observational material is given by Pettersen [P6]. The rapidly rotating, early M dwarf Gliese 890 is the best case, with a well-defined modulation having a period of 0.43 days, in phase with the Hα emission (minimum light occurs at minimum Hα equivalent width) [P5], [Y1]. Spots of moderate amplitude have been observed on AU Mic [T5] and EV Lac [C4]. M dwarfs that show periodic variations in one observing run may appear non-varying in subsequent months or years [R5], [C4]. Figure 5.11 illustrates a time series of light curves for the dMe star AU Mic (note that the data entering this figure are all from unpublished sources!). A clear variation with a period of ∼4.8 days is seen for 1971 and 1981 (although the phase of light maximum appears to have shifted). Observations in 1979 and 1980 show little or no periodic modulation.

M dwarfs that have variable light levels but no sign of periodic rotational variation may have long periods similar to Proxima Centauri, whose ∼ 80 day rotational modulation with an amplitude of ∼0.01 magnitudes would surely not have been discovered were it not the subject of intense scrutiny by the HST. They might also represent cases where spots are appearing and disappearing on short time-scales, or are distributed uniformly over the surface, so that little rotational modulation is produced. The discussion in Section 5.2.1 regarding turbulent dynamos suggests that uniform surface coverage is a feature of that model; the spot observations (or lack of them) on M dwarfs support that prediction. Further support for uniform surface coverage is found in the constant polarisation measurements of the magnetic field (Section 5.2.2). An apparent wish to cling to a solar analogy (a few large, isolated spots rather than many small spots), has led many to invoke 'saturation' to explain the lack of convincing observations of spots. In this context, 'saturation' refers to a surface which is completely covered by large sunspot-like spots, such that no unspotted surface exists [V5]. While the observations do not rule this out, it is also not required, since uniform (and not necessarily complete) coverage of small spots would produce the same unvarying effect on the light curve.

Photometric observations of young open clusters provide additional support for the view that large, isolated starspots which contrast sharply with the surrounding photosphere are not common on M dwarfs, although G and K dwarfs in the same clusters do show evidence of such behaviour. Only two M dwarfs show rotational modulation in IC 2391 [P4], while spots were found only in stars with $M > 0.5\,M_\odot$ in the Pleiades [K4]. A large study of IC 2602 [B4] revealed low-level photometric modulation in most of the confirmed members, including several M dwarfs. Further studies of VLM cluster stars found low-amplitude modulation for one star in α Per [M4] and two stars in the Pleiades [T1]. Apparently spot activity does exist on some M dwarfs, but in most cases at a low level (amplitudes of ~0.02 magnitudes), which presents a challenge in detection.

An absence of rotational modulation is expected if spots are located only near the rotational poles, where they would be visible during the entire rotation period of the star. Both Doppler imaging and photometric studies of RS CVn stars seem to require polar spots ([H3], [B9], but see [B13] for an opposing viewpoint), primarily to explain stationary features in line profiles and changing levels of maximum brightness in long-term photometric light curves. Young et al. [Y1] argued that polar spots were required to model their data for Gliese 890. Theoretical support for polar spots on early-type M dwarfs was found by Buzasi [B12], who performed solar dynamo calculations on $0.4\,M_\odot$ stars. He found that reasonable values of the rotation velocity led to flux tube emergence at much higher latitude in these stars, compared to the Sun (see also [S6] for a discussion of a similar effect in rapidly rotating solar-type stars).

If detection of starspots on M dwarfs is difficult, finding spot cycles poses an even greater challenge. Some tenuous evidence exists for long-term photometric variability on timescales of decades [B7], but there are as yet no convincing cases. The solitary M dwarf in the Mount Wilson Ca II K survey does not exhibit cyclic behaviour. If the dynamo operates similarly in M dwarfs as in the solar-type stars, the dMe stars could be examples of the young, active stars from the Mount Wilson study that showed strong activity but little cyclic behaviour. On the other hand, the lack of cycles could be a natural feature of a turbulent, rather than a shell, dynamo dominating the production of the magnetic field, as suggested by the Durney model [D8].

When a star that previously showed a periodic variation stops varying, it often has a *fainter* mean light level. This has been interpreted as an indication that more of the surface is covered by spots that are darker than the surrounding photosphere – the star has become 'more active'. Alternatively, the star may have been initially covered with bright plage areas during the period of variation, and the underlying photosphere may be revealed during non-varying periods. Pettersen *et al.* [P7], in their 'Zebra effect' paper on BY Dra stars, highlighted the difficulty in distinguishing between a bright star with dark spots and a dark star with bright spots. Support for the latter interpretation may be found in recent analyses of solar luminosity variations over the course of an activity cycle, which show that the Sun is brightest at activity maximum despite having a larger coverage of sunspots [F7], [H17]. Presumably this represents a stronger contribution from the plage areas (facular network) in the active state. However, the Mount Wilson survey [R1] provided evidence that, while relatively inactive solar-type stars (like the Sun) are brighter at the maximum of the Ca II K activity cycle, the opposite is true for active G–K stars – those stars are darker at activity maximum. Hence, the relative importance of bright plage areas compared to dark spot areas appears to vary systematically with the activity level. If late-type dwarfs mimic the behaviour of active solar-type stars, they would be expected to be darker when more active, in contrast to results for BY Dra stars [P7] and Gliese 890 [Y1], although some authors do claim to find a correlation between lower mean light level and increased flare activity [M7], [C4]. More data are required to sort out the dependence of photometric light variations on activity level in M dwarfs, and hence to clarify the interpretation of observations in terms of cyclic behaviour.

5.4.2   Flares

While spots on M dwarfs remain elusive, flares are obvious and easily observed through even a small telescope. They are distinguished by blue and ultraviolet continuum emission which can increase the brightness of the star by several magnitudes in a manner of seconds. Imagine the feeling of the photometric observer, monitoring a count rate of some few hundred counts/sec on an M dwarf flare star, and suddenly finding that the count rate has become a few hundred *thousand* counts/sec! (Experienced observers say that the adrenaline rush accompanying such observations is the only reason to tolerate the other 99% of the time when the star stubbornly refuses to flare.) The flares on M dwarfs can be remarkably energetic in comparison to the bolometric luminosity of the star, and will often completely dominate the blue, ultraviolet and X-ray emission of the quiet atmosphere. Indeed while the flux in the U band filter may increase by a factor of 100–1,000, the ultraviolet and X-ray emission may increase by up to 10^4–10^5 times in brightness for periods of a few minutes or more.

The observations were described briefly in Chapter 2. Flares are identified by the impulsive phase emission in the blue and ultraviolet continuum and by exceedingly

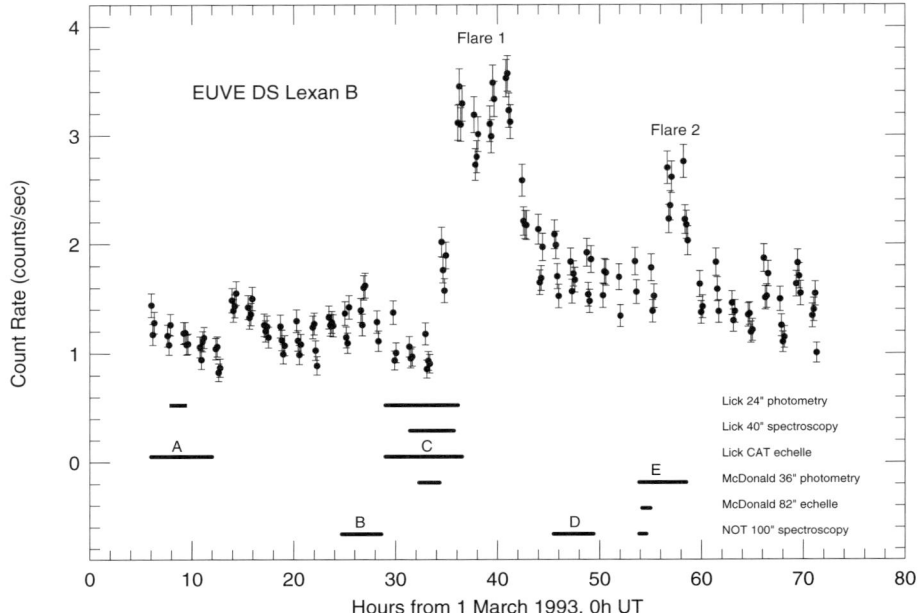

Figure 5.12. Soft X-ray light curve observed with EUVE during a flare on AD Leo. Several ground-based observatories were obtaining photometric and spectroscopic observations at the times indicated. (Data from [H7].)

strong emission lines in the optical and ultraviolet (see Figure 2.23(b)). The continuum emission is typically that of a black-body with a temperature of $\sim 9{,}000{-}10{,}000$ K [M8], [K1], [H5]. The optical emission lines consist of the same lines seen in the quiescent (dMe) spectrum – the hydrogen Balmer lines and Ca II H and K – with the addition of many short-lived species with higher excitation, such as He I and He II, the Ca II infrared triplet, and numerous singly and doubly ionised metals, all indicating a significant increase in the amount of emitting material at chromospheric, or slightly higher, temperatures. Ultraviolet emission lines such as C IV and N V become much stronger during flares, which also indicates an increase in the amount of material at transition region temperatures. The soft X-ray emission rises more slowly than the optical continuum and emission lines, peaks during the gradual phase when the optical emission has already begun to decay, and has longer rise and decay times than the impulsive phase emission. Figure 5.12 shows the soft X-ray light curve during two flares on AD Leo as observed with the NASA EUVE satellite. Optical (U-band) light curves of the impulsive and extended gradual phases of Flare 1 are illustrated together with the EUVE data and the line flux in the Balmer Hγ line in Figure 5.13. Clearly the continuum brightening and optical emission lines precede the soft X-ray emission in this flare.

Impulsive phase time-scales are generally seconds to a few minutes, while the gradual phase may last from minutes to hours. More energetic flares typically last longer in both phases. It is noteworthy that the Ca II H and K lines generally act

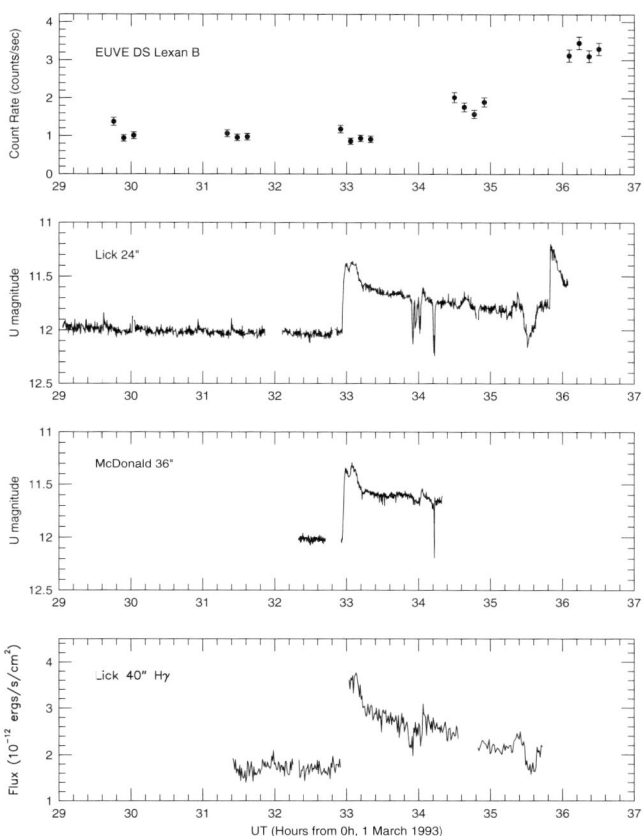

Figure 5.13. The photometric light curves in the U-band filter from two observatories, together with the soft X-ray and Hγ line flux data. The U-band continuum and the Hγ line flux have a similar response to the impulsive flare heating, while the X-ray emission is delayed. These observations support the chromospheric evaporation model described in the text. (Data from [H4].)

more like a gradual phase indicator than the hydrogen Balmer lines (see Figure 2.23(a)), and have been used as a proxy for the soft X-ray emission. The difficulty in modelling the hydrogen Balmer lines and Ca II lines simultaneously has already been described, and the same is true in flare models; their differing evolution during the flare may be a clue to the suspected lateral inhomogeneity in their formation on the stellar surface.

There are correlations between the total energy (summed over the flare) released in the emission lines, the continuum and the X-rays, such that $E_X \sim E_U$, $E_{H\gamma} \sim E_{CaIIK}$, and $E_{H\gamma} \sim 0.05 E_U$ [B11], [H4]. Figure 5.14 illustrates these relations, which exist for flares on stars with a wide range of spectral types, and

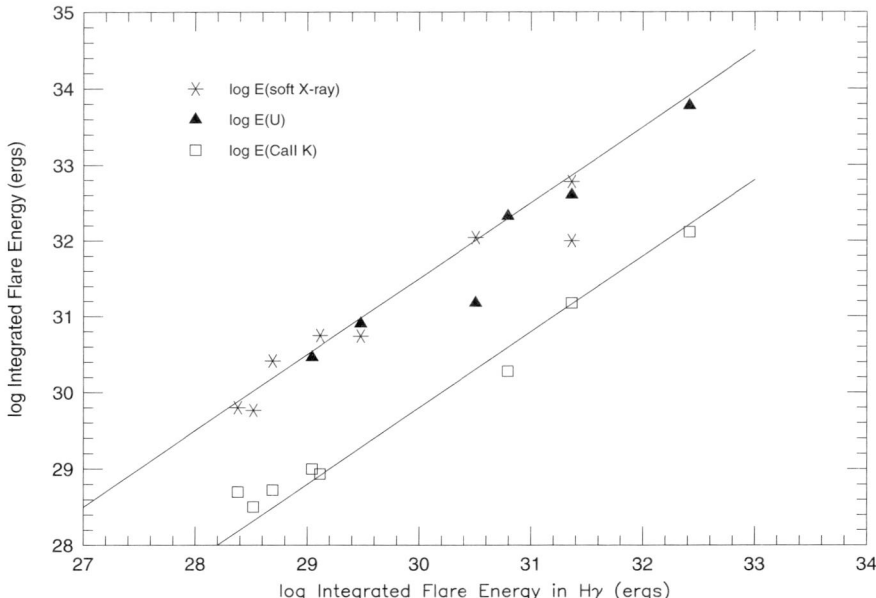

Figure 5.14. Correlations between the energy emitted in chromospheric emission lines (Ca II, Hγ), the white-light continuum (U-band) and the thermal corona (soft X-ray emission) are characteristic of both solar and dMe flares. (Data from [H4].)

over a broad range of total flare energy ($E_U \sim 10^{26}$–10^{34} ergs), suggesting (from an empirical standpoint) that the heating rates and mechanisms for producing the flares may be similar. Area coverage may then be the principal discriminator in determining the total energy produced by a given flare.

Note that the blue continuum, $H\gamma$ (and higher-order H Balmer lines), and Ca II K line are typically observed during M dwarf flares, since these are the most dramatically affected. $H\alpha$ often shows relatively little change during flares on early to mid-M dwarfs (although it can vary tremendously in late M dwarf flares, see Section 5.6.1). In contrast, $H\alpha$ is the principal diagnostic in solar flares, and the blue continuum radiation is rarely strong enough to be observed on the Sun, which has a much higher photospheric background. Comparison between solar and stellar flares has been hampered by the lack of similar observational material. This is not necessarily the fault of the observers; even if the underlying mechanism causing the flares is the same, the M dwarf atmosphere responds differently to the influx of energy, making it necessary to observe the flare radiation where it chooses to appear, rather than in a more convenient, solar-type diagnostic. A step toward obtaining comparable observations has been made by Johns-Krull et al. [J3], who observed a solar flare using an innovative experimental set-up allowing solar observations through a stellar spectrograph. They found differences in the hydrogen Balmer line formation during solar flares compared to the stellar counterparts. More observations are needed to understand the extent to which stellar flares are analogues of solar flares, as is generally assumed.

For the Sun, an enormous body of work has led to a generally accepted model for the production of solar flares, as summarised in [L6], [D3]. The flare energy is released into the atmosphere following a catastrophic event between magnetic flux tubes (or coronal loops, as they are called when they have reached coronal height). Magnetic reconnection occurs, which produces a current sheet, and electrons are rapidly accelerated along the sheet (essentially down the magnetic field lines) towards the lower atmosphere. The electrons collide with ions in the ambient plasma and emit hard X-rays via non-thermal bremsstrahlung, from which a power-law distribution can be inferred for the electron energy spectrum. These hard X-rays have been directly observed on the Sun with satellites such as ASCA, SOHO, and Yohkoh, but are not yet observable from stars, as the hard X-ray flux is many orders of magnitude smaller than the chromospheric and thermal (soft X-ray) emission observed later in the flare. White light (broadband optical) continuum emission is sometimes observed from small kernels at the footpoints of solar flare loops, and is temporally and spatially correlated with the hard X-ray emission [N1], implying that it is emitted at or near the sites where the accelerated electrons impact the pre-flare chromosphere and, in some cases, penetrate into the photosphere. The white-light emission typically emanates from hydrogen recombination radiation, showing a strong Balmer jump [F1]. The spiralling motion of the energetic electrons around the magnetic field lines results in gyrosynchrotron emission at microwave frequencies, which is also observed directly in solar flares (and a few stellar flares [J1], [A1]). The non-thermal electrons lose most of their energy in the lower atmosphere, resulting in increased chromospheric emission and the heating of chromospheric plasma to coronal temperatures. The term *evaporation* is used to describe the process of heating the cool chromospheric plasma, and the subsequent flow of hot plasma up into the corona. The evaporation is accompanied by a downward moving shock front which forms a *chromospheric condensation* – a relatively cool and quite dense region that is propelled toward the photosphere, and which is the site of much of the chromospheric emission during the flare [F2]. As might be expected, the emission is strongly Doppler-shifted; redshifted Hα emission is observed from the condensation, and blueshifted emission in transition region lines marks the evaporating plasma. Downflow (condensation) velocities of tens of km s^{-1} and upflow (evaporation) velocities of a few hundred km s^{-1} are typically observed [I1], [Z1].

As the corona increases in density (and hence column depth) due to chromospheric evaporation, it becomes effective at stopping energetic electrons, which now heat the corona directly while the amount of heat deposited in the lower atmosphere diminishes. During this time the corona is dense and hot, and is emitting copious thermal soft X-rays, which in turn provide a secondary source of heating for the chromosphere. When the coronal heating ceases, the material in the loop gradually condenses back to its pre-flare configuration, and the soft X-ray flare emission dies away.

Empirical support for this model is found through observations of the 'Neupert effect' [N3]. Chromospheric evaporation, which is responsible for increasing the

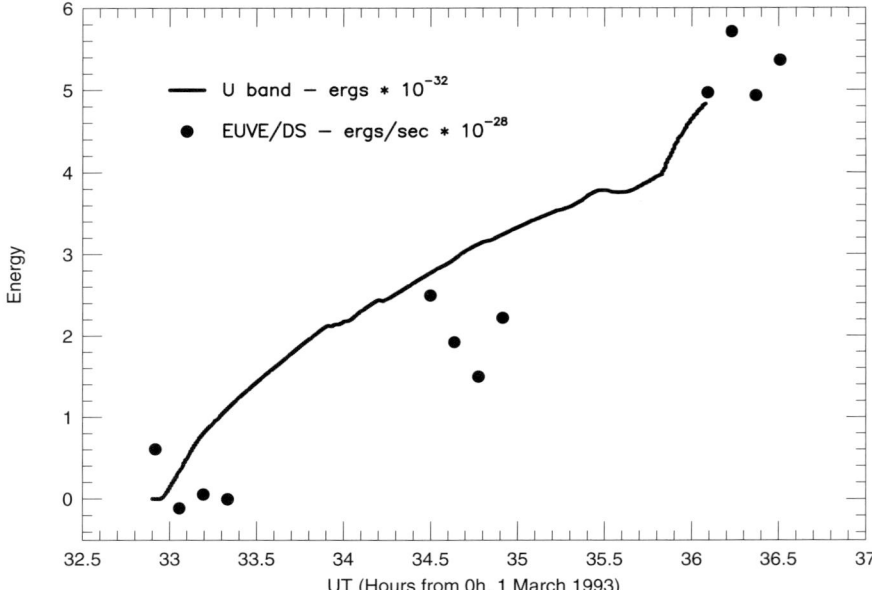

Figure 5.15. The first example of the Neupert effect measured in a stellar flare. The time integral of the U-band emission (solid line) is proportional to the instantaneous thermal soft X-ray emission observed with EUVE (filled circles) for the 1993 flare on AD Leo. The U-band (optical white light) emission is used as a proxy for the hard X-ray flux which measures the flare heating rate, while the thermal emission measures the current amount of material that has been evaporated into the corona by the flare heating. (Data from [H7].)

coronal emission measure and hence the soft X-ray flux, is proportional to the flare heating rate, measured by the hard X-ray flux. The instantaneous emission measure represents a time integral of the evaporation (heating) process, implying that the soft X-ray flux should be a time integral of the hard X-ray flux. Dennis and Zarro [D4] found that more than 80% of large solar flares exhibit the Neupert effect. For stellar flares, white-light continuum emission [H7] or radio gyrosynchrotron emission [G10] have been used as a proxy for the hard X-rays to estimate the flare heating rate. Both studies found evidence for the Neupert effect in M dwarfs, indicating that the general features of the solar model may be applicable on low-mass stars. Figure 5.15 illustrates the first observation of the Neupert effect during an M dwarf flare by Hawley *et al.* [H7].

The exact details of the flare emission are clearly different for M dwarfs; the white-light continuum is much stronger, and does not often show evidence for hydrogen recombination radiation as the production mechanism. Figure 2.23(b) is a spectrum from the impulsive phase of a large flare on AD Leo, with no sign of a Balmer jump [H4]. Flare models have also been largely unable to produce the white-light continuum with any realistic heating mechanisms [H5], although Houdebine [H12] showed that semi-empirical flare models with very large coronal and transition region pressure could reproduce some of the continuum properties. Models incorporating the dynamics in the chromosphere and the formation of the chromospheric condensation, together with realistic treatment of the heating and radiation, are only now being developed for the Sun [A2], and have not yet been applied to M dwarfs. It is ironic that the production of the white light during M dwarf flares remains a mystery, although it is the most obvious and spectacular manifestation of these events.

The line profiles observed during flares also remain unexplained. As shown in Figure 2.23(b), extensive broadening – up to 30 Å at the base of the H Balmer lines – is a common feature during the impulsive phase of flares, but the mechanism has been variously attributed to the Stark effect, turbulence, and flows. Models incorporating the Stark effect and turbulence have not been successful at reproducing the observed line broadening [H5], [H12]. Since the lines are often symmetric, flows are not a promising candidate mechanism, as they would require equal amounts of material emitting at the same temperature to be flowing in both directions. When asymmetries are observed [G11], they offer a probe of the mass motions that are occurring. High time and spectral resolution observations of line profiles are required for further progress in this area.

A final, elusive feature of stellar flares is the presence of diminutions in the continuum radiation just before an outburst [C7], [G1], [D6], [H7]. An example is shown in Figure 5.13 at time 35.5 hours, just before the second impulsive peak, and was seen in both photometric and spectroscopic observations. Speculation about the production of these diminutions centres on enhanced opacity mechanisms [G9], or dark absorbing material from a prominence passing in front of the emitting regions. A theoretical prediction of dimming at the beginning of solar flares has been made by [A2], but the mechanism relies on the emission being due to hydrogen recombination radiation, which may not be applicable for M dwarf flares.

While the radiated flare energy has been relatively well-observed and catalogued for many flares, little work has been carried out on determining the mechanical energy dissipated through flows or coronal mass ejections in M dwarfs. The kinetic energy and mass momentum have been computed for a flare on AD Leo [H14], and Cully et al. [C8] proposed a model incorporating solar-like coronal mass ejections to explain the EUVE observations of a flare observed on AU Mic. The mass loss that M dwarfs will undergo through flaring has been estimated at a few times $10^{-13} M_\odot \, yr^{-1}$ [C3], which, when considering the large number of M dwarfs in the Galaxy, could contribute significantly to the mass and energy balance in the interstellar medium. Recently, Mullan et al. [M10] found that a surprisingly large value of the mass-loss rate, a few times $10^{-10} M_\odot \, yr^{-1}$, was permitted by their radio observations, which were, however, at the detection limit of their instrumentation. This value includes both a steady wind and flare mass ejections. In response, Lim and White [L5] pointed out that such a strong wind would completely absorb the radio emission observed during some stellar flares, and found that mass-loss rates must be less than $\sim 10^{-13} M_\odot \, yr^{-1}$. van den Oord and Doyle [V2] subsequently provided a theoretical basis for calculating properties of stellar winds in cool stars, and found that the maximum mass-loss rates were on the order of $10^{-12} M_\odot \, yr^{-1}$. Radio and submillimetre observations of dMe stars will in future be important for obtaining direct measurements of the mass-loss rate. If mass loss from M dwarfs does play an important role in the mass balance of the ISM, there may be implications for the measured interstellar deuterium abundance as well as other light elements produced via spallation reactions in the corona during flare events.

Flare data for low-mass members of open clusters are sparse, consisting mostly of reports of detections, but with little in the way of reliable statistics. IAU Symposium No. 137 on *Flare Stars in Stellar Clusters, Associations and the Solar Vicinity*, was held in Armenia in 1989; the proceedings from that meeting together with the review by Pettersen [P8] comprise most of the older work. Newer work is mostly contained in IBVS bulletins (see also [H2], [J4]). Studies to determine the number of flare stars in clusters, and the change in flaring rate with age, would be valuable.

5.5   GLOBAL PROPERTIES OF CHROMOSPHERIC ACTIVITY IN M DWARFS

5.5.1   Activity effects on colours, magnitudes and bandstrengths

Magnetic activity, and the accompanying outer atmospheric heating, produce subtle effects on the colours, magnitudes and bandstrengths of active stars. Evidence of these effects has been known for many years: dMe stars are brighter (in absolute magnitude) than dM stars of the same colour [V3]; the V–K colours of active stars are redder than inactive stars of the same spectral type [M11]; and excess ultraviolet and blue emission is often observed in active stars [H15]. However, it was only with the availability of large, statistically complete samples – notably the PMSU spectroscopic survey of the nearby stars [H8] – that the extent of these effects became apparent.

First, field dMe stars lie systematically above and to the red of the main sequence defined by the inactive (dM) stars, particularly at early spectral type (M0–M3). The same effect is present among active cluster stars. Figure 5.16 shows data for the 8-parsec sample, other field stars with good colours and *Hipparcos* parallaxes, and the nearby clusters NGC 2516, the Hyades and IC 2602. The active stars are clearly offset from the dM main sequence. The data are not sufficient to determine whether the active stars are redder at the same luminosity, or brighter at the same colour, than the inactive stars.

Second, individual molecular bandstrengths are affected: the various sub-bands of the $\gamma(0,0)$ TiO bandhead near 7000–7100 Å change their behaviour, depending on the activity level of the star. Figure 5.17 shows the sub-bands plotted against one another for the active stars, with the solid lines again representing the mean relations for the inactive stars (see [H8] for a description of the TiO indices). The dMe stars lie below the mean dM relation in TiO2, while the opposite is true in TiO4 – in other words, the $\lambda7050$ Å subband is stronger and the $\lambda7120$ Å subband is weaker in active stars than in inactive stars. The overall bandstrength, TiO5, is only slightly affected by the activity and therefore still serves as an indicator of spectral type. Plotting

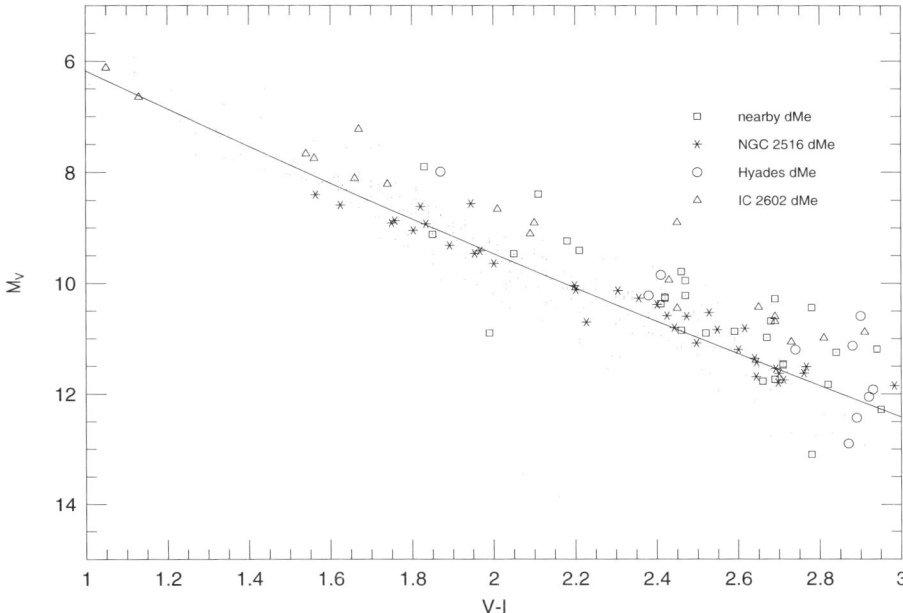

Figure 5.16. Active stars in the field and in clusters lie systematically above and to the red of the main sequence defined by the inactive stars. Inactive field M dwarfs are shown as the small dots, with the solid line being a fit.

TiO2 against TiO4 shows the effect most strongly, and the extent of the offset is positively correlated with the activity strength measured by $L_{H\alpha}/L_{bol}$ [H8].

The bandhead near 6230 Å, comprised primarily of TiO in M dwarfs earlier than M4, is also affected, appearing shallower in active stars at the same V–I colour as inactive ones [H9]. When the λ6230 Å bandstrength is substituted for V–I in Figure 5.16, the luminosity difference between the dM and dMe stars disappears! Evidently the colour is most sensitive to the presence of activity. Detailed models of M dwarf atmospheres (including the chromosphere) are required to provide insight into the physical processes that lead to these activity-related effects.

5.5.2   Empirical relationships between age and activity

Observations of solar-type dwarfs in open clusters show that these stars exhibit a well-defined age–activity relationship. The emission in the Ca II K line is used to represent the activity, and this emission decays with age, such that a young solar-type star has a large Ca II K line flux, while an older star (like the Sun) has a rather small line flux. Skumanich [S8] was the first to quantify the relation as $F(Ca) = At^{-1/2}$, the now famous '*t* to the half' law; Soderblom *et al.* [S9] have since shown that the relation is probably more complicated than a simple power-law fit of emission flux to age.

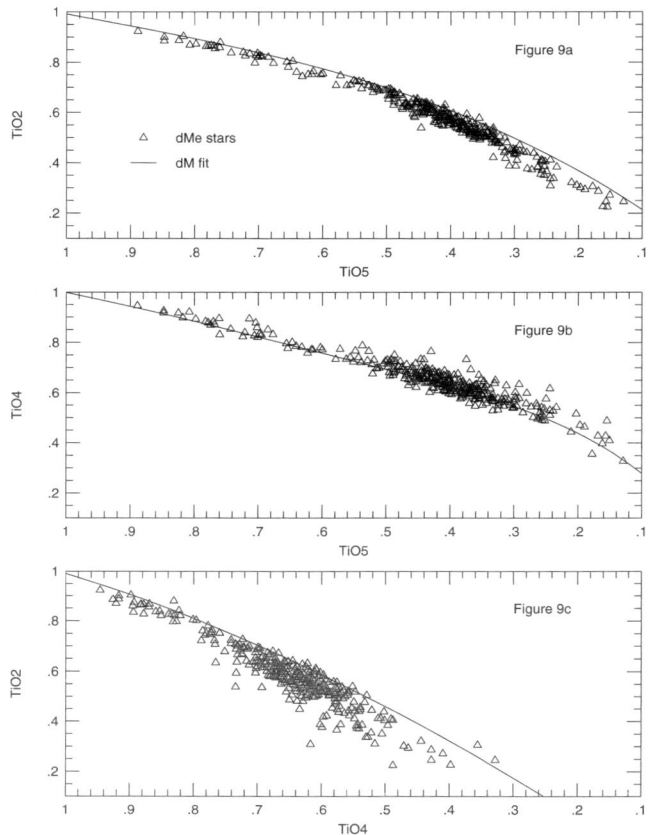

Figure 5.17. Activity effects on the TiO bandstrengths. (Data from [H8].)

Observations of M dwarfs, however, do not obviously fit a solar-type relation (see [E1] for an opposing viewpoint). Activity in these stars appears to be a function of both age and mass, with more prolonged lifetimes in lower-mass stars. This is demonstrated both by variations in the mean kinematics – which, as described in Chapter 6, can be used as statistical age estimators – and through observations of stars in clusters.

Wielen [W1] originally demonstrated that dMe stars as a group had smaller velocity dispersions and a smaller asymmetric drift than the dM stars, both indicative of a younger population. More detailed recent analysis of the PMSU survey shows that the dMe dwarfs have a smaller vertex deviation ('V' velocity) and smaller velocity dispersions than the complete dM sample [H8]. The principal axes also are not diagonal in Galactic co-ordinates – an indication of a sample that has not had time to relax. The dM sample appears to have slightly higher velocity dispersion than the dM + dMe sample (as it should if the dMe stars are younger and hence contribute lower velocity dispersion in the cumulative sample). When the dM stars are subdivided into early- and late-type subsamples (at TiO5 = 0.5 or $M_V \sim 11$, spectral type $\sim$M3), the later-type stars have somewhat higher velocity dispersions than the earlier stars, which is attributed to an age effect: if M dwarfs of late spectral type retain their activity properties for longer than those of early spectral type, then the removal of the dMe stars from the early sample will leave some relatively young stars, while the removal of dMe stars from the late sample will leave only quite old stars. These kinematic results therefore provide a clue that the later-type M dwarfs retain their activity for a longer time than the early-type M dwarfs. Such an effect provides a natural explanation for the larger percentage of active stars seen at late spectral types, described in Chapter 2.

Observations of open clusters provide additional information on age effects, since the age of the cluster is independently determined; for example, from main-sequence fitting (Chapter 6). M dwarfs in the Pleiades and Hyades have received the most extensive scrutiny. Stauffer and collaborators [S11] showed that cluster members exhibit strong Hα emission at late spectral types in both clusters, and that the difference between them is the earliest spectral type where emission is detected. In the Pleiades the emission is prevalent among the late K stars, while in the Hyades it does not appear until the early M stars. However, the mean activity strength, measured by $L_{H\alpha}/L_{bol}$, lies at the same level in both clusters. Several clusters – including the important older cluster M67 [H10] – have now been surveyed, and all show the same effect: the spectral type (mass, colour, absolute magnitude) where the activity 'turns on' is what changes with age, rather than the strength of the emission at a given spectral type. This result is in accord with the kinematic result above; it explains the prevalence of dMe stars at late spectral types in the field (most of them have not yet turned off), and also allows the maintenance of nearly constant activity strength at all spectral types, as shown in Figure 2.21.

The 'Hα limit mass' where the emission turns on is illustrated in Figure 5.18 for the $\sim$150-Myr-old cluster NGC 2516. Using similar data for several other clusters, Hawley *et al.* [H10] constructed the age–activity relationship shown in Figure 5.19. Approximately linear relationships in log age provide an adequate fit to the available data. The age–activity relationship for M dwarfs does not indicate decay of chromo-spheric emission with age, as in the solar-type stars, but instead shows that activity depends on the mass of the star at different ages. This empirical relationship between age and activity provides an important (although as yet unexplained) clue concerning the physical processes governing the production of the magnetic fields underlying the chromospheric heating, and perhaps the heating mechanisms themselves.

5.5.3   The connection between rotation and activity

The shell dynamo mechanism which produces the magnetic field in the Sun and solar-type stars depends critically on the stellar rotation. Empirically, studies of

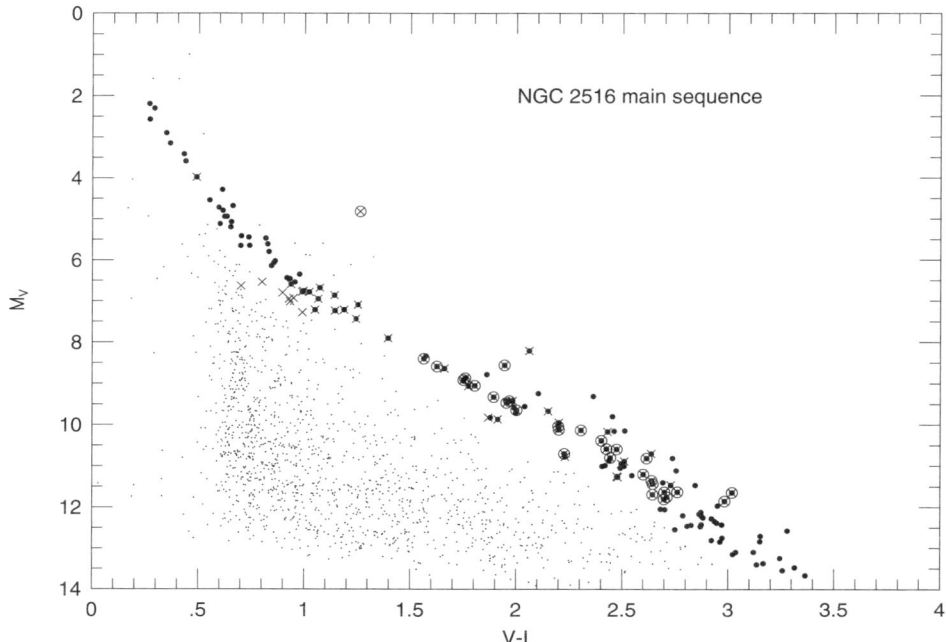

Figure 5.18. Colour-magnitude diagram for the open cluster NGC 2516, illustrating the concept of the Hα-limit mass. Candidate cluster members are the enlarged dots, stars with spectroscopic observations are marked with crosses, and stars that showed Hα emission are circled. The activity turns on at (V–I) ~ 1.5 in this cluster. The object at V–I ~ 1, M_V ~ 5 is a foreground dMe star, not a cluster member. (Data from [H9].)

solar-type stars show good correlation between rotation velocity and activity indicators such as Ca II K emission, and an even stronger correlation between activity and the Rossby number, the ratio of the rotation period to the convective turnover time [N4]. The dynamo number N_D, which characterises the strength of an $\alpha\Omega$ dynamo, is proportional to the inverse square of the Rossby number, thus providing a physical explanation for the rotation–activity correlation.

Rotation velocities may be measured by modelling the effect of rotation on the spectral line profiles. All other line-broadening mechanisms – such as pressure broadening, turbulence and magnetic field effects – must also be modelled, and allowance must be made for the limb-darkening [G7]. The result of the analysis is the line-of-sight component of the rotation, $v \sin i$, where i is the inclination of the rotational polar axis. An alternative method is to measure the photometric rotation period; for example, by observing rotational modulation due to starspots (see Section 5.4.1). The period, together with an estimate of the stellar radius, allows a determination of the rotation velocity at the spot latitude (equatorial spots are generally assumed). If both spectroscopic and photometric rotation estimates are available, the inclination can be determined.

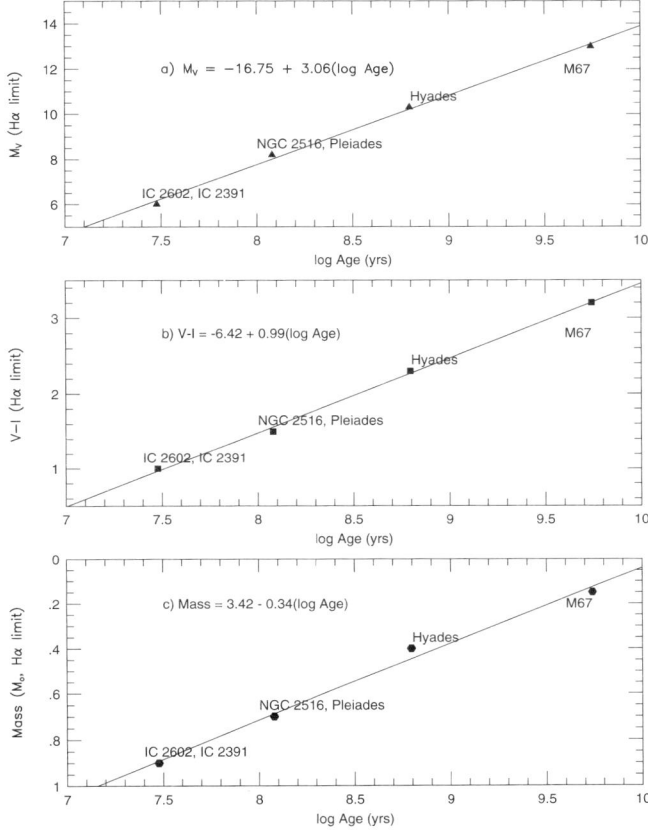

Figure 5.19. Age–activity relationships for M dwarfs using Hα limit measurements in clusters. (a) M_V against log age; (b) V–I against log age; (c) mass ($M_\odot$) against log age. (Data from [H10].)

Early efforts to measure the spectroscopic rotation of low-mass stars were hampered by instrumental limitations. The stars were too faint to observe at high resolution in order to measure $v \sin i$ from the line-broadening; in addition, there was a growing realisation that rotation velocities in M dwarfs were nearly always very low compared to earlier-type stars. A notable exception is the dM0e star Gliese 890 [Y1], [P5], with $v \sin i = 70\,\mathrm{km\,s^{-1}}$ and a measured rotation period of just 0.43 days, but it is the exception rather than the rule; the surveys by [S10] and [M2] found few M dwarfs above their detection limits of $10\,\mathrm{km\,s^{-1}}$ and $3\,\mathrm{km\,s^{-1}}$ respectively.

Recently, a large survey has been carried out by Delfosse and collaborators [D2], who detected rotation above $2\,\mathrm{km\,s^{-1}}$ in 24 out of 99 nearby stars. They found that the later-type M dwarfs (the study included stars as late as spectral type M6.5) were

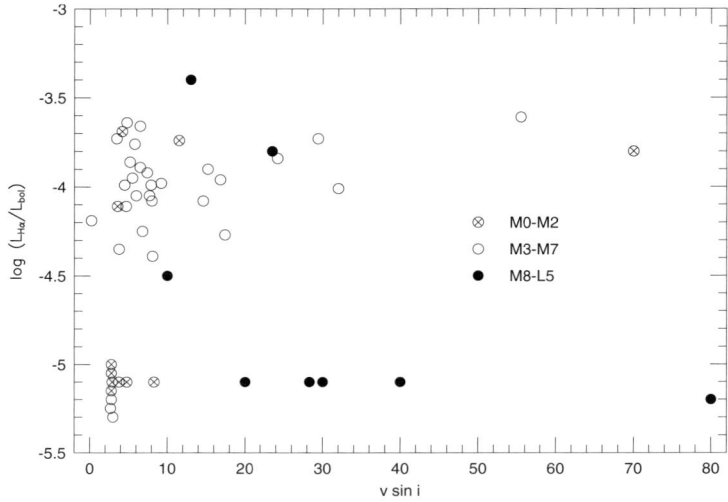

Figure 5.20. The activity strength is nearly uncorrelated with rotation, showing evidence only for a possible threshold value below $2\,\mathrm{km\,s^{-1}}$. Very low or non-detections are shown at $\log L_{H\alpha}/L_{bol} \leqslant -5$; the early-type stars are arbitrarily offset to show them individually at velocities near $2\,\mathrm{km\,s^{-1}}$.

more likely to have measureable rotation, and that the early-type stars with measureable rotation were nearly all members of a young kinematic population, as determined by their space velocities. A picture thus emerges in which rotation lasts longer in lower-mass stars, such that in a fairly old field sample only a few early-type stars have rotation (the young ones), while most of the later-type stars have rotation (because it lasts longer in those stars). These are characteristics very much like those described for the age–activity relationship above, and would seem, at face value, to indicate that the longer time-scale for the rotation velocity to decay in later-type stars is responsible for the continued activity to late types in those stars. However, Figure 5.20 plots activity strength ($\log L_{H\alpha}/L_{bol}$) against rotation velocity, showing that while it is true that the later-type stars with measureable rotation velocity do have activity, there is no strong correlation between the magnitude of the velocity and the activity strength; stars can be equally active with velocities ranging from $0.2\,\mathrm{km\,s^{-1}}$ to $>20\,\mathrm{km\,s^{-1}}$.[5] The data indicate a rather low threshold for rotation velocity necessary to maintain magnetic activity in M dwarfs ($<2\,\mathrm{km\,s^{-1}}$). The threshold effect has also been termed 'saturation', but note that the mean activity level shows large scatter, and by no means all stars achieve the highest observed activity strengths illustrated in Figure 2.21. The latest M and L dwarfs show very little activity, except for the interesting object PC0025+0447 (at $\log L_{H\alpha}/L_{bol} = -3.4$), and the two active late M dwarfs LHS 2397a and LHS 2924. The late-type objects at high velocity are nearly all brown dwarfs, as evidenced by the presence of lithium in their spectra.

[5] It should be noted that these are $v \sin i$ measurements for the most part; we assume that inclination effects are statistically insignificant when using a large, unbiased sample.

Bopp and Fekel [B8] originally noted that active close binaries (BY Dra variables) with synchronous rotation attained a rotational velocity above a threshold value of $\sim 5\,\mathrm{km\,s^{-1}}$, and postulated that active single stars rotated at least this fast. Several recent investigations also show that binarity can enhance activity in late-type subdwarfs [G5], [P3]. Rotation therefore plays some role in the production of magnetic fields and magnetic activity – at least in early M dwarfs.

The lowest mass dwarfs, with spectral types later than about M7, begin to deviate significantly from even the 'threshold' rotation–activity relationship in the earlier M dwarfs. The first VLM dwarf to be found with very rapid rotation was LP 655-4 (BRI 0021-0214). [B5], using the HIRES echelle spectrograph on the Keck telescope, found $v \sin i = 40\,\mathrm{km\,s^{-1}}$, but no measureable H$\alpha$ emission (Hα was later measured at a low level by [T2]). Subsequently, several more VLM objects with similarly high rotation and little or no activity have been found. These stars are discussed further in the next section.

5.6   ACTIVE PROPERTIES OF THE LOWEST-MASS STARS AND BROWN DWARFS

The extreme faintness of the very low-mass (VLM) stars and brown dwarfs, coupled with the small size of the available samples, has hampered progress in the understanding of their magnetic activity. Even the choice of magnetic diagnostic may change – as between the G–K dwarfs and the early M dwarfs, where the chromospheric properties switch from favouring Ca II K to favouring Hα. A few very young ($\sim 1\,\mathrm{Myr}$) objects have been observed in X-rays, but most of the well-known low-mass dwarfs were not found in a systematic search of the ROSAT database [N2]. The expected X-ray emission is near or below the detection limits in most cases, and the next generation of X-ray satellites will be required to investigate the coronal properties of these stars. A search for non-thermal radio emission also produced non-detections for the brightest and most active nearby objects [K5]. The presently available data thus consist of a limited number of Hα observations, and these are discussed together with evidence for spots and flares on VLM dwarfs.

5.6.1   Hα emission

Results for the early to mid-M dwarfs (M0–M6) suggest that the incidence of magnetic activity, as measured by the presence of Hα emission, increases monotonically towards later spectral type (lower mass); by M5–M6 more than half of M dwarfs exhibit activity. This increase has been attributed to the longevity of the activity in later-type stars, so that if all stars are born active, in a disk sample only the youngest M0 stars will still be active, while nearly the entire age span of the M6 stars will be active. Recent data on the 'ultracool' M dwarfs (later than M6.5) demonstrates that this trend continues to spectral type M7, where there are about 20 stars known, all of them with Hα in emission [G6].

Stauffer *et al.* [S12] were the first to question whether the Hα emission properties changed in the lowest-mass dwarfs. Their data for VLM Pleiads showed a decrease in activity level at the latest types measured. A second indication of anomalous behaviour was the discovery of several VLM field dwarfs with no (or very low) activity, but rapid rotation [B5], [M3], [T3]. The latter study found that the VLM population as a group had similar kinematics to the M0–M6 nearby dwarfs in the PMSU survey, and thus there was no kinematic signature of a younger (or older) age. Instead, they observed a smooth decrease in activity with decreasing mass in objects across the stellar and substellar regimes, and postulated that the onset of dust formation (see Chapter 4) could be affecting the production of magnetic activity in the lower-mass dwarfs. A notable exception to the decreasing activity in the latest-type M dwarfs is the remarkable dM9.5e dwarf PC0025+0447 [M9]. This object has a Balmer emission spectrum that is very similar in activity strength to earlier-type M dwarfs, and hence exhibits enormous equivalent widths, since the stellar continuum is so depressed at this late spectral type. Martin and collaborators have suggested that PC0025 is a brown dwarf based on their detection of lithium. However, Liebert has proposed an alternative scenario, where the M dwarf is the brighter member of a VLM-dwarf/brown dwarf contact binary, with material accreted in Roche lobe overflow supplying both the high level of Hα emission and intermittent lithium absorption.

The comprehensive study by Gizis *et al.* [G6] confirmed a general decline in activity strength at spectral types later than M7. Figure 5.21 illustrates their results for late M dwarfs and L dwarfs. They found a possible correlation of Hα EW with tangential velocity in their sample, such that the more active objects had larger velocities, possibly indicating an older population. Based on this correlation, they divided their sample into 'very low-mass stars of age similar to the Galactic disk' (which had reasonably strong magnetic activity, similar to that in the earlier type field M dwarfs), and 'young brown dwarf field stars' (low magnetic activity, some having rapid rotation measurements). Figure 5.22 shows that the M8–L0 spectral type (temperature) range is populated by these two groups at different times during their evolution. Gizis *et al.* thus advocate using the presence of activity as an age indicator in the opposite sense from the earlier-type stars (Section 5.2); the older, low-mass stars have maintained their activity, and represent the tail of the distribution found for the earlier-type stars, while the younger brown dwarfs are unable to produce magnetic activity, even with rapid rotation and young age. This explanation would attribute the decline in activity to a gradual increase in the number of brown dwarfs at later spectral types, in contrast to the proposition that it is the spectral type (temperature) of the object that controls its activity, regardless of the mass (that is, if dust formation in the atmosphere is controlling the magnetic field production and emergence). The disentanglement of the relative importance of mass and surface temperature will be the next step in understanding the Hα emission properties; but again, it should be remembered that it is not yet known whether Hα is the best diagnostic to use for measuring magnetic activity in these lowest-mass objects.

No magnetic activity has yet been observed in the majority of the even cooler

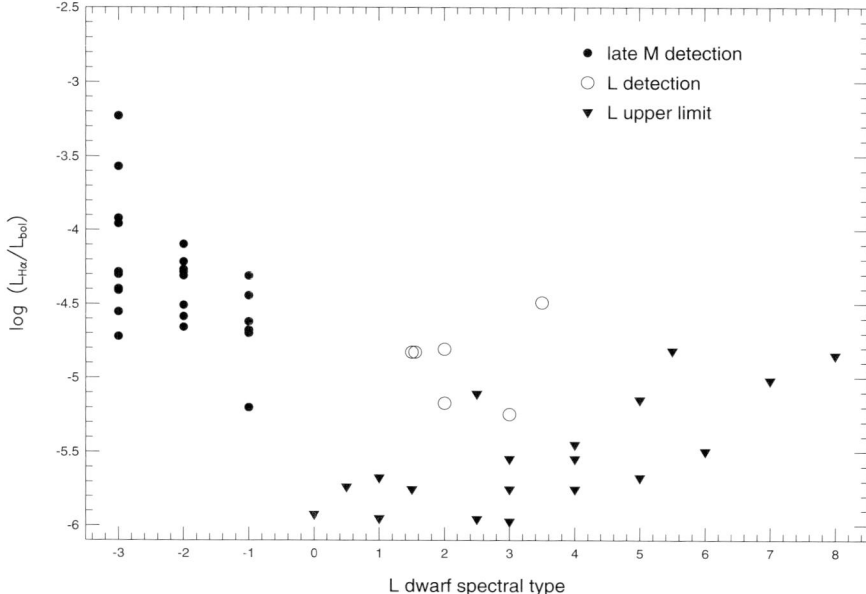

Figure 5.21. Activity strength against spectral type for the latest M and L dwarfs. The activity strength declines toward later types. Note that the PC0025 object would lie at M9, $\log L_{H\alpha}/L_{bol} = -3.4$. (Figure courtesy of J. Gizis.)

methane (T) dwarfs. If these objects are similar to Jupiter, they might have significant magnetic fields leading to auroral activity, but the emission line fluxes would be too low to be detected outside the solar system. However, observations by the 2MASS Rare Object Team indicate weak, but persistent emission from one T dwarf, 2MASS 1237+65. Current speculative analysis centres on the possibility that this might be an interacting binary, a cooler analogue of PC0025.

5.6.2   Spots and flares

As described in Section 5.4.1, spots have been observed on a few VLM objects, including brown dwarfs in the Pleiades and α Per at spectral types M6–M7. However, field objects at later spectral types show little sign of spot modulation [T1], although recent results indicate a possible detection in one object [B2]. There is a complication with 'weather' in dwarfs that have dusty atmospheres. Large weather systems such as the Great Red Spot on Jupiter may produce a rotational modulation signal similar to that expected from magnetic active regions, and thereby introduce ambiguity into the interpretation of the results [T4].

Observations of flares provide further evidence of magnetic activity in VLM stars, although as yet there is no evidence of this activity in brown dwarfs. The well-known dM8e star VB 10 has been observed to flare in Hα [H11], the ultraviolet [L8] and the

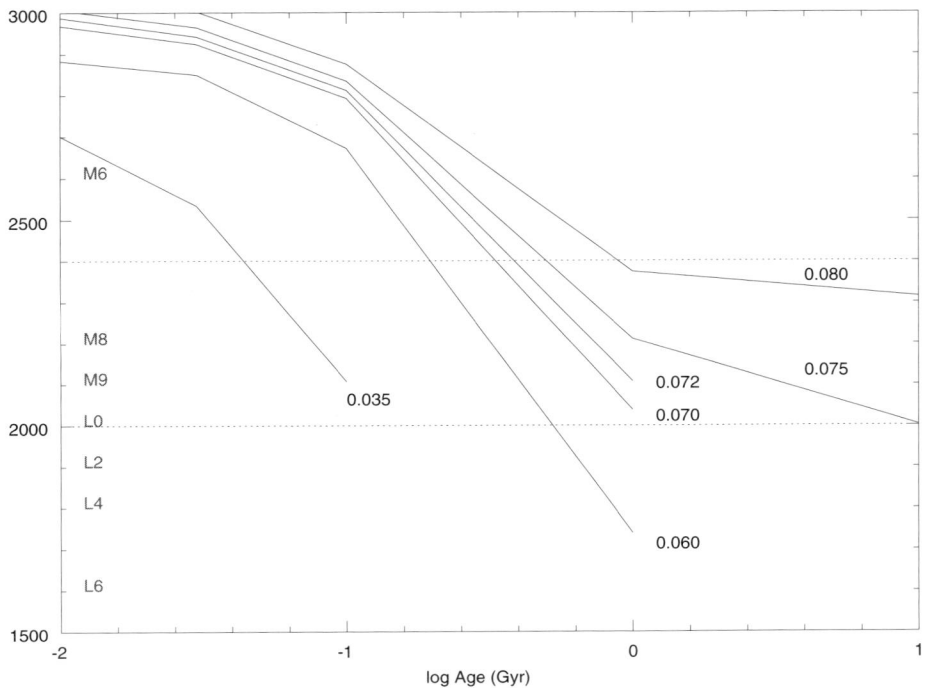

Figure 5.22. Evolutionary models for VLM dwarfs. At a given temperature (spectral type) both old, low-mass stars and young, brown dwarf interlopers can co-exist, making it difficult to disentangle the magnetic activity properties, which may be quite different for the two groups.

X-ray [F6]. Both the UV and X-ray flares were quite energetic compared to the (undetected) quiescent emission, lying at the upper end of the flare-energy/quiescent luminosity relationship given in [L1]. A flare on a dM9.5e object discovered with 2MASS showed noticeable enhancements in Hα and other optical emission lines, and in optical continuum radiation [L4]. Figure 5.23 depicts the flare spectrum, which appears quite similar to those of earlier type dMe flares. (The PC0025 object is shown for comparison; it clearly does not exhibit a flare signature). The flare attained a very high luminosity – perhaps nearly equal to L_{bol} – during the impulsive phase. The frequency of flaring in these objects is not yet documented, but will be essential in deciding whether a significant fraction of their energy is emitted during flares, as suggested by these studies.

5.7   COMPILATION AND SYNTHESIS OF ACTIVITY RESULTS

Table 5.3 provides a compilation of the various results described throughout this chapter. The activity features are organised into three categories: solar-type dwarfs,

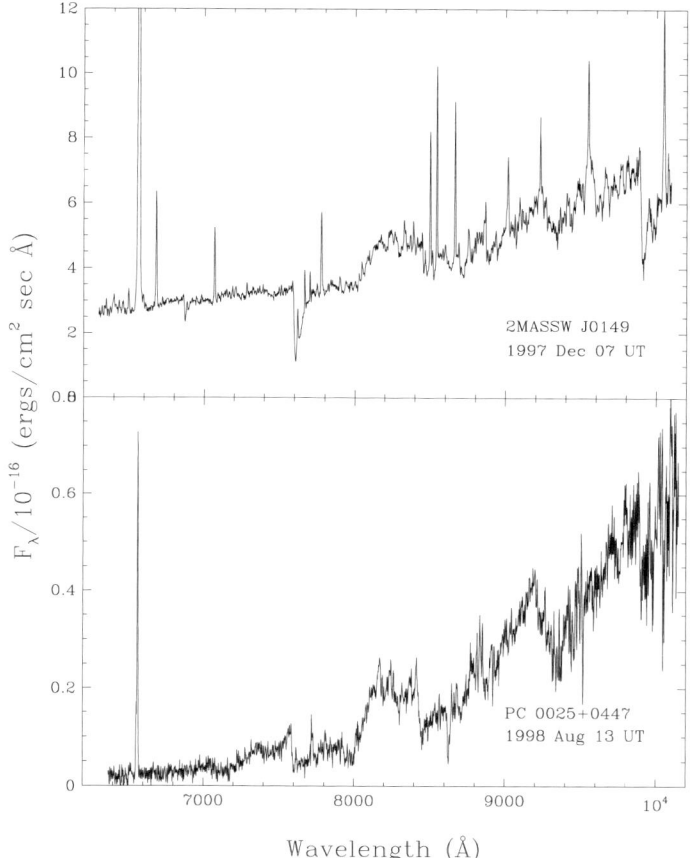

Figure 5.23. Flare spectrum of a dM9.5e 2MASS dwarf, together with a spectrum of the interesting dM9.5e dwarf PC0025. The greatly enhanced emission lines and continuum in the flare spectrum are typical of flares on earlier-type M dwarfs, while the PC0025 spectrum is clearly not of flare origin. (Figure courtesy of J. Liebert and the *Astrophysical Journal*.)

early to mid-M dwarfs, and late M–L dwarfs. Figure 5.24 is a schematic attempt to synthesise these features into an overall picture of magnetic activity as manifested in the different objects. It is only a snapshot in time; the relative importance of the various activity components may also evolve. The essential points are:

- The $\alpha\Omega$ (shell) dynamo operates in solar-type stars, and explains the activity dependence on rotation and youth. A turbulent component may be partially responsible for early chaotic behaviour, before the shell dynamo begins to dominate and produce cyclic behaviour as seen in the Sun.
- In the early M dwarfs (M0–M3), before the onset of complete convection, the shell dynamo probably still plays a role, with the turbulent dynamo becoming

Table 5.3. Magnetic activity properties of low-mass dwarfs.

Feature	Solar type	Early–mid M	Late M–L
Magnetic fields	Yes	Yes	?
B, filling factor	1,000 G, 1%	3,000 G, 50%	?
Spots	Yes	Yes (some, low amp)	Yes (some, low amp)
Cycles	Yes	No	?
Flares	Yes	Yes	Yes
Chromospheric radiation	Ca II	Hα	Hα?
Fraction that are active	Small	Increasing → all	Decreasing → none
Activity strength	Weak	Strong, large scatter	Weakening → none
Rotation	Correlated	Weak/none (threshold?)	None/anti-correlation? fast rotation inhibits?
Radio, X-rays	Yes	Yes	Only in very young stars
Age behaviour	Decays with time	Lasts longer at lower masses	Only occurs in young BD or old stars?
Dynamo	Shell	Turbulent	Turbulent → none (primordial?)

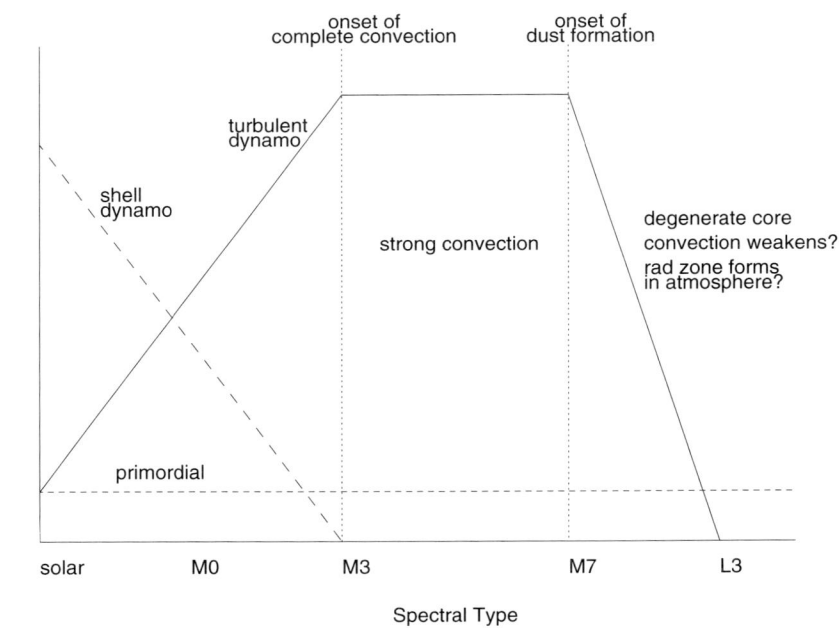

Figure 5.24. Schematic representation of magnetic activity in low-mass dwarfs.

increasingly important at later spectral types. Moderate dependence on rotation (at least requiring a threshold level) and age, as in the solar-type stars, is expected. The lack of cycles and infrequent spot observations may be explained by the increased role of the turbulent dynamo.

- The fully convective M dwarfs (M3–M7) have the activity characteristics proposed for a turbulent dynamo: few spots, chaotic coverage with no global properties such as cycles or organised dipole fields, and little or no dependence on rotation. The longevity of the activity, which increases in stars of lower mass, must be related to the efficiency of the dynamo production and/or heating mechanisms, but no theory yet exists to explain the observations in detail.

- The latest-type dwarfs (M8–L) show a significant decline in activity. This could be attributed to a loss of efficiency in the turbulent dynamo at low mass, perhaps because the convection becomes less vigorous in the cooler objects. An alternative explanation involves the formation of radiative zones in atmospheres with significant dust opacity, perhaps impeding the emergence of magnetic flux in the surface layers. A complication in these objects is the overlap in spectral type of low-mass stars and very low-mass brown dwarfs, making it difficult to disentangle dependences on mass, age and surface temperature. A further caveat is that it has not yet been established that $H\alpha$ is the best diagnostic for magnetically heated outer atmospheres in these cool dwarfs.

- The role of the primordial field – particularly in very young brown dwarfs – has yet to be explored. The decay of the primordial field may provide an explanation for the differences in activity observed between brown dwarfs in young star-formation regions (strong in $H\alpha$ and X-ray emission) and brown dwarfs in older clusters such as the Pleiades and in the field (weak or no activity).

5.8 SUMMARY

We have discussed the dynamo production of magnetic fields, and direct detection techniques in solar-type and lower-mass dwarfs. A few M dwarfs have strong magnetic fields with larger filling factors than are observed on the Sun. Properties of solar and M dwarf outer atmospheres have been reviewed, together with general modelling techniques. The evidence for spots and flares on M dwarfs has been described: spots are less prevalent than in active solar-type dwarfs, but exist at low amplitude in some objects. Flares are much more energetic compared with their bolometric luminosity in M dwarfs than in the Sun, but a model for their production that follows a solar analogy matches most of the observable phenomena. The mechanisms for producing the blue continuum emission and very broad lines remain elusive. Flare effects on the local stellar neighbourhood through mass loss and energetics may be important.

Global properties of active M dwarfs have been described: their colours, molecular bandstrengths, luminosities, and particularly their Hα emission behaviour. Magnetic activity remains at a similar level, but lasts longer, in later type (lower-mass) M dwarfs through spectral type M7, and then begins to decline toward later types. There is no obvious correlation with rotation, as in solar-type stars, other than a possible requirement of a 'threshold' value of $< 2\,\text{km}\,\text{s}^{-1}$. In the latest type M and L dwarfs, it is common to find rapid rotation and little or no activity. These latest objects may lose their activity because of changes in the dynamo production induced at low mass (such as reduced convection), or because the atmospheric properties preclude magnetic flux emergence in regions of significant dust formation. The role of the primordial field is as yet unexplored. Understanding the properties of magnetic activity in the mid-M dwarfs, and the decline in magnetic activity in the lowest mass stars and brown dwarfs, will be a rich topic in the coming years. In particular, we do not yet know if the effects of magnetic activity on the internal structure and evolution of these objects is inconsequential or profound.

5.9   REFERENCES

A1   Abada-Simon, M., Aubier, M., 1997, *A&AS*, **125**, 511.
A2   Abbett, W. P., Hawley, S. L., 1999, *ApJ*, **521**, 906.
A3   Allard, F., Hauschildt, P. H., 1995, *ApJ*, **445**, 433.
A4   Arnaud, M., Rothenflug, R., 1985, *A&AS*, **60**, 425.
A5   Ayres, T. R., 1979, *ApJ*, **228**, 509.
A6   Ayres, T. R., 1981, *ApJ*, **244**, 1064.
A7   Ayres, T. R., Rabin, D., 1996, *ApJ*, **460**, 1042.
A8   Ayres, T. R., Marstad, N. C., Linsky, J. L., 1981, *ApJ*, **247**, 545.
B1   Babcock, H. W., 1961, *ApJ*, **133**, 572.
B2   Bailer-Jones, C. A. L., Mundt, R., 1999, *A&A*, **348**, 800.
B3   Baliunas, S. L. *et al.* (26), 1995, *ApJ*, **438**, 269.
B4   Barnes, S. A., Sofia, S., Prosser, C. F., Stauffer, J. R., 1999, *ApJ*, **516**, 263.
B5   Basri, G., Marcy, G. W., 1995, *AJ*, **109**, 762.
B6   Benedict, G. F., *et al.* (12), 1998, *AJ*, **116**, 429.
B7   Bondar, N. I., 1995, *A&AS*, **111**, 259.
B8   Bopp, B. W., Fekel, F., 1977, *AJ*, **82**, 490.
B9   Bromage, G. E., Phillips, K. J. H., Dufton, P. L., Kingston, A. E., 1986, *MNRAS*, **220**, 1021.
B10 Bruls, J. H. M. J., Solanki, S. K., Schuessler, M., 1998, *A&A*, **336**, 231.
B11 Butler, C. J., Rodono, M., Foing, B. H., 1988, *A&A*, **206**, L1.
B12 Buzasi, D. L., 1997, *ApJ*, **484**, 855.
B13 Byrne, P. B., 1996, in 'Stellar surface structure', *Proceedings of IAU Symposium 176*, ed. Strassmeier, K. G. and Linsky J. L., Kluwer, Dordrecht, p. 299.
C1   Canfield, R. C. *et al.*, 1986, in *Energetic Phenomena on the Sun*, ed. M. Kundu and B. Woodgate, NASA Conf. Publ. 2439, chapter 3.
C2   Carlsson, M., Stein, R. F., 1997, *ApJ*, **481**, 500.

C3   Coleman, G. D., Worden, S. P., 1976, *ApJ*, **205**, 475.

C4   Contadakis, M. E., 1995, *A&A*, **300**, 819.

C5   Craig, I. J. D., McClymont, A. N., Underwood, J. H., 1978, *A&A*, **70**, 1.

C6   Cram, L. E. 1982, *ApJ*, **253**, 768.

C7   Cristaldi, S., Gershberg, R. E., Rodono, M., 1980, *A&A*, **89**, 123.

C8   Cully, S. L., Fisher, G. H., Abbott, M. J., Siegmund, O. H. W., 1994, *ApJ*, **435**, 449.

C9   Cully, S. L., Fisher, G. H., Hawley, S. L., Simon, T., 1997, *ApJ*, **491**, 910.

D1   DeLuca, E. E., Gilman, P. A., 1991, 'The solar dynamo', in *The Solar Interior and Atmosphere*, University of Arizona Press, p. 275.

D2   Delfosse, X., Forveille, T., Perrier, C., Mayor, M., 1998, *A&A* **331**, 581.

D3   Dennis, B. R., Schwartz, R. A., 1989, *Solar Phys.*, **121**, 75.

D4   Dennis, B. R., Zarro, D. M., 1993, *Solar Phys.*, **146**, 177.

D5   Dere, K. P., Landi, E., Mason, H. E., Monsignori Fossi, B. C., Young, P. R., 1997, *A&AS*, **125**, 149.

D6   Doyle, J. G., Butler, C. J., Byrne, P. B., van den Oord, G. H. J., 1988, *A&A*, **193**, 229.

D7   Durney, B. R., 1997, *ApJ*, **486**, 1065.

D8   Durney, B. R., De Young, D. S., Roxburgh, I. W., 1993, *Solar Phys.*, **145**, 207

E1   Eggen, O. J., 1990, *PASP*, **102**, 166.

F1   Fang, C., Ding, M. D., 1995, *A&AS* **110**, 99.

F2   Fisher, G. H., 1989, *ApJ*, **346**, 1019.

F3   Fisher, G. H., Hawley, S. L., 1990, *ApJ*, **357**, 243.

F4   Fisher, G. H., Canfield, R. C., McClymont, A. N., 1985, *ApJ*, **289**, 425.

F5   Fisher, G. H., Fan, Y., Howard, R. F., 1995, *ApJ*, **438**, 463.

F6   Fleming, T. A., Giampapa, M. S., Schmitt, J. H. M. M., 2000, *ApJ*, in press.

F7   Foukal, P & Lean, J. 1988, *ApJ*, **328**, 347.

G1   Giampapa, M. S., Africano, J. L., Klimke, A., Parks, J., Quigley, R. J., Robinson, R. D., Worden, S. P., 1982, *ApJ*, **252**, 39.

G2   Giampapa, M. S., Africano, J. L., Klimke, A., Parks, J., Quigley, R. J., Robinson, R. D., Worden, S. P., 1982, *ApJ*, **252**, L39.

G3   Giampapa, M. S., Rosner, R., Kashyap, V., Fleming, T. A., Schmitt, J. H. M., Bookbinder, J. A., 1996, *ApJ*, **463**, 707.

G4   Giampapa, M. S., Worden, S. P., Linsky, J. L., 1982, *ApJ*, **258**, 740.

G5   Gizis, J. E., 1998, *AJ*, **115**, 2053.

G6   Gizis, J. E., Monet, D. G., Reid, I. N., Kirkpatrick, J. D., Williams, R. J., 2000, *AJ*, in press.

G7   Gray, D. F., 1992, *The Observation and Analysis of Stellar Photospheres*, 2nd edition, Cambridge University Press, p. 375.

G8   Griffiths, N. W., Jordan, C., 1996, in *The Ninth Cambridge Workshop on Cool Stars, Stellar Systems and the Sun*, ed. R. Pallavicini & A. Dupree, ASP Conf. Ser. 109, p. 647.

G9   Grinin, V., 1983, in *Activity in Red Dwarf Stars*, ed. P. B. Byrne and M. Rodono, Reidel, Dordrecht, p. 223.

G10 Guedel, M., Benz, A. O., Schmitt, J. H. M. M., Skinner, S. L., 1996, *ApJ*, **471**, 1002.

G11 Gunn, A. G., Doyle, J. G., Mathioudakis, M., Houdebine, E. R., Avgoloupis, S., 1994, *A&A*, **285**, 489.

H1 Hale, G. E., 1908, *ApJ*, 28, 315.

H2 Hambaryan, V. V., 1998, in *The Tenth Cambridge Workshop on Cool Stars, Stellar Systems and the Sun*, ed. R. Donahue and J. Bookbinder, ASP Conf. Ser. 154, p. 1492.

H3 Hatzes, A. P., Vogt, S. S., Ramseyer, T. F., Misch, A., 1996, *ApJ*, **469**, 808.

H4 Hawley, S. L., Pettersen, B. R., 1991, *ApJ*, **378**, 725.

H5 Hawley, S. L., Fisher, G. H., 1992a, *ApJS*, **78**, 565.

H6 Hawley, S. L., Fisher, G. H., 1992b, *ApJS*, **81**, 885.

H7 Hawley, S. L., *et al.* (10 others), 1995, *ApJ*, **453**, 464.

H8 Hawley, S. L., Gizis, J. E., Reid, I. N., 1996, *AJ*, **112**, 2799.

H9 Hawley, S. L., Tourtellot, J. G., Reid, I. N., 1999, *AJ*, **117**, 1341.

H10 Hawley, S. L., Reid, I. N., Tourtellot, J. G., 1999, in *Very Low Mass Stars and Brown Dwarfs*, ed. R. Rebolo, Cambridge University Press, in press.

H11 Herbig, G., 1956, *PASP*, **68**, 531.

H12 Houdebine, E. R., 1992, *IrAJ*, **20**, 213.

H13 Houdebine, E. R., Doyle, J. G., 1994, *A&A*, **289**, 169.

H14 Houdebine, E. R., Foing, B. H., Doyle, J. G., Rodono, M., 1993, *A&A*, **278**, 109.

H15 Houdebine, E. R., Mathioudakis, M., Doyle, J. G., Foing, B. H., 1996, *A&A*, **305**, 209.

H16 Houdebine, E. R., Stempels, H. C., 1997, *A&A*, **326**, 1143.

H17 Hudson, H. S., 1988, *ARA&A*, **26**, 473.

H18 Hudson, H. S., Acton, L. W., Hirayama, T., Uchida, Y., 1992, *PASJ*, **44**, L77.

I1 Ichimoto, K., Kurokawa, H., 1984, *Solar Phys.*, **93**, 105.

J2 Jackson, P. D., Kundu, M. R., White S. M., 1989, *A&A*, **210**, 284.

J3 Johns-Krull, C. M., Valenti, J. A., 1996, *ApJ*, **459**, L95.

J4 Johns-Krull, C. M., Hawley, S. L., Basri, G., Valenti, J. A., 1997, *ApJS*, **112**, 221.

J5 Jones, K. L., Page, A. A., 1991, *PASAu*, **9**, 277.

K1 Kahler, S. *et al.* (30), 1982, *ApJ*, **252**, 239.

K2 Kalkofen, W., 1991, 'The heating of the solar chromosphere', in *The Solar Interior and Atmosphere*, University of Arizona Press, p. 911.

K3 Katsova, M. M., Boikko, A. Ya., Livshits, M. A., 1997, *A&A*, **321**, 549.

K4 Krishnamurthi, A. *et al.* (15), 1998, *ApJ*, **493**, 914.

K5 Krishnamurthi, A., Leto, G., Linsky, J. L. 1999, *AJ*, **118**, 1369.

L1 Lacy, C. H., Moffett, T. J., Evans, D. S., 1976, *ApJS*, **30**, 85.

L2 Lanza, A. F., Catalano, S., Cutispoto, G., Pagano, I., Rodono, M., 1998, *A&A*, **332**, 541.

L3 Leighton, R. B., 1969, *ApJ*, **156**, 1.

L4 Liebert, J., Kirkpatrick, J. D., Reid, I. N., Fisher, M. D., 1999, *ApJ*, in press.

L5 Lim, J., White, S. M., 1996, *ApJ*, **462**, L91.

L6   Lin, R. P., Hudson, H. S., 1976, *Solar Phys.*, **50**, 153.

L7   Linsky, J. L., Bornmann, P. L., Carpenter, K. G., Hege, E. K., Wing, R. F., Giampapa, M. S., Worden, S. P., 1992, *ApJ*, **260**, 670.

L8   Linsky, J. L., Wood, B. E., Brown, A., Giampapa, M. S., Ambruster, C., 1995, *ApJ*, **455**, 670.

L9   Lockwood, G. W., Skiff, B. A., Radick, R. R., 1997, *ApJ*, **485**, 789.

M1   Marcy, G. W., 1984, *ApJ*, **276**, 286.

M2   Marcy, G. W., Chen, G. H., 1992, *ApJ*, **390**, 550.

M3   Martin, E. L., Basri, G., Delfosse, X., Forveille, T., 1997, *A&A*, **327**, L29.

M4   Martin, E. L., Zapatero-Osorio, M. R., 1997, *MNRAS*, **286**, L17.

M5   Mathioudakis, M., Fruscione, A., Drake, J. J., McDonald, K., Bowyer, S., Malina, R. F., 1995, *A&A*, **300**, 775.

M6   Mauas, P. J. D., Falchi, A., 1994, *A&A*, **281**, 129.

M7   Mavridis, L. N., Avgoloupis, S., 1993, *A&A*, **280**, L5.

M8   Mochnacki, S., Zirin, H., 1980, *ApJ*, **239**, L27.

M9   Mould, J., Cohen, J., Oke, J. B., Reid, I. N., 1994, *AJ*, **107**, 2222.

M10 Mullan, D. J., Doyle, J. G., Redman, R. O., Mathioudakis, M., 1992, *ApJ*, **397**, 225.

M11 Mullan, D. J., Stencel, R. E., Backman, D. E., 1989, *ApJ*, **343**, 400.

N1   Neidig, D. F., Kiplinger, A. L., Cohl, H. S., Wiborg, P. H., 1993, *ApJ*, **406**, 306.

N2   Neuhauser, R. *et al.* (9), 1999, *A&A*, **343**, 883.

N3   Neupert, W. M., 1968, *ApJ*, **153**, L59.

N4   Noyes, R. W., Hartmann, L. W., Baliunas, S. L., Duncan, D. K., Vaughan, A. H., 1984, *ApJ*, **279**, 763.

O1   Oranje, B. J., 1986, *A&A*, **154**, 185.

P1   Parker, E. N., 1955, *ApJ*, **122**, 293.

P2   Parker, E. N., 1975, *ApJ*, **198**, 205.

P3   Pasquini, L., Lindgren, H., 1994, *A&A*, **283**, 179.

P4   Patten, B. M., Simon, T., 1996, *ApJS*, **106**, 489.

P5   Pettersen, B. R., Lambert, D. L., Tomkin, J., Sandmann, W. H., Lin, H., 1987, *A&A*, **183**, 66.

P6   Pettersen, B. R., 1983, in *IAU Colloquium 71, Activity in Red-Dwarf Stars*, Reidel, Dordrecht, p. 17.

P7   Pettersen, B. R., Hawley, S.L., Fisher, G. H., 1992, *Solar Phys.*, **142**, 197.

P8   Pettersen, B. R., 1989, *Solar Phys.*, **121**, 299

R1   Radick, R. R., 1991, in *The Sun in Time*, University of Arizona Press, p. 787.

R2   Raymond, J. C., Smith, B. W., 1977, *ApJS*, **35**, 419.

R3   Rice, J. B., 1996, in 'Stellar surface structure', *Proceedings of IAU Symposium 176*, ed. Strassmeier, K. G. and Linsky J. L.,Kluwer, Dordrecht, p. 19.

R4   Robinson, R. D., 1980, *ApJ*, **239**, 961.

R5   Rodono, M. *et al.* (21), 1986, *A&A*, **165**, 135.

R6   Rosner, R., 1980, in *The First Cambridge Workshop on Cool Stars, Stellar Systems and the Sun*, p. 79.

R7   Rosner, R., Tucker, W. H., Vaiana, G. S., 1978, *ApJ*, **220**, 643.

S1   Saar, S. H., 1996, in *IAU Colloquium 153, Magnetodynamic Phenomena in the Solar Atmosphere – Prototypes of Stellar Magnetic Activity*, eds. Y. Uchida, T. Kosugi, H. S. Hudson, Kluwer, Dordrecht, p. 367.

S2   Saar, S. H., Linsky, J. L., 1985, *ApJ*, **299**, L47.

S3   Saar, S. H., Linsky, J. L., Beckers, J. M., 1986, *ApJ*, **302**, 777.

S4   Schou, J. *et al.* (23), 1998, *ApJ*, **505**, 390.

S5   Schrijver, C. J., 1987, in *The 5th Cambridge Workshop on Cool Stars, Stellar Systems and the Sun*, p. 89.

S6   Schuessler, M., Solanki, S. K., 1992, *A&A*, **264**, L13.

S7   Sheeley, N. R. Jr., Nash, A. G., Wang, Y.-M., 1987, *ApJ*, **319**, 481.

S8   Skumanich, A., 1972, *ApJ*, **171**, 565.

S9   Soderblom, D. R., Duncan, D. K., Johnson, D. R. H., 1991, *ApJ*, **375**, 722.

S10   Stauffer, J. R., Hartmann, L., 1986, *ApJS*, **61**, 531.

S11   Stauffer, J. R., Giampapa, M. S., Herbst, W., Vincent, J. M., Hartmann, L. W., Stern, R. A., 1991, *ApJ*, **374**, 142.

S12   Stauffer J. R., Liebert, J., Giampapa, M., Macintosh, B., Reid, I. N., Hamilton, D., 1994, *AJ*, **108**, 160.

T1   Terndrup, D. M., Krishnamurthi, A., Pinsonneault, M. H., Stauffer, J. R., 1999, *AJ*, **118**, 1814.

T2   Tinney, C. G., Delfosse, X., Forveille, T., 1997, *ApJ*, **490**, L95.

T3   Tinney, C. G., Reid, I. N., 1998, *MNRAS*, **301**, 1031.

T4   Tinney, C. G., Tolley, A. J., 1999, *MNRAS*, **304**, 119.

T5   Torres, C. A. O., Ferraz-Mello, S., 1973, *A&A*, **27**, 231.

V1   Valenti, J. A., Johns-Krull, C. M., Piskunov, in *The Tenth Cambridge Workshop on Cool Stars, Stellar Systems and the Sun*, ASP Conf. Ser. 154, p. 1357.

V2   Van den Oord, G. H. J., Doyle, J. G., 1997, *A&A*, **319**, 578.

V3   Veeder, G. J., 1974, *AJ*, **79**, 1056.

V4   Vernazza, J. E., Avrett, E. H., Loeser, R., 1981, *ApJS*, **45**, 635.

V5   Vilhu, O. 1984, *A&A*, **133**, 117.

W1   Wielen, R. 1974, in *Highlights of Astronomy*, Vol. 3, ed. Contopoulos, G., Reidel, Dordrecht, p. 395.

W2   Wilson, O. C., Bappu, M. K. V. 1957, *ApJ*, **125**, 661.

Y1   Young, A., MacGregor, Skumanich, 1990, *ApJ*, **349**, 608.

Z1   Zarro, D. M., Canfield, R. C., Strong, K. T., Metcalf, T. R., 1988, *ApJ*, **324**, 582.

Z2   Zirin, H. 1988, *Astrophysics of the Sun*, Cambridge University Press.

6

A Galactic structure primer

6.1 INTRODUCTION

The first chapters of this book consider M dwarfs as individual stars, and study the underlying astrophysics through observations of their atmospheric properties. Low-mass stars are also potentially highly effective as probes of Galactic structure, since they are the most populous constituent of each stellar population. Until recently, their utility in this role has been restricted both by difficulties in interpreting their emergent spectra in terms of luminosities, temperatures and abundances, and by their low intrinsic luminosities, which limit observations to relatively nearby stars. However, with the development of the atmosphere models described in Chapter 4, together with substantial improvements in instrumental efficiency and the consequent availability of extensive, statistically well-defined samples, it is now becoming possible to better exploit their potential.

Detailed observations of late-type M dwarfs are limited to stars within 100 parsecs of the Sun, but early-type M dwarfs are accessible at distances of 2–3 kiloparsecs. The latter observations reach the transition region between the two main stellar populations of the outer Galaxy – the disk and the halo. Moreover, even though surveys of late-type stars may be confined to M dwarfs which are *currently* within a few hundred parsecs of the Sun, the majority of those stars have completed many Galactic orbits, and may have been formed at Galactic radii well separated from the current location of the Sun. Thus, local samples are still representative of global Galactic properties.

Our goal in this chapter is to provide context for the statistical studies of M dwarfs described in the following chapters. We do so by first reviewing the origins of the stellar populations concept and its contribution to our understanding of Galactic structure. Subsequent sections outline the basic properties of the disk and halo populations. The succeeding chapters describe how M dwarfs are used to probe aspects of Galactic structure, notably the underlying mass distribution.

6.2   STELLAR POPULATIONS

The realisation that stars in our Galaxy could be divided into stellar populations – groups of stars with well-characterised kinematic, spatial and chemical abundance distributions – marks one of the crucial steps in our understanding of galaxy formation. We now recognise that those distinct properties reflect significant differences in conditions prevalent during the formation and subsequent evolutionary histories of those groups of stars. Presenting the Galaxy as a combination of a small number of basic building blocks, rather than as a collection of 100 billion suns, is a striking simplification which opens the way toward theoretical investigation of the mechanisms underlying galaxy formation. However, like all simplifications, the population concept can obscure rather than illuminate, either through the broad, large-scale picture hiding important small-scale details, or through its overuse in a proliferation of physically almost indistinguishable subcomponents.

The population concept owes its origin to Baade [B1], [B2], and excellent reviews of its development are given by Blaauw [B13] and Sandage [S5]. Sandage provides an illuminating account of how Baade's hypothesis grew through synthesis of results obtained over a period of thirty years and drawn from two areas of observational research: studies of high-velocity Galactic stars, originating with Charlier [C9]; and spectroscopic and photometric observations of extragalactic systems by Curtis, Hubble, Humason and others. The latter observations revealed significant differences in the structure of spiral and elliptical galaxies, with the ellipticals staunchly refusing to resolve into the individual stars clearly evident in the disks of the nearer spirals. It seemed that the brightest stars in the early-type systems (elliptical) were both fainter and redder than the blue stars visible in the arms of spiral systems.

The identification of large numbers of high-velocity stars led to the realisation that there was a preferred direction of motion relative to the Sun – a consequence, we now understand, of the relative mean rotation of the disk and halo. Analysing data for several types of stars, Strömberg [S19], [S20] codified this velocity difference as the 'asymmetric drift' (preferred motion towards $l = 270°$, $b = 0°$, the direction opposite to Galactic rotation), and also showed that there was a significant correlation between the square of the U-velocity dispersion (radial from the Galactic centre) and the extent of the Strömberg drift. Lindblad [L5] and Oort [O2] demonstrated that the latter feature is a natural consequence of the distribution of kinetic energy between ordered rotation and velocity dispersion (pressure support) in kinematically-separate sub-systems: the lower the energy in rotation, the higher the velocity dispersion, and *vice versa*. The correlation is now usually plotted as rotational lag versus the total velocity dispersion (Figure 6.1).

Sandage has emphasised the key role played in this saga by Shapley's [S14], [S15] discovery of the nearby Sculptor and Fornax dwarf ellipticals. These were not only resolved into individual stars, but Baade and Hubble [B4] were able to identify RR Lyrae stars and, in Fornax, globular clusters associated with the system. This marked the first step towards drawing comparisons between the stellar content in Galactic globulars and ellipticals. The clinching observations came with Baade's wartime photographic studies of the Andromeda spiral galaxy (M31) and its satellite

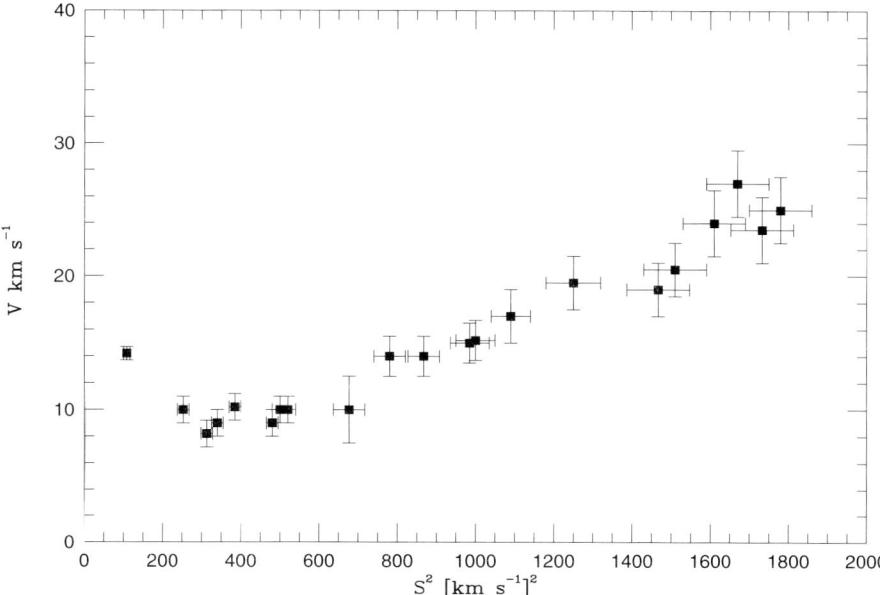

Figure 6.1. Strömberg asymmetric drift: V is the mean rotational lag relative to the Sun and S^2 the total velocity dispersion ($\sqrt{\sigma_U^2 + \sigma_V^2 + \sigma_W^2}$) for groups of stars from the Hipparcos catalogue. (Data from [D2].)

dwarfs, M32, NGC 205, NGC 147 and NGC 185, made using the 100-inch Hooker telescope on Mount Wilson. For the first time, the brightest stars were resolved in both the central bulge of M31 and in the dwarf elliptical companions. Those stars proved to be K giants, with luminosities comparable to the brightest stars in Galactic globular clusters. Based on the similarity in the Strömberg drift derived for globulars, RR Lyrae stars, short-period (P < 210 days) Mira stars and the high-velocity stars, Baade concluded that those Galactic objects are representative of the dominant stellar population in both spiral bulges and elliptical systems. In his own words [B2]: 'This leads to the further conclusion that the stellar populations of the galaxies fall into two distinct groups, one represented by the well-known H–R diagram of the stars in the solar neighbourhood (the slow-moving stars), the other by that of the globular clusters. Characteristic of the first group (type I) are highly luminous O- and B-type stars and open clusters; the second (type II), short-period Cepheids [RR Lyrae stars] and globular clusters. Early-type nebulae (E-Sa) seem to have populations of the pure type II. Both types seem to co-exist in the intermediate and late-type nebulae. The two types of stellar populations had been recognised among the stars of our own Galaxy by Oort as early as 1926.'

From the outset, Baade realised that studies of nearby high-velocity stars represented the only available method of probing the characteristics of the lower-luminosity dwarf members of his Population II. Spectroscopic observations of the so-called subdwarfs and evolved high-velocity stars, such as RR Lyrae stars, had already revealed peculiarities, notably line-strengths weaker than expected for typical stars of the appropriate spectral type. These abnormalities were generally assumed to reflect different atmospheric conditions rather than different chemical composition. With neither a well-developed theory of stellar evolution, nor understanding of nucleosynthesis, it is not surprising that the standard paradigm in the early twentieth century was that all stars had abundances similar to those observed in the Sun.

Sandage identifies two papers which solidified the dichotomy in the characteristics of the disk and halo populations. Chamberlain and Aller's [C8] spectral analysis of the two archetypal subdwarfs, HD 19445 and HD 140283, produced the first metal abundances that were substantially below those found in the Sun;[1] and Roman [R8] demonstrated that subdwarfs share the common photometric characteristic of having bluer (U–B) colours (an 'ultraviolet excess') for a given (B–V) than disk dwarfs. Moreover, the extent of the ultraviolet excess was apparently dependent on the size of the Strömberg drift [R5]. Combined with observations of stars in globular clusters, these results provided a stronger link between the field subdwarfs and the Galactic globular cluster system, characterising them as members of a low-rotation, high-velocity-dispersion, metal-poor stellar population – the Galactic halo.

Baade originally proposed only two distinct stellar populations. However, the expanding empirical database accumulated through the 1950s revealed complexities, particularly amongst the local Population I stars, which were difficult to accommodate in this simple picture. The result was the development of a more detailed classification scheme at the 1957 Vatican conference on stellar populations [O1]. Three populations and five sub-populations were defined, each with its own distinct kinematics and abundance distribution, as summarised in Table 6.1. The original Population II and Population I were both subdivided, into respectively, the 'Halo' and 'Intermediate Pop II'; and the 'Extreme' (young) and 'Older Pop I'. The 'Disk' population was added as a component between the 'Intermediate Pop II' and the 'Older Pop I'. Various types of stars which were regarded as representative tracers of each component are included in Table 6.1.

These revisions represented a philosophical transition from the 'either/or' dichotomy of Baade's original scheme to a series of sub-populations with continuously varying properties, reminiscent of Lindblad's [L5], [L6] division of the Galaxy into kinematic sub-systems. Moreover, for the first time this classification was cast in terms of an evolutionary scenario, with halo stars marking the first stage of star formation in the Galaxy, and extreme Population I the most recent. The remaining components represent intermediate stages of galaxy formation, with a broad trend between age and properties such as velocity dispersion and spatial distribution.

The revised, more gradualistic classification scheme devised at the Vatican

---

[1] Chamberlain and Aller's published results indicate abundances of 0.1–0.03 solar for the various elements studied, while both stars are actually less than one-hundredth the abundance of the Sun. Sandage reports that the initially-derived metallicities were indeed much lower, but given the prevailing climate of opinion of the time (as expressed by the journal referee), the temperature of each star was adjusted slightly, moved by either internal or external prompting, to give somewhat more palatable results (see also Chapter 8 in [C13]). HD 140283 is now known to be a halo subgiant, rather than a subdwarf.

Table 6.1. The Vatican stellar population scheme.[1]

	Halo Pop. II	Intermediate Pop. II	Disk
Tracers	Subdwarfs Globulars	Stars with $|z| > 30\,\mathrm{km\,s^{-1}}$ Long-period variables Periods <250 days	Galactic nucleus stars Planetary nebulae
	RR Lyrae stars, $P > 0.4$ days		RR Lyrae stars, $P < 0.4$ days Weak-line stars
$\langle|z|\rangle$ (parsecs)	2,000	700	400
$\langle|Z|\rangle$ (km s^{-1})	75	25	17
Axial ratio, c/a	0.5	0.2	0.04?
Central concentration	Strong	Strong	Strong?
Distribution[2]	Smooth	Smooth	Smooth?
Age (10^9 years)	6	5–6	1.5–5

	Older Pop. I	Extreme Pop. I
Tracers	A-type stars Strong-line stars dMe stars	Gas Spiral structure Supergiants Cepheids, T Tauri
$\langle|z|\rangle$ (parsecs)	160	120
$\langle|Z|\rangle$ (km s^{-1})	10	8
Axial ratio, c/a	?	0.01
Central concentration	Weak	Weak
Distribution[2]	Patchy Spiral arms	Very patchy Spiral arms
Age (10^9 years)	0.1–1.5	<0.1

[1] From [B13].
[2] The total mass associated with each component in this model is $1.6 \times 10^{10}\,M_\odot$ for the halo; $4.7 \times 10^{10}\,M_\odot$ for the intermediate Population II and disk combined; $5 \times 10^9\,M_\odot$ for the older Population I; and $2 \times 10^9\,M_\odot$ for extreme Population I.

conference provides a better approximation to the overall distribution of stellar properties. Paradoxically, however, repercussions from the codification into five sub-populations have led to almost as much confusion as illumination, particularly in recent years.

Dividing and classifying objects, establishing order in the face of apparent disorder, plays a prominent role during the early stages of many scientific disciplines. Classification schemes lay the foundation for subsequent understanding of why differences and similarities occur. Stellar spectral classification exemplifies the success possible with this approach.

Problems arise, however, when boundaries between adjacent sub-groups are less than clearly defined, as is the case with the Vatican system. While mean properties such as abundance and rotational velocity are different for each component, the distributions overlap, particularly among the disk and population I components. The result is that many individual stars cannot be assigned uniquely to one component. Applying slightly different criteria, such as emphasising kinematics over abundance, can lead to different effective definitions of a population, even though each is identified by name as the same component. All four of the spatially-flattened sub-populations in the Vatican scheme are subject to these vagaries of interpretation.

Given those concerns, the current best option in stellar population studies is to return to the simplicity of Baade's scheme, dividing the Galaxy at moderate ($\sim$1 kpc) to large radii into two major components, the disk and the halo, with the Galactic Bulge as a centralised, possible third component. The disk is less homogeneous than the halo, including stars spanning a significant range of age, kinematics and abundance, but whether this dispersion is the product of Galactic gradualism (that is, continuous evolutionary processes), or reflects the presence of discrete sub-populations, remains a subject of some debate.

The simplest method of appreciating the relative extent of the different Galactic components is through observations of nearby external galaxies. Figure 6.2 presents an image of the edge-on spiral galaxy NGC 4565, an Sb system generally regarded as similar to our own Milky Way. The central bulge and the disk, with a prominent absorbing dust lane, are obvious; the halo population is not. While deep images have identified moderate numbers of globular clusters in this and other spirals, the field halo population lies below the current detection level in all external spiral galaxies. Based on observations of tracers such as metal-poor RR Lyrae stars within our own Galaxy, it probably describes a spherical distribution extending to radii beyond the edge of the NGC 4565 disk.

6.3 THE GALACTIC HALO

The stellar halo is an old, metal-poor population, characterised by little or no current star formation. The most prominent members of the halo are globular clusters, $\sim$100 systems each consisting of 10^5 to 10^6 stars concentrated within radii of 10–30 parsecs. Figure 6.3 shows a typical example, the intermediate abundance cluster M13 in the constellation of Hercules. The sheer number of stars in these systems renders them excellent templates of the sparse field halo population, which comprises only $\sim$0.2% of the stars in the Solar Neighbourhood.

Figure 6.4 plots colour–magnitude diagrams for two globulars: M68 (from [W1]) at [Fe/H] = −2.0 [C6], one of the lower-abundance systems; and M5 [R2], an intermediate-abundance system with [Fe/H] = −1.1. In both clusters, the main-sequence turn-off lies between M_V = +3.5 and +4, corresponding to a mass of $\sim$0.8 $M_\odot$ and ages of 11–13 Gyr. Evolved stars traverse well-populated subgiant and red giant branches (hydrogen shell-burning) before dropping onto the horizontal branch after the ignition of helium core-burning at the tip of the first giant branch (the helium flash). Depending mainly on mass and abundance, a horizontal branch

Figure 6.2. The edge-on spiral galaxy NGC 4565. (From the STScI Digital Sky Survey scans of a plate taken as part of the second Palomar Sky Survey, courtesy of Palomar Observatory and STScI.)

star may evolve through the instability strip and pulsate as an RR Lyrae variable before initiating double shell-burning (hydrogen and helium burning). Subsequent evolution takes the star up the second, or asymptotic, giant branch, a phase terminated by envelope ejection and planetary nebula formation. (Further details on the evolution of these metal-poor, solar-type stars are given in the textbooks referenced in Chapter 3.)

Halo stars with masses approaching the hydrogen-burning limit can be identified within the Solar Neighbourhood. These are M subdwarfs, discussed in detail in Chapter 11. Studies of the halo population at large heliocentric distances are, of necessity, based on the more luminous population tracers, notably horizontal branch (HB) stars, particularly RR Lyrae stars, and globular clusters.

6.3.1   The metal-poor main sequence

The majority of halo stars in the solar neighbourhood are hydrogen-burning main-sequence stars which have intrinsically lower luminosities than solar-abundance

Figure 6.3. The globular cluster M92, from a plate taken with the 48-inch Oschin Schmidt telescope. (Courtesy of Palomar Observatory.)

dwarfs of the same (B–V) colour. The term 'subdwarf' was coined by Kuiper [K5] to describe stars which lay between disk dwarfs and white dwarfs in the H–R diagram. The existence of such objects was first suggested by Adams and Joy [A1], who identified three weak-lined 'A-type' stars – HD 19445, HD 219617 and HD 140283 – as having unusual absolute magnitudes for their spectral properties. The latter two stars are now known to be extremely metal-poor subgiants.

By the 1950s, several other subdwarfs had been identified, but it remained unclear whether these stars were distinct in both the observational $((M_V, (B–V)))$ and theoretical $((M_{bol}, T_{eff}))$ planes. Sandage and Eggen [S3] demonstrated that this was the case. As discussed in Chapter 4, decreased line and continuum absorption with decreased metal abundance leads to a star of given mass having a higher effective temperature and emitting a larger proportion of its flux in the optical régime (that is, a smaller bolometric correction, and a brighter M_V). Thus, while the subdwarf sequence is often characterised as lying 'below' the solar-metallicity main sequence in the colour–magnitude diagram, the true offset is primarily in colour (temperature), with lower abundance stars lying progressively further blueward of the disk dwarf sequence.

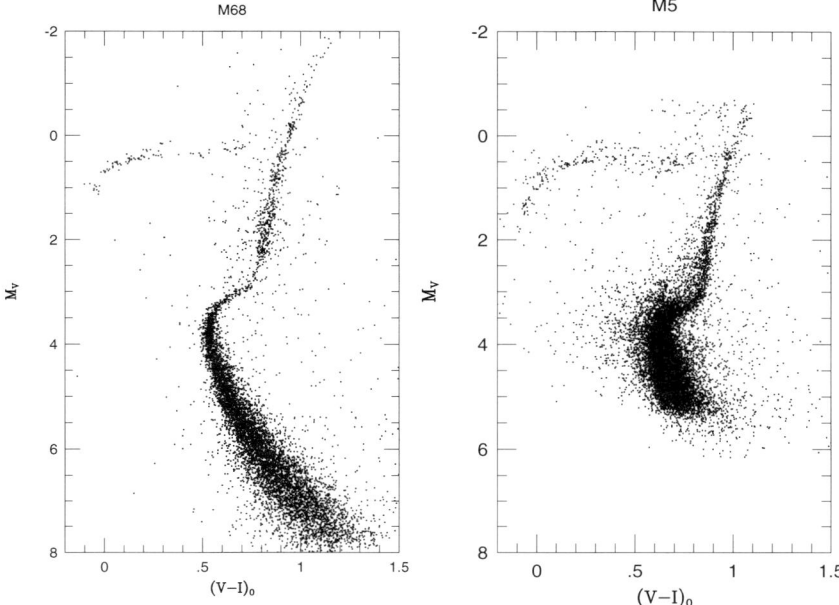

Figure 6.4. Colour–magnitude diagrams for the globular clusters M5 and M68. (M68 courtesy of A. Walker.)

Theoretical models predict that the main-sequence absolute magnitude of normal, single stars of a given colour (temperature) is a monotonic function of metallicity: the Russell–Vogt theorem outlined in Section 2.4. Until recently, few subdwarfs had both abundance determinations and parallaxes measured with sufficient accuracy to provide a strong test of that basic hypothesis. Observations from the ESA Hipparcos astrometric satellite project [E7] have improved the situation for F, G and early K-type subdwarfs. Figure 6.5 plots the $(M_V, (B-V))$ distribution described by subdwarfs and subgiants with $[Fe/H] < -0.7$ and parallaxes measured to a precision of at least 15% $(\sigma_\pi/\pi < 0.15)$. The chemical abundances are derived from analysis of high-resolution echelle spectra [A5], [G3]. The mean $(M_V, (B-V))$ relationship defined by disk dwarfs in the immediate Solar Neighbourhood is plotted as a reference. By and large, data for FGK subdwarfs meet the theoretical expectations of a monotonic decrease in M_V with decreasing $[m/H]$.

6.3.2   The density distribution

The extended nature of the halo has been evident since Shapley's analysis of the globular cluster distribution. Globular clusters are large, luminous objects, and while a few distant $(R > 30\,\mathrm{kpc})$ or highly-obscured systems were identified in the last twenty years, most have been known since Dreyer's compilation of the New General Catalogue in 1885 (and many since Messier's 1785 catalogue of non-comets). The

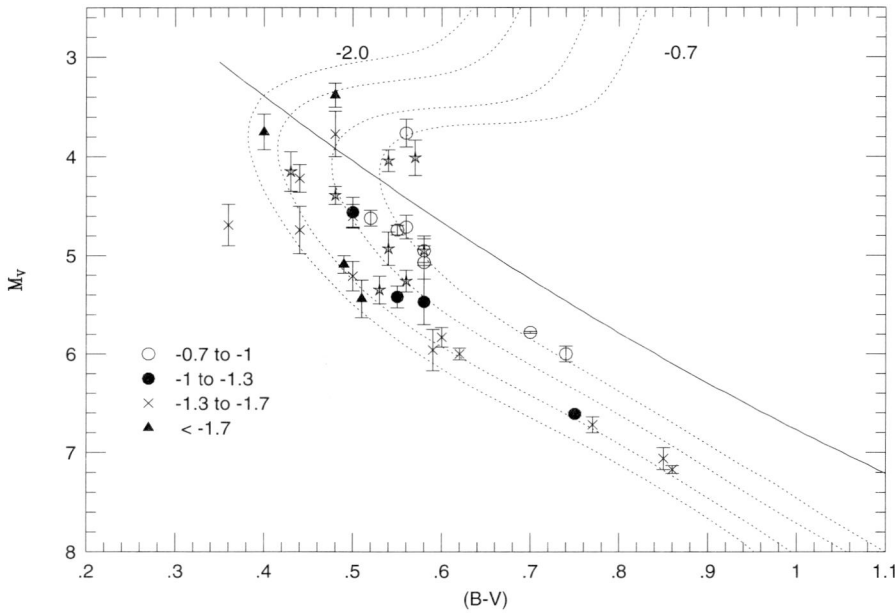

Figure 6.5. $(M_V, (B–V))$ data for single subdwarfs with abundance determinations by [G3] or [A5], compared with 12-Gyr isochrones predicted by [D1] for [Fe/H] $= -0.7, -1.0, -1.5$ and -2.0.

projected distribution in Galactic co-ordinates shows a clear concentration toward the Galactic centre, which led Shapley to formulate his Great Galaxy hypothesis. Distances can be estimated using a number of techniques, notably main-sequence fitting, or calibrating the mean magnitude of either non-variable horizontal branch stars or RR Lyrae stars. The latter variables had been identified in several clusters by Bailey in the late 1880s [B6], and Shapley's initial distance-scale analysis was based on what we now know to be an erroneous RR Lyrae absolute-magnitude calibration. The inferred distance to the centre of the cluster distribution was $\sim 20\,\mathrm{kpc}$.[2]

Modern analyses of the inferred three-dimensional spatial distribution by Harris [H1] and Zinn [Z1] lead to the conclusion that the metal-poor ([Fe/H] < -1) clusters describe a near-spheroidal distribution, with the space density following a radial variation, $\rho(R) \propto R^{-3.5}$, where R is Galactocentric radius. Higher-abundance systems describe a much flatter spatial distribution (Figure 6.6), and have higher net rotation than the classical metal-weak clusters [Z2]. This apparent dichotomy has been confirmed by more recent observations [A3], and the two subsystems are often referred to as 'halo' (metal-poor, $[m/H] < -1$) and 'disk' clusters respectively. There

---

[2] Shapley's Galaxy had a diameter about 30 times larger than the diameter of Kapteyn's Universe, the prevailing model of the time. However, Shapley's Milky Way was envisaged as a loose aggregation of sub-systems, with Kapteyn's Universe being the local system, rather than as a coherent structure. In the so-called Great Debate, Shapley's Great Galaxy was more accurate in scale, but Curtis' Island Universe model was more accurate in morphology.

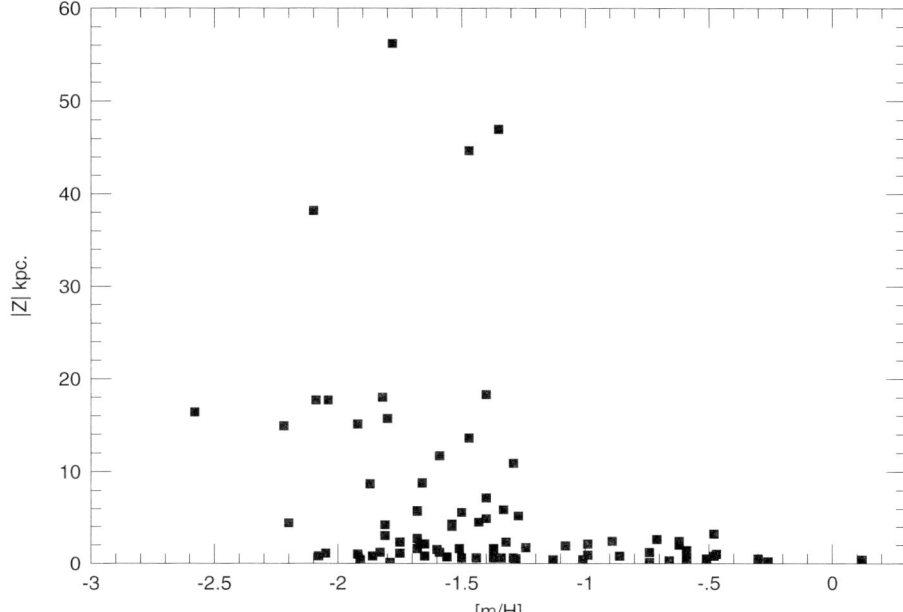

Figure 6.6. The present-day distribution of distance above the Galactic Plane of globular clusters, plotted as a function of abundance (from [Z2]). The higher-abundance systems are more closely confined to the disk.

are, however, good grounds, notably the strong spatial concentration near the centre of the Galaxy, for associating the 'disk' system with the Galactic bulge, rather than with the field-star disk population.

Globular clusters have completed as many as 100 Galactic orbits during their lifetimes. It is therefore inevitable that some systems, probably a substantial fraction of the initial population, have been disrupted through tidal interactions. The surviving systems are likely to be objects which have spent a small fraction of the time in high-density regions near the Galactic Centre or close to the Galactic Plane. Thus, there is no guarantee that the present distribution of $\sim 10^5 \, M_\odot$ stellar aggregations provides an accurate reflection of the underlying field-star distribution, and it is important to supplement the cluster data with observations of field stars.

Blue horizontal branch (BHB) stars, especially RR Lyrae variables, are well suited as density tracers, since they are luminous, readily identifiable through their colours and/or variability; and their absolute magnitudes can be calibrated with some reliability. A complication is that the HB morphology depends on the core mass following red giant branch evolution, which varies as a function of metallicity and age [L3]. Metal-rich clusters, such as 47 Tucanae ($[m/H] = -0.7$), possess only a short, red horizontal branch, with few RR Lyrae stars. Moving to lower abundances, theoretical HB tracks for ages of 10 Gyrs or more intersect the instability strip, and clusters such as M3 and M5 have significant numbers of RR Lyrae variables.

Decreasing the abundance further leads to a continued migration towards higher temperatures until, at $[m/H] = -1.7$ to -1.9, the horizontal branch lies fully blueward of the instability strip (an example being NGC 6397, $[m/H] = -1.8$). That trend reverses at lower abundances, mainly since lower mass-loss rates on the giant branch lead to higher core masses, and clusters such as M15 and M68 ($[m/H] \sim -2.1$) have nearly symmetric HB morphologies and well-populated instability strips.

In addition to the overall trend in morphology with metallicity, global analyses must address the 'second parameter problem': clusters of the same abundance may have significantly different HB morphologies. This is exemplified by the inter-mediate-abundance pair M3 and M13, where M13 has an extended blue horizontal branch and almost no RR Lyrae variables, and by M92 and M68, where M92 has a more extended BHB than does M68. The origin of these differences remains a subject of debate (see, for example, [S8]), but age variations are usually indicted as responsible.

Regardless of the underlying mechanism, these variations in HB morphology suggest that the density distribution of BHB stars may not match $\rho(r)$ for the underlying stellar population if the halo is chemically inhomogeneous. Fortunately, significant inhomogeneities do not seem to be present: the halo population appears to be well mixed at all radii. Most density analyses derive density distributions $\rho(r) \propto r^{-3.5}$ [K2], generally consistent with the globular cluster distribution. With regard to the three-dimensional shape, results favour a near-spherical distribution at large radii, slightly flattened along the vertical axis ($c/a \sim 0.9$). The halo may be more complex near the disk, with perhaps half of the local subdwarf population drawn from a flatter component (axial ratio $c/a \sim 0.5$: [H2], [S17]). There is no clear evidence for differences in chemical composition, and the origin of these two possible halo components remains under investigation.

6.3.3   Halo kinematics

Globular clusters, RR Lyrae variables, field horizontal branch stars, metal-poor giants and local subdwarfs have all been used to probe halo kinematics. In the case of the more luminous and more distant tracers, such as RR Lyrae variables and globular clusters, transverse motions are poorly determined. However, statistical techniques can be used to reconstruct the full three-dimensional motions from radial velocities given a sample well-distributed on the sky.

Table 6.2 lists results derived by recent studies, expressed as a Schwarzschild velocity ellipsoid in the (U, V, W) co-ordinate system described in Chapter 1. Apparently the mean rotational velocity of the halo lags that of the Sun by 180–230 km s^{-1} which, since the net rotational velocity at the Solar Circle is estimated as between 200 and 220 km s^{-1}, implies that the halo has negligible mean rotation. There are also indications that the net rotation decreases with increasing height above (or below) the Galactic Plane, and that the outer halo has significant retrograde rotation [M4]. This may provide a clue to the formation and early history. The typical halo star has a substantial velocity relative to that of the Sun and

Table 6.2. The kinematics of the stellar halo.

	$\bar{U}^a$	$\bar{V}$	$\bar{W}$	σ_U	σ_V	σ_W	N_*	$\langle V_{\tan}\rangle^b$
	$(\mathrm{km\,s^{-1}})$			$(\mathrm{km\,s^{-1}})$				$(\mathrm{km\,s^{-1}})$
Clusters[1]		-160		118	118	118	66	172
$\Delta S \geq 5$ RR Lyrae stars[2]		-220		210	119	91	33	228
Halo RR Lyrae stars[3]	9	-210	-12	168	102	97	162	209
Metal-poor stars[4]		-200		153	93	107	887	201
Metal-poor stars[5]	16	-217	-10	161	115	108	180	215
Subdwarfs[6]		-180		130	105	85	452	177
Subdwarfs[7]		-195		133	98	94	420	189
Disk dwarfs[8]	-9	-22	-7	41	27	21	311	39

[a] Typical uncertainties for the halo kinematics are 10–$15\,\mathrm{km\,s^{-1}}$.
[b] The median tangential velocity for each sets of kinematics.
[1] Frenk and White [F3].
[2] Metal-poor RR Lyrae stars, from Oort [O3].
[3] The Halo-3 statistical parallax analysis by Layden *et al.* [L2].
[4] [Fe/H] <-1.5 field stars, Beers and Sommer-Larsen [B8].
[5] Metal-poor giants and RR Lyrae stars, Hipparcos data, Chiba and Yoshii [C10].
[6] Kinematically unbiased sample of field subdwarfs with [Fe/H] <-1.2, Norris [N3].
[7] Kinematically unbiased sample of field subdwarfs with [Fe/H] <-1.6, Morrison *et al.* [M16].
[8] Nearby stars, single population, Hawley *et al.* [H3]: uncertainties $\pm 5\,\mathrm{km\,s^{-1}}$.

subdwarfs make a disproportionately large contribution(relative to their space density) to catalogues of high proper-motion stars. As a result, proper motion surveys offer one of the most effective methods of identifying local members of the halo population (see Chapter 11).

6.3.4   The abundance distribution of the halo

Heavy elements originate from stellar nucleosynthesis. The first halo stars formed from protogalactic gas which was almost entirely H and He. The halo abundance distribution reflects how those stars and succeeding generations polluted their environment. Chapter 4 outlined the physical principles underlying the measurement of stellar metallicities, $[m/H]$. This section summarises the applications of those principles to study the chemical composition of halo subdwarfs.

Abundances are often expressed as [Fe/H] – the iron-to-hydrogen abundance relative to the solar value expressed in logarithmic units. The equivalence $[m/H] \equiv [Fe/H]$ is valid for the majority of disk stars, where the relative abundances of individual elements remain approximately constant: that is, if iron is deficient by a factor of two relative to the Sun ([Fe/H] $= -0.3$), then the same holds for Mg, Ca, O

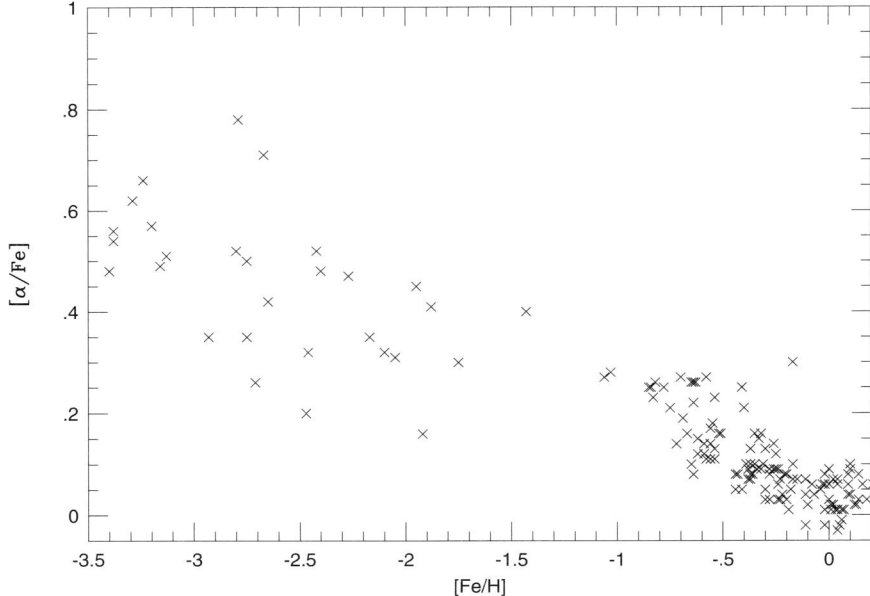

Figure 6.7. The variation in the α-element/iron abundance ratio as a function of iron abundance. (Data from [E1].)

and so on. That is not the situation for most halo stars. A number of species, notably the so-called α-elements (O, Ca, Ti, Mg, Si, S), have abundances which are enhanced above the scaled-solar values in most subdwarfs [3] with [Fe/H] < -1 (Figure 6.7). This has implications for halo formation, as discussed in Section 6.3.6.

Since most halo stars have non-solar abundance ratios, the terms 'metallicity' and '[Fe/H]' cannot be used interchangeably, as with disk dwarfs. Abundances are therefore usually written as [m/H], which can be approximately related to the iron abundance by

$$Z_M = Z_{Fe}(0.638 \times f_\alpha + 0.362) \tag{6.1}$$

where Z_{Fe} is the iron abundance (in linear units), f_α the α-element enhancement and Z_M the overall effective metal abundance [S1]. This is a useful relation, since many abundance analyses are based specifically on measuring the strength of Fe absorption lines.

Measuring abundances

Most studies of the halo abundance distribution are based on observations of either

[3] As with most statistical correlations, this behaviour does not hold for *every* star. A number of subdwarfs are known with both [Fe/H] < -1 and solar element ratios, and with [Fe/H] ~ -0.5 and enhanced [O/Fe] [E1].

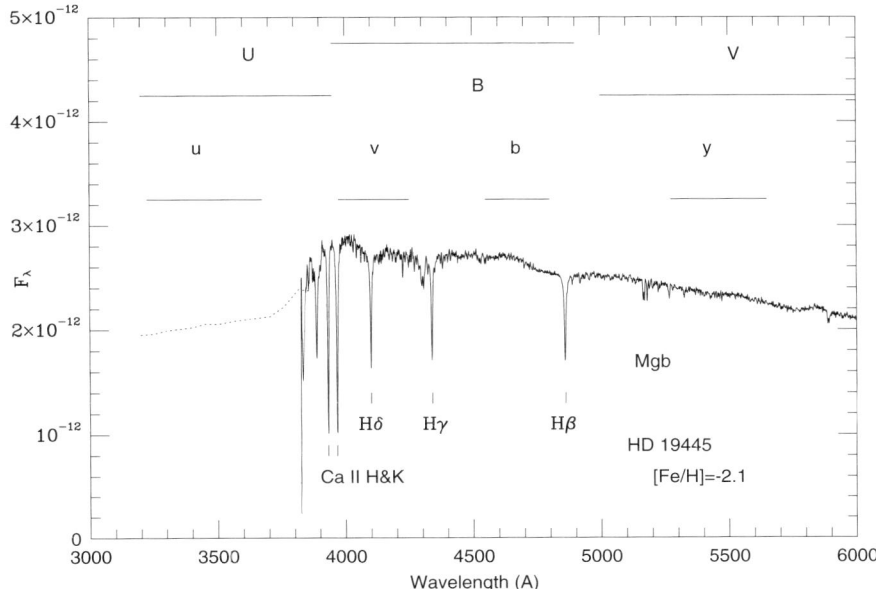

Figure 6.8. The UBV and uvby photometric passbands matched against the spectral energy distribution of the F-type subdwarf, HD 19445. The solid line is the measured spectrum; the dotted line shows the extension to shorter wavelengths.

F, G subdwarfs or K-type red giants. In both cases, a variety of photometric and spectroscopic techniques are used to estimate [m/H]. The first studies [S2] centred on Roman's nearby high-velocity F, G subdwarfs, and used UBV photometry to quantify ultraviolet excess as

$$\delta(U–B) = (U–B)_{obs} – (U–B)_{std} \tag{6.2}$$

where $(U–B)_{std}$ is the colour of a solar-abundance dwarf with the same (B–V) colour. Figure 6.8 illustrates the basis for this technique: the (U–B) colour measures the relative flux emitted over the 3300–3900 Å and 4100–4900 Å regions. Metal-poor stars have weaker absorption lines, less line blanketing, and therefore more ultraviolet flux than solar-abundance dwarfs with the same (B–V) colour (temperature). Figure 6.9 shows the ((U–B),(B–V)) diagram outlined by photometric standards supplemented by data for subdwarfs with abundance of [m/H] < −1.5.

The Strömgren *uvby* photometric system [S21], [C12] provides a finer-tuned measurement of the same spectral characteristic, with the index

$$m_1 = (u – b) – (b – y) \tag{6.3}$$

measuring line blanketing in the ultraviolet relative to the blue–visual. As with $\delta(U–B)$, relative abundances are determined by comparing the metallicity index against the standard value at that temperature:

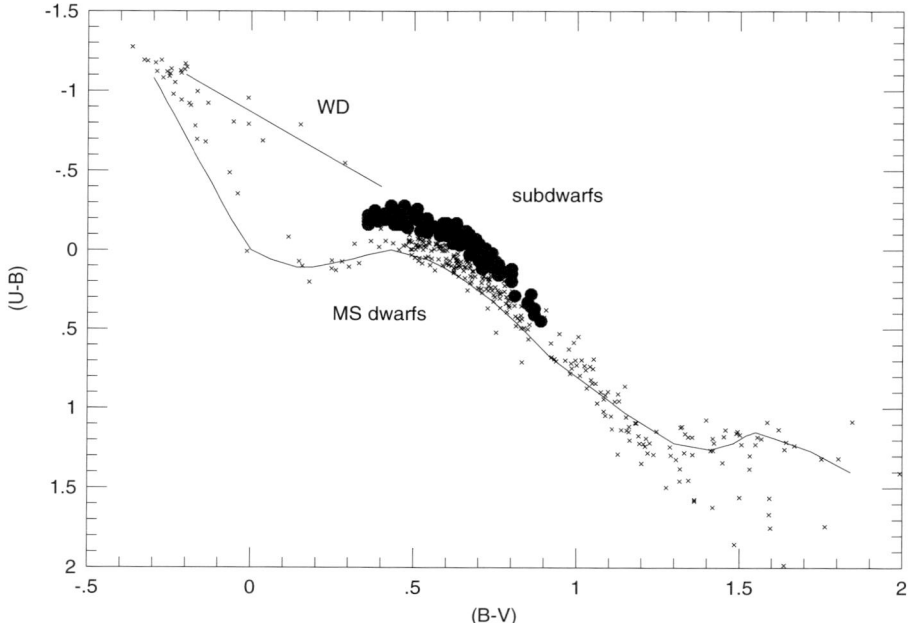

Figure 6.9. The $((U–B), (B–V))$ diagram. Solid points are photometric standards, predominantly disk stars; open circles plot data for extreme subdwarfs. (From Carney *et al.* [C5].)

$$\delta m_1 = m_1(obs) - m_1(std) \tag{6.4}$$

In this case $(b–y)$ serves as the temperature indicator and the $m_1(std)$ calibration is defined by the disk dwarf $(m_1, (b–y))$ relation [N1]. Strömgren indices have been measured for several thousand F, G and early K stars in both disk and halo.

In a similar manner, intermediate-band photometric systems have been devised to measure relative abundances of K-type stars. The two systems used most frequently are the DDO system [M1] and the Washington system [C3]. The former has six filters of bandwidth ~ 100 Å, centred at $\lambda\lambda 3500, 3800, 4100, 4200, 4500$ and 4800 Å, usually supplemented by the '51' filter, centred on the Mgb lines (Figure 6.10). The latter system employs four filters, $CMT_1 T_2$, with bandwidths close to 1000 Å and centred at $\lambda\lambda 3910, 5085, 6330$ and 8050 Å. As with UBV and uvby data, relative abundances are determined by using one colour as a temperature estimator (usually DDO colour $C(45–48)$ and the Washington colour $(M–T_2)$) and matching an abundance-sensitive index against a standard relationship. The two systems are often used in conjunction, with temperature estimated from the broader-band Washington data and abundances from DDO indices. As Figure 6.10 shows, the DDO colour difference $C(41–42)$ provides a measure of the strength of CN absorption, while $C(48–51)$ measures Mgb/MgH strength.

Considered in isolation, photometric indices provide only a relative ranking of stellar abundances. Setting those measurements on an absolute abundance scale

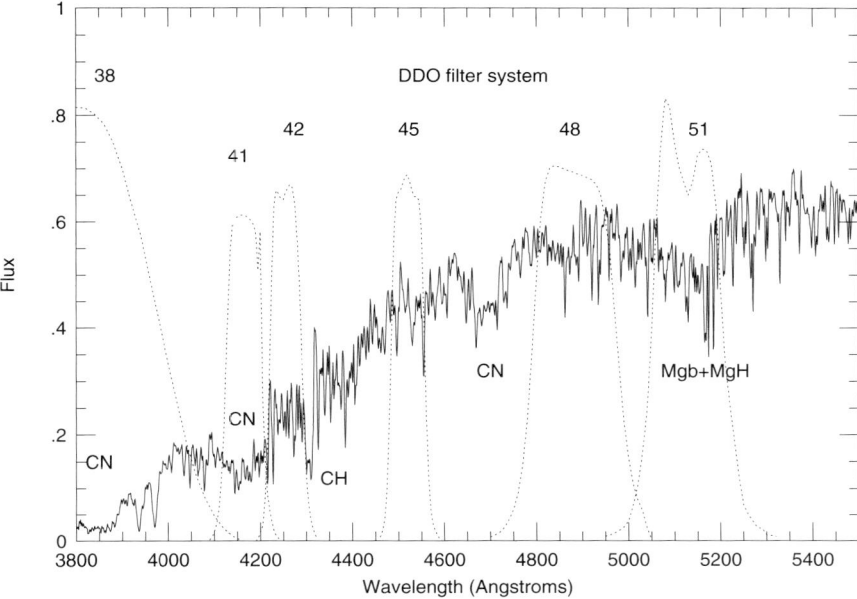

Figure 6.10. The passbands of the DDO photometric system matched against the spectrum of a K0 disk giant.

requires detailed analysis of observations at higher spectral resolution. G dwarfs have a wealth of features at blue/visual wavelengths, notably Ca II H and K, the G-band (CH) and the Mg I Fraunhofer b triplet, as well as numerous atomic lines due to Fe. All vary significantly with decreasing abundance (Figure 6.11). The larger variation in linestrengths between $[Fe/H] = -2$ and -1 (linear régime) than between -1 and solar abundance (flat or damped régime) illustrates the principle of curve of growth analysis (Figure 4.4). Narrowband spectrophotometric indices can be devised to measure specific features in low resolution spectra [R10], or individual line profiles can be analysed at spectroscopic resolutions >20,000 [A5], [G3], [C4]. Modern abundances are derived either through direct line analyses or by spectrum synthesis, matching the observed spectrum against theoretical predictions for a range of [m/H]. Stars calibrated in this manner serve as standards for lower-resolution spectroscopic and photometric systems.

The halo abundance distribution

High spectral-resolution abundance analyses provide the most accurate metallicity measurements, but require high signal-to-noise spectra and a substantial investment of time on 4-m class (or larger) telescopes. Photometric and spectrophotometric observations require fewer resources, and those techniques therefore provide the majority of metallicity estimates. Measuring uncertainties are typically given as

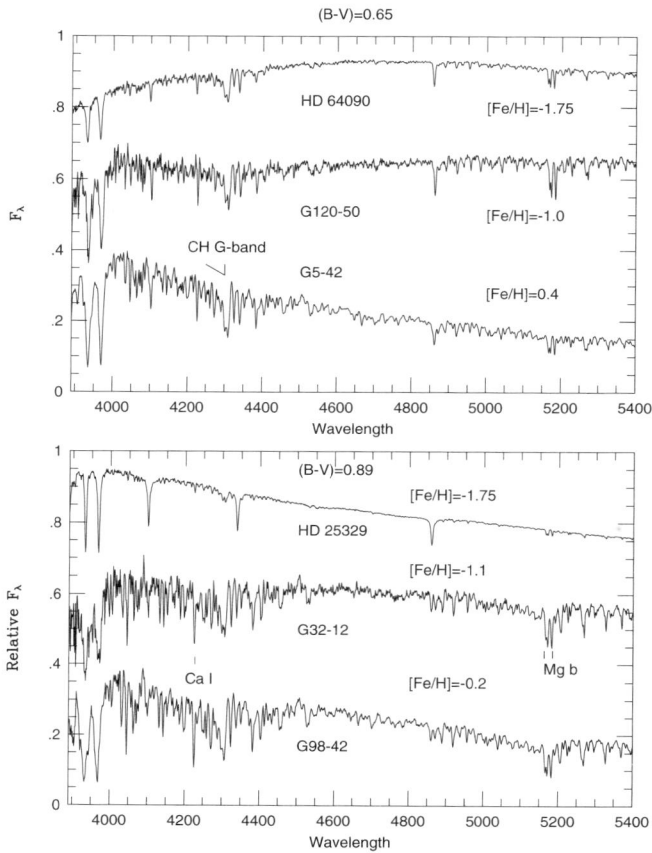

Figure 6.11. Spectra of two sets of G-type dwarf with identical (B–V) colours and abundances of $[Fe/H] = 0, -1$ and -2.

$[m/H] = \pm 0.2$, but systematic offsets of similar magnitude in both zero-point and scale exist between different systems or different analyses based on the same technique [R4], [C11]. The latter often reflect the particular choice of standard stars as primary calibrators.

With an increasing number of stars having high-resolution spectroscopic observations, lower-resolution systems will be calibrated in a self-consistent manner in the near future. At present, analyses rest to a substantial extent on older observations, which often remain uncorrected for systematic errors. In particular, the fundamental reference point – the iron abundance of the Sun – was revised in the early 1990s, rescaling $[Fe/H]$ by definition. The pre-1991 reference value for the solar iron abundance was Fe = 7.67, where the abundance is expressed in logarithmic units with H = 12.0; the revised value [B12] is Fe = 7.52. Abundance scales defined using standards with high-resolution analyses pre-dating 1991 (such as the extensive

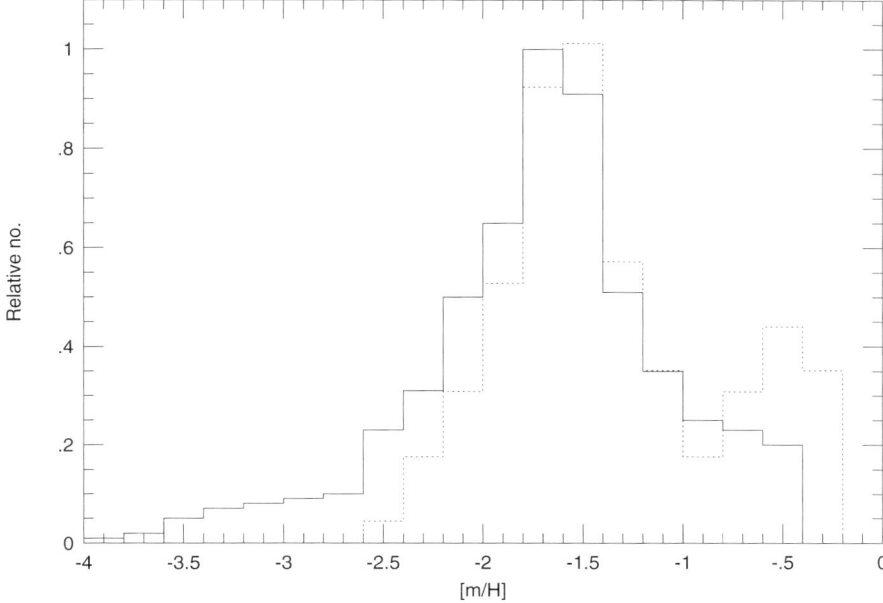

Figure 6.12. The halo abundance distribution. The solid line plots the field-star distribution from [L1]; the dotted line plots the cluster abundance distribution from [Z2].

surveys undertaken by [N1] and [C4]) therefore underestimate $[Fe/H]$ for *every* star by 0.15.

Figure 6.12 compares the abundance distribution of globular clusters [Z2] against data for nearby subdwarfs [L1]. Both abundance scales probably underestimate $[m/H]$ by 0.1 to 0.3 for abundances exceeding -2. Deciphering the detailed distribution at $[m/H] > -1$ is complicated by the presence of 'disk' globular clusters and metal-weak disk dwarfs respectively. However, both distributions indicate a modal abundance of $[m/H] \sim -1.5$, or a metallicity $\sim \frac{1}{30}$ that of the Sun.

The field subdwarf distribution in Figure 6.12 extends to lower abundances than the cluster distribution, probably reflecting the destruction of the oldest clusters and their consequent absence from the present-day census. Few extremely metal-poor stars were known until recently, but large-scale objective prism surveys, aimed specifically at detecting stars with very weak lines, have now succeeded in discovering over 50 objects with $[Fe/H] < -3.5$ [B7]. The star with the lowest measured abundance is the red giant CD-38:245, with $[m/H] = -4.5$ [B10]. Extremely metal-poor stars are expected to be rare: a mass of only several $\times 10^{-5}\, M_{\odot}$ in 'metals' per star is all that is required to raise abundances to $\frac{1}{1000}$ solar ($[m/H] = -3$) in a star-forming cloud. The mean abundance of the interstellar medium increased rapidly to $[m/H] = -2$ with the first generation of evolved stars; reaching higher abundances requires proportionately more mass in stellar ejecta, accumulated over longer timescales, which, in turn, allows more star formation.

Thus, the number of stars increases with increasing abundance for $[m/H] < -1.4$. The decrease at higher abundances presumably marks the cessation of star formation in the halo. There is no evidence for significant variation in the mean abundance as a function of position in the halo.

6.3.5   The age of the Galactic halo

This issue has two aspects: absolute age and the range of ages spanned by members of the halo. The absence of OBA main-sequence stars at large distances from the Galactic Plane indicates that star formation ceased at least 1 Gyr ago. Further quantifying that statement demands comparisons between observations and theoretical stellar models. The most effective method of deriving absolute ages is the determination of the mass of stars currently evolving onto the subgiant branch; that is, the mass at the main-sequence turnoff. Although simple in principle, this technique remains difficult to implement in practice. In contrast, more easily-calibrated methods are available for estimating the age distribution of halo constituents. We summarise those before discussing absolute ages.

Relative ages

Elemental abundances in field subdwarfs provide an important clue to the overall time-scale of halo formation. At halo metallicities ($[m/H] < -1$), iron abundance (the abscissa in Figure 6.7) can be taken as a (non-linear) time-dependent variable. The α-elements form through nuclear reactions involving capture of α particles, which occur in type II supernovae – that is, supernovae with massive ($>8-9\,M_\odot$) progenitors. In contrast, iron originates predominantly in type I supernovae, which form through thermonuclear runaway in accreting white dwarf stars in binary systems (I2). The latter systems have evolutionary timescales of ~ 1 Gyr, since the intermediate-mass white dwarf progenitor must evolve through the red giant, Cepheid, asymptotic giant branch and planetary nebula phases, before forming a white dwarf; massive stars evolve to type II supernovae in a matter of only a few million years. Thus, the high $[\alpha/\text{Fe}]$ ratios measured for halo stars suggest that most formed before substantial numbers of type I supernovae could contribute to the Galactic iron abundance; as type I supernovae became more common, the $[\alpha/\text{Fe}]$ ratio was reduced toward the solar value. This hypothesis implies that most of the Galactic halo formed in a span of only 1–2 Gyr.

   Globular cluster analyses suggest a similarly short formation period. Relative ages for clusters of similar abundance can be estimated using several techniques – by notably measuring the offset in colour between the turn-off and the base of the giant branch ($\Delta(\text{B–V})_{TO}$ [V1], [S6]). The (B–V) colour at the base of the giant branch is primarily a function of metallicity in old stellar systems (Figure 6.13). As the system ages, the main-sequence turn-off moves to lower luminosities and redder colours, but the base of the giant branch maintains a constant colour, so the colour difference decreases. $\Delta(\text{B–V})_{TO}$ can be measured directly from the observed colour–magnitude diagram, without requiring corrections for foreground reddening or knowledge of

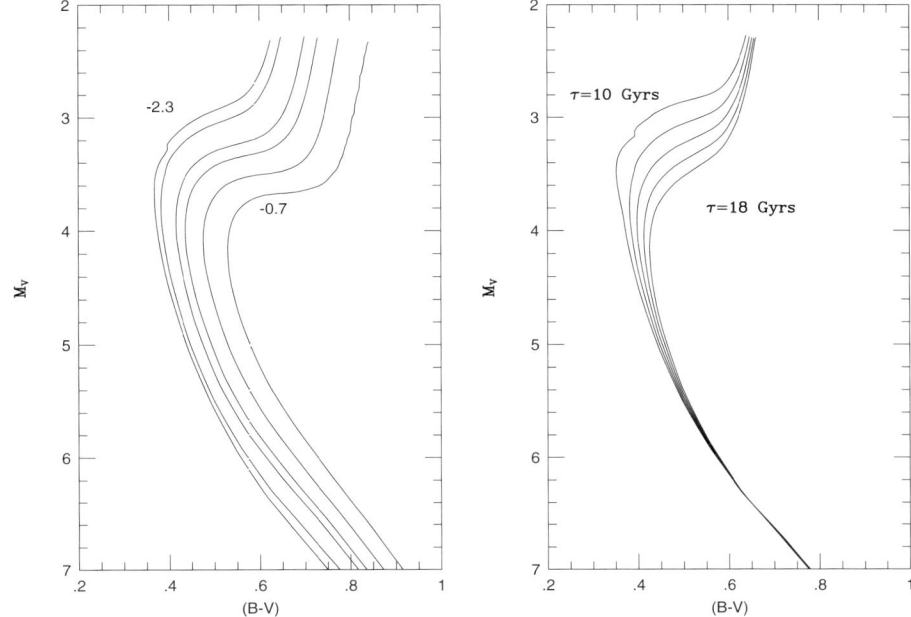

Figure 6.13. Theoretical isochrones for metal-poor globular clusters. (*Left*) 12-Gyr isochrones for abundances of [Fe/H] = −0.7, −1.0, −1.3, −1.5, −2.0 and −2.3; (*right*) shows the evolution of an [Fe/H] = −2.0 system, plotting isochrones for age 10, 12, 14, 16 and 18 Gyr.

the exact cluster distance. Thus, this parameter provides a robust measurement of the cluster age *distribution* at a given metal abundance. Recent studies (see [S7]) indicate that the majority of clusters have similar ages, with a few (∼ 10%) younger by 3–5 Gyr. Many of the younger clusters lie at Galactocentric radii exceeding 10 kpc (for example, Palomar 14 [S8]), a matter discussed further below.

Absolute age determinations

Absolute age measurements rest almost exclusively on globular cluster data, primarily comparison of the luminosity at the main sequence turnoff against evolutionary models. This procedure places stringent demands on both observation and theory. Observationally requirements are accurate photometry; accurate measurement of foreground reddening, allowing determination of intrinsic colours and magnitudes; accurate abundance determination; and, above all, a reliable distance estimate. Transforming the measured absolute magnitude, M_V(TO) to an age requires accurate modelling of convection and helium diffusion; reliable opacities; an appropriate mix of chemical abundances; and the definition of the appropriate equation of state. As yet, no models fulfil all of these goals, but recent years have seen substantial improvements. Figure 6.13 illustrates the (M_V, (B–V)) colour–magnitude diagram for stars of various age and abundance.

The task of obtaining precise colour–magnitude data for globular clusters has been rendered straightforward by the availability of large-format CCDs on 4 m-class telescopes. Foreground reddening can be estimated to reasonable precision using far-infrared maps from either the IRAS or COBE satellites. Cluster distances are more problematic. The primary method of distance estimation is main-sequence fitting, [S4]: a fiducial sequence derived from the reddening-corrected cluster colour–magnitude diagram (that is, $(V_0, (B–V)_0)$ is matched against either model isochrones (the theoretician's route to distances) or an empirical $(M_V, (B–V)_0)$ sequence based on subdwarfs with accurate parallax measurements (the observer's route to distances). In either case, the offset $(V_0–M_V)$ provides the distance estimate.

Both empirical and theoretical approaches to main-sequence fitting must deal with significant obstacles to obtaining accurate distance estimates: in the former case, the calibration rests on only a handful of subdwarfs; the latter demands accurate models and accurate transformation from the theoretical to observational plane. Given these concerns, many studies (for example, [B15]) opt for semi-empirical methods, adjusting theoretical isochrones to match the few subdwarfs with accurate data. The availability of Hipparcos astrometry, and better parallaxes for a larger number of metal-poor stars, has eased matters to some extent.

Prior to 1995, most analyses favoured ages of 16 Gyrs or more for low-abundance clusters such as M92 and M68, with intermediate-abundance systems, such as M5 or M13, 2 to 3 Gyr younger. Subsequently, improved stellar evolution models have become available which incorporate revised opacities and an improved equation of state [D1], [C7]. Compared to previous studies, the new models predict lower luminosities for main-sequence stars of given mass, implying higher masses and younger ages for a given $M_V(TO)$. Combined with Hipparcos-based main-sequence fitting analyses [R4], which indicate an increase of $\sim 5\%$ in cluster distances (i.e. brighter $M_V(TO)$), these new models reduce globular cluster age estimates to ~ 11 to 14 Gyr (Figure 6.14). There is no obvious trend towards younger ages with increasing abundance.

Summarising ages

Current analyses indicate that the bulk of the stellar halo was formed in a 1 to 2 Gyr burst of star formation $\sim 12.5 \pm 1.5$ Gyr ago. The presence of a small number of younger globular clusters suggests that star formation continued at a reduced level for the succeeding 3 to 5 Gyr, primarily in the outer halo.

6.3.6   Forming the halo

The conclusions summarised in the previous subsections, concerning the kinematics, density distribution, abundance distribution and age distribution of the halo, provide baseline constraints for the two main models which have been proposed to describe the formation of the halo: the monolithic collapse model, originating with Eggen, Lynden-Bell and Sandage [E2]; and the fragmentary accretion model proposed by Searle and Zinn [S13].

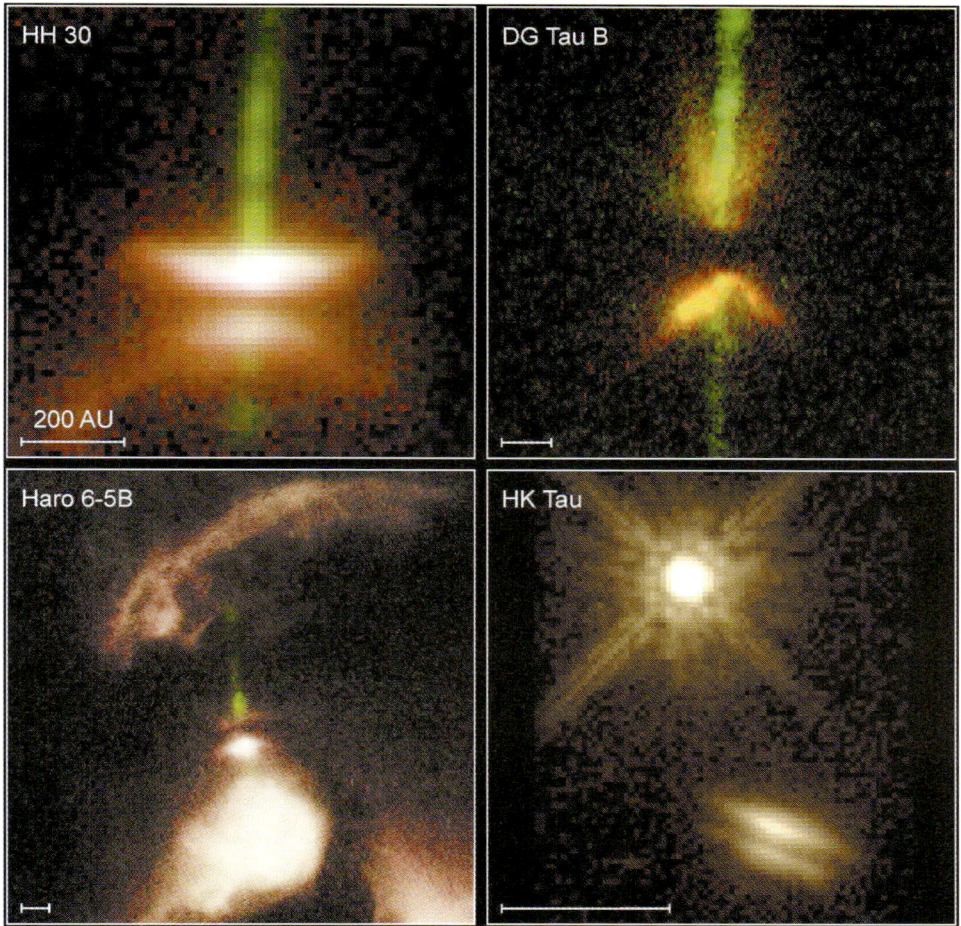

Figure 3.18. Hubble Space Telescope images of gaseous jets from Herbig–Haro objects. The protostellar disk is evident as the dark lane in HH 30. (Courtesy of STScI.)

Figure 3.19. A protoplanetary disk, or 'proplyd', set against the Orion Nebula. (Courtesy of STScI.)

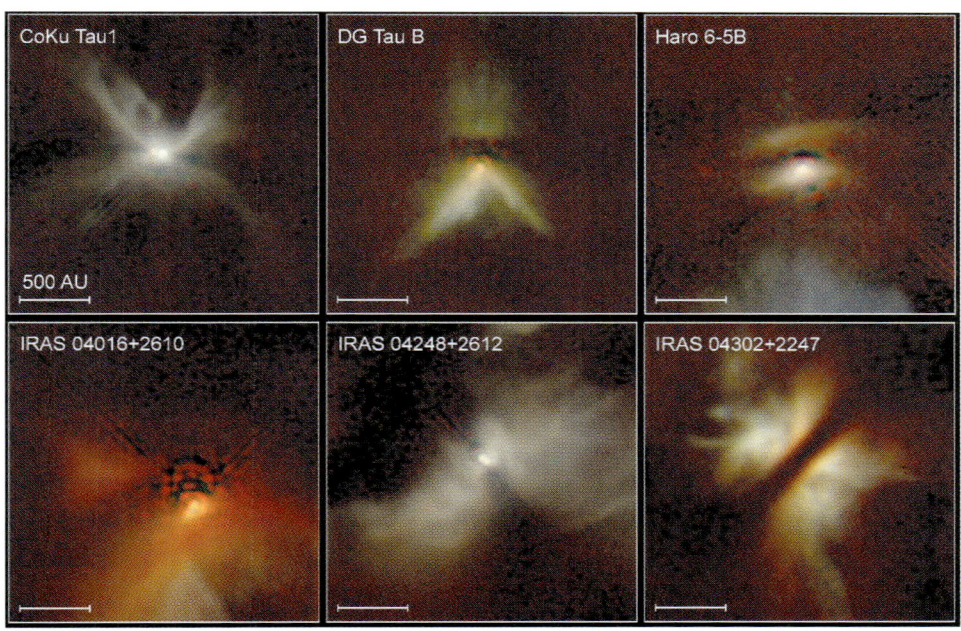

CoKu Tau1   DG Tau B   Haro 6-5B

500 AU

IRAS 04016+2610   IRAS 04248+2612   IRAS 04302+2247

Figure 3.20. Hubble Space Telescope images of disks around young stars. (Courtesy of STScI.)

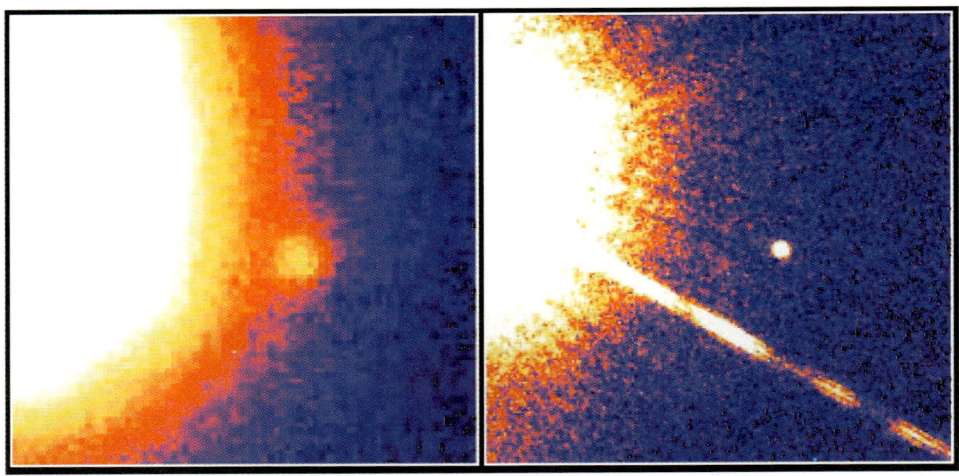

Figure 9.5. Images of Gl 229B: (*left*) discovery image, Palomar Observatory 60-inch reflector, 27 October 1994; (*right*) Hubble Space Telescope image, 17 November 1995. (Courtesy of STScI.)

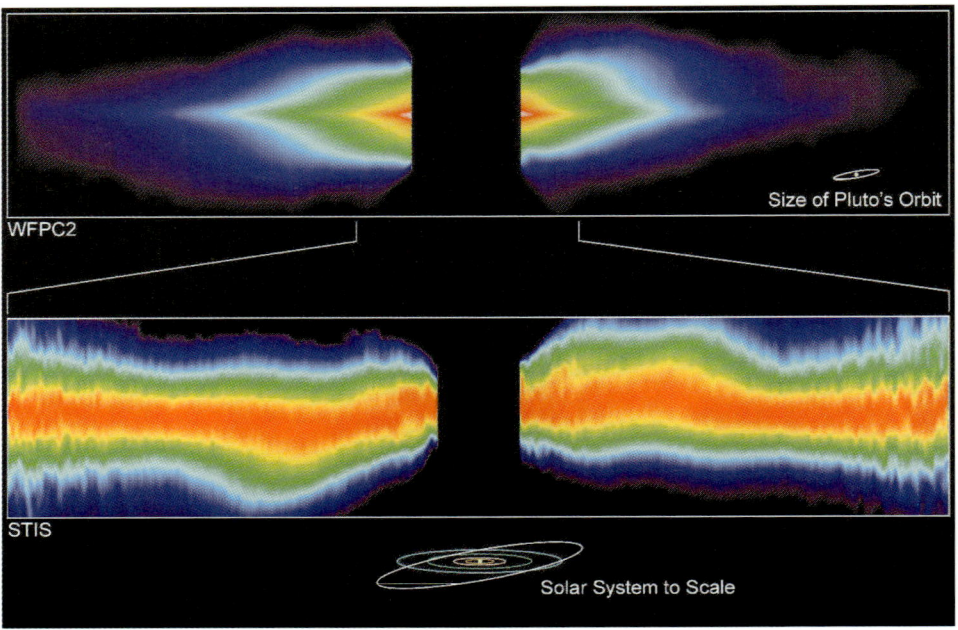

Figure 10.4. Hubble Space Telescope images of the dusty debris disk around β Pictoris. The warps in the light distribution suggest the presence of massive objects – planetesimals. (Courtesy of STScI.)

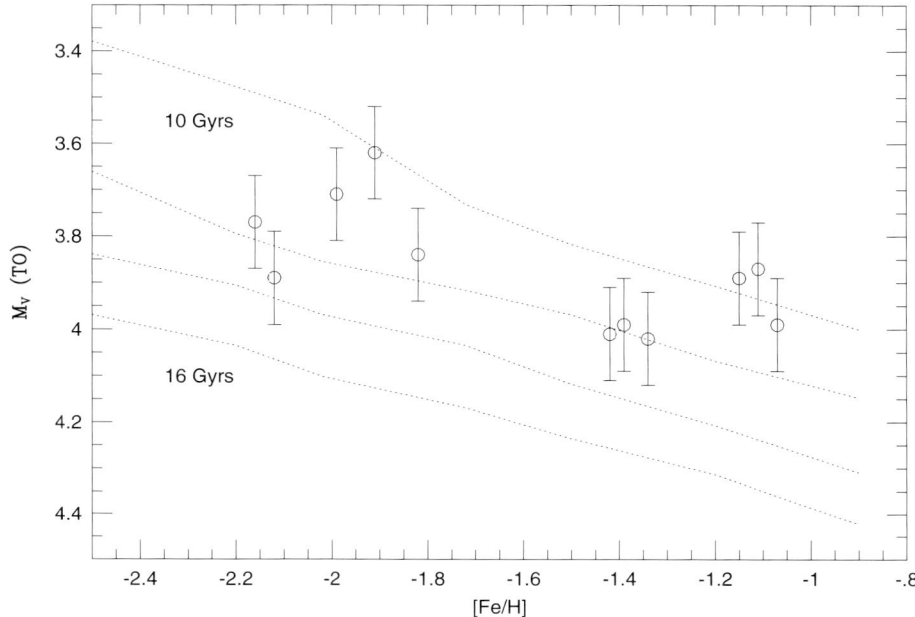

Figure 6.14. Age calibration for globular clusters: turn-off absolute magnitudes, deduced from main-sequence fitting against subdwarfs with Hipparcos parallaxes, are matched against the predictions of the [D1] set of models. Ages in the range 11–13 Gyr are indicated.

The ELS monolithic collapse model

The Eggen, Lynden-Bell and Sandage (ELS) model marks a landmark as the first serious analysis of galaxy formation. ELS combined kinematic and photometric data for 221 stars, drawn in almost equal numbers from two catalogues of, respectively, nearby stars and high-velocity stars. Reconstructing Galactic orbits from the space motions, the results showed that both orbital eccentricity and $|W|$, the velocity perpendicular to the Plane, were correlated strongly with the ultraviolet excess, $\delta(U–B)$: stars with high velocities and high eccentricities had stronger ultraviolet excess (that is, lower metallicity). ELS hypothesised that the observed relationships stemmed from an underlying correlation between age and abundance, with metal-poor stars having formed during the initial collapse of the largely gaseous protogalaxy. High-angular-momentum, metal-rich stars on nearly circular orbits (like the Sun) are denizens of the rotating disk which was the result of that collapse, chemically enriched by the metals produced by nucleosynthesis in massive stars of the first generation. ELS envisaged collapse by more than a factor of 10 in radius over a timescale of $\sim 10^8$ years or less. Since most subdwarfs form at large radius in this model, they are predicted to follow predominantly radial orbits, with large motions perpendicular to the Plane. This produces the observed $(\delta(U–B)$, eccentricity) and $(\delta(U–B), |W|)$ correlations.

In their original analysis, ELS proposed a rapid, dissipationless (non-collisional) collapse, which implies that there should be no abundance gradient in the resulting stellar halo. However, in a later analysis of a larger sample of subdwarfs, Sandage [S3] found evidence for a continuous increase in $|W|$ with decreasing δ(U–B). This, he argued, 'shows that a chemical gradient exists in the Galactic halo such that stars with the largest $\langle|Z|\rangle$ (maximum attainable height above the Plane) have the lowest metal abundance. This gradient appears to be a natural consequence of the collapse of the halo towards the Plane, with the metal enrichment taking place as the collapse proceeds.' That is, as the protogalaxy collapses, chemical enrichment from stellar mass-loss leads to increasing abundance in the interstellar medium, while star formation is confined within a volume of continuously decreasing radius. Each succeeding generation of star-forming clouds acquires higher angular momentum as the collapse proceeds, leading to a smooth correlation between rotation and abundance.

This model, however, requires that the collapse be sustained over a time that is sufficient to allow the stellar ejecta from one generation of stars to become well-mixed in the gas clouds which form the succeeding generation. This demands a longer timescale than the $\sim 10^8$ years envisaged by ELS in their original model. Prompted by the observational results outlined by [S3], modified Galaxy-formation models were proposed where the collapse is slowed down by pressure support provided by winds from high mass OB stars and supernovae (see [Y1], for example). These models can prolong the collapse phase for several Gyrs.

The SZ fragmentary accretion model

Searle and Zinn's [S13] model was conceived partly to address the complications encountered by the ELS model and partly as a result of more extensive observations of globular clusters in the outer halo. The latter revealed variations in the morphology of the colour–magnitude diagram (notably the horizontal branch) amongst clusters with similar metal abundance: the 'second parameter' problem, discussed in Section 6.3.2. This effect is particularly pronounced amongst the clusters at large Galactocentric radii. SZ proposed age as the second parameter, suggesting that the formation of the halo consisted of chaotic accretion of individual dwarf galaxy-sized (10^7–$10^9\,M_\odot$) gas clouds, rather than as a smooth monolithic collapse. Each accreted fragment has its own star-formation history, leading to a dispersion of properties amongst clusters and field stars once those fragments coalesce to form the halo. While most fragment accretion occurs during the initial 2 to 3 Gyr of the formation history, the Galaxy can continue to absorb low-mass systems at later epochs, providing a natural explanation for the presence of younger clusters at large radii.

A comparison of the ELS and SZ models

These models make different predictions about the expected properties of the halo population. The modified ELS model predicts a radial abundance gradient and a correlation between rotational velocity and abundance in response to the slow,

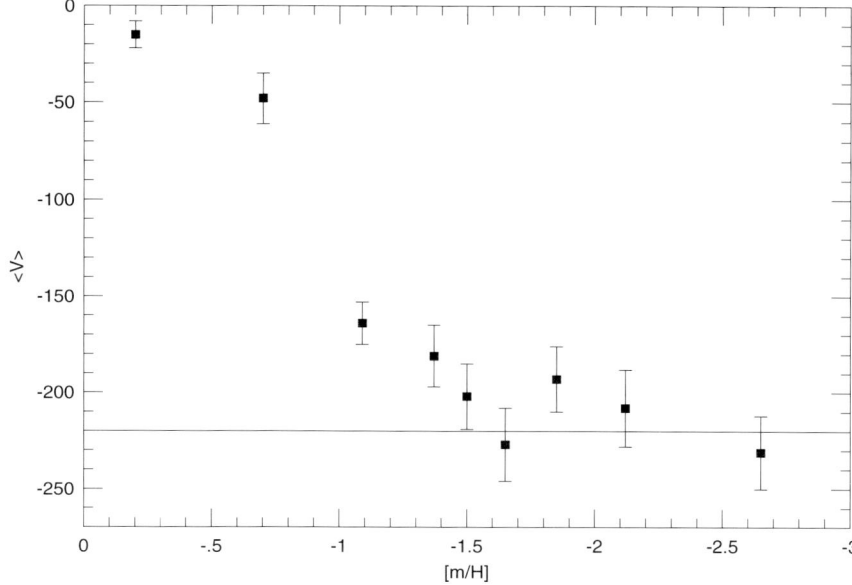

Figure 6.15. The correlation between rotation and abundance, from [C10] and [B8]. The near-constant rotation for $[m/H] < -1.5$ contradicts the predictions of halo models involving rapid, monolithic collapse.

unidirectional collapse. On the other hand, Searle and Zinn predict a dispersion in age- and abundance-dependent properties with radius, with no monotonic abundance gradient and no expectation of a correlation between rotation and abundance.

Observations tend to support the latter scenario. Halo star surveys undertaken subsequent to the [S3] analysis have resulted in the identification of halo stars with near-circular, high-angular-momentum orbits. Those stars escaped previous detection since they have low proper motions; their presence adds significant scatter to the smooth (velocity, $[m/H]$) correlations which underlie the ELS hypothesis. With the addition of these new data, the mean rotational velocity of halo subdwarfs appears to be independent of abundance for $[m/H] < -1.5$ (Figure 6.15). The weak correlation evident between $[m/H] = -1$ and -1.5 may stem from contamination by disk stars. Most significantly, observations suggest that the outer halo has a net rotational velocity which is retrograde with respect to the motion of the Sun about the Galactic centre [M4]. Clearly, this result cannot be accommodated within a monolithic collapse model. There is also direct evidence for the importance of accretion. Recent large-scale kinematic surveys have identified possible star streams in the halo, presumably remnants of disrupted fragments [M5], [J2].

In the final analysis, it is likely that both processes played a significant role in the formation of the Galaxy. Sandage has described the SZ scenario as 'ELS with noise'. A plausible composite model might consist of a high-mass ($> 10^{10}\,M_\odot$) system undergoing monolithic ELS collapse to form the disk and inner halo, while simultaneously accreting smaller SZ fragments to form the less homogeneous outer halo. Indeed, the discovery of many interacting and merging extragalactic systems and the current engorging of the Sagittarius dwarf by our own Galaxy [I1] illustrates that the latter process continues to this day.

6.4   THE GALACTIC BULGE

M dwarfs in the Galactic Bulge have apparent magnitudes too faint to be accessible to detailed observations. However, for completeness, we include a brief summary of the large-scale properties of that subsystem in our Galactic survey. A recent review of observational and theoretical studies is presented by Wyse *et al.* [W7].

Observational studies of the Bulge are complicated by the fact that the line-of-sight passes along the length of the disk and through substantial foreground obscuration. Far-infrared observations by the IRAS and COBE satellites provided the first clear picture of the large-scale structure of the Bulge, but questions remain about whether the central component is more closely related to the halo or the disk. In halo models, the Bulge forms during the initial collapse of the protogalaxy, either as the central component of the large-scale collapse in the ELS model, or as part of the inhomogeneous SZ merging-fragment model. The alternative is post-collapse formation. For example, Pfenniger and Norman [P4] have proposed that a strong central bar could first accumulate a concentration of stars, and then develop a bulge as gravitational resonance interactions drive stars out of the Galactic Plane.

Deciding amongst the various formation models is not straightforward. What are the constituents of the Galactic Bulge? The radically different colours of stars in the disk and Bulge of M31 provided the main stimulus for Baade's two-population hypothesis. His original identification of the Bulge as a metal-poor population was prompted by the discovery of RR Lyrae stars in the low-absorption field at ($l = 1°$, $b = -4°$) now known as 'Baade's window' [B3]. Arp's [A4] subsequent colour-magnitude work appeared to confirm that result, with the detection of a well-populated giant branch, red horizontal branch and a main-sequence that was essentially truncated at G-type stars – morphological features clearly compatible with globular cluster characteristics. However, by the time of the Vatican conference, Morgan had shown that a significant component in the Bulge had strong-lined spectra, consistent with a near-solar abundance [M7], [M14], [M15]. Recent higher-resolution spectroscopic data confirm that the abundance distribution is comparable to that in the Solar Neighbourhood stars, although some stars have abundance ratio anomalies [M2]. In particular, the [Eu/Fe] ratio is enhanced in the Bulge stars, suggesting that Type II supernovae (sources of Eu) played a prominent role in the metal enrichment, and implying, as in the halo, a formation timescale of only ~ 1 Gyr.

The most extensive recent study of the Bulge is Minniti's [M11], [M12] survey of the radial velocities and abundances of several hundred K giants in two fields at intermediate radii. His data show a strong correlation between these two parameters, with the [Fe/H] < -1 stars having no net rotation but a large line-of-sight velocity dispersion ($\sigma \sim 109$ km s^{-1}), while the higher abundance stars have lower dispersion and significant mean rotation. The implication is that the former stars represent the innermost extension of the halo population, which also contributes Baade's RR Lyrae stars, while the metal-rich stars form a chemically- and kinematically-distinct component, perhaps related to the 'disk' globular clusters.

Synthesising these observations, it is likely that 'the bulge' is a composite entity, including both the inner halo and a higher-abundance, rotationally-supported population. The latter probably formed rapidly, and may be a near-contemporary of the halo. The halo stars form during collapse, and therefore have radial orbits and little net rotation; enriched gas sinks towards the centre of the potential well during the earliest phases of Galaxy formation, and, through dissipative collisions, forms a rotating system. Star formation within those rotating gas clouds produces Bulge stars. The issue, however, is far from settled.

6.5 THE GALACTIC DISK

The overwhelming majority of stars in the Solar Neighbourhood are members of the disk population. The Sun lies in the outer disk between the Scutum and Perseus spiral arms, with the nearest star-forming regions (in Taurus) ~ 150 parsecs distant. The typical density of the inter-arm interstellar medium is ~ 0.5 atoms cm^{-3}, but the Sun resides within the 'Local Bubble', diameter ~ 100 parsecs, where the average density is lower by over a factor of 10 [F6], [P2]. Several other similar structures lie within 500 parsecs of the Sun, and are probably due to the effects of both winds from massive OB stars and blast-waves from supernovae [B15]. General star-counts (that is, the number of stars as a function of apparent magnitude) show that there are $\sim 10\%$ more stars towards the South Galactic Pole (SGP) than towards the North Galactic Pole (NGP). The simplest explanation for this asymmetry is that the Sun lies 20–30 parsecs above (north of) the mid-Plane of the disk. Finally, the Sun's distance from the Galactic Centre (the Solar Radius) is usually taken as $R_\odot = 8.5 \pm 1.5$ kpc.

6.5.1   The kinematics of disk stars

The rapid rotation of the Galactic disk is the result of conservation of angular momentum during Galactic collapse. The original protogalaxy had net rotation. Collisions between gas clouds during collapse dissipated energy in motion perpendicular to the present-day Galactic Plane, but preserved motion parallel to the Plane, producing ordered rotation. Disk stars show variation in mean kinematics, primarily velocity dispersion, as a function of spectral type. Table 6.3 presents results from a recent analysis of astrometric data obtained by the Hipparcos satellite [D2]. Stellar kinematics are usually expressed in the (U, V, W) velocity system (see Section 1.3.3), but there can be misalignments between the principle axes of the best-fit ellipsoid and the Galactic co-ordinate system. The parameters given in Table 6.3 for each group of stars are the velocity dispersions for the principal axes of the ellipsoid, with σ_1 closest to σ_U, σ_2 closest to σ_V, and σ_3 closest to σ_W. The vertex deviation, l_v, corresponding to the Galactic longitude of the direction of σ_1, is also given. Finally, velocity dispersions in U, V and W for nearby M dwarfs (from [H2]) are listed for comparison [H3]. The solar motion derived from the Hipparcos analysis is

$$U_\odot = 10.00 \pm 0.32 \text{ km s}^{-1}$$
$$V_\odot = 5.23 \pm 0.62 \text{ km s}^{-1}$$
$$W_\odot = 7.17 \pm 0.38 \text{ km s}^{-1}$$

Table 6.3 shows that early-type stars have substantially lower velocity dispersions than GKM dwarfs. This behaviour was pointed out originally by Parenago [P1], and is referred to as 'Parenago's discontinuity'. Its origin lies in the dynamical evolution experienced by disk stars coupled with the decrease in main-sequence lifetimes with increasing mass. Young stars emerge from the parent star-forming region with the space motion of the cloud, which generally lies within 10 km s^{-1} of the local circular velocity. As those stars orbit the Galaxy, they undergo gravitational interactions, particularly with massive (10^4–$10^6\ M_\odot$) molecular clouds, leading to random changes in motion, that is, scattering [J1]. Spitzer and Schwarzschild [S18] demonstrated that under such circumstances the overall velocity dispersion, σ_t, is expected to increase with $\sigma_t \propto \tau^{1/3}$, where τ is the age (see also [W3], [W4]). The correlation between spectral type and σ_t follows: early-type stars have shorter main-sequence lifetimes,

Table 6.3. Velocity dispersions of disk stars.

Spectral type	$(B-V)_{min}$	$(B-V)_{max}$	σ_1	σ_2	σ_3	l_v
B–A5	−0.24	0.14	14.4	9.6	5.5	30°3
B5–F0	0.14	0.31	20.2	9.6	8.1	22°8
F0–F2	0.31	0.41	22.4	11.9	9.4	19°8
F3–F5	0.41	0.47	26.3	15.9	12.2	10°2
F5–F7	0.47	0.53	30.5	18.5	13.4	6°8
F7–G0	0.53	0.58	33.0	21.9	15.1	1°9
G0–G3	0.58	0.64	37.7	23.6	21.3	10°2
G3–G5	0.64	0.72	38.2	24.0	20.9	7°6
>G5	0.72	1.55	37.3	26.1	18.3	13°1
M dwarfs	1.1	1.6	41	27	21	
Me dwarfs	1.1	1.6	27	20	15	
≤M3	1.1	1.4	39	24	22	
>M3	1.4	1.6	42	29	22	
M dwarf g1	1.1	1.6	35	21	20	
M dwarf g2			52	36	32	

Data for spectral types B to G are from [D2]. M dwarf velocity dispersions are $\sigma_U, \sigma_V, \sigma_W$, from [H3]. g1 and g2 are velocity dispersions for the 2-Gaussian match to the latter dataset.

and therefore, as a population, are subject to less scattering than GKM dwarfs with main-sequence lifetimes comparable with, or exceeding the age of, the Galactic disk.

The velocity dispersion/age dependence is also illustrated in comparing the kinematics of dM and dMe dwarfs. Since chromospheric activity declines relatively rapidly with time for early- and mid-type M dwarfs (Section 5.5.2), a sample of dMe dwarfs is biased towards stars younger than the average age of the disk. As Table 6.3 shows, those stars also have lower velocity dispersions.

Table 6.3 characterises the velocities of disk stars as a velocity ellipsoid, with Gaussian dispersion in each co-ordinate. That model is a simplification, as illustrated in Figure 6.16, where velocity data for a volume-complete sample of M dwarfs are matched against the best-fit dispersions from [H3]. The observations are poorly matched by single Gaussians; probability plots indicate that composite distributions give a better fit. In a probability plot, the cumulative distribution of a variable (for example, U-velocity) is graphed in units of the standard deviation (σ_U from Table 6.3). A single Gaussian distribution results in a straight line in this plane, slope σ; sums of Gaussians, g_i, produce line segments, slope σ_i. Considering the data for nearby M dwarfs (Figure 6.16(d)–(f)), both U and W distributions can be represented as three line segments, with the outer segments having the same slope. This is consistent with these distributions being the sum of two Gaussian components (g1 and g2 in Table 6.3), with the higher-velocity component accounting for ~10% of the total sample. The distribution in V is more complex, due largely to the extended tail toward negative velocities reflecting the Strömberg asymmetric drift).

Representing the stellar velocity distribution as a sum of two components is reminiscent of the division of disk stars into several sub-populations in the Vatican Galaxy model (Table 6.1). However, it is likely that even these multi-component models are simplified representations of the local kinematics. Detailed studies of *Hipparcos* proper-motion data for nearby stars [D3] reveals clumping in velocity space. Those clumps are 'moving groups': probably stars which formed in clusters associated with a particular molecular cloud-complex. The Orion I association, for example, may give rise to such a moving group several hundred Myrs in the future. Amongst the more prominent systems are groups associated with the Hyades, Pleiades and Coma clusters, and with Sirius [E3], [E4], [E5], [E6], which suggests that larger groups can persist as coherent kinematic structures for at least 1 Gyr. While fewer than 10% of disk stars are likely to be members of moving groups, further analysis is required to determine their full importance in Galactic dynamics.

Finally, an important consequence of the substantial velocities acquired by older disk dwarfs is that those stars migrate to regions in the disk far from their birthplace. With an overall velocity dispersion of $\sigma_U \sim 30\,\mathrm{km\,s^{-1}}$, M dwarfs which are currently within the immediate Solar Neighbourhood may have originated from star formation regions at Galactic radii of 4–12 kpc. As a result, a *local* sample of disk dwarfs can provide a fair sampling of the properties of the Galactic disk population as a whole.

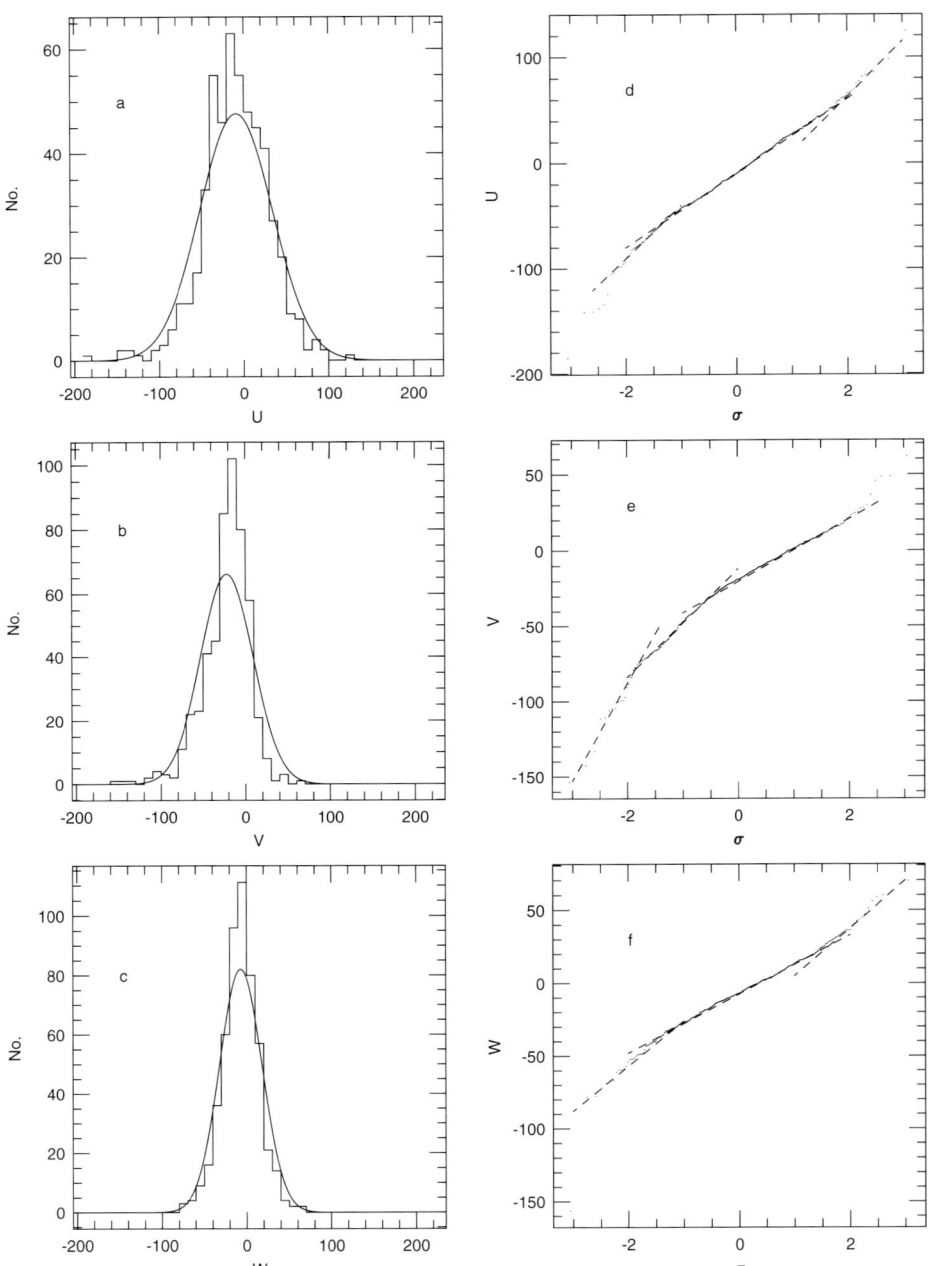

Figure 6.16. The velocity distribution of a volume-limited sample of nearby M dwarfs. Panels a–c display the (U, V, W) histograms for 514 systems with $8 < M_V < 15$; panels d–f show the corresponding probability plots, with the linear relations indicating the best-fit Gaussian distributions. (From [R3].)

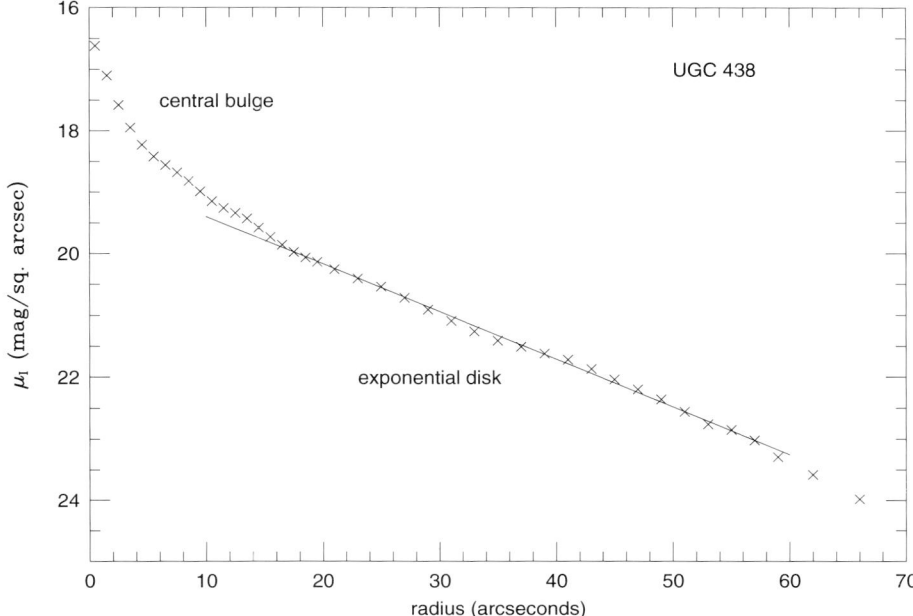

Figure 6.17. The radial surface brightness profile of the disk in the face-on spiral galaxy UGC 438 (NGC 214). (From [J3].)

6.5.2   The disk density distribution

The radial density law

The determination of the radial density distribution of the Galactic disk is complicated by our vantage point within the disk. Observations of other spiral galaxies (such as NGC 4565, Figure 6.2) provide a clearer picture of the overall structure. The stellar density distribution in the disk of an external system is derived through surface photometry, making due allowance for contributions from the central bulge, star formation and dust absorption in spiral arms. The first extensive analysis was undertaken by Freeman [F2], who found that radial intensity profiles of several nearby spirals could be represented as exponential distributions, $I(r) \propto e^{-\frac{r}{r_0}}$, where r_0 is the scale-length, estimated as 2–3 kpc. As an illustration of this technique, Figure 6.17 shows the radial surface-brightness profile, in magnitudes per square arcsecond (μ_I) of the face-on spiral galaxy NGC 214 (also known as UGC 438): the distribution is almost linear for $r > 15$ arcsec corresponding to an exponential distribution. The rise in brightness near the centre is due to the bulge component in the galaxy.

Adopting the reasonable assumption that the Galactic disk is similar to external spirals, models can be used to predict the expected surface brightness distribution and star-counts from our viewpoint within the disk. Those predictions are matched against observations. Estimates of the disk scalelength, r_0, range from 2 to 6 kpc

[R6], [K6]. However, the longer estimates are based on optical photometry (see [K3], for example) and are therefore vulnerable to bias due to inadequate compensation for reddening. Infrared data, less susceptible to interstellar absorption, indicate scalelengths between 2 and 3 kpc, consistent with $r_0 = 2.5$ kpc deduced from direct star-counts towards the anticentre [R6]. The latter data also show evidence for a sharp decrease in the number of disk stars at a distance of ~ 6 kpc beyond the Solar Radius: an edge to the Galactic disk (although a few open clusters are known at larger radii). Similar truncations in the density distribution have been detected in other spiral galaxies [K4].

The vertical density law

The density distribution perpendicular to the Plane, $\rho(z)$, can be derived from direct star-counts, either by inverting the apparent magnitude–colour (or spectral type) distribution or by building Galaxy models. In the former case, the observations are used to infer absolute magnitudes, and hence distances, for each star, providing direct estimates of $\rho(z)$; in the latter, the Galactic stellar populations are para-meterised (density laws, luminosity functions, colour–magnitude relationships) and predicted star-counts matched against observations spanning a range of directions (l, b). The model parameters are adjusted until suitable agreement is achieved. These two techniques can be viewed as complementary, using deductive and inductive methods of analysis.

The thickness of the disk depends on the vertical velocity dispersion, σ_W, which, as described in the previous section, increases with increasing age. Most molecular clouds and star-forming regions lie within 60 parsecs of the mid-Plane, and the short main-sequence lifetimes of early-type (OBA) stars leads to their distributions being similarly restricted in z. Thus, Parenago's discontinuity in $(\sigma_t$, spectral type) is reflected in $\rho(z)$, with few unevolved A-type stars reaching distances of $z > 500$ parsecs.

As with the radial density law, $\rho(z)$ is often modelled as an exponential, $\rho(z) \propto \exp(-z/z_0)$, with scaleheights of $z_0 \sim 100$ parsecs for OBAF stars, and 250–350 parsecs for later-type dwarfs [M10], [B5]. This parameterisation has its origin in Oort's summary of the Vatican conference reinforced by Schmidt's [S11], [S12] analysis of classical star-count data. Schmidt derived 'equivalent widths' for the thickness of the disk, where the equivalent width is 'the quantity with which the spatial densities in the Plane must be multiplied to give the surface densities in the Plane.' If the vertical density law is an exponential distribution, then the exponential scale-height is half the equivalent width. Hence, Schmidt's equivalent width estimate of 600 parsecs for stars with $M_V > +5$ is consistent with (but does not require) an exponential scaleheight of 300 parsecs.

Direct analyses of $\rho(z)$ reveal more complex behaviour. Gilmore and Reid [G1] reconstructed $\rho(z)$ from starcounts toward the South Galactic Pole, using an $(M_V, (V-I))$ relation to derive photometric parallaxes (see Section 7.5.2) for G and K stars. The resulting density law shows a clear change in slope at $z \sim 1.5$ kpc (Figure 6.18). Such a feature can be expected in the transition between the highly

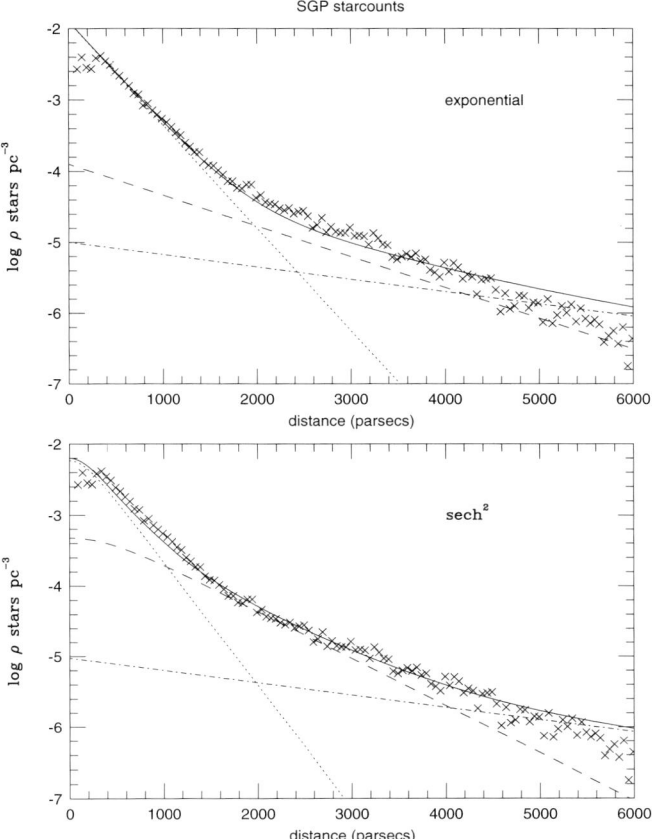

Figure 6.18. The stellar density distribution perpendicular to the Galactic plane. The upper panel models the distribution as two exponentials, scale-heights 300 and 1,000 pc, with 70 : 1 density normalisation; the lower, as two sech^2 distributions, $h_0 = 500$ and 1,400 pc, and a 30 : 1 normalisation. The dotted and dashed lines shows the density distribution of the individual components; the solid line plots the combined $\rho(z)$.

flattened disk and the halo, but at heights of 5 kpc or more and at lower space densities. The additional stars between 1.5 and 5 kpc were interpreted by Gilmore and Reid as representives of a separate stellar population, termed the 'thick disk', with a local density $\sim 1.5\%$ of the 'classical' old disk and a scale-height $\sim 1,100$ parsecs. Subsequent observations [F4], [R1], [R5] have confirmed the existence of these additional stars, although the relationship between them and the old disk, whether a separate population [G2] or the tail of a distribution [N4], remains to be established.

The overall density law can be modelled using a double exponential. However, exponential distributions have a sharp cusp at $z = 0$. An alternative representation is to follow the approach outlined by Camm [C1], [C2], who demonstrated that an isothermal (single-Gaussian velocity dispersion) population follows a sech^2

distribution. Writing

$$\rho(z) = \rho_0 \operatorname{sech}^2 \left(\frac{z}{h_0} \right) \tag{6.5}$$

the density distribution tends to an exponential, scale-height $h_0/2$ at large z,

$$\rho(z) \propto \exp(-2z/h_0) \tag{6.6}$$

At small z, the gradient of the sech^2 function is less steep than an exponential, giving a physically more realistic density distribution.

The composite, perhaps two-component, density law, $\rho(z)$, may be related to the similarly composite, perhaps two-component, velocity distributions $f(U)$ and $f(W)$ (Figure 6.16). At present, however, the comparison is no more than intriguing. As Figure 6.18 shows, the density distribution for the 'thick disk' component is constrained mainly by observations at $z > 1.5\,\mathrm{kpc}$, or 1–2 scale-heights above the Plane. Extrapolating to $z = 0$ to estimate the local density normalisation (the parameter that might be deduced from the velocity distributions) is neither straightforward nor unambiguous. The observed densities at large height can be matched by a range of models, with an anti-correlation between the values adopted for h_0 and ρ_0. In the case of a sech^2 distribution, current analyses are consistent with $500 < h_0 < 700$ parsecs for the old disk, and $1200 < h_0 < 1800$ parsecs for the 'thick disk', with local density normalisations (old disk : 'thick disk') between 70 : 1 and 10 : 1. As with most areas of Galactic structure, further analysis is required. Nonetheless, all of these estimates suggest that, integrated over its vertical distribution, the 'thick disk' contributes $\sim 10\%$ of the total mass of the disk, making those stars a substantive Galactic component.

6.5.3   The abundance distribution

Metallicities of disk stars are determined using techniques similar to those used in studying halo stars. In contrast to the halo, high-resolution spectroscopic analyses show that most disk dwarfs have elemental abundance ratios close to scaled-solar values. The most extensive catalogues of metallicity determinations are based on Strömgren photometry of F and G dwarfs [E1], [W6] and analyses of the Mgb region of the spectrum [C4]. These studies place more than 70% of disk stars in the abundance range $-0.3 < [m/H] < 0.15$, with the remainder forming a metal-poor tail extending to $[m/H] \sim -1$ (Figure 6.19(a)). The latter stars are often associated with the thick disk, although it is not established that these lower-abundance stars are identical with the higher-velocity component in the Solar Neighbourhood and the more extended component in the $\rho(z)$ distribution.

There are two important caveats in considering these results: first, as described in Section 3.4, systematic errors are present in some low-resolution abundance calibrations, usually in the sense of underestimating $[m/H]$; second, none of the analyses of the disk abundance distribution published to date is based on a reliable volume-complete sample. Most studies, including the results plotted in Figure 6.19, have samples selected from proper-motion catalogues or from pre-Hipparcos versions of the Gliese nearby-star catalogue. Figure 6.19(b) illustrates the potential problems invoked by that approach: nominally a sample of single G dwarfs within 10 parsecs of the Sun, plotting the $(M_V, (B–V))$ colour–magnitude distribution using Hipparcos parallax data shows that the sample not only includes previously unrecognised binaries, but also a number of subgiants. The abundance distribution of such an heterogeneous sample is not a reliable indicator of the underlying distribution of the Galactic disk population.

Figure 6.19(b) also illustrates the calibration problems present in some metallicity scales. Each star in the [W6] sample has an abundance estimate based on Strömgren photometry (using the [N1] calibration). The overall trend in the colour–magnitude diagram is in line with expectations: lower-abundance stars lie at fainter M_V for a given colour. However, comparison with theoretical isochrones shows that the metal-poor dwarfs are more luminous than expected, consistent with the Strömgren calibration underestimating $[m/H]$. A thorough analysis of the abundance distribution of local stars remains to be undertaken, but it is likely that at least 90% of disk dwarfs have abundances $[m/H] > -0.4$.

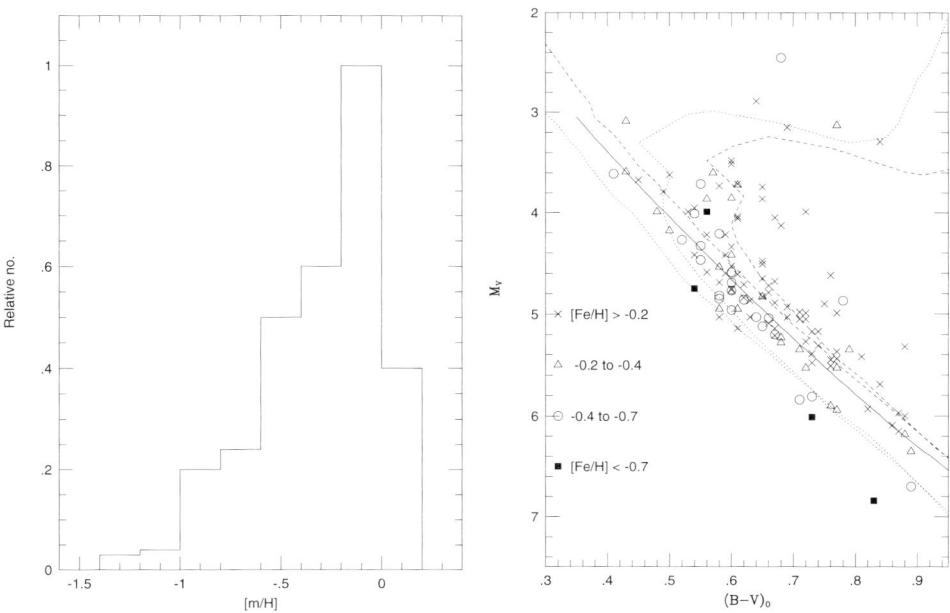

Figure 6.19. (a) The abundance distribution of the Galactic disk, inferred from Strömgren photometry of a pre-Hipparcos sample of nearby G dwarfs (from [W6]); (b) The post-Hipparcos colour–magnitude diagram for the same stars; several are subgiants or unresolved binaries. The dashed lines are 1 and 5 Gyr isochrones for $[m/H] = 0.0$; dotted lines plot similar data for $[m/H] = -0.4$ (from [B6]); the solid line is the mean relationship for nearby stars.

6.5.4   The age of the Galactic disk

The rapid rotation and high metallicity of the Galactic disk point to its formation after the halo. In principle, the absolute age of the oldest disk stars can be determined by matching the position of the main-sequence turn-off against theoretical isochrones, as in studies of halo globular clusters. In practice, the fact that star formation has continued throughout the existence of the disk complicates the task of identifying the oldest stars in the field. In general, results are consistent with an age of ~ 10 Gyr.

A variation on this theme is to estimate the age of the oldest disk open cluster. This has the advantage of isolating a coeval, single-abundance group of stars, simplifying the age determination; on the other hand, open clusters are fragile, loosely-bound objects which are easily disrupted. Few old open clusters survive, and only a lower limit can be set on the age of the disk. At present, Berkeley 13 [P3] is the oldest known, with an age $\tau \sim 12$ Gyr, while NGC 7789 and NGC 188 have $\tau \sim 8$ Gyr [F5].

Finally, the white dwarf luminosity function, $\Phi_{WD}(M_V)$, can be used as an age estimator. Virtually all single stars (and most binary components) with masses less than 7–$8\,M_\odot$ become white dwarfs during their final, dying phase of evolution. Mestel's [M8] early analytical models of these degenerate stars predicted that $L \propto (M/\tau_c)^{7/5}$, where τ_c is the cooling time, the time since the star became a white dwarf. If the disk has a finite age, $\Phi_{WD}(M_V)$ must drop to zero at a particular luminosity set by the finite cooling time available. Schmidt [S11] originally suggested this as a method of measuring the age of the disk. The main complication is the identification of a complete sample of low-luminosity degenerates, which are photometrically similar to main-sequence K dwarfs. The most recent studies [W5], [L4] indicate ages of 8–12 Gyr. Overall, current results indicate that the oldest stars in the Galactic disk are 2–3 Gyr younger than a typical star in the Galactic halo.

6.5.5   The star-formation history of the disk

At present, we lack an accurate determination of the detailed star formation history of the disk, the stellar birth-rate, $B(t)$. The most obvious constituents of the disk population are open clusters. Their age distribution [F5] shows that star formation has been taking place continuously over the ~ 10-Gyr lifetime of the Galactic disk. However, these structures have masses of $\sim 10^3$–$10^4\,M_\odot$, with correspondingly low binding energy, and therefore can be disrupted by gravitational interactions on timescales of 1 Gyr or less (see Section 7.6.2). The majority of old $(\tau > 3$ Gyr) clusters which have survived to the present day have orbits which are significantly inclined to the Galactic Plane, leading to their spending the majority of their existence undisturbed by gravitational perturbations. Thus, the number of open clusters as a function of age is not a reliable guide to the overall stellar birth-rate.

A variety of techniques has been used to attempt to disentangle the star-formation history through observations of field stars. Assuming that there is a single-valued relationship between age and metallicity, the abundance distribution of disk stars

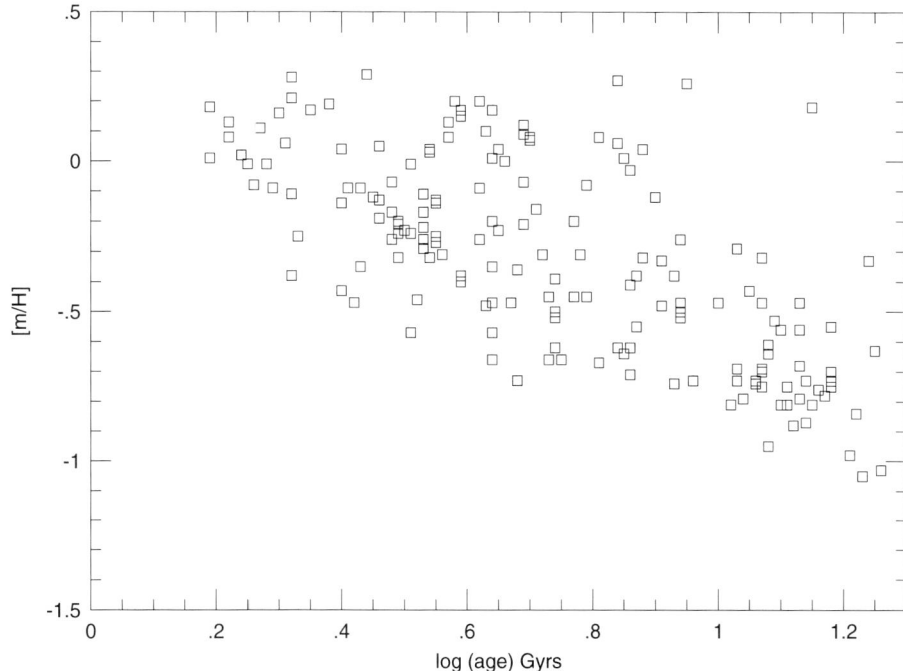

Figure 6.20. The age–metallicity relationship for stars in the Galactic disk (data from [E1]).

can be used to infer the past star-formation activity [T1], [R7]. Unfortunately, detailed analyses show that there is significant variation in the abundances of stars formed at any given time in the disk. Figure 6.20 plots results for F and G dwarfs in the field, where ages have been estimated by matching the observed Strömgren colours against theoretical isochrones [E1]. The broad distribution negates the use of metallicity as an age indicator at abundances higher than $[m/H] \sim -1$.

Attempts have also been made to use the white dwarf luminosity function, Φ_{WD}, to probe the star-formation history. Consider the case where there is a burst of star formation, duration τ_D between times τ_1 and τ_2, superimposed on a nearly-constant background rate. White dwarf progenitors have masses from 1 to 7–8 $M_\odot$, with corresponding main-sequence lifetimes of 10 Gyrs to 5×10^7 yr (0.05 Gyr). Thus, burst stars first make a contribution to Φ_{WD} at time $\tau_1 + 0.05$ Gyr, with the contribution increasing as lower-mass burst stars evolve through the red giant phase. Models [N2] show that the net effect is the appearance of a 'bump' in Φ_{WD} whose location can, in principle, be used to estimate both the duration of the starburst and the time since its occurrence. In practice, Φ_{WD} is defined to only moderate precision [W5], and while there is marginal evidence for a mild, recent (< 0.3 Gyr old) burst of star formation, the data are generally consistent with little variation in the average birth-rate over the last 4 Gyr.

Both lithium abundance and chromospheric activity have been employed as age indicators in late-F and G-type dwarfs. In those stars, the temperature at the base of the convective envelope is sufficiently high to allow partial lithium burning through the (p, α) process (see Section 3.3.3), with the result that the lithium abundance decreases with time. Early analyses [D4] favour a mildly decreasing SFR over the last 4–5 Gyr, but the mass-dependence of the age/(Li abundance) relationship complicates the analysis. Moreover, the surprising detection of lithium at low abundance levels in K dwarfs, where deep convection should have resulted in rapid lithium depletion, raises questions about the reliability of the technique [F1].

Chromospheric activity in solar-type stars is characterised through measurement of Ca II K emission strength, the time dependence of which is generally approximated by the $\tau^{-1/2}$ Skumanich relationship (Section 5.5.2). Given accurate data for an unbiased sample of field stars, that age calibration can be used to reconstruct the disk star-formation history. Barry [B7] suggested that there have been significant bursts in activity 4–5 Gyr before the present and within the last 10^9 years. Soderblom *et al.* [S16], however, have criticised that analysis, and argue that the available data [H4] can be represented by a constant star-formation rate. The distribution of chromospheric activity in M dwarfs is also consistent with only minor departures from constant star-formation over the history of the disk. In short, a reliable method for determining the detailed disk star-formation history remains to be identified, but there is no evidence for substantial variation.

6.6   SUMMARY

Our aim in this chapter has been to provide an overview of current understanding of Galactic structure. We have

- reviewed the historical development of Galactic structure, with particular emphasis on the development of the concept of stellar populations;
- summarised current understanding of the kinematics, spatial distribution and abundances of the Galactic halo, and considered how those properties are interpreted in terms of the age and the initial stages of formation of the Galaxy;
- outlined the most significant features of the Galactic Bulge, and considered its relation to the disk and halo;
- summarised the main properties of the Galactic disk, particularly the complex spatial distribution perpendicular to the Plane and the results recently derived from analysis of Hipparcos data.

Summarising the overall picture, the Galactic halo is a metal-poor, non-rotating population, with a near-spherical distribution; the disk is a rapidly-rotating, metal-rich population. A plausible formation scenario has most halo stars forming within a 1–2 Gyr period, $\sim 12.5 \pm 1.5$ Gyrs ago, during the initial collapse of the protoGalaxy, with residual star formation continuing in accreting fragments for a further 3–6 Gyrs. Most of the bulge stars form during, or shortly after, the halo phase of Galactic evolution. The disk forms as a result of dissipative collapse, with nucleosynthetic products from the preceding generation of halo stars rapidly driving the abundance to higher levels. The oldest disk stars have ages of $\sim$10 Gyrs, and the net rotation is $\sim$200 km s^{-1} at the Solar Radius. The density distribution perpendicular to the Plane is complex, and of uncertain origin.

The following chapters are concerned primarily with a simple quantity: the stellar mass function, the number of stars and substellar-mass brown dwarfs as a function of mass. Although simple in concept, this parameter has wide-reaching consequences for both Galactic dynamics and the theory of star-formation. As we will describe, its derivation is also far from straightforward, and has absorbed substantial observational and theoretical effort throughout the latter half of the twentieth century.

6.7   REFERENCES

A1   Adams, W. S., Joy, A. H., 1922, *ApJ*, **56**, 242.
A2   Adams, W. S., Joy, A. H., Humason, M. L., Brayton, A. M., 1935, *ApJ*, **81**, 187.
A3   Armandroff, T. E., 1989, *AJ*, **97**, 375.
A4   Arp, H. C., 1955, *AJ*, **60**, 1.
A5   Axer, M., Fuhrmann, K., Gehren, T., 1994, *A&A*, **291**, 895.
B1   Baade, W., 1944, *ApJ*, **100**, 137.
B2   Baade, W., 1944, *ApJ*, **100**, 147.
B3   Baade, W., 1951, *Publ. Univ. Mich. Obs.*, **10**, 7.
B4   Baade, W., Hubble, E.. 1939, *PASP*, **51**, 40.
B5   Bahcall, J. N., Soneira, R. M., 1980, *ApJS*, **44**, 73.
B6   Bailey, S. I., 1917, *Harvard Annals*, **78**, pt. 2.
B7   Barry, D. C., 1988, *ApJ*, **334**, 436.
B8   Beers, T. C., Sommer-Larsen, J., 1996, *ApJS*, **96**, 175.
B9   Beers, T. C., Rossi, S., Norris, J. E., Ryan, S. G., Molaro, P., Rebolo, R., 1998, *Sp. Sci. Rev.*, **84**, 139.
B10 Bessell, M. S., Norris, J., 1984, *ApJ*, **28**(5), 622.
B11 Biémont, E., Badoux, M., Kurucz, R. L., Ansbacherm W., Pinnington, E. H., 1991, *A&A*, **249**. 539.
B12 Binney, J. E., Tremaine, S., 1998, *Galactic Dynamics*, W. H. Freeman & Company, San Francisco.
B13 Blaauw, A., 1965, in *Galactic Structure*, Vol. 5, *Stars and Stellar Systems*, ed. A. Blaauw and M. Schmidt, p. 435.
B14 Bolte, M., Hogan, C., 1995, *Nature*, **376**, 399.
B15 Breitschwerdt, D., Egger, R., Freyberg, M. J., 1996, *Sp. Sci. Rev.*, **78**, 183.
C1   Camm, G., 1950, *MNRAS*, **110**, 305.
C2   Camm, G., 1952, *MNRAS*, **112**, 155.
C3   Canterna, R., 1976, *AJ*, **81**, 228.
C4   Carney, B. W., Laird, J. B., Latham, D. W., Kurucz, R. L., 1987, *AJ*, **94**, 1066.
C5   Carney, B. W., Latham, D. W., Laird, J. B., Aguilar, L. A., 1994, *AJ*, **107**, 2240.
C6   Carretta, E., Gratton, R. G., 1997, *A&AS*, **121**, 95.
C7   Cassisi, S., Castellani, V., Degl'innocenti, S., Weiss, A., 1998, *A&AS*, **129**, 267.

C8   Chamberlain, J. W., Aller, L. H. 1951, *ApJ*, **114**, 52.

C9   Charlier, C. V. L., 1913, *Medd. Lund Obs. Ser. II*, No. 9, p. 92.

C10 Chiba, M., Yoshii, Y., 1998, *AJ*, **115**, 168.

C11 Clementini, G., Carretta, E., Gratton, R., Merighi, R., Mould, J. R., McCarthy, J. K., 1995, *AJ*, **110**, 2319.

C12 Crawford, D. L., 1979, in 'Problems of calibration of multicolour photometric systems', ed. A. G. Davis Philip, *Dudley Obs. Report* No. 14, p. 23.

C13 Croswell, K., 1995, *The Alchemy of the Heavens*, Doubleday, New York.

D1   D'Antona, F., Caloi, V., Mazzitelli, I., 1997, *ApJ*, **477**, 519.

D2   Dehnen, W., Binney, J. J., 1998, *MNRAS*, **298**, 387.

D3   Dehnen, W., 1999, *AJ*, **115**, 2384.

D4   Duncan, D. K., 1981, *ApJ*, **248**, 651.

E1   Edvardsson, B., Andersen, J., Gustafsson, B., Lambert, D. L., Nissen, P. E., Tomkin, J., 1993, *A&A*, **275**, 101.

E2   Eggen, O. J., Lynden Bell, D., Sandage, A. R., 1962, *ApJ*, **136**, 748.

E3   Eggen, O. J., 1983, *AJ*, **88**, 642.

E4   Eggen, O. J., 1986, *PASP*, **98**, 423.

E5   Eggen, O. J., 1996, *AJ*, **111**, 1615.

E6   Eggen, O. J., 1998, *AJ*, **116**, 1810.

E7   ESA, 1997, *The Hipparcos Catalogue*, ESA SP-1200.

F1   Favata, F., Micela, G., Sciortino, S., 1996, *A&A*, **311**, 951.

F2   Freeman, K. C., 1970 *ApJ* **160**, 811.

F3   Frenk, C. S., White, S. D. M., 1980, *MNRAS*, **198**, 173.

F4   Friel, E. D., 1987, *AJ*, **93**, 1388.

F5   Friel, E. D., 1995, *ARA&A*, **33**, 381.

F6   Frisch, P. C., York, D. G., 1983, *ApJ*, **271**, L59.

G1   Gilmore, G. F., Reid, I. N., 1983, *MNRAS*, **202**, 1025.

G2   Gilmore, G., Wyse, R. F. G., Kuijken, K., 1989, *ARA&A*, **27**, 555.

G3   Gratton, R. G., Carretta, E., Castelli, F., 1997a, *A&A*, **314**, 191.

G4   Gratton, R. G., Fusi Pecci, F., Carretta, E., Clementini, G., Cosri, C. E., Lattanzi, M., 1997b, *ApJ*, **491**, 749.

H1   Harris, W. E., 1976, *AJ*, **81**, 1095.

H2   Hartwick, F. D. A., 1987, in *The Galaxy*, ed. R. Carswell and G. Gilmore (Cambridge University Press, Cambridge).

H3   Hawley, S. L., Gizis, J. E., Reid, I. N., 1996, *AJ*, **112**, 2799.

H4   Henry, T. J., Soderblom, D. R., Donahue, R. A., 1996, *AJ*, **111**, 439.

I1   Ibata, R., Gilmore, G., Irwin, M. J., 1995, *MNRAS*, **277**, 781.

I2   Iben, I., Tutukov, A. V., 1984, *ApJS*, **54**, 335.

J1   Jenkins, A., 1992, *MNRAS*, **257**, 620.

J2   Johnston, K. V., 1998, *ApJ*, **495**, 297.

J3   de Jong, R. S., van der Kruit, P. C., 1994, *A&AS*, **106**, 451.

K1   Kent, S. M., Dame, T. M., Fazio, G., 1991, *ApJ*, **378**, 131.

K2   Kinman, T. D., Suntzeff, N. B., Kraft, R. P., 1995, *AJ*, **108**, 1722.

K3   van der Kruit, P. C., 1986, *A&A*, **157**, 230.

K4   van der Kruit, P. C., Searle, L., 1982, *A&A*, **110**, 61.

K5   Kuiper, G. P., 1939, *ApJ*, **89**, 549.

L1   Laird, J. B., Rupen, M. P., Carney, B. W., Latham, D. W., 1988, *AJ*, **96**, 1908.

L2   Layden, A. C., Hanson, R. D., Hawley, S. L., Klemola, A. R., Hanley, C. J., 1996, *AJ*, **112**, 2110.

L3   Lee, Y. W., Demarque, P., Zinn, R., 1994, *ApJ*, **423**, 248.

L4   Leggett, S. K., Ruiz, M. T., Bergeron, P., 1998, *ApJ*, **294**.

L5   Lindblad, B., 1925, *Upsala Medd.*, No. 3.

L6   Lindblad, B., 1926, *Ark. Mat. Astron. Fys.*, **19B**, No. 7.

L7   Lutz, T., Upgren. A. R., 1980, *AJ*, 85, 1390.

M1   McClure, R. D., 1979, in *Problems of Calibration of Multicolour Photometric Systems*, Dudley Observatory Report No. 13, ed. A. G. Davis Philip (Schenectady, New York), p. 83.

M2   McWilliam, A., Rich, R. M., 1994, *ApJS*, **91**, 749.

M3   McWilliam, A., 1997, *ARA&A*, **35**, 503.

M4   Majewski, S. R., 1993, *ARA&A*, **31**, 575.

M5   Majewski, S. R., Munn, J. A., Hawley, S. L., 1996, *ApJ*, **459**, L73.

M6   Matteucci, F., Greggio, L., 1986, *A&A*, **154**, 279.

M7   Mayall, N. U., 1946, *ApJ*, **104**, 290.

M8   Mestel, L., 1952, *MNRAS*, **112**, 583.

M9   Mihalas, D., Binney, J. J., 1981, *Galactic Astronomy*, W. H. Freeman and Company, San Francisco.

M10 Miller, G. E., Scalo, J. M., 1979, *ApJS*, **41**, 513.

M11 Minniti, D., 1995, *ApJ*, **459**, 175.

M12 Minniti, D., 1995, *ApJ*, **459**, 579.

M13 Minniti, D., Olszewski, W. E., Liebert, J., White, S. D. M., Hill, J. M., Irwin, M. J., 1995, *MNRAS*, **277**, 1293.

M14 Morgan, W. W., 1956, *PASP*, **68**, 509.

M15 Morgan, W. W., 1959, *AJ*, **64**, 432.

M16 Morrison, H. L., Flynn, C., Freeman, K. C., 1990, *AJ*, **100**, 1191.

N1   Nissen, P. E., Schuster, W. J. 1989, *A&A*, **222**, 69.

N2   Noh, H.-R., Scalo, J., 1990, *ApJ*, **352**, 605.

N3   Norris, J., 1986, *ApJS*, **61**, 667.

N4   Norris, J., 1987, *ApJ*, **314**, L39.

O1   O'Connell, D. J. K., ed., 1958, *Specola Astronomica Vaticana*, Vol. 15, Specola Vaticana, Vatican City.

O2   Oort, J. H., 1928, *Bull. Astr. Inst. Neth.*, **159**, 273.

O3   Oort, J. H., 1965, in *Galactic Structure*, Vol. 5, *Stars and Stellar Systems*, ed. A. Blaauw and M. Schmidt, p. 435.

P1   Parenago, P. P., 1950, A. Zh., **27**, 150.

P2   Paresce, F., 1984, *AJ*, **89**, 1022.

P3   Phelps, R., 1998 *ApJ*, **483**, 823.

P4   Pfenniger, D., Norman, C., 1990, *ApJ*, **363**, 391.

R1   Reid, I. N., Majewski, S. R., 1993, *ApJ*, **409**, 635.

R2   Reid, I. N., 1996, *MNRAS*, **278**, 367.

R3   Reid, I. N., Hawley, S. L., Gizis, J. E., 1995, *AJ*, **110**, 1838.

R4   Reid, I. N., 1998, *AJ*, **115**, 204.

R5   Robin, A. C., 1994, *IAU Symposium 161*, ed. H. T. MacGillivray *et al.*, p. 403.

R6   Robin, A. C., Crézé, M., Mohan, V., 1992, *A&A*, **265**, 32.

R7   Rochapinto, H. J., Maciel, W. J., 1997, *MNRAS*, **289**, 882.

R8   Roman, N. G., 1954, *AJ*, **59**, 307.

R9   Roman, N. G., 1955, *ApJS*, **2**, 195.

R10 Rose, J. A., 1991, *AJ*, **101**, 937.

S1   Salaris, M., Chieffi, A., Straniero, O., 1993, *ApJ*, **414**, 580.

S2   Sandage, A., Eggen, O. J., 1959, *MNRAS*, **119**, 278.

S3   Sandage, A. R., 1969, *ApJ*, **158**, 1115.

S4   Sandage, A., 1970, *ApJ*, **162**, 841.

S5   Sandage, A., 1986, *ARA&A*, **24**, 421.

S6   Sarajedini, A., Demarque, P., 1990, *ApJ*, **365**, 219.

S7   Sarajedini, A., 1997, *AJ*, **113**, 682.

S8   Sarajedini, A., Chaboyer, B., Demarque, P., 1997, *PASP*, **109**, 1321.

S9   Scalo, J. M., Miller, G. E., 1980, *ApJ*, **239**, 153.

S10 Schuster, W. J., Nissen, P. E., 1989, *A&A*, **221**, 65.

S11 Schmidt, M., 1959, *ApJ*, **129**, 243.

S12 Schmidt, M., 1963, *ApJ*, **137**, 758.

S13 Searle, L., Zinn, R., 1978, *ApJ*, **225**, 357.

S14 Shapley, H., 1938, *Bull. Harvard Coll. Obs.*, No. 908.

S15 Shapley, H., 1938, *Nature*, **142**, 715.

S16 Soderblom, D. R., Duncan, D. K., Johnson, D. R. H., 1991, *ApJ*, **375**, 722.

S17 Sommer-Larsen, J. Zhen, C. 1990, *MNRAS*, **242**, 10.

S18 Spitzer, L., Schwarzschild, M., 1953, *ApJ*, **118**, 106.

S19 Strömberg, G., 1922, *ApJ*, **56**, 265.

S20 Strömberg, G., 1925, *ApJ*, **61**, 363.

S21 Strömgren, B., 1966, *ARA&A*, **4**, 433.

T1   Twarog, B. A., 1980, *ApJ*, **242**, 242.

V1   Vandenberg, D. A., Bolte, M., Stetson, P. B., 1990, *AJ*, **100**, 445.

W1   Walker, A. R., 1994, *AJ*, **108**, 555.

W2   Wielen, R., 1974, *Highlights in Astronomy*, Vol. 3, p. 395, ed. G. Contopolous (Reidel, Dordrecht).

W3   Wielen, R., 1977, *A&A*, **60**, 263.

W4   Wielen, R., Fuchs, K., 1983, in *Kinematics, Dynamics and Structure of the Milky Way*, ed. W. L. H. Shuter, Reidel, Dordrecht, p. 81.

W5   Wood, M. A., Oswalt, T. D., 1998, *ApJ*, **497**, 870.

W6   Wyse, R. F. G., Gilmore, G., 1995, *AJ*, **110**, 2771.

W7   Wyse, R. F. G., Gilmore, G., Franx, M., 1997, *ARA&A*, **35**, 637.

Y1   Yoshii, Y., Saios, H., 1979, PASJ, **31**, 339.

Z1   Zinn, R., West, M. J., 1985, *ApJS*, **55**, 45.

Z2   Zinn, R., 1985, *ApJ*, **293**, 424.

7

The stellar luminosity function

7.1 INTRODUCTION

The stellar luminosity function, $\Phi(M)$ – the number of stars per unit absolute magnitude per unit volume – is one of the fundamental quantities required for understanding star formation and investigating the structure of our Galaxy. In recent years, most emphasis has been placed on determining the shape of $\Phi(M)$ at faint magnitude – and hence the mass spectrum, with the ultimate goal of assessing the likely dark-matter contribution of VLM stars and brown dwarfs. However, it was Galactic structure analyses, rather than an interest in the relative number of luminous and faint stars *per se*, which prompted the first derivation of the luminosity function.

   The study of the stellar luminosity function has a long and chequered history of contentious debate, reviewed in the first part of this chapter. Latter sections synthesise results from recent surveys to provide an estimate of the luminosity function defined by field stars in the Galactic disk. Chapter 8 discusses the transformation of these results from luminosity to mass.

7.2 THE EARLY YEARS

William Herschel [H10] inaugurated the study of Galactic structure through his programme of 'star-gaging' – counting the total number of stars that drifted through the 15-arcmin diameter field of view of his 20-foot reflecting telescope[1] within a set time for a grid of reference fields distributed across the celestial sphere. His observations yielded the average surface density of stars brighter than the limiting magnitude of his telescope (about 12th magnitude), for Herschel had no means of

[1] The focal length of the telescope was 20 feet; the speculum mirror had a diameter of 18 inches.

obtaining accurate photometry nor, lacking distance estimates, had he any means of estimating intrinsic luminosities. Herschel adopted the assumption that the stars in his survey had the same average luminosity in each field, and that there was no significant intervening obscuration, and inferred the three-dimensional structure of the Galaxy from the observed surface densities. On that basis, he deduced that the Sun lay in a flattened system which bifurcated towards the Great Rift in Cygnus, a feature which we now know to be due to absorption within the Plane of the Galaxy.

The opportunity to apply an approximate distance scale to these observations arrived with the first measurements of stellar trigonometric parallaxes. In 1847 John Herschel undertook observations of apparent brightnesses of 191 southern stars, a small number of which had measured trigonometric parallaxes. By the middle years of the nineteenth century, measurements had been obtained of stars with luminosities ranging from $20 L_{\odot}$ (Sirius) to $0.1 L_{\odot}$ (61 Cygni, or Gl 820). These results provided the first quantitative demonstration of the substantial range in intrinsic brightness spanned by stars in the Milky Way galaxy.

The main tool of Galactic structure studies in the late nineteenth century was stellar number–magnitude counts – indeed, this remains one of the most effective means of probing the stellar density distribution. Star-count analysis aims to invert the observed number counts as a function of apparent magnitude to infer the underlying density distribution. To do so, due allowance must be made for the range of intrinsic luminosities. The formalism for this analysis – the fundamental equation of stellar statistics – was developed by von Seeliger [S7].

The observed number–magnitude distribution of stars, $A(m)$, can be expressed as an integral equation involving a density law, $D(r)$, a volume element, and the stellar luminosity function, $\Phi(M)$, corrected for line-of-sight absorption, $a(r)$,

$$A(m) = \omega \int \Phi(M - a(r))D(r)r^2 \, dr \qquad (7.1)$$

where ω is the solid angle covered by the observations.

von Seeliger's equation demonstrates that an estimate of $\Phi(M)$ is essential in determining the structure of the Galaxy, $D(r)$, the main focus of early star-count studies. The most influential proponent of this research was the Dutch astronomer, Jacobus Kapteyn, who oversaw the compilation and analysis of one of the most extensive photographic star catalogues yet undertaken, and, through his association with Hale and Mount Wilson Observatory, laid the foundations for the immensely fruitful influx of Dutch astronomers (including van Maanen, Luyten, Kuiper and Bok) into US observatories. Elected to the Chair of Astronomy and Theoretical Mechanics at Gröningen in 1878, Kapteyn found himself in charge of an observatory with staff, but little in the way of instrumentation to allow him to assemble the extensive data essential for the pursuit of what he called the 'sidereal problem' – the distribution of stars in space.

Fortunately for Kapteyn, David Gill, Her Majesty's Astronomer at the Cape of Good Hope, soon found himself in entirely the opposite situation. Gill discovered the astronomical potential of photography in 1882, when he assisted a local amateur – Mr Allis – to obtain a picture of the Great Comet of that year, mounting a camera on one of the observatory telescopes. Gill noticed that the photograph not only captured detailed structure in the cometary tail, but also revealed numerous faint stellar images. Realising the potential for stellar cartography, Gill circulated copies of this and other astronomical photographs amongst various leading astronomers, including Admiral Mouchez at Paris Observatory who, in turn, encouraged the brothers Henry to produce lenses suitable for astrophotography. The latter circumstance led to the initiation of the all-sky *Carte du Ciel* in 1887. Gill had started his own southern photographic survey – a project discussed in letters to Kapteyn in 1885 [M9]. Kapteyn responded by offering to help with the measurement and reduction of these plates – an offer which eventually led to the production of the *Cape Photographic Durchmusterung* in 1896. (In the meantime, the Board of Visitors of the Royal Observatory, Greenwich, had cancelled all funding for the project, and Gill was forced to devote almost a third of his salary for several years to the completion of the project. Further details are given by Murray [M9] and references therein.)

Kapteyn published one of the first observational determinations of the 'luminosity curve' of stars. While the techniques used in those early analyses have been superseded, the results not only track our evolving understanding of $\Phi(M)$, but also offer insight into some of the problems and errors encountered in more recent investigations. Before considering these results, a brief mention of the treatment of interstellar absorption is in order. Seeliger explicitly used the term $a(r)$ in the star-count equation, but few early studies corrected observations for that effect. Kapteyn, in particular, noted [K3]: 'Undoubtedly one of the greatest difficulties, if not the greatest of all, in the way of obtaining an understanding of the real distribution of the stars in space lies in our uncertainty about the amount of loss suffered by the light of the stars on its way to the observer.' Kapteyn generally ignored absorption in his star-count analysis, and the effects are of little importance in luminosity function determinations where the majority of stars lie within 100 pc of the Sun. The question of general interstellar absorption (as opposed to the small, dark nebulae observed by Barnard and others) was not laid to rest until Trumpler's [T5] studies of the distribution of open clusters.

7.2.1   The method of mean parallaxes

The most straightforward method of deriving $\Phi(M)$ is to count the number of stars as a function of absolute magnitude in a known volume. However, this method requires a complete sample, and it is only in recent years that statistically complete stellar samples can be compiled within even modest volumes. With accurate trigonometric parallaxes, and even photometry, available for only a handful of stars in the early years of this century, luminosity function derivations relied on less direct, statistical techniques.

The method of mean parallaxes – used by Kapteyn in his initial work on the luminosity function [K1] and one of the techniques included in van Rhijn's [R16] analysis – assumes that one can use the sparse sample of stars with known trigonometric parallaxes as unbiased indicators of the distance distribution amongst the general number–magnitude–proper motion distribution, $N(m, \mu)$. Taking known stars with measured parallaxes, Kapteyn calculated the mean parallax for a grid of points in apparent magnitude and μ, $\bar{\pi}(m, \mu)$. Within each bin, the distribution of true parallaxes is fitted to a Gaussian distribution[2] in $\log \pi$. Given the distribution of π for each $N(m, \mu)$ cell, one can integrate over μ to estimate the distance distribution of stars of a given apparent magnitude, and hence derive the absolute magnitude distribution. Summing the number of stars in an absolute magnitude bin within given parallax limits determines the luminosity function.

Figure 7.1(a) shows the luminosity curve derived by Kapteyn from his 1902 analysis (curve V from [K1]). This paper is also notable for establishing the ordinate used in early plots of the luminosity function (the logarithm of the local space density in stars $\mathrm{mag}^{-1}\,\mathrm{pc}^{-3} + 10$), and for defining absolute magnitude as the apparent magnitude a star would have at a distance of 10 parsecs (parallax of $0.''1$),

$$\mathrm{M} = m + 5 + 5\log\pi \qquad\qquad (7.2)$$

Kapteyn commented, [K1]: 'The curve appears to reach a maximum about the absolute magnitude 10.5. Whether for fainter stars it will descend as rapidly or more rapidly, and whether it will soon reach a limit, below which no luminous stars exist, are questions for which a knowledge of the number of large proper motion stars of fainter than the ninth magnitude is required.'

Kapteyn's method of presenting $\Phi(\mathrm{M})$ as the number of stars per unit absolute magnitude prompted one of the first controversies. G. Comstock, director of Washburn Observatory (Madison, Wisconsin), interpreted the densities as given per unit luminosity (that is, a linear, rather than logarithmic, scale). With that interpretation, Kapteyn's results appeared to disagree sharply with Comstock's own analysis of trigonometric parallax data [C2]. This prompted an extended and somewhat acrimonious exchange of articles in the journals, setting a prophetic precedent for the field.

7.2.2   The method of trigonometric parallaxes

The second method devised by Kapteyn, this was adopted subsequently by van Rhijn [R15], [R16], Bok and Macrae [B5], Trumpler and Weaver [T6], Starikova [S11] and is even discussed at length by Mihalas and Binney [M8]. Given the observed distribution $N(m, \pi)$ for a set of stars $\pi > \pi_{\lim}$, the aim is to correct both for errors due to random observational uncertainties and for systematic effects due to incompleteness in the sample.

Consider the number–magnitude–proper motion distribution, $N(m, \mu)$, for a set of stars with $\mu > \mu_{lim}$, a subset of which have measured parallaxes. We can construct $N_\pi(\mathrm{M}, \mu)$, the number–absolute magnitude–proper motion distribution for the latter. If those stars with measured parallax are an unbiased selection from the

---

[2] Kapteyn gave no explicit justification for the choice of fitting function, but with a total of only 58 parallax stars, a Gaussian distribution is as reasonable a choice as any.

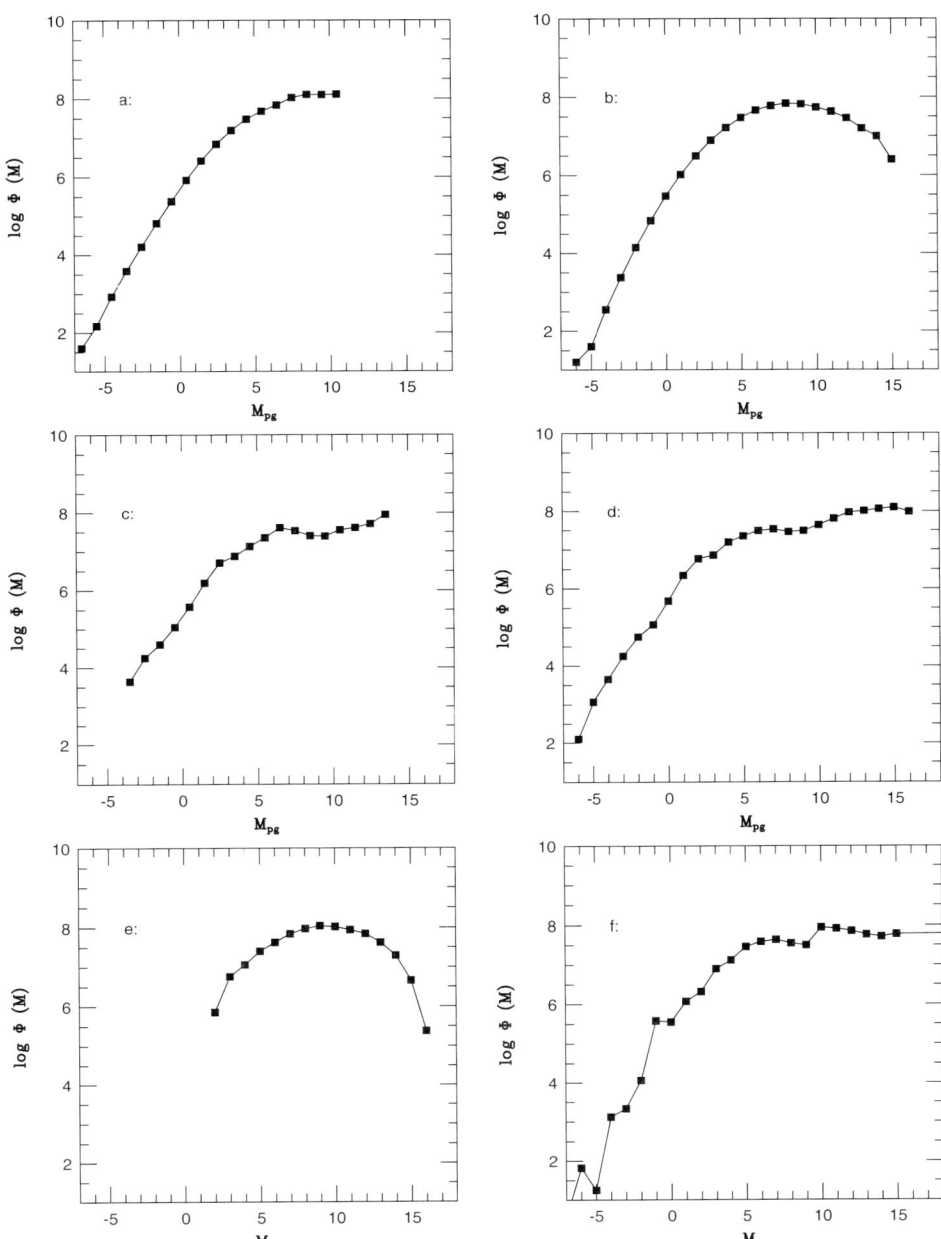

Figure 7.1. The changing form of the luminosity function. The results shown here are derived either using the method of mean parallaxes or the method of mean trigonometric parallaxes and are from: (a) Kapteyn [K1]; (b) Kapteyn and van Rhijn [K4]; (c) van Rhijn [R15]; (d) van Rhijn [R16]; (e) van Maanen [M3]; and Starikova's [S11] analysis. Following the convention of the time, the ordinate is the logarithm of the space density (in stars $pc^{-3} mag^{-1} + 10$).

whole, then each star with known π represents f stars in total, where

$$f = \frac{N(m, \mu)}{N_\pi(m, \mu)} \tag{7.3}$$

Generally, analysis is limited to the range of $N(m, \mu)$, where f is small (<2).

Two further systematic corrections are required. First, the uncertainty in π is symmetric, but the volume of a shell of given width increases with π^{-2}. The result is that more stars with smaller parallaxes (larger distances) are scattered into a parallax-limited sample than are scattered out (Malmquist bias, further discussed below). The net result is that the average parallax in a given interval $\pi \pm \delta\pi$ is smaller than would be the case if all measurements were error-free; that is,

$$\langle \pi \rangle < \langle \pi_0 \rangle$$

where π is the observed parallax, and π_0 is the true parallax.

The second correction takes account of stars within the distance limit, $\pi > \pi_{\lim}$, but with proper motions below the sample limit, $\mu < \mu_{\lim}$. These stars have a tangential velocity below the critical value for inclusion in the sample, either through projection effects (that is, most of the relative motion is in the radial velocity) or through their having a space velocity closely matching the solar motion. In either case, the number of such stars can be estimated by adopting a particular model for the local stellar kinematics (see [T6] and [M8]).

Figures 7.1(b)–7.1(e) show luminosity functions derived using this technique. van Maanen's [M3] study was based on analysis of 2,380 stars with $\mu > 0\overset{''}{.}5$ per year, 651 of which had parallax measurements – a significant improvement over the sparse data originally available to Kapteyn in 1902. In general, the later surveys, analysing data taken to faint limiting magnitudes, show the luminosity function peaking at fainter absolute magnitudes than the initial studies, with the space densities constant or slowly declining (except in van Maanen's $\Phi(M_{vis})$).

7.2.3   The method of mean absolute magnitudes

Luyten's analyses

This third statistical method was used extensively by Luyten [L7], [L8], [L11], primarily in the analysis of catalogues of high proper-motion stars. Following a suggestion by Hertzsprung, Luyten combined apparent magnitude and proper motion to produce a quantity analogous to absolute magnitude. At first Luyten referred to this quantity as the 'Hertzsprung index', M_μ, but later adopted the term 'reduced proper motion', H, defined as

$$H = m + 5 + 5 \log \mu \tag{7.4}$$

Luyten calibrated this index against absolute magnitude using observations of stars with known trigonometric parallax. Initially, a magnitude term was included in the equation, although Luyten later settled on a linear equation,

$$\bar{M}_H = a + bH \tag{7.5}$$

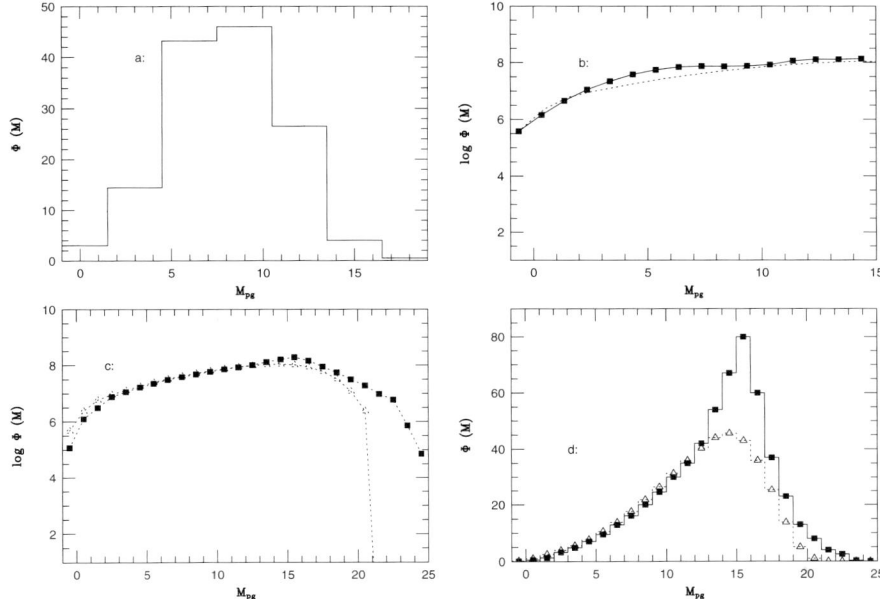

Figure 7.2. Luminosity functions derived using the method of mean absolute magnitudes. The data are from: (a) Luyten [L8]; (b) Seares [S6] and (dotten line) Luyten [L9]; (c) Luyten ([L9], dotted line; [L11], solid line) – logarithmic representation; and (d) Luyten [L9], [L11], presented as the number of stars predicted within 10 parsecs.

Luyten characterised the distribution of residuals about the mean relation as a Gaussian with a r.m.s. dispersion of ~ 1.2 magnitudes in absolute magnitude. Thus, H is useful only as a statistical predictor of absolute magnitude for a group of stars, rather than as a provider of accurate estimates for individual cases. Nonetheless, Luyten applied this calibration to transform the observed distribution N(H) from his proper-motion surveys to a distribution in absolute magnitude, M, using estimates of the velocity distribution to allow for stars with proper motions below the survey limit.

Luyten's initial calibration [L7] of equation (7.5) gave coefficients of $a = -0.69$ and $b = 0.54$ and, analysing data for 749 stars with $\mu > 0\overset{''}{.}5$ yr^{-1}, he derived a distribution $\Psi(H)$ which reached a maximum at $H \sim 12.5$, implying a maximum in the luminosity function at $M_{pg} \sim 8.5$ (Figure 7.2(a)). This was in broad agreement with the luminosity function derived some three years earlier by Kapteyn and van Rhijn. In contrast, Seares [S6] derived higher densities in his analysis combining proper-motion stars and a larger sample of stars with spectroscopic parallaxes. This result stems primarily from Seares' use of two linear relations to calibrate H against M, with coefficients $b \sim 0.67$ at $H < 8$, and $b \sim 1$ at larger H. The latter calibration places a larger number of stars at fainter absolute magnitudes than in Luyten's analysis, with a corresponding effect on $\Phi(M)$ (Figure 7.2).

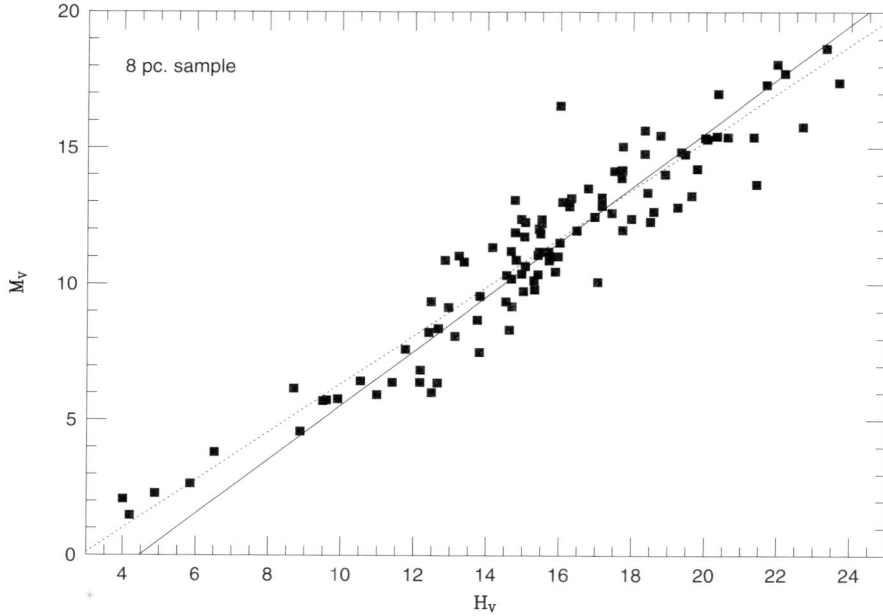

Figure 7.3. The relationship between reduced proper motion, H, and absolute magnitude, M_V, defined by modern observations of stars in the nearby-star catalogue with trigonometric parallaxes accurate to better than 10%. The dotted line is the best-fit relation, equation (7.9); the solid line is equation (7.10), where the slope is set to unity.

Luyten [L8] disputed the validity of Seares' revised calibration, but his own luminosity function analyses evolved in the same direction. The final calibration adopted by Luyten is

$$\bar{M}_H = -3.5 + 0.86H \tag{7.6}$$

and this equation was used in his analysis of both the Bruce proper motion survey, covering the southern skies to $m_{pg} \sim 17.7$, and of the Palomar proper motion surveys [L5]. The latter extend from the north celestial pole to $\delta = -42°$ and have a limiting magnitude of $m_{pg} \sim 20$. The luminosity functions from those surveys peak at $M_{pg} = 14.5$ and 15.7 magnitudes respectively, and are shown in Figure 7.2(c) and 7.2(d).

Reconsidering mean absolute magnitudes

Since accurate photometry is now available for many nearby stars with well-determined parallaxes, we can re-examine the calibration of $\bar{M}_H$. First, note that since

$$\mu = \frac{V_T}{\kappa r} \tag{7.7}$$

where V_T is the transverse velocity in km s^{-1}, r is the distance in parsecs and $\kappa = 4.74$, equation (7.4) can be rewritten as

$$\langle H \rangle = m + 5 - 5 \log r + 5 \langle \log V_T \rangle - 3.38$$

$$= M + 5 \langle \log V_T \rangle - 3.38 \tag{7.8}$$

This relationship implies that the coefficient 'b' in equation 7.5 should be 1.0, as in Seares' calibration, unless the mean transverse velocity varies with M_V.

Figure 7.3 plots H_V against M_V for stars within 8 parsecs of the Sun. The best-fit relation is

$$M_V = -2.56 + 0.89 H_V, \qquad \sigma_M = 1.24 \, \text{mag} \tag{7.9}$$

which implies that the average tangential velocity rises from $\sim$ 18 km s^{-1} at $M_V \sim 0$ to 35 km s^{-1} at $M_V \sim 12$. This variation is not unreasonable, given the short main-sequence lifetimes of early-type stars. However, a more reliable calibration of M_V versus H_V for later-type stars – where the main-sequence lifetime exceeds the age of the Galactic disk – follows if we substitute the appropriate value of $\langle \log V_T \rangle$ in equation (7.8). Averaging $\log V_T$ for stars in the 8-parsec sample gives

$$M_V = -4.46 + H_V \tag{7.10}$$

There is a further effect to consider in interpreting mean absolute magnitude analyses of proper-motion samples. Equation (7.8) shows that the relationship between H and M is single-valued only for a population with a simple uni-modal velocity dispersion. This circumstance holds for a volume-limited sample of nearby stars, almost all of which are members of the Galactic disk (see Figure 6.16). Halo stars, however, have a higher velocity dispersion and negligible net rotation (a substantial solar motion), leading to a higher average tangential motion. As discussed further in Chapter 11, the number of stars contributed by a stellar population to a proper-motion sample is given by

$$N(\mu > \mu_{lim}) \propto \rho_0 V_T^3 \tag{7.11}$$

where ρ_0 is the local space density. Thus, high-velocity halo stars make a disproportionately large contribution to proper-motion selected samples. Moreover, equation (7.10) does not provide the appropriate absolute magnitude calibration [H5], [R4]. Since

$$\langle V_T(\text{halo}) \rangle \approx 6 \times \langle V_T(\text{disk}) \rangle$$

then

$$M_V(\text{halo}) = -8.3 + H_V \tag{7.12}$$

That is, applying an $\bar{M}_H$ relationship calibrated against disk dwarfs to a sample of halo subdwarfs leads to absolute magnitudes which are too faint.

This miscalibration is unimportant for proper-motion surveys with bright magnitude limits (such as the Bruce survey), but becomes increasingly important for surveys extending to fainter magnitudes, such as the Luyten Palomar Schmidt surveys. As will be discussed further in Chapter 11, this bias can be turned to advantage by using broadband colours to isolate halo subdwarfs and degenerate white dwarfs in proper-motion surveys. With photometry of only modest accuracy, however (as in Luyten's catalogues) the segregation between disk and halo is not evident, and an analysis which assumes that a single calibration will necessarily produce an inflated estimate of the number of low-luminosity dwarfs. This effect is evident in Luyten's Palomar Schmidt analysis, where the low-luminosity tail extends to beyond $M_{pg} = 20$ (Figure 7.2(d)): follow-up spectroscopy has shown that these stars are almost exclusively halo subdwarfs.[3]

7.2.4 Kuiper's nearby-star luminosity function

In the late 1920s and early 1930s, the main direction of Galactic structure studies shifted towards determining the distribution of stars within the disk. In part this reflected limitations imposed by the available technology – low-sensitivity photographic plates and hand-driven measuring engines. More emphasis was placed on understanding the behaviour of the stellar luminosity function at brighter magnitudes, than on identifying stars at the lowest luminosities. The final pre-1950 study which is relevant to determining the form of the luminosity function at faint magnitudes is Kuiper's analysis of a sample of 254 stars with $\pi \geq 0\rlap{.}''095$ [K8] – a precursor of the more recent work by Wielen [W5], [W6] and Reid *et al.* [R11].

While many of the stars in Kuiper's sample had trigonometric parallax measurements, a number of the lower-luminosity stars also had distances determined from spectroscopic parallaxes. Based on those data, Kuiper derived a luminosity function similar to Luyten's results, with a maximum at $M_V \sim 10$ (Figure 7.4). Spectroscopic observations only were available for a larger number of M dwarfs, and Kuiper presented preliminary analysis in addition to the luminosity functions derived from only trigonometric parallaxes (a) and trigonometric plus spectroscopic data (b). This curve (c) is plotted in Figure 7.4, and Kuiper commented that: 'The difference c–b shows how much remains to be discovered even for the nearest stars.' This passing remark sets the scene for the *cause célebre* of M dwarfs and the missing mass.

7.3 THE 1970S: A PLETHORA OF M DWARFS?

'If there is a generalisation to be drawn from this whole discussion ... it is that the principles are easy but the practice is difficult.'

I. King, 1977 [K6]

---

[3] Luyten was aware of potential problems with his analysis, particularly concerning the Palomar data. There were very few determinations of accurate parallaxes for stars fainter than $\sim$ 16th magnitude even in 1968, so Luyten had no means of checking the validity of his calibration at those magnitudes; nor were the parameters of the halo population well characterised at that time. Nonetheless, it is surprising that no attempt was made to understand why the coefficient, b, was not unity in any of the calibrations except Seares' two-segment fit.

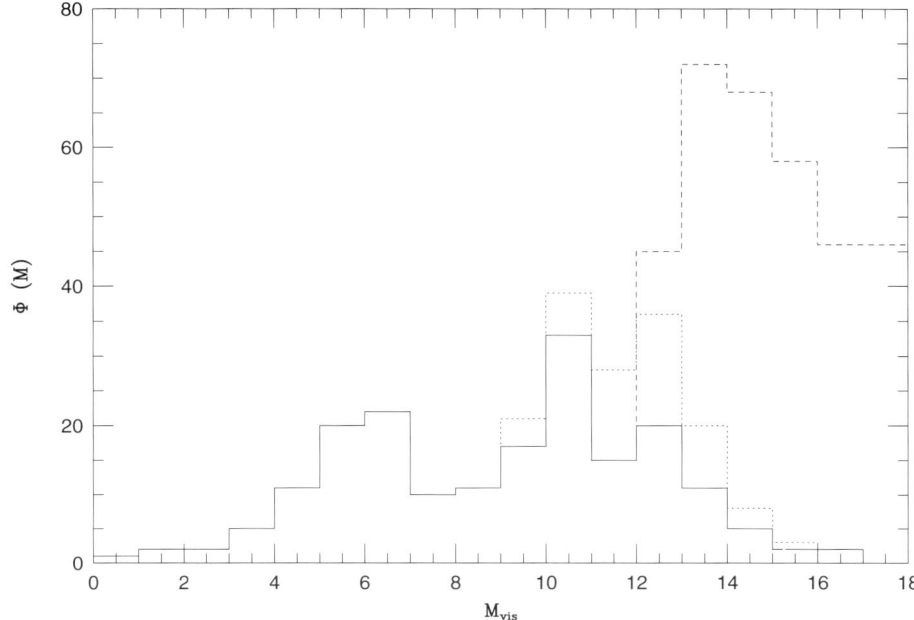

Figure 7.4. The three luminosity functions derived by Kuiper [K8] from the analysis of nearby stars. Curve a (solid line) is based only on trigonometric parallaxes; curve b (dotted line) includes stars with spectroscopic parallax distance estimated; and curve c (dashed line) is from a 'larger sample'.

In his 1966 review of the stellar luminosity function, McCuskey [M1] compared the results from available studies, and found, with two exceptions, excellent agreement. The first exception lay between the van Rhijn [R16] and Luyten [L9] functions for $7 < M_{pg} < 10$, with the van Rhijn $\Phi(M)$ dipping below Luyten's data. This discrepancy arises because Luyten smoothed his data to give a continuously increasing function [L8]. The second discrepancy lay at fainter magnitude, and had more far-reaching effects. Sanduleak [S1] had compiled a sample of $\sim 1,200$ late-type dwarfs from an objective prism survey of the north Galactic cap. Determining distances to those stars using spectroscopic parallaxes, Sanduleak – like Kuiper before him – derived volume densities up to a factor of five higher than Luyten's results $11 < M_B < 15$ (Figure 7.5(a)).

This result was not confirmed by a similar high-latitude survey by Klare and Schaifers [K8], and inevitably drew criticism from Luyten. However, one of the main astrophysical problems of the time concerned the search for the 'missing mass' in the Galactic disk. Oort's [O1], [O2] analyses, and others, of the motions of stars perpendicular to the Galactic plane implied that the gravitational force required a local mass density, $\rho_{disk}(0)$, of $0.15\,M_{\odot}\,\mathrm{pc}^{-3}$. Summing the contributions to $\rho_{disk}(0)$ from the visible constituents of the solar neighbourhood gave a mass density of $\sim 0.06\,M_{\odot}\,\mathrm{pc}^{-3}$ from stars, and $0.03\,M_{\odot}\,\mathrm{pc}^{-3}$ from gas and dust, leaving $\sim 0.06\,M_{\odot}\,\mathrm{pc}^{-3}$ unaccounted for. Low-luminosity M dwarfs have a high

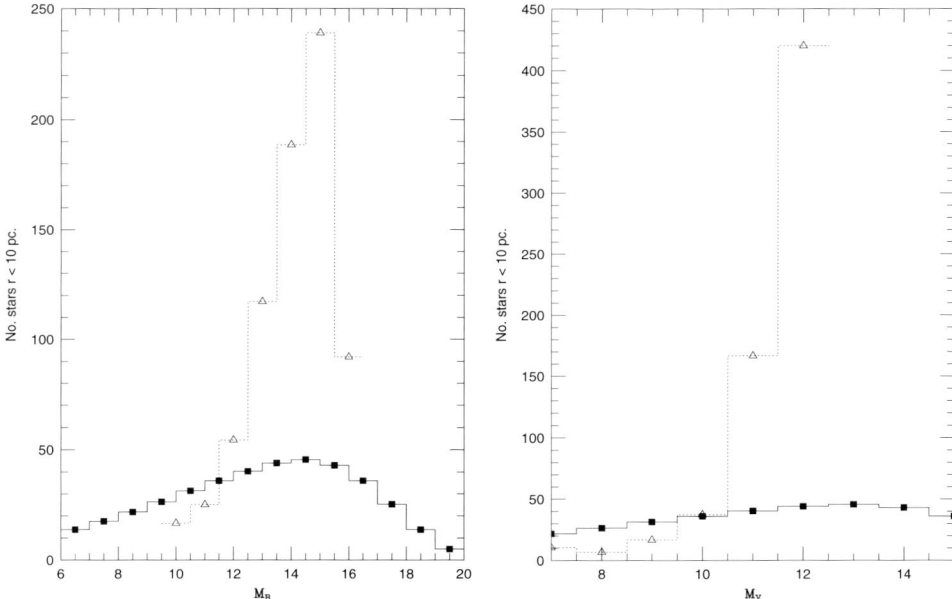

Figure 7.5. (a) A comparison of Luyten's 1939 luminosity function and Sanduleak's analysis of the north polar objective prism survey; (b) a comparison of Luyten's 1968 results and Weistrop's luminosity function.

mass-to-light ratio ($M/L > 10$), so numerous late-type M dwarfs could hide substantial mass without adding significantly to the total light in a system.

Support for Sanduleak's $\Phi(M)$ appeared to come from results based on an entirely different technique. Weistrop [W2] had painstakingly undertaken iris diaphragm photometry of $\sim 13{,}500$ stars brighter than V ~ 18 within a 13.5 square-degree field towards the north Galactic pole (NGP) – the same region surveyed by Sanduleak, although Weistrop's data consisted of direct U, B and V plates taken with the 48-inch Oschin Schmidt at Palomar. Calibrating the iris photometry against a sequence of stars with known magnitudes (from photoelectric observations), she used the (M_V, (B–V)) relationship (Figure 2.13) to estimate photometric parallaxes (that is, given (B–V), estimate M_V), and hence distances and space densities. Her results suggested that the Luyten function underestimated the number of faint M dwarfs by a factor of between 5 and 10 (Figure 7.5(b)). Further apparent corroboration was supplied by the identification of numerous very red (colour class +4) stars from Giclas et al. [G2] visual scans of blue and red plates from the Lowell Observatory survey, and in new objective prism surveys by Smethells [S9] and Sanduleak [S2].

The complication facing this hypothesis lay in explaining why these late M dwarfs were not included in Luyten's proper-motion surveys. Assuming that the absolute magnitude estimates (from spectroscopy and (B–V)) were reliable, the only reasonable explanation seemed to be that those stars formed a low velocity-dispersion component in the Galactic disk. With low space motions, the stars would also have low proper motions, and would therefore not contribute to a sample selected by proper motion. This hypothesis seemed to be consistent with proper-motion measurements of a subset of the Sanduleak M dwarfs by Pesch and Sanduleak [P4], Pesch [P3], Gliese [G5] and Murray and Sanduleak [M10]. Although each study was based on only ~ 20 stars, all had small proper motions, consistent with very low space motions *if the M dwarfs were actually at the distances indicated by the spectroscopic parallaxes.* Murray and Sanduleak, for example, estimated a transverse velocity dispersion of only $10\,\mathrm{km\,s^{-1}}$, from which they inferred a vertical dispersion $\sigma_W < 8\,\mathrm{km\,s^{-1}}$. The latter result implies a vertical density distribution similar to B stars, and a local mass density of $\sim 0.05\,M_\odot\,\mathrm{pc^{-3}}$ – sufficient to account for almost all of Oort's 'missing mass'. Gliese [G5] reached similar conclusions from comparable analysis of the motions of 75 Lowell colour-class +4 stars towards the SGP.

While these results appeared to answer the 'missing mass' question, they posed conundra of their own. First, the existence of a substantial, low velocity-dispersion population of M dwarfs implies either a correspondingly large recent burst of star formation, confined, moreover, to low-mass stars, since an excess of solar-type stars would be very hard to hide; or that velocity dispersion and age were not correlated parameters [O3]. Second, while the presence of so much mass in a very thin disk might help explain the Oort 'missing mass', it raised severe problems for maintaining the stability of the disk as a whole; essentially, there appeared to be insufficient kinetic energy to stop gravitational collapse [S3]. Finally, Luyten – who vehemently opposed the revised luminosity functions in a series of inimitable papers ([L12] *et seq.*) – pointed out that the number densities derived at faint absolute magnitudes implied at least 800 stars within 5 pc of the Sun (as compared with 56 known) and 29 stars within 2 pc, where only α Cen AB, Proxima Cen and Barnard's star are known.

'There is no trusting appearances.'
R. B. Sheridan

The M dwarf missing mass hypothesis started unravelling in the mid-1970s. Jones [J3] used narrowband photometry to estimate photometric parallaxes for 19 of the 21 stars in the [M10] SGP sample. He derived a mean parallax of $0\rlap{.}{''}009$ while [M10] had estimated $0\rlap{.}{''}021$, implying over a factor of two increase in the derived velocity dispersions. These results – later confirmed by broadband BVRI photometry by Weistrop (W3) – were not consistent with the Sanduleak stars being members of a low velocity-dispersion population, while the larger distances implied not only brighter absolute magnitudes than indicated by the spectroscopic parallaxes, but a lower space density.

The most significant blow, however, came from a reassessment of Weistrop's photometry by Faber *et al.* [F1] and by Weistrop [W3] herself. This revealed the presence of systematic errors in both the (B–V) colours of the standard stars and in the photographic (B–V) colours used to infer absolute magnitudes and photometric parallaxes. The sense of the error was a colour term such that colours were systematically too red for the reddest stars. The error was only ~ 0.2 magnitudes at $(B–V) \sim 1.7$ magnitudes but the $(M_V, (B–V))$ colour–magnitude diagram has a slope of ~ 8–10 for M dwarfs, so this error led to an inferred M_V too faint by 1.5–2 magnitudes. Thus, a 15th magnitude M dwarf with $M_V = 11$ at $\sim 60\,\mathrm{pc}$ was being misidentified as an $M_V = 13$ dwarf at a distance of only 25 pc. Since the result of this error was both to overestimate the absolute magnitude (too faint) and underestimate the distance, the result was a substantial overestimate of the local density of late-type M dwarfs. Once the calibration was corrected, the number densities fell in line with those expected from a luminosity function like Luyten's $\Phi(M)$.

Similar problems beset the objective prism surveys. Luyten [L12] was amongst the first to suggest that, given the steep relation between M_V and spectral type (Figure 2.10), a small systematic error in classification could account for high inferred number densities. Indeed, Pesch (P3) had noted already that: 'there seems to be a tendency to assign too late a type to objective-prism spectra of M stars near the plate limit.'

Accurate broadband photometry confirmed the presence of these systematic errors. Weistrop [W4] and Reid [R3] both obtained photoelectric BVRI data of the 186 M dwarfs catalogued by Smethells. A comparison between (V–I) photometric parallaxes and Smethells' $(M_V, \text{spectral type})$ relation shows that the spectroscopic absolute magnitude estimates are too faint by ~ 0.5 magnitudes at type K7, and by ~ 3 magnitudes for M5 and later. A number of these stars have since been observed by the astrometric satellite Hipparcos, and the resulting trigonometric parallaxes confirm the brighter absolute magnitudes of the (V-I) calibration. Re-calculating the luminosity function using the photometric parallaxes gives results indistinguishable from the Luyten or van Rhijn $\Phi(M)$ (Figure 7.6).

In a similar manner, BVRI photometry of stars from NGP [S2] and Pesch and SGP [P4] objective prism surveys demonstrated that only $\sim 10\%$ of the stars are actually fainter than $M_V \sim 13$, rather than the originally posited $\sim 50\%$ [R10], [P5]. Other surveys are similarly affected [T2], in spite of their use of (R–I) photometry to supplement the photographic data [R10]. Finally, catalogues of red dwarfs derived from visual estimates of colour classes (as in the Lowell survey) proved to be over-optimistic in the classification. Luyten originally pointed out that not only had he classified many of the Lowell +4 stars as type K and K–M in his surveys, but that a substantial number of the stars which lay in the overlap region between two Lowell fields were classified as colour class +3 (red), as opposed to class +4 (extremely red), on the second plate. Accurate photometry showed that nearly all of these stars were early-type M dwarfs at distances of 50–100 pc, rather than >M5 dwarfs within a few parsecs of the Sun.

'Everything's got a moral, if you can only find it.'

Lewis Carroll

Following the resolution of what Luyten unkindly termed the 'Weistrop Watergate', the IAU devoted a joint discussion session of the 1976 General Assembly to the matter of the local density of M dwarfs. Opening that session, Gliese enquired rhetorically whether all of the efforts of the last few years had produced any worthwhile result – or had they simply turned full circle, returning the subject to

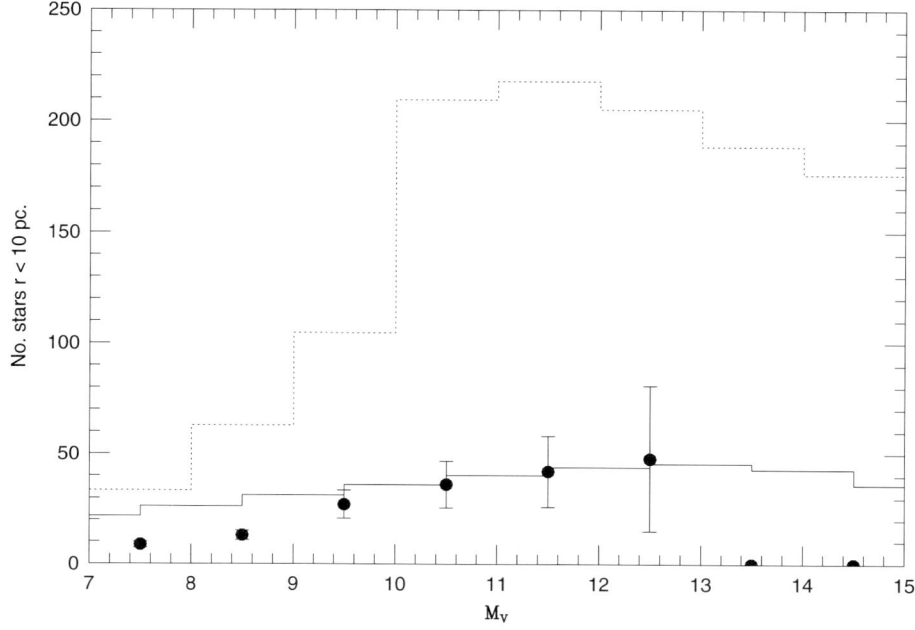

Figure 7.6. A comparison of the luminosity function derived by Smethells (dotted histogram); Reid's [R4] re-analysis of the same data, using (V–I) photometry rather than low-resolution prism spectroscopy (solid points); and the Luyten [L11] $\Phi(M)$ (solid histogram).

the position held in 1963? Answering his own question, Gliese argued that there had been some progress, notably the many observational programmes which had been prompted by the controversy, several of which used innovative techniques. Weis-trop's survey, for example, was the first to attempt to apply faint, accurate number–magnitude star-counts to Galactic structure work on the wide-field scale made possible by 48-inch Schmidt plates.

One can also point to this incident as an excellent example of the scientific method actually at work – messier and more subjective than the idealised objective comparison of observation and theory, but achieving the same result. New observations appeared to solve a theoretical problem (in this case, disk 'missing mass'), and cast doubt on an established viewpoint (the van Rhijn/Luyten $\Phi(M)$). These initial results were apparently supported by independent data, prompting a certain amount of uncritical bandwagoning, and a growing acceptance of the new hypothesis, despite problems (some valid, some histrionic) raised by the old guard. As time passes, the new result attracts more detailed scrutiny, and either withstands those more rigorous tests or, as in the present case, reveals an increasing number of inconsistencies and is rejected. Yet the attention generated by the now-invalidated hypothesis leads to better understanding of the subject as a whole.

A prime lesson learned from this affair is that systematic and random errors generally conspire to enhance the apparent number of unusual objects, particularly in large-scale surveys. One must be careful in sifting through such data, and highly sceptical of potentially exciting scientific discoveries. In short: be wary if your survey finds too much of what it was looking for.

7.4   SYSTEMATIC BIASES

Before summarising results from more recent luminosity function studies, two sources of systematic error should be considered.

7.4.1   Lutz–Kelker corrections

Any trigonometric parallax measurement has an associated uncertainty, usually assumed to be drawn from a Gaussian distribution, rms dispersion σ. The measured parallax is the best estimate of the distance to an individual star, but uncertainties can lead to systematic biases when data are combined in a sample based on the value of the parallax. Writing the true parallax as π_0, the distribution of the observed parallax, π, for a given true value is

$$g(\pi_0|\pi) = \frac{1}{\sqrt{2\pi}\sigma} \exp\left(\frac{-(\pi_0 - \pi)^2}{2\sigma^2}\right) \qquad (7.13)$$

We require the inverse distribution; that is, the distribution of *true* parallaxes given the observed distribution of π, $g(\pi, \pi_0)$. To calculate this, the overall parallax distribution, $N(\pi)$ must be taken into account. If the number of stars in the range $\pi_0 \pm d\pi$ were constant for all π, then $g(\pi|\pi_0) = g(\pi_0, \pi)$. However, if the sample has uniform density, then the number of stars within a given shell, r to $r + dr$, is given by

$$N(r) = V \, dr \qquad (7.14)$$

where we have defined $V = 4\pi = 4(3.14159)$ to avoid confusion. Hence, the number of parallaxes in the range π_0 to $\pi_0 + d\pi$ is

$$N(\pi_0) = V \frac{d\pi}{\pi_0^3} \qquad (7.15)$$

The number of stars increases with decreasing parallax, so if we consider a sample defined by $\pi \pm d\pi$, a larger number of more distant (smaller parallax than $\pi - d\pi$) stars will be scattered into the sample than are either scattered out of the sample or are scattered into the sample from smaller distances (larger parallax than $\pi + d\pi$). If we select stars from this sample based on $\pi > \pi_{lim}$, then $\langle\pi\rangle > \langle\pi_0\rangle$: the average distance is underestimated, and the true mean absolute magnitude is brighter than the formal value calculated directly from the individual parallaxes.

This effect was recognised by Russell and by Trumpler and Weaver, but was first quantified by Lutz and Kelker [L6], who showed that the resulting bias is a strong function of the accuracy of the parallax measurement. Writing $Z = \pi/\pi_0$, then for a

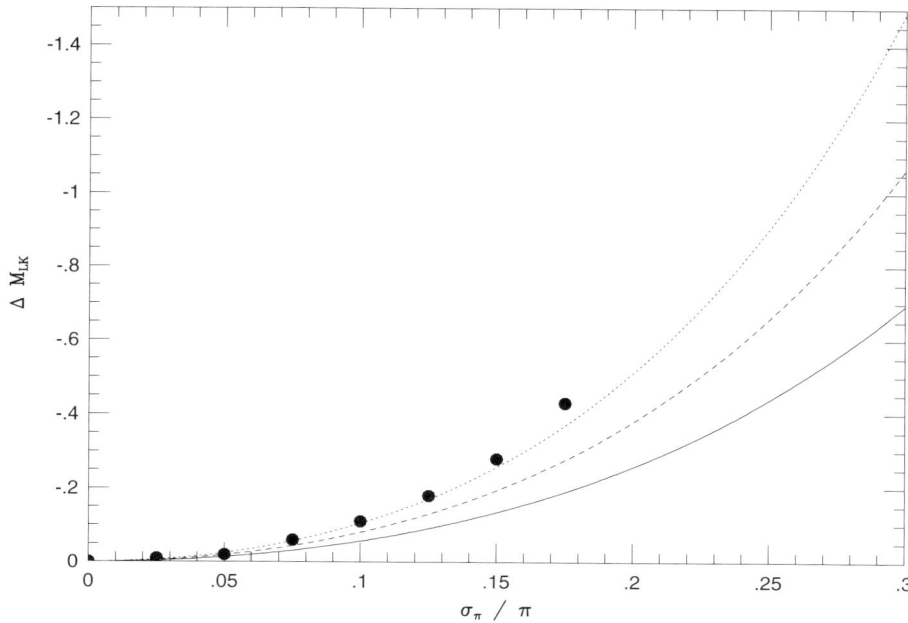

Figure 7.7. Lutz–Kelker corrections. The solid points mark the original [L6] calculations for a uniform density distribution; the solid line is Hanson's approximation for a power-law parallax distribution, $N_D(\pi) \propto \pi^{-n}$, with $n = 2$, the dotted line his results for $n = 3$, and the dashed line for $n = 4$.

uniform density distribution

$$g(\pi|\pi_0) \propto G(Z) = Z^{-4} \exp\left(\frac{-(Z-1)^2}{2(\sigma/\pi_0)^2}\right) \tag{7.16}$$

Integrating this distribution produces the mean offset as a function of σ/π_0, the Lutz–Kelker bias (Figure 7.7)

$$\Delta M_{LK} = \langle \bar{M}_t - M_{obs} \rangle = \left\langle 5 \log \frac{\pi}{\pi_0} \right\rangle \tag{7.17a}$$

where $\bar{M}_t$ is the best estimate of the true absolute magnitude and the appropriate value to use in combining data statistically – for example, when deriving a mean colour–magnitude relationship.

Lutz and Kelker's original calculations were specifically for a parallax-selected sample drawn from a uniform distribution, $N_D(\pi) \propto \pi^{-4}$ or $N_D(r) \propto r^3$. Hanson [H4] showed that selection effects such as proper-motion limits or a magnitude limit modify the distribution N_D to give less steep power laws. These can be understood as parallax-independent effects which reduce the number of stars at larger distances, and therefore also reduce the likelihood of parallax measurement errors scattering those stars into a π-selected sample. Hanson derives appropriate expressions for the general case $N_D(\pi) \propto \pi^n$:

$$\Delta M = -2.17\left[(n+0.5)\ \frac{\sigma_\pi^2}{\pi} - \left(\frac{6n^2 + 10n + 3}{4}\right)\frac{\sigma_\pi^4}{\pi}\right] \qquad (7.17b)$$

The corrections are less pronounced for smaller n (Figure 7.7), but still rise sharply with increasing σ_π/π for parallax precision worse than $\sim 15\%$.

Lutz–Kelker corrections have been misused extensively in the literature. Four important points should be made:

- LK corrections are appropriate only for a sample selected based on the value of the measured parallax. For example, if one selects all Cepheid variables in the Hipparcos catalogue with parallaxes measured to an accuracy of 20%, Lutz–Kelker bias is present; if the sample includes all Cepheids brighter than 10th magnitude, LK bias can be avoided.
- ΔM_{LK} depends on $N_D(\pi)$ of the parent sample from which the parallax-selected subset is drawn, not the parallax distribution of the underlying Galactic population. The density distribution can be estimated from the cumulative proper-motion distribution of the parent sample: since $\mu \propto r^{-1}$, a cumulative distribution $N_C(\mu) \propto \mu^n$ implies differential number counts $N_D(r) \propto r^{n-1}$ [H4].
- If a sample is modified based on criteria which are independent of π (such as an apparent magnitude limit, or spectroscopic line-strengths), then ΔM_{LK} is modified. There is no unique value for a given star – ΔM_{LK} depends on context.
- the uncertainties inherent in LK correction are minimised if a sample is restricted to stars with parallaxes measured to high precision.

7.4.2   Malmquist bias

Classical Malmquist bias

The most frequently applied statistical correction in astronomy is the classical version of Malmquist bias, which is important not only for number counts and luminosity function work – both Galactic and extragalactic – but also for calibrating distance determinations. Malmquist's [M4] original formulation was designed to estimate the frequency distribution of absolute magnitudes for stars of a given apparent magnitude and given spectral type (or colour). Consider the case of stars of spectral type S, which have a mean absolute magnitude M_0. Since there is an intrinsic dispersion to the (M, S) relationship, stars with apparent magnitudes in the range $m \pm \delta m$ magnitudes are drawn from a range of absolute magnitudes, and hence the magnitude limit m $m + \delta m$ corresponds to a larger effective distance limit for the intrinsically more luminous stars. If the convolution of the density law and the volume element is such that the number of stars increases with distance, as is usually the case, then this leads to a magnitude-limited sample including a higher proportion of stars with $M < M_0$ than with $M > M_0$. As a result, the mean absolute magnitude of the observed sample is biased to a brighter absolute magnitude than M_0.

Malmquist quantified the extent of the bias under the assumption that the intrinsic luminosity function for stars of a given spectral type could be characterised as a Gaussian, dispersion σ about a mean magnitude, M_0,

$$\Phi_S(M) = \frac{1}{\sigma\sqrt{2\pi}} \exp\left(-\frac{(M-M_0)^2}{2\sigma^2}\right) \tag{7.18}$$

If $A(m,r)$ is the joint distribution of stars as a function of apparent magnitude and distance, and $M(m,r)$ is the absolute magnitude distribution, then the mean absolute magnitude of the observed sample is given by

$$\bar{M}(m) = \frac{\int A(m,r)M(m,r)\,dr}{\int A(m,r)\,dr} \tag{7.19}$$

and $M(m,r) = m + 5 - 5\log r$ (Malmquist assumes no interstellar absorption, but the equations can be re-written to take that into account; see [M8]). The stellar number distribution is

$$A(m,r)\,dr = \omega\Phi_S(M)D_S(r)r^2\,dr = dA_S(m) \tag{7.20}$$

where $D_S(r)$ is the density law and $A_S(m)$ the number-magnitude distribution of stars of spectral type S. Substituting into equation (7.19) produces

$$\bar{M}(m) = \frac{\omega\int(m + 5 - 5\log r)\Phi_S(M)D_S(r)r^2\,dr}{A_S(m)} \tag{7.21}$$

with

$$A_S(m) = \omega\int\Phi_S(M)D_S(r)r^2\,dr \tag{7.22}$$

Since $\Phi_S(M)$ is given by equation (7.18), we can write

$$\begin{aligned}\frac{dA_S(m)}{dm} &= \omega\int\frac{d\Phi}{dm}D_S(r)r^2\,dr \\ &= -\frac{\omega(M-M_0)}{\sigma^2}\int\Phi_S(M)D_S(r)r^2\,dr\end{aligned} \tag{7.23}$$

Combining equations (7.23), (7.22) and (7.21), we have

$$\sigma^2\frac{dA_S(m)}{dm} = -A_S(m)\bar{M}(m) + M_0A_S(m) \tag{7.24}$$

and so

$$\begin{aligned}\bar{M}(m) &= M_0 - \frac{\sigma^2}{A_S(m)}\frac{dA_S}{dm} \\ &= M_0 - \frac{\sigma^2}{\log_{10}e}\frac{d\log A_S}{dm}\end{aligned} \tag{7.25}$$

For a uniform density distribution, star-counts increase as $10^{0.6m}$, so this becomes

$$\bar{M}(m) = M_0 - 1.38\sigma^2 \tag{7.26}$$

The mean absolute magnitude of the sample, $\bar{M}(m)$, is therefore brighter than the mean absolute magnitude of all stars of spectral type S, M_0, by the amount $-1.38\sigma^2$. At high Galactic latitude ($b > 40$) the density law of the Galactic disk leads to a shallower slope in the number counts. For a single exponential distribution,

$$A(m) \propto \left(\frac{z}{h}\right)^3 \exp\left(\frac{-z}{h}\right) \tag{7.27}$$

where h is the scale-height, and z is the height above the plane, giving a Malmquist bias [S15] of

$$\bar{M}(m) = M_0 - \frac{0.2}{\log_{10} e}\sigma^2\left(3 - \frac{z}{h}\right) \tag{7.28}$$

Thus, in the case of a simple exponential density law, $\rho(z)$, the Malmquist bias falls to zero at 3 scale-heights. However, $\rho(z)$ in the disk is best represented as a convolution of several components (Chapter 6). Star-counts at high Galactic latitude have a typical logarithmic slope closer to 0.4 for magnitudes $13 < V < 20$, and so

$$\bar{M}(m) = M_0 - 0.92\sigma^2 \tag{7.29}$$

In photometric parallax work, the dispersion in the main-sequence colour–magnitude relation is typically ~ 0.4 magnitudes, giving Malmquist bias of ~ 0.15 magnitudes. That is, if we consider stars within a given range in colour, classical Malmquist analysis implies that the mean absolute magnitude within that range is 0.15 magnitudes brighter than would be inferred from the direct colour–magnitude relation.

Malmquist bias in a continuous distribution

Classical Malmquist analysis treats the luminosity function as if it were a series of discrete intervals, each characterised by a given spectral type or colour. The analysis allows only for uncertainties due to dispersion in the absolute magnitude, whereas the mean absolute magnitude is also affected by observational uncertainties in the measured colour. Stobie *et al.* [S15] have demonstrated that applying the classical Malmquist method to a photometric parallax survey does not provide an unbiased estimate of the luminosity function.

A direct consequence of Malmquist bias is that, by biasing the mean absolute magnitude to brighter values, one effectively samples a larger volume, leading to an increase in the number of stars, ϕ, assigned to a given bin in $\Phi(M)$. If we adopt the assumption of a uniform space density, then the fractional increase in the luminosity function is given by

$$\frac{\Delta\phi}{\phi} = \left(\frac{0.6\sigma}{\log_{10} e}\right)^2 - \left(\frac{0.6\sigma^2}{\log_{10} e}\right)\frac{\phi'}{\phi} \tag{7.30}$$

where ϕ' is $d\phi/dM$. The first term arises from the additional volume sampled due to the underestimate in the mean absolute magnitude, and the second term measures the effect of sampling the luminosity function at a different value, $M_m < M_0$. The latter term is zero for a flat (uniform density) luminosity function.

The above correction is appropriate for standard Malmquist analysis, but allowing for uncertainties introduced by errors in the measured colour reduces the effective Malmquist bias [S15]. The fractional change in $\Phi(M)$ becomes

$$\frac{\Delta\phi}{\phi} = \frac{1}{2}\sigma^2\left[\left(\frac{0.6}{\log_{10} e}\right)^2 - \left(\frac{1.2}{\log_{10} e}\right)\frac{\phi'}{\phi} + \frac{\phi''}{\phi}\right] \qquad (7.31)$$

The final term is negligible and can be ignored, while the second term, in ϕ'/ϕ, is identical to that in equation (7.30). However, the first term, tied to the change in sampling volume, is half the value derived from the classical analysis. The net result is that directly applying standard Malmquist methods will lead to *over-correction* of the observed densities, producing an *underestimate* of $\Phi(M)$ by a factor of $\frac{1}{2}\left(\frac{0.6}{\log_{10} e}\right)^2\sigma^2$, or $\sim 15\%$ for $\sigma = 0.4$ magnitudes.

7.5   MODERN TIMES: FIELD-STAR SURVEYS

> 'If the fool would persist in his folly he would become wise.'
>
> William Blake

Recent investigations of $\Phi(M)$ for low-mass dwarfs rely on one of three techniques: star counts in the immediate Solar Neighbourhood; application of photometric parallaxes to wide-field field-star surveys; and surveys for low-luminosity main-sequence stars in open clusters. Each method offers its own advantages and disadvantages. This section concentrates on the two former techniques. Open-cluster studies are reviewed in the following section, while other specialised techniques are outlined in the final section.

7.5.1   A census of the Solar Neighbourhood

Concentrating on the nearest stars to the Sun seems parochial at first sight. However, as discussed in Section 6.5.1, the gradual diffusion of Galactic orbits and the position of the Sun between spiral arms means that a local sample should include stars of all ages since the formation of the Galactic disk ($\tau \sim 10\,\mathrm{Gyr}$), formed in molecular clouds at Galactic radii of $\sim 4\text{--}15\,\mathrm{kpc}$. Hence these stars provide a fair sampling of the average luminosity function of main-sequence stars (average age, 5 Gyr) in the Galactic disk population. Nearby stars also offer the best opportunity of discovering companions in multiple systems. As John Faulkner expresses it: 'search locally, think globally.'

The most difficult step is to identify a complete sample of the nearest stars, particularly at faint magnitudes and low Galactic latitudes where the star density exceeds 1,000 stars deg^{-1}. Consequently, these analyses are restricted to the immediate Solar Neighbourhood. Kuiper's catalogue (Section 7.2.4) was limited to $\pi > 0\!\!\!.''095$, or $r < 10.5\,\mathrm{pc}$, and consisted mainly of stars brighter than 15th magnitude, identified as candidate nearby stars based on their having substantial annual proper motions. Larger telescopes, more sensitive detectors and new techniques led to more discoveries, and in the 1950s Wilhelm Gliese took on the task of compiling and maintaining nearby-star catalogues.

Gliese's first compilation – published in 1957 as the *Catalogue of Nearby Stars* (CNS1 [G3]) – is limited to stars identified as being within 20 pc, and includes astrometric and photometric data for 1,094 stars in 915 systems (Gl 1 to Gl 915). By 1969 and the publication of the second catalogue (CNS2 [G4]), the list had expanded to include 1,890 stars in 1,529 systems, with a parallax limit of $\pi > 0\!\!\!.''040$ ($r < 25\,\mathrm{pc}$). Prompted by the publication of the CNS2, Woolley (at that time the Astronomer Royal) oversaw the compilation of a rival RGO catalogue [W7], which includes a further 259 stars (in 216 systems) from diverse (and, in some cases, dubious) sources. Many of these additional stars, together with newer discoveries were included in a supplement to the CNS2, published by Gliese and Jahreiss [G6]. The most recent catalogue – the CNS3 [G7], includes data for approximately 3,820 stars in 3264 systems (some binary companions are not listed separately in the current version of the new catalogue). Thus, the Solar Neighbourhood stellar census more than tripled over a 40-year period.

The stars in the CNS3 are drawn from many sources – primarily proper motion surveys and, in particular, Luyten's catalogue of stars with $\mu > 0\!\!\!.''5\,\mathrm{yr}^{-1}$ (the LHS catalogue). Other stars originate from objective prism surveys, wide-field photometric surveys, and, most recently, X-ray surveys, notably the ROSAT all-sky survey [S5]. Barely 60% of the systems had trigonometric parallax measurements when the catalogue was compiled, and of those only half were measured to a precision $\sigma_\pi/\pi < 15\%$. Most of the remaining stars had photometric or spectroscopic parallaxes, although some could be said to have entered the catalogue more through hearsay and rumour than through direct observational evidence of proximity. Uniform distance estimates were clearly required, and have been supplied from two sources: a spectroscopic survey of late-type dwarfs and astrometry by the Hipparcos satellite.

The PMSU spectroscopic survey ([R11], [H8]) includes 2,075 of the 2,200 candidate K and M dwarfs in the CNS3. 70 of these stars (mainly from the RGO catalogue) prove to be either misclassified earlier-type (AFG) stars or M giants – in either case lying at distances of 50–1,000 pc. Similarly, spectroscopic parallax estimates place many of the genuine K and M dwarfs beyond the nominal 25-pc limit of the CNS3, while Hipparcos astrometry of brighter FGK dwarfs also eliminates almost one-third as outwith the distance limits. This is not unexpected, given that both Malmquist and Lutz–Kelker bias tend to produce an overestimate in π and an underestimate in distance. Jahreiss is currently compiling the CNS4 based on these new data.

It should be recognised that both Hipparcos and PMSU are more effective in removing misclassified stars than in adding previously unrecognised nearby stars. The Hipparcos satellite measured parallaxes of around 118,000 stars to V $\sim$ 12, including almost every star brighter than V $=7.5$, but only a subset of fainter stars. The latter were pre-selected in 1982, so any star not suspected of being within 25 pc at that time was not observed by Hipparcos. Similarly, the PMSU sample was drawn from the CNS3 and did not add new candidates. Given these limitations, how complete is the nearby star sample?

Completeness

Statistics are only as reliable as the parent sample. As already discussed in Chapter 2, the southern sky is less well surveyed than the north, as is evident when considering the distribution with declination of systems in the CNS3. Table 7.1 lists the total numbers for four equal areas, considering stars brighter than and fainter than $M_V = 7.5$. The incompleteness amongst faint stars at southern declinations is obvious.

Table 7.1. The declination distribution of CNS3 stars.

	$> +30°$	$+30°–0°$	$0°– -30°$	$< -30°$
$M_V < 7.5$	256	218	236	228
$M_V \geq 7.5$	681	722	507	351

The simplest method of assessing completeness is to determine where the cumulative distribution function of stellar systems, $(\log(N), \log(r))$, drops below a straight line, slope 3 – the relationship expected for a uniform distribution. This technique was applied by Wielen and collaborators [W5], [W6] in analysing the CNS2 and its supplement. Those studies are notable for identifying what has become known as the 'Wielen dip' at $M_V = +7$, a feature now recognised as introduced by the physics underlying the (mass, M_V) relationship (Sections 3.5.2 and 8.3). As noted in Section 7.2.3, this feature is present in van Rhijn's analysis [R16], but was removed intentionally by Luyten. The completeness limits adopted in determining $\Phi(M_V)$ are listed in Table 7.2.

Table 7.2. Completeness limits for nearby-star analyses.

Wielen/CNS2			PMSU/CNS3		
M_V	$\delta_{\lim}$	$r_{\lim}$	M_V	$\delta_{\lim}$	$r_{\lim}$
≤ 7.5	All sky	20 pc	≤ 7.0	$\geq -30°$	22 pc
7.5–9.5	$\geq -30°$	20 pc	7.0–9.0	$\geq -30°$	22 pc
9.5–11.5	$\geq -30°$	10 pc	9.0–10.0	$\geq -30°$	20 pc
11.5–17.5	$\geq -30°$	5 pc	10.0–13.0	$\geq -30°$	14 pc
			13.0–14.0	$\geq -30°$	12 pc
			14.0–15.0	$\geq -30°$	10 pc
			15.0–18.0	$\geq -30°$	5 pc

An alternative method of testing completeness is to use the V/V_{max} statistic devised by Schmidt [S4]. For each star, the ratio r^3/r^3_{max} is calculated, where r is the observed distance, and r_{max} is the maximum distance that allows the star to remain in the sample. If the sample is uniformly distributed, then as many stars are expected within the inner half of the volume as within the outer, and the average value of V/V_{max}, $\langle V/V_{max} \rangle = 0.5$. If the sample is incomplete at large distances,[4] then $\langle V/V_{max} \rangle < 0.5$. This method can also be adapted to construct a luminosity function, since each star makes a contribution of $1/V_{max}$ to the total density.

The V/V_{max} method was used by Upgren and Armandroff (U1) to test whether the 'Wielen dip' at $M_V \sim 7$ could be explained by incompleteness in the nearby-star sample. They found no evidence for significant differential incompleteness. Armandroff [A2] later extended this analysis to consider general star-counts at high Galactic latitude, and provided the first quantitative indication that the 'Wielen dip' is a global feature of the disk luminosity function. Reid *et al.* [R11] use both V/V_{max} and $(\log(N), \log(r))$ measurements to determine completeness limits for the PMSU survey (see Table 7.2).

Both of the $(\log(N), \log(r))$ and V/V_{max} tests identify distance limits within which a sample of stellar *systems* can be taken as statistically complete; they do not address the related question of whether every component in each system has been identified. This question is not amenable to simple analysis, since not every system has been subjected to the same degree of attention from high-resolution imaging and/or radial-velocity surveys. This is the main reason why the full Hipparcos dataset cannot be incorporated in luminosity function analyses: while every G dwarf within 30 parsecs has an accurate parallax measurement, the multiplicity fraction is unknown. A NASA programme (RECONS – the Research Consortium for Observing Nearby Systems) is currently engaged in the extensive (and necessary) follow-up observations.

The luminosity function

Figure 7.8 compares the luminosity functions derived by [W6], [R11], and based on statistics of the 8-parsec sample in the Appendix. In each case the star-by-star luminosity function is plotted, counting each component separately. There is reasonable agreement amongst these three determinations, as expected, since there is substantial overlap between the underlying samples. Each shows a maximum in number density near $M_V \sim 12$. The peak is strongest in the Wielen *et al.* analysis; their sample includes a number of M dwarfs since removed to larger distance (or different classifications) by [R11]. Data for a 5.2-parsec distance limit shows a strong spike in number density at $M_V = +16$, but a deficit at $M_V = +11$. None of the datasets is large, but the 5.2-pc sample includes only 52 stars in total and the discrepancy probably reflects small number statistics. The probable average star density implied by the other datasets is ~ 0.09 stars parsec^{-3}, corresponding to an average separation of 2.25 parsecs.

---

[4] Note that the statistic is differential: a uniformly incomplete sample has $\langle V/V_{max} \rangle = 0.5$.

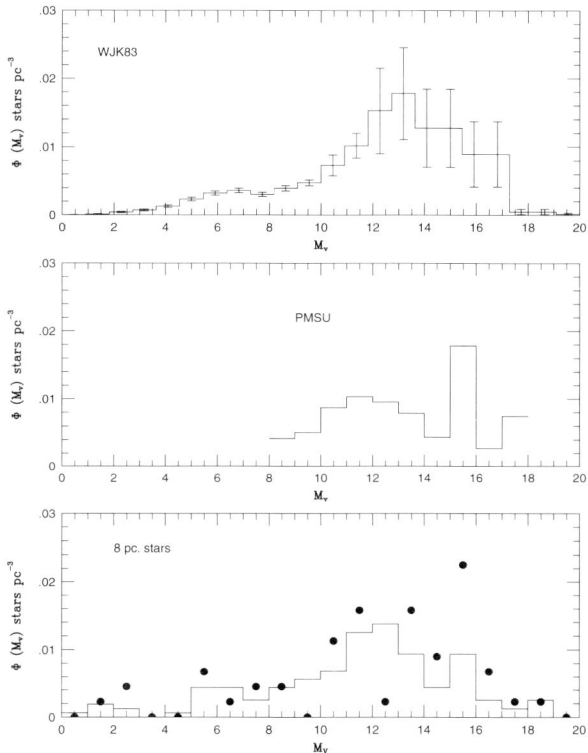

Figure 7.8. The V-band luminosity function derived from nearby-star samples. The uppermost diagram plots $\Phi(M_V)$ from the Wielen *et al.* analysis of the CNS2 and its supplement; the middle panel shows the Reid *et al.* analysis of the CNS3 supplemented by the PMSU spectroscopic survey; and the lowest panel shows $\Phi(M_V)$ for the 8-parsec sample listed in the Appendix (solid line), and an $r < 5.2$-parsec subset (solid points). Note that only nine stars contribute to the $M_V = 15$ 5.2-parsec sample bin. All of the luminosity functions are shown on the same scale.

7.5.2   Photometric parallax surveys of field stars

While Weistrop's Schmidt survey suffered from unfortunate systematic errors, her work revealed the potential of wide-field imaging as a means of identifying field M dwarfs. Selection by colour avoids the inherent kinematic biases of proper-motion surveys, besides providing a more accurate and objective absolute magnitude estimator than spectral type. (B–V) is not the colour of choice for M dwarf photometric parallaxes, given both the steep gradient $dM_V/d(B–V)$ and the faint B-band luminosities of late-type dwarfs. M8 dwarfs like VB10 are barely detectable beyond 10 parsecs on the POSS I O-band plates. Finally, hand-measurement of

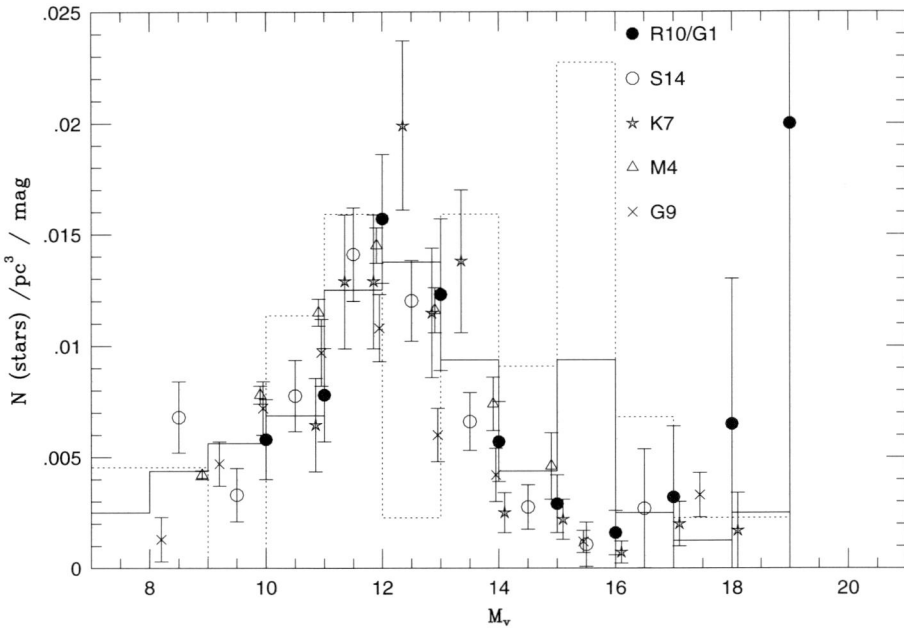

Figure 7.9. V-band luminosity functions derived from the photometric parallax surveys identified in the figure legend. The solid histogram plots $\Phi(M_V)$ from the 8-parsec sample, and the dotted line marks the 5.2-parsec densities.

positions and magnitudes for tens of thousands of objects presents a daunting task in logistics.

These problems were ameliorated significantly by the development of both automated plate-measuring machines and the near-infrared sensitive IVN photographic emulsion in the late 1970s. Combined with suitable filters [B1], the latter provides wide-field I-band data, permitting surveys based on (V–I) or (R–I) colours. Schmidt plates, covering over 30 square degrees, can be scanned and digitised in 8–10 hours by machines such as COSMOS [S14] and APM [K5]. Those scans provide positions accurate to ∼1 arcsec, photometry to ±0.1 magnitudes, and morphological information allowing separation of stars and galaxies for 20,000–100,000 objects per plate.

Comparing the predicted absolute magnitude, $M_V(\text{phot})$, against the observed apparent magnitude, V, gives the photometric parallax, π_{pp}, where

$$\pi_{pp} = 10^{0.2(M_V(\text{phot})-V-5)} \tag{7.32}$$

Initial low-mass star surveys were based on V (IIaD emulsion) and I-band plates [R10], [G1], [S15], while later studies combined R (IIIaF) and I-band material [H7], [L4], [T3]. The latter gain in depth ($R \sim 20.5$ against $V \sim 20$) at the expense of larger uncertainties in the absolute magnitude calibration ($\sigma_{V-I} \sim 0.34$ magnitudes in M_I against $\sigma_{R-I} \sim 0.44$ magnitudes). In each case the analysis involved combining data from several plates in each passband (to eliminate photographic defects); determining mean colours for each object; selecting candidate M dwarfs and estimating absolute magnitudes and distances; setting an appropriate distance limit (typically 100 pc); and deriving space densities, with due allowance for the disk density distribution, $\rho(z)$. While we refer to these observations as 'wide-field' surveys,[5] most analyses cover only a single ~ 30 square degree field, although Tinney's [T3] dataset spans nine fields and ~ 270 square degrees.

Photographic analyses have been supplemented by three recent studies based on wide-field CCD imaging: a CCD-transit survey by Kirkpatrick *et al.* [K7], covering a strip of ~ 27.3 square degrees to a limiting magnitude of R = 19.0 magnitudes[6]; analysis of HST WFPC data by Gould *et al.* [G8], [G9], including 53 fields encompassing 0.28 square degrees to I ~ 27 magnitudes; and Martini and Osmer's [M5] UBVRI$_{75}$I$_{85}$ survey of 0.83 square degrees to V = 22, a by-product of a multicolour search for high-redshift QSOs. Those datasets provide more accurate photometry for individual objects, although the intrinsic dispersion in the calibrating photometric parallax relationship dominates the uncertainties in both photographic and CCD surveys. Both the deep WFPC observations and Martini and Osmer's analysis include M dwarfs at heights of 500–2,000 pc above the plane, requiring simultaneous solution for $\Phi(M_V)$ and $\rho(z)$.

Despite the different provenance of these various surveys, the results are in broad agreement. There is, however, a discrepancy between the photometric surveys and nearby-star data: the former indicate significantly lower space densities at $M_V \sim 13$–16. Both Dahn *et al.* [D1] and Kroupa [K10] have suggested that this circumstance might arise through low-mass companions being omitted from the photometric surveys: the stars in those samples are at typical distances of 50–100 parsecs (much further for the HST data), with a corresponding reduced efficiency in detecting companions. However, unrecognised binarity also add stars to a photometric sample. Consider a sample selected to have $\pi_{pp} > \pi_{lim}$. Binaries are brighter than single stars of the same colour, but have identical M_V(phot). Hence, an unresolved binary with $\pi_0 < \pi_{lim}$ can have $\pi_{pp} > \pi_{lim}$. This effect compensates to a considerable extent for the missing 'hidden' companions [R5]; more importantly, there simply are not enough M dwarf binaries to make a difference [R8].

The explanation for this discrepancy is more mundane: the colour–magnitude relationships used in the photometric-parallax analyses plotted in Figure 7.10 fail to match the observed lower main sequence. Figure 7.11 compares several of those calibrations against data for the 8-parsec stars. None of the former adequately match the latter – particularly the 'kink' at $(V–I) \sim 2.9$ (see Section 3.5). The spline fit adopted originally by [R10] comes closest, but underestimates M_V by +0.4

---

[5] An alternative description is 'pencil-beam' surveys – much deeper than their width.

[6] The [K7] scans cover 19.4 square degrees in the northern Galactic hemisphere, and 7.9 in the south, but the eight reddest M dwarfs ($M_I > 13$) lie in the latter portion of the survey. Since the Sun lies above (north of) the mid-point of the Galactic Plane, Kirkpatrick *et al.* suggested that this imbalance might indicate that the latest-type M dwarfs are tightly confined to the Plane, as one would expect for young, substellar objects. As discussed in more detail in the following chapter, this hypothesis is not supported by subsequent studies.

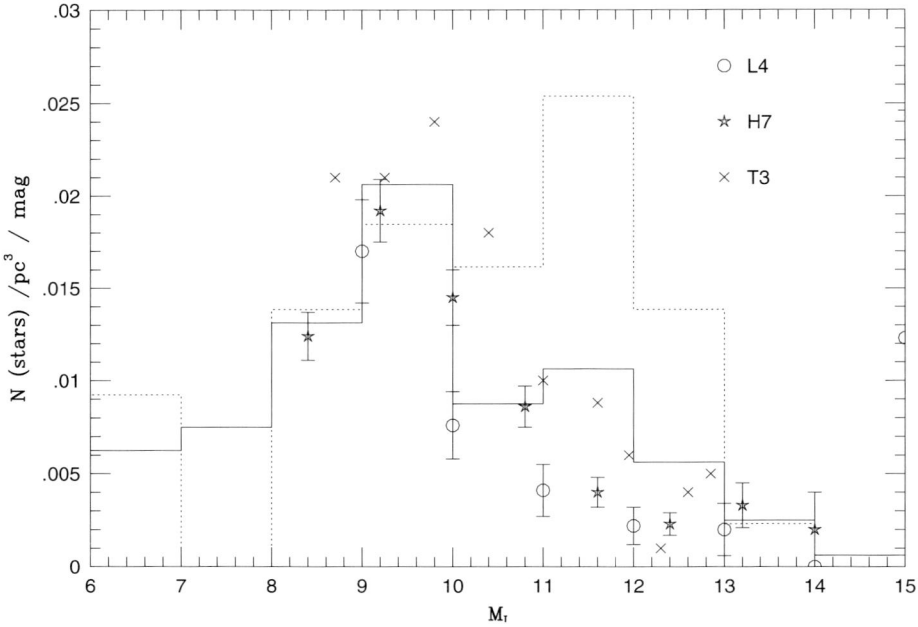

Figure 7.10. I-band luminosity functions derived from photometric parallax surveys. The histograms have the same meaning as in Figure 7.9. The points come from [L4], [H7], [T3].

magnitudes at (V–I) ~ 3.7. Most studies adopt a linear relationship close to that proposed by [S14]:

$$M_V^{SIP} = 2.89 + 3.34(V–I), \quad (V–I) > 1.0 \tag{7.33}$$

The net result of applying this relationship (or the calibration adopted by [K7]) is that stars with $3 < (V–I) < 4$ are assigned absolute magnitudes that are too bright and distances that are too large; this leads to an underestimate in $\Phi(M_V)$ for $13 < M_V < 15$. In addition, the added Malmquist bias due to the effective increase in σ_M near $(V–I) = 2.9$ produces an enhancement in the inferred number densities at $M_V = 12$ (Figure 7.12) – exactly the form of the observed discrepancy between the photometric parallax studies and the nearby-star analysis.

7.5.3   Photometric parallax analyses versus a local census

The original impetus for using photometric techniques to study the luminosity function was the fact that those methods provide a kinematically-unbiased way to identify the nearby stars. Any low space-motion component excluded systematically from proper-motion surveys would be detected by the photometric technique. The existence of structure in the lower main sequence complicates the photometric analysis, but in general space densities derived from these surveys are consistent with the results from the local census. There is no evidence for any additional

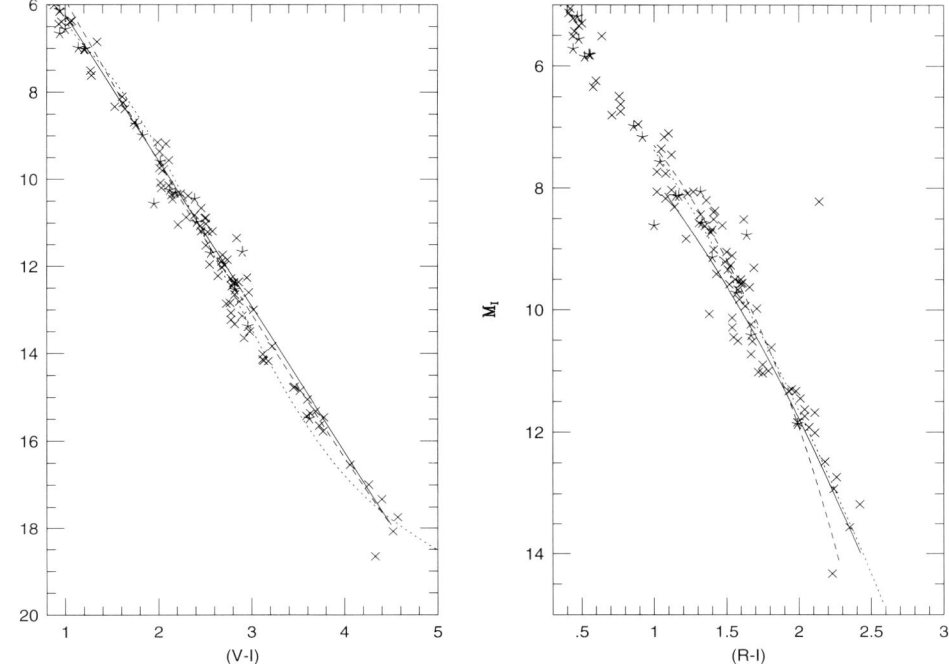

Figure 7.11. The $(M_V, (V–I))$ and $(M_I, (R–I))$ colour–magnitude diagrams for individual stars in the the 8-parsec sample compared with the mean colour-magnitude relations used to infer photometric parallaxes. The three (V–I) relations plotted are from [R10] (dotted line), [S15] (solid line) and [K7] (dashed line); the (R–I) calibrations are from [H4] (solid line), [L4] (dotted line) and [T3] (dashed line).

component of late-type dwarfs which might make a substantial contribution to the local mass-density. Photometric techniques are most effective when there is a low dispersion in the relationship between colour and magnitude, as is the case at (V–I) > 3.2 (Figure 7.11). As a result, these techniques continue to provide a powerful tool in searching for VLM stars and substellar-mass brown dwarfs. In recent years, emphasis has shifted towards surveys at near-infrared wavelengths, as will be discussed in more detail in Chapter 9.

7.6 M DWARFS IN OPEN CLUSTERS

Open clusters are potent tools for investigating evolutionary phenomena in stars, since cluster members are products of a single star-forming event of known metallicity and of measurable age. In particular, open clusters off the prospect of testing for variations in $\Phi(M_V)$ in different environments.

Cluster ages and distances are generally determined by comparing the location of the observed main-sequence turnoff against theoretical models (see Figure 6.13). The accuracy of the derived results rests on the accuracy of the models. Recent

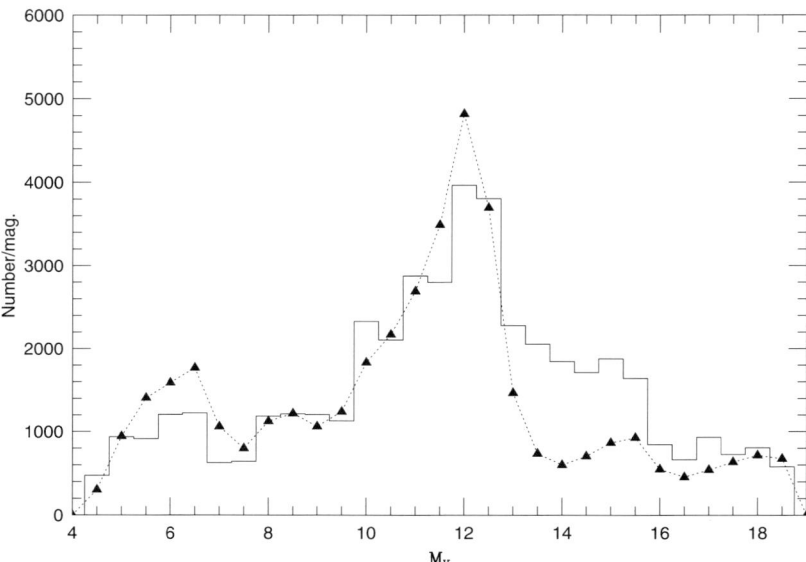

Figure 7.12. Systematic biases introduced in $\Phi(M_V)$ by using a linear $(M_V, (V–I))$ relationship. The histogram outlines the 'true' luminosity function in a model simulation; the triangles, the results of deriving distances and luminosities using a linear $(M_V, (V–I))$ relationship.

observational results have highlighted the importance of the level of convective overshoot adopted in models of intermediate-mass stars (those near the turnoff): that is, the extent to which a convective element travels beyond the point of neutral buoyancy (see equation (3.49)). This parameter determines the amount of fuel available for hydrogen-burning; the greater the overshoot, the longer-lived the star and the older the age associated with a given turn-off luminosity. Until recently, cluster ages were derived (at least in the USA) assuming little or no convective overshoot. The latter half of the 1990s, however, saw a revision in the age scale, stimulated largely by the results from applying the 'lithium test' to VLM stars and brown dwarfs in clusters (see Section 9.7). For example, the age of the Pleiades has been revised from 70 Myr to ∼125 Myr. The older ages are broadly consistent with those derived from model calculations that include significant convective overshoot (see [M2], and references therein). The cluster ages cited here are linked to the revised scale.

Only a limited number of Galactic clusters are suitable for studies of the luminosity function of lower main-sequence stars for the following reasons:

● Detecting stars at or below the hydrogen-burning limit is practical only in the nearest (< 150 pc) clusters – not only because VLM dwarfs are intrinsically very faint, but also because sifting through the foreground and background stars to identify the cluster members is significantly more difficult in more distant clusters.

- Stochastic dynamical effects due to gravitational interactions with other massive objects gradually disrupt open clusters, stripping lower-mass stars preferentially and changing the luminosity function.
- Low-mass stars in open clusters younger than ~ 20 Myrs are still evolving onto the main sequence, with luminosities which exceed the main-sequence values. Thus, while well-suited to searching for substellar-mass brown dwarfs (Chapter 9), a comparison with field-star data is problematic; reconstructing the mass function is highly dependent on the accuracy of evolutionary models (as discussed in the following chapter).

Given these caveats, the prime targets are the three best-known clusters – the Hyades, the Pleiades and Praesepe – and these clusters have received the most attention. Before discussing the results, we review techniques used to segregate cluster members from the dominant background of Galactic field-stars.

7.6.1 Cluster membership criteria

Most open clusters lie close to the Galactic plane, and an average cluster has a total of only a few hundred members. Consequently, the contrast between cluster and field can fall to below $1:1,000$ at faint apparent magnitudes. Identifying those few low-luminosity members demands the application of a series of astrometric, photometric and kinematic tests, and the exclusion of stars failing any one criterion.

Proper motions and radial velocities

The velocity dispersion of stars in an open cluster is extremely low; Gunn *et al.* [G10] measure $\sigma \sim 0.23\,\mathrm{km\,s}^{-1}$ for the Hyades. This is a necessary requirement for the cluster to survive as a discernible entity. The potential energy of a star, mass m_i, in a cluster with n stars, is given by

$$\Omega_i = -Gm_i \sum \frac{m_j}{r_{ij}} \tag{7.34}$$

so the kinetic energy required for that star to escape the cluster is $-\Omega_i$ ([C1], Chapter 5)). The average kinetic energy required to escape the cluster is

$$\bar{E}_\infty = -\frac{2}{n}\Omega = \frac{Gm^2 n}{R} \tag{7.35}$$

where $\bar{R}$ is the average radius of the cluster, and m is the average stellar mass. Since $\bar{E}_\infty = \frac{1}{2}m\bar{v}_{\mathrm{esc}}^2$, this reduces to

$$\bar{v}_{\mathrm{esc}}^2 = 4\sigma^2 \tag{7.36}$$

where σ is the rms velocity dispersion in the cluster. Hence, stars with a velocity only twice the observed velocity dispersion are expected to escape from the cluster in a matter of a few Myr. Surviving cluster members occupy a correspondingly reduced velocity range.

The practical consequence of the limited range in velocities is that kinematic criteria can be used to identify cluster members. Radial velocities present one possible means of selection, but large-scale observations of faint ($V > 16$) stars at the required precision of $< 1 \, \text{km s}^{-1}$ remain impractical, while in some cases (such as the Pleiades) the cluster velocity lies within the field star distribution (see [S13]). Proper motions, readily measureable from wide-field photographic plates, offer a more tractable option as the first step towards distinguishing cluster from field.

Open clusters have typical diameters of 15–20 pc, corresponding to angular diameters of $10°$–$30°$ in the three nearest systems. The proper motion of an individual star is the projection of the space velocity onto the plane perpendicular to the line of sight, so while the cluster stars have identical space velocities, the changing perspective across the cluster leads to different motions, with the individual vectors directed towards a particular point on the sky – the convergent point (CP) [B6].[7] Individual stellar motions are resolved into two components directed towards ($\mu_{u'}$), and perpendicular to (μ_t), the direction of the CP. The dispersion in the latter component, σ_{μ_t}, provides a measure of the intrinsic velocity dispersion of cluster members. If the CP has equatorial co-ordinates (α_P, δ_P), then the expected direction of motion, θ (measured north through east) of a star at position (α_i, δ_i) is given by

$$\cos \theta = \frac{(\sin \delta_P - \sin \delta_i \cos \lambda_i}{\cos \delta_i \sin \lambda_i} \tag{7.37}$$

where λ_i is the angular separation between the star and the CP.

The changing perspective across a cluster also leads to a change in the magnitude of the proper motion, since

$$\mu_{u'}^i = \frac{V_S \sin \lambda_i}{\kappa r_i} \tag{7.38}$$

where V_S is the cluster space motion, $\kappa = 4.74$, and r_i is the distance. To allow for this projection effect, the observed motions, $\mu_{u'}$, are adjusted to the cluster centre:

$$\mu_u^i = \mu_{u'} \frac{\sin \lambda_C}{\sin \lambda_i} \tag{7.39}$$

where λ_C is the angular separation of the cluster centre and the CP.

Equation (7.38) shows that $\mu_{u'} \propto r^{-1}$. As a result, individual values of μ_u^i span a range of values due to the intrinsic depth of the cluster along the line of sight. This is analogous to the 'finger of God' effect produced by galaxy clusters in redshift-based maps of the local Universe: in the galaxies, velocity variations of objects at the same distance lead to an apparent elongation in distance; in open clusters, distance variations in objects with the same velocity lead to an elongation in μ_u. Indeed, inverting the process, if V_S is known, r_i can be determined from $\bar{\mu} - \mu_u^i$. The amplitude of this effect decreases with cluster distance: for the Hyades,

[7] The intrinsic cluster velocity dispersion means that the proper-motion vectors for all cluster members actually intersect in a small area, rather than a unique point [P2]. This is measureable only with high-accuracy astrometry such as the Hipparcos Hyades data, and can be taken into account.

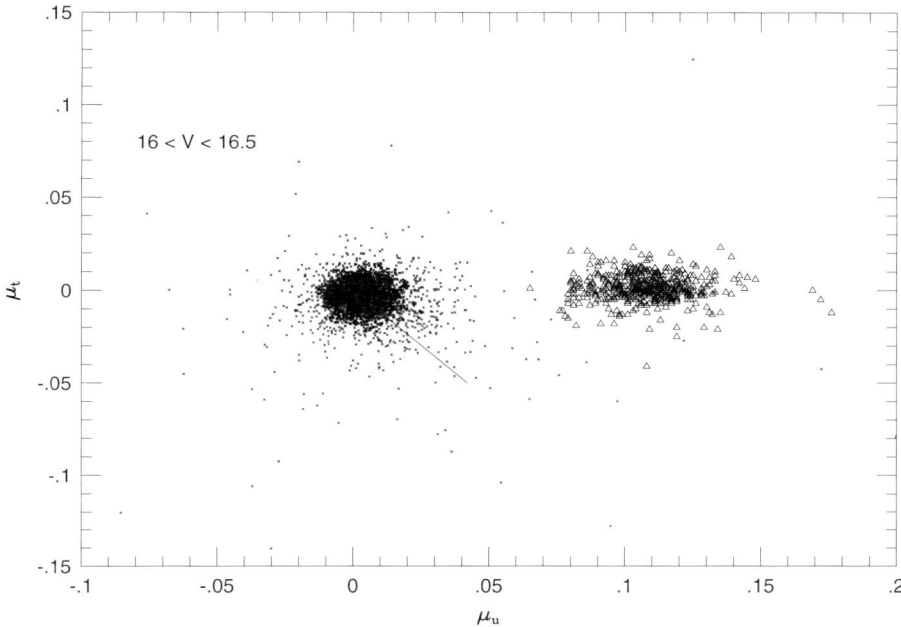

Figure 7.13. The proper motion distribution in (μ_u, μ_t) for stars with $16 < V < 16.5$ in a 50 square degree region near the Hyades. Candidate cluster members $(8 < V < 20)$ are plotted as open triangles. The line indicates the proper-motion vector due to reflex solar motion. (Data from [R6].)

$0\rlap{.}''08 \leq \mu_u^i \leq 0\rlap{.}''15$ (Figure 7.13); for the more distant Pleiades, $0\rlap{.}''035 \leq \mu_u^i \leq 0\rlap{.}''05$. In fact, defining $\Delta\mu_u = \mu_{max}^i - \mu_{min}^i$, then $\Delta\mu_u / \bar{\mu}_u$ measures the ratio between cluster diameter and mean distance.

Proper-motion selection is effective if the mean cluster motion is distinct from that of the local field-stars. This circumstance holds for the Hyades (Figure 7.13) and, to a lesser extent, the Pleiades ($\bar{\mu}_\alpha = 0\rlap{.}''02\,\mathrm{yr}^{-1}$, $\bar{\mu}_\delta = -0\rlap{.}''04\,\mathrm{yr}^{-1}$) and Praesepe ($\bar{\mu}_\alpha = 0\rlap{.}''02\,\mathrm{yr}^{-1}$, $\bar{\mu}_\delta = -0\rlap{.}''04\,\mathrm{yr}^{-1}$). In the case of the Hyades, an initial sample of $\sim400{,}000$ stars with $V < 20$ reduced to ~400 candidate members [R6]. Some field-stars, however, will have proper motions consistent with cluster membership, and the smaller the offset between cluster and field, the greater the degree of contamination. Other techniques must be employed to eliminate these interlopers.

Photometric selection

By definition, we expect open cluster members to outline a well-defined main-sequence in the H–R diagram. Given accurate multicolour photometry, and distance estimates based on either individual measurements or mean cluster proper-ties, a colour–magnitude diagram can be constructed for the candidate cluster members. Figure 7.14 shows the results of applying this procedure to the proper-motion selected Hyades candidates from Figure 7.13. Approximately two-thirds of

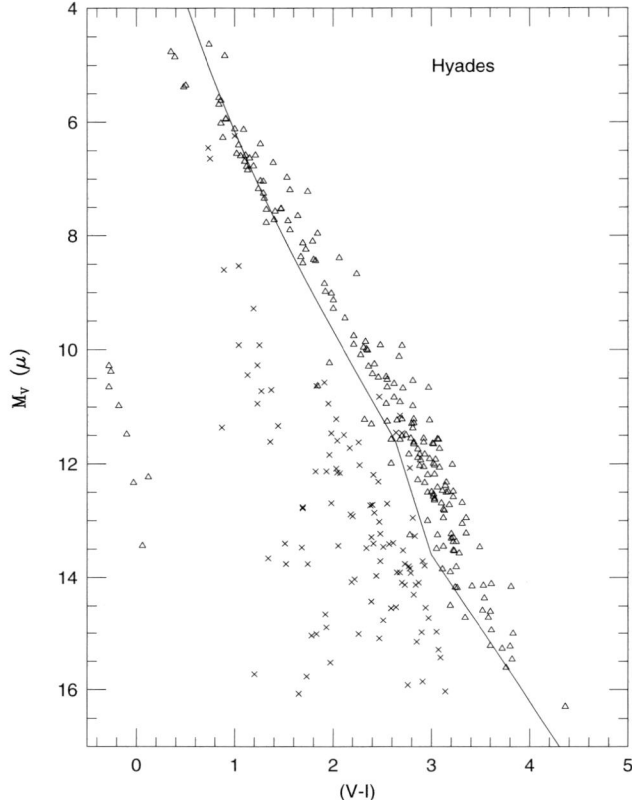

Figure 7.14. The $(M_V, (V–I))$ colour–magnitude diagram for the Hyades proper-motion candidate members. Open triangles identify stars with photometry consistent with cluster membership; crosses indicate background field-stars. The solid line marks the nearby-star mean colour–magnitude relationship.

the stars lie on a main sequence, which is somewhat redder than the field-star relationship. Most stars lying above the Hyades sequence (brighter in M_V at a given colour) are binaries.

Over a quarter of the proper-motion candidates, however, lie well below the main sequence. Eggen [E1] originally suggested that these stars might constitute 'Hyades subdwarfs' – an hypothesis which we now know to be incompatible with stellar evolution theory. Hanson and Vasilevskis [H6] demonstrated that none of the apparently sub-luminous stars have radial velocities consistent with cluster member-ship, and they are undoubtedly background stars with velocities which are sufficiently high to produce proper motions matching Hyades members.[8]

[8] One star with discrepant photometry in Figure 7.14 is a Hyades member: HZ9, a white dwarf/red dwarf binary, $M_V = 10.6$, $(V–I) = 2.0$.

Spectroscopic membership criteria

Proper-motion astrometry and photometry are techniques which use integrated starlight, and are therefore efficient methods of examining large samples of faint stars. Even after applying these criteria, however, some field-stars will remain misclassified as cluster members. Eliminating these stars demands more detailed observations, but the sample size is now sufficiently small that such observations are feasible. Radial velocity provides an additional criterion; measurement of stellar activity, either chromospheric (Hα emission) or coronal (X-ray emission), offers another option for late-type stars in relatively young clusters such as Praesepe, the Hyades and the Pleiades [P8], [P9], [R12], [S13].

7.6.2   Dynamical effects and present-day luminosity function in clusters

Once a membership list has been compiled for a cluster of known distance, construction of the luminosity function is straightforward. This represents the *present-day luminosity function* for the cluster, and two time-dependent processes must be taken into account before comparing results with other studies: stellar evolution and dynamical evolution of the cluster. The former is a systematic change in the mass–luminosity relationship with age, which can be taken into account using theoretical models (see Chapter 3). Dynamical evolution, in contrast, is a stochastic effect which increases with time and is less susceptible to analysis.

Large open clusters, such as the Pleiades or the Orion star-forming region, have masses of no more than $\sim 2,000\,M_\odot$ at formation. With diameters of ~ 10 parsecs, individual stars are bound only weakly to the cluster. Most clusters have orbits which lie close to the Galactic Plane, bringing the system into close proximity with massive giant molecular cloud complexes (GMCs). Stars on the near-side of the cluster receive a greater gravitational impulse from the GMC, and the additional energy may unbind stars from the cluster potential. The net result of this tidal effect is that an open cluster is gradually pulled apart over the course of time as stars 'evaporate' from the cluster, leading to the observed scarcity of old open clusters [F3]. A cluster which has an orbit with low inclination to the Galactic Plane is likely to survive for only a few orbits (perhaps several $\times 10^8$ years), while those few clusters in high-inclination orbits can survive for several Gyr.

If stellar evaporation were to occur at the same rate for stars of all masses, then the only effect on the luminosity function would be a change in the zero-point. However, this is not the case. As stars in a cluster relax to match the gravitational potential, we expect equipartition of energy; that is, the kinetic energy, $\frac{1}{2}mv^2$, should be the same for all groups of stars. This implies that lower-mass stars must have a higher velocity dispersion and, with higher σ_V, these stars have a more extended spatial distribution. Mass segregation occurs, with more massive stars and binaries confined to the central regions of the cluster.[9] Since lower-mass stars spend more

---

[9] There are also indications that high-mass stars form preferentially near the cluster core; for example, the Trapezium stars in the Orion Nebula Cluster and the BA stars in the Pleiades.

time at large distances from the cluster centre, they are more likely to be stripped from the cluster by tidal interactions with massive bodies. Thus, we expect the ratio between the number of low-luminosity and high-luminosity stars to decrease with time.

Both Spitzer [S10] and Chandrasekhar [C1] derive analytical expressions for the relaxation time-scales for star clusters. Spitzer gives the following relationship:

$$\tau_R = 8.0 \times 10^5 \frac{N^{0.5} R^{1.5}}{\langle m \rangle^{0.5} (\ln N - 0.5)} \text{ yr} \tag{7.40}$$

where N is the total number of stars, R is the cluster radius in parsecs, and $\langle m \rangle$ is the average mass in solar units. This implies relaxation times of $\sim 10^8$ yr for rich open clusters such as the Pleiades.

Spitzer's calculations, however, consider only internal dynamical evolution (that is, gravitational interactions between cluster members) and are more appropriate to $10^5 M_\odot$ globular clusters in the galactic halo than $10^3 M_\odot$ open clusters within the disk. The latter clusters are more loosely bound gravitationally, and external encounters can have a proportionately more significant effect. More recent calculations rely on N-body simulations, which include at least 10^3 particles and must model both the mean Galactic gravitational potential and short-term effects due to encounters with GMCs. Details of the computational techniques used to model the evolving cluster potential can significantly influence the final results [Z2]. Moreover, the binary fraction adopted is also important, since binaries can affect cluster dynamics by transferring energy to other stars during close encounters. In a binary–single star encounter, the binary becomes more tightly bound while the single star acquires additional orbital energy, and therefore spends more time in the cluster halo where it is more vulnerable to tidal stripping. Available models [T1], [F4], [Z2] indicate typical lifetimes of 2–4 Gyr for clusters with 1,000–2,000 particles, with significant mass segregation occurring in a few hundred Myr – a slightly longer time-scale than derived in analytical calculations. Further detailed theoretical work on this complicated problem is required.

Empirically, there is clear evidence for mass segregation in clusters as old as, or older than, the Hyades ($\tau \sim 600$ Myr). Figure 7.15 (*top*) plots the line-of-sight distance distribution of stars with $M_V < 8.5$ and $8.5 \le M_V \le 12$ for Hyades members from Reid [R7]. The higher-mass stars are more concentrated towards the cluster core. The effect is even more pronounced in older clusters, such as the $\sim 3.5 \times 10^9$-yr-old NGC 2420, lying at $(m-M)_0 = 11.95$ magnitude. The other diagrams in Figure 7.15 plot colour–magnitude and luminosity functions for two fields in NGC 2420, the first near the cluster centre, the other offset by 10 arcmin (~ 7 parsecs). This cluster has a particularly rich binary population, evident in the doubling of the main sequence in the central regions. In the outer field, however, there are few binaries and no stars within two magnitudes of the turnoff. Even in the central field, $\Phi(M_V)$ reaches a maximum at $M_V \sim 6$ – six magnitudes brighter than in the field. Clusters as young as the Pleiades (~ 120 Myr) are less affected by dynamical evolution, but are unlikely to be entirely immune from such effects.

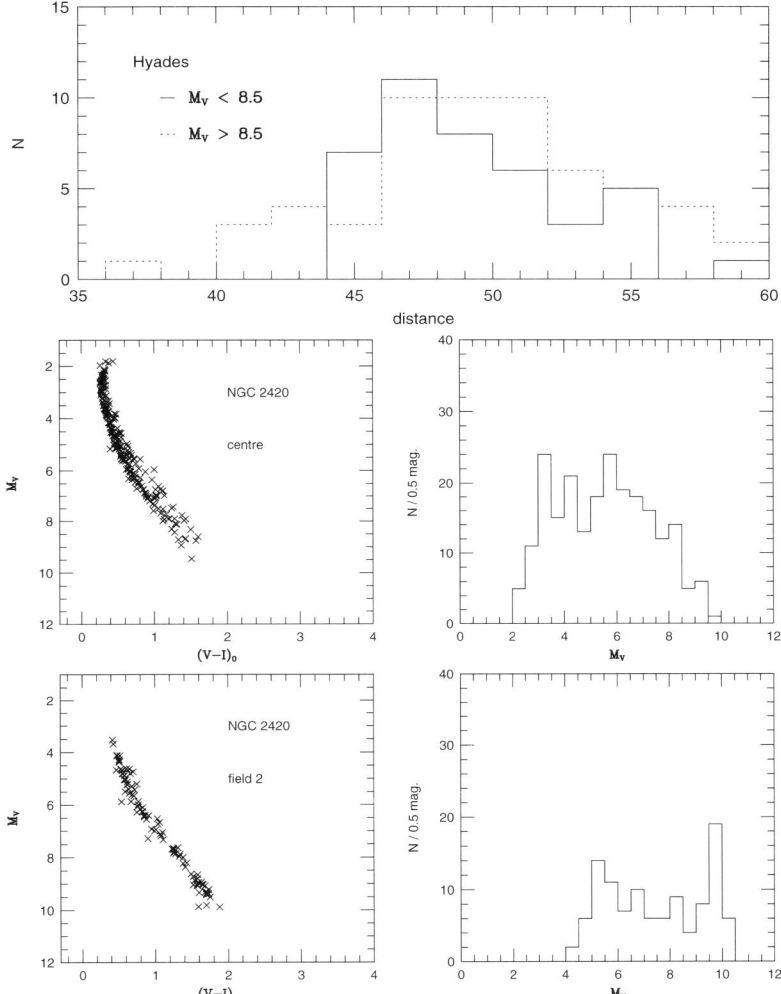

Figure 7.15. Mass segregation in older open clusters. The *top* panel compares the radial density distribution of Hyades stars with $M_V < 7$ and $7 \leq M_V \leq 12$; the *lower* panels compare colour–magnitude
diagrams and $\Phi(M_V)$ for two regions in NGC 2420. Note the scarcity of bright stars and binaries in
the outer field.

7.6.3 $\Phi(M_V)$ for nearby clusters

The Hyades

The nearest substantial open cluster – the Hyades – was first identified as a coherent
moving group by Boss [B6]. With a mean motion of $\bar{\mu} = 0\rlap{.}''11\,\mathrm{yr}^{-1}$, the cluster has
been the target of numerous proper-motion surveys [B8], [A1], [H3], [P1], [L14], [R6].

Most recently, 150 of the brightest cluster members were observed by the Hipparcos satellite, and Perryman *et al.* [P2] combine the measured space motion of $V_S = 45.72\,\mathrm{km\,s^{-1}}$ with astrometric data for 134 stars within the 10-parsec diameter cluster core to derive a distance of $46.34 \pm 0.27\,\mathrm{pc}$ $((m\text{–}M)_0 = 3.33)$ to the centre of mass. Matching the colour–magnitude diagram against the appropriate theoretical models (the cluster is mildly metal-rich, $[\mathrm{Fe/H}] = +0.11$ [B4]), they infer an age of $625\pm25\,\mathrm{Myr}$.

Due to its proximity, the Hyades subtends a substantial solid angle, and most surveys cover only a fraction of the cluster. Luyten *et al.* Palomar data ([L14]) are the exception, but the candidate list of 929 members lacks adequate supplementary photometry and spectroscopy to eradicate contaminating field stars. The most useful wide-field survey combines data from POSS I with UK Schmidt $B + V$ plates, covering $\sim$112 square degrees (one-third of the cluster) to $V \sim 19.5$ [R6]. Follow-up VRI photometry [R7], [L3] and spectroscopy [R12], [S13] serves to eliminate field-stars from the sample, while Hubble Space Telescope images [R9] have been used to search for binary companions at separations of $0''\!.1$ or more.

Figure 7.16(a) compares $\Phi(M_V)$ for the Hyades cluster against data for the 8-parsec sample, matching the two at $M_V = +7$. The younger age of the Hyades leads to the cluster having proportionately more stars with $M_V < +4$. As with field-stars, the Hyades luminosity function peaks at $M_V \sim +12$, but the maximum is less

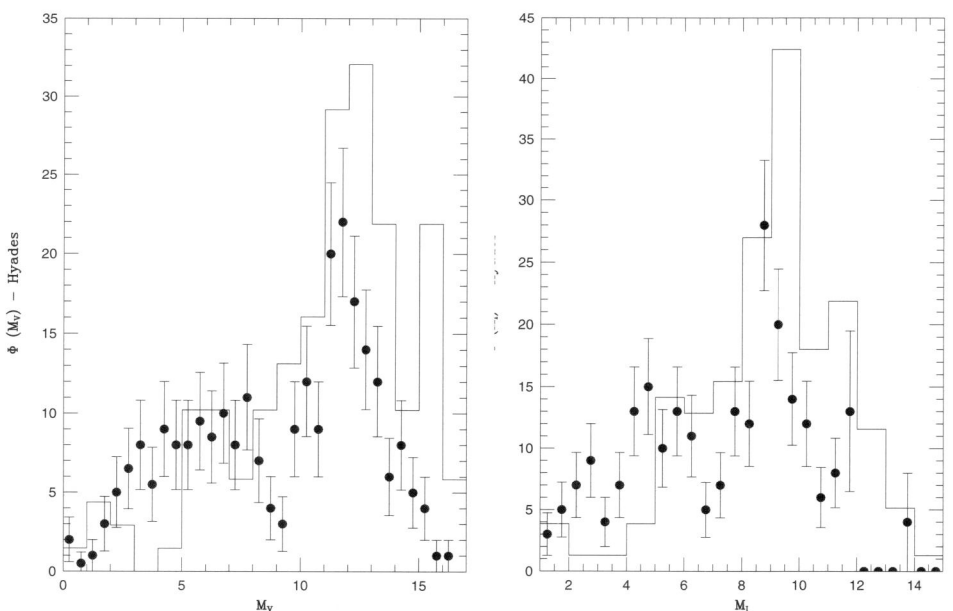

Figure 7.16. (a) $\Phi(M_V)$ for the Hyades, from [R6] and [R12]; (b) $\Phi(M_I)$, including fainter stars from [B7] and [L5]. Both datasets include low-mass binary companions identified by HST imaging. The histogram plots $\Phi(M)$ for the 8-parsec data, scaled to match at $M_V = +7$ and $M_I = +5$.

pronounced. In the field, there are three times as many stars with $M_V = 12 \pm 0.5$ as with $M_V = 6 \pm 0.5$, that is $\Phi(M_V = 12) : \Phi(M_V = 6) = 3 : 1$. In the Hyades, $\Phi(M_V = 12) : \Phi(M_V = 6) = 3 : 2$. The difference is probably due to dynamical evolution in the cluster. The lower-mass stars have suffered more attrition through tidal stripping.

The observed number of Hyades stars drops sharply for $M_V > 12$. The [R6] survey becomes incomplete at $M_V > 16$, but Leggett and Hawkins [L5] and Bryja *et al.* [B7] have pushed the search for low-mass members to fainter luminosities, in the central 25 square degrees of the cluster. [B7] combined scans of POSS I E plates with Luyten's 1962 Palomar plates to search for proper motion members fainter than $m_r = 20$, but, after spectroscopic follow-up observations, added only two new members at $M_V \sim 16.5$. Leggett and Hawkins used multiple R- and I-band UK Schmidt plates to search for late-type M dwarfs with colours and magnitudes consistent with membership, and identified 12 candidates. One of these stars, LH0418+13, has Hα emission and a radial velocity consistent with membership, and is probably the faintest known cluster member at $M_I \sim 13.85$ [R13]. Five non-members from the [L5] survey are discussed in more detail in Chapter 9.

Dynamical models predict that stars ejected from a cluster remain in the vicinity for a significant period of time. Those stars contribute to the Hyades moving group [E2] – stars with Hyades-like properties (colours and Hα emission), but with space motions differing by 2–$5\,\mathrm{km\,s^{-1}}$ from members of the cluster proper. Analysing the current white dwarf population in the cluster, Weidemann *et al.* [W1] suggest that up to two-thirds of the original protocluster of higher-mass ($>1\,M_\odot$) stars have been ejected. With a present-day mass of $\sim 400\,M_\odot$, this implies that the original cluster may have exceeded $1{,}200\,M_\odot$ with $3{,}000$–$4{,}000$ members, comparable to the Orion Nebula Cluster.

The Pleiades

The Pleiades is younger than the Hyades and approximately three times more distant, with a diameter of $\sim 10°$. The B-type stars on the upper main sequence form a compact and distinctive group, identified by both Greek and Chinese civilisations over 3,000 years ago. The cluster was recognised as a bound system in the late nineteenth century, and the first thorough proper-motion surveys were carried out by Trumpler [T4] and Hertzsprung [H11], who together catalogued ~ 500 candidate members brighter than $m_{pg} \sim 15$ within $2°$ of the cluster centre. As with the Hyades, there have been numerous subsequent studies [L1], [S12], [H1], [M7]. The [L1] survey is limited to brighter stars, $B < 15$, but both the [S9] and [M10] catalogues extend to $M_V = +12$ over 16 and 16.5 square degree areas respectively. Many of the [S12] candidate Pleiades members have spectroscopic or X-ray observations confirming membership. The [H1] survey, based on POSS I E and UK Schmidt R- and I-band plates, covers 25 square degrees to $R = 20.5$. Smaller regions of the Pleiades have been searched to much fainter magnitudes by surveys aiming specifically at detecting substellar-mass brown dwarfs, and these results are discussed in Chapter 9.

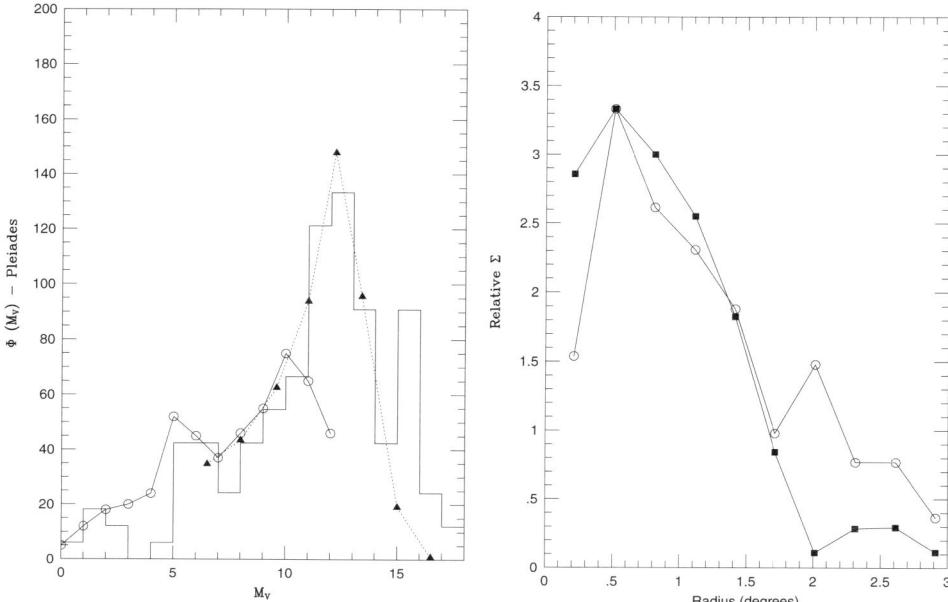

Figure 7.17. (a) $\Phi(M_V)$ derived from the Pleiades surveys by [S12] (open circles) and [H1] (solid triangles) compared with the 8-parsec data (histogram); (b) the separate radial density profiles of G dwarfs (solid squares) and mid-type M dwarfs (open circles). There is little evidence for significant mass segregation at radii $<1.8°$; there may be a more extended halo of low-mass M dwarfs.

Traditional main-sequence fitting and convergent-point analyses placed the cluster at a distance of $r \sim 132 \pm 10$ pc [L1]. This result was challenged by initial analysis of Hipparcos parallax data, which appeared to favour a significantly larger parallax, $\pi = 8{.}''6 \pm 0{.}''24$, or $r = 116$ pc [L2], [M6]. This revision would pose severe problems for stellar evolutionary theory, since the solar-abundance Pleiades would be subluminous with respect to the Sun. More recent analyses [P6] suggest that the discrepancy rests with systematic bias at the $0{.}''001$ level in the *Hipparcos* data. The older value for the distance of 132 pc $((m-M)_0 = 5.60)$ appears more reliable.

Figure 7.17(a) compares luminosity function data from two Pleiades surveys against $\Phi(M_V)$ for the 8-parsec sample, with all of the results scaled to match at $M_V \sim 6$. The maximum in $\Phi(M_V)$ lies at a slightly brighter magnitude than in the field, as expected given the younger age. The cluster does not show the deficit of low-mass stars present in the Hyades. Moreover, while BA-type stars and binaries are more centrally concentrated than the cluster G dwarfs [R1], the M dwarf radial density profile matches the G dwarf distribution to radii of ~1.8 degrees. There may be a more extended, low-density halo of the lower-mass stars (Figure 7.17(b)). However, few of the M dwarfs at large radii have been confirmed as cluster members.

Praesepe

Praesepe is similar to the Hyades in both age (~ 650 Myr) and abundance

([Fe/H] = +0.12), but lies at a distance of ~165 parsecs. As a result, the cluster has received less observational attention than its two neighbours in Taurus. The first extensive proper-motion surveys were carried out by van Rhijn [R14] and Klein-Wassink [K9], with the latter identifying ~160 likely members brighter than 14th magnitude. The more massive stars and binaries are concentrated towards the central regions of the cluster, but the degree of mass segregation appears to be less pronounced than in the Hyades. The most recent astrometric survey are by Jones and Stauffer [J2], covering the central ~16 square degrees in V and I to V ~ 17.5 magnitudes, and by Hambly *et al.* [H2], who used UK Schmidt R- and I-band plates to reach R ~ 20 magnitude over an area of 19 square degrees. While both surveys apply proper-motion and photometric criteria to identify likely cluster members, neither set of candidates has extensive spectroscopic follow-up observations to confirm membership. The I-band luminosity function derived from the [H2] survey peaks at $M_I \sim 9$, matching results for the Hyades, but remains flat at fainter magnitudes. In part, this probably reflects field-star contamination, but there may also be differences between the present-day mass functions in the two clusters. More detailed observations of the candidate low-luminosity Praesepe members are required.

7.6.4   Summary

Results from detailed observations of the Hyades reveal the potential pitfalls of using open clusters as probes of $\Phi(M_V)$. Even at the relatively young age of 600 Myrs, dynamical evolution can modify the stellar distribution to the extent that recovering the original attributes lies outside the bounds of current analytical techniques. The Pleiades studies, however, show that rich, 100–150-Myr-old clusters may be dynamically intact, and therefore suitable for statistical analysis. As yet, only the Pleiades has received the observational attention necessary to test for dynamical evolution and to derive reliable luminosity (and mass) functions. The Pleiades results are in broad agreement with data for the general field. A number of more distant southern clusters – notably NGC 2516 ($\tau \sim 100$ Myr), IC 2602 and IC 2391 (both $\tau \sim 55$ Myr), are currently under study by several groups. Initial analyses of IC 2602 [F2] and IC 2391 [R17] suggest a scarcity of even mid-type M dwarfs, as illustrated in Figure 7.18. This implies either much more drastic dynamical evolution than in the Pleiades, or a different initial luminosity function. However, studies are still restricted to relatively small areas within each cluster, and more extensive observations currently being undertaken by the Harvard–Smithsonian group (Stauffer, Barrado y Navuescés and collaborators) will provide a more thorough test of the universality (or otherwise) of $\Phi(M_V)$.

7.7   SPECIALISED SURVEY METHODS

7.7.1   M dwarfs in front of dark clouds

Most techniques used to search for late-type M dwarfs are effective at moderate to high Galactic latitudes. Herbst and Dickman [H9], however, devised a variation on

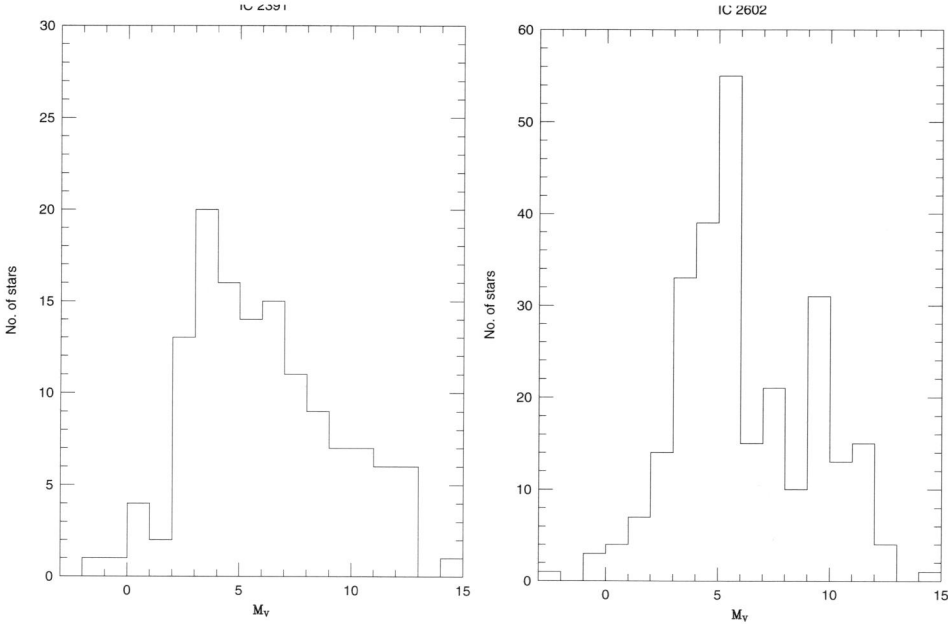

Figure 7.18. Visual luminosity functions for the young open clusters IC 2391 and IC 2602. Current datasets are far from complete, but note the scarcity of even early-type M dwarfs.

the photometric parallax method which not only works best at low latitudes, but also, in a further contradiction of convention, relies on the presence of substantial interstellar absorption along the line of sight. The principle is simple: a relatively nearby, high-density molecular cloud is identified using, for example, millimetre-wavelength CO observations. Accurate, multicolour photometry is obtained for stars lying within the central high-density region of the cloud. Since the cloud has substantial absorption at visual wavelengths, most of those stars can be expected to lie in front of the cloud, while luminous stars visible through the cloud will have colours that are severely reddened. The cloud serves as a backdrop to provide a catalogue of main-sequence stars within the foreground volume. Absolute space density measurements demand a distance estimate to the cloud, but relative densities (determining the shape of $\Phi(M_V)$) do not, since the same effective volume is sampled at all absolute magnitudes.

   The main limitations of this method are twofold. First, obscuration within a molecular cloud complex is patchy, and observations obtained along sight-lines with moderate reddening run the risk of including background, reddened stars in the catalogue. This limits analysis to dense cloud cores where $A_V > 10$ magnitudes. This restriction leads to the second, and crucial problem: the total solid angle subtended by high-density, high-obscuration clouds is very small. [H9] discuss results for a single cloud in the Scorpius–Centaurus region, covering an area of 220 square arcmin and enclosing only 25 cubic parsecs (against $1{,}608\,\mathrm{pc}^{-3}$ enclosed by the northern 8-parsec sample). More recently, Jarrett *et al.* [J1] used CCD photometry to probe to fainter magnitudes towards clouds in the Taurus and Ophiuchus star-forming regions ($r \sim 150\,\mathrm{pc}$), covering 2,000 square arcmin and ~ 200 cubic pc. A total of only 22 foreground stars with $9 < \mathrm{M}_V < 18$ contribute to the luminosity function derived in the latter study. Thus, while it is an interesting technique, this is not an effective method of determining $\Phi(\mathrm{M})$.

7.7.2   M dwarf companions to white dwarfs

White dwarf stars serve as targets for two types of low-mass companion searches: wide common proper-motion pairs, or unresolved companions at smaller separations, identified through infrared observations. For all except the first few million years of their existence, white dwarfs have bolometric luminosities comparable with those of late-K and M dwarfs. The scarcity of metals in their high-gravity atmospheres leads to energy distributions close to the appropriate black-body curve. Since the cooling time to reach a temperature of even 5,000 K is several Gyr, most white dwarfs have flux distributions which peak at optical wavelengths. If such a star is in a binary with a low-mass main-sequence companion, the white dwarf contributes the majority of the flux in the blue, but the M dwarf dominates at near-infrared and longer wavelengths. Indeed, even a low-luminosity VLM companion ($\mathrm{M}_V \sim +18$) produces a substantial infrared excess as compared with that of an isolated white dwarf of the same luminosity (Figure 7.19).

   White dwarf/red dwarf binaries can be identified by their having unusual optical/infrared colours (for example, (V–K)). The first extensive binary survey was undertaken by Probst [P7], who obtained K-band data for 100 white dwarfs, identifying seven composite systems, only two of which were previously undetected. Zuckerman and Becklin [Z1] extended observations to a larger sample, with the principle aim of detecting VLM dwarfs at or below the hydrogen-burning limit. (Two objects of particular interest in the latter context, G29-38 and GD 165B, are discussed in more detail in Chapter 9.) Figure 7.20 compares the K-band luminosity function for their complete sample against the nearby-star data: note that only 20 M dwarfs were discovered amongst a sample of ~ 150 targets, suggesting a low binary frequency for M dwarf companions to intermediate-mass ($> 1.5\,M_\odot$) stars. While the statistics are sparse, the [Z1] luminosity function rises more steeply towards low luminosities than either the 8-parsec or 5.2-parsec samples. However, this may simply be a selection effect: the targets [Z1] were all identified as white dwarfs based on their blue colours. A white dwarf/mid-M dwarf binary is likely to have colours closer to a G or early-K dwarf (such as HZ 9 in the Hyades), and is therefore excluded *a priori* from the catalogues of candidate white dwarf proper-motion stars which underlie the [Z1] sample. Further data – particularly on common proper-motion wide systems – should provide a more definitive answer for $\Phi(\mathrm{M}_k)$.

7.7.3   M dwarfs as wide companions

Finally, brief mention should be made of an older technique used to identify low-

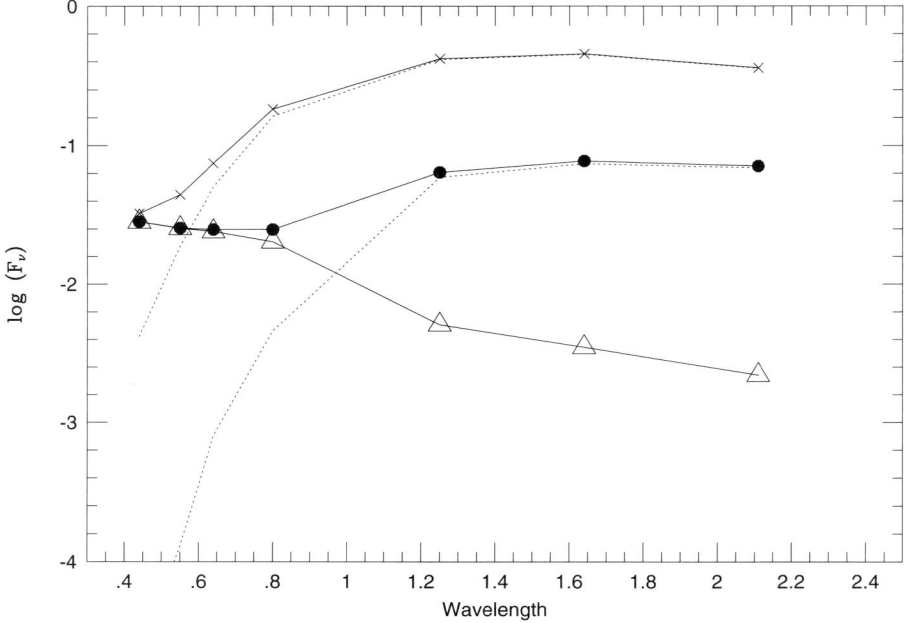

Figure 7.19. The spectral energy distribution of binary systems including a white dwarf and a later-type M dwarf. The white dwarf data (open triangles) are for GJ 1037 ($M_V = 12.9$, $T_{eff} \sim 9,800$ K); the composite distributions include Barnard's Star (crosses) and VB 10 (solid points). The dotted lines show the energy distributions of the M dwarfs alone.

luminosity companions: searching for common proper-motion companions of known nearby stars. Numerous such systems were identified in the course of the Bruce and Palomar proper-motion surveys, with Luyten paying particular attention to white dwarf/red dwarf pairs. The most effective exponent of this technique was van Biesbroeck, who concentrated his studies on stars known to be within 5–10 pc of the Sun. Searching specifically for very low-luminosity companions. van Biesbroeck identified 24 such stars [B3], the most famous being Gl 752B (VB10), which held the record as the star of lowest known luminosity from its discovery in 1944 [B2] until the mid-1980s. Recent surveys by Simons *et al.* [S8] have failed to add any additional companions amongst the stars within 8 parsecs of the Sun. We return to these matters in Chapter 9.

7.8 THE STELLAR LUMINOSITY FUNCTION IN THE GALACTIC DISK

This chapter has reviewed the methods used to study $\Phi(M_V)$, and has summarised recent results. The two main techniques currently employed in these studies are: first, surveying the stars within the immediate vicinity of the Sun; and, second, deep

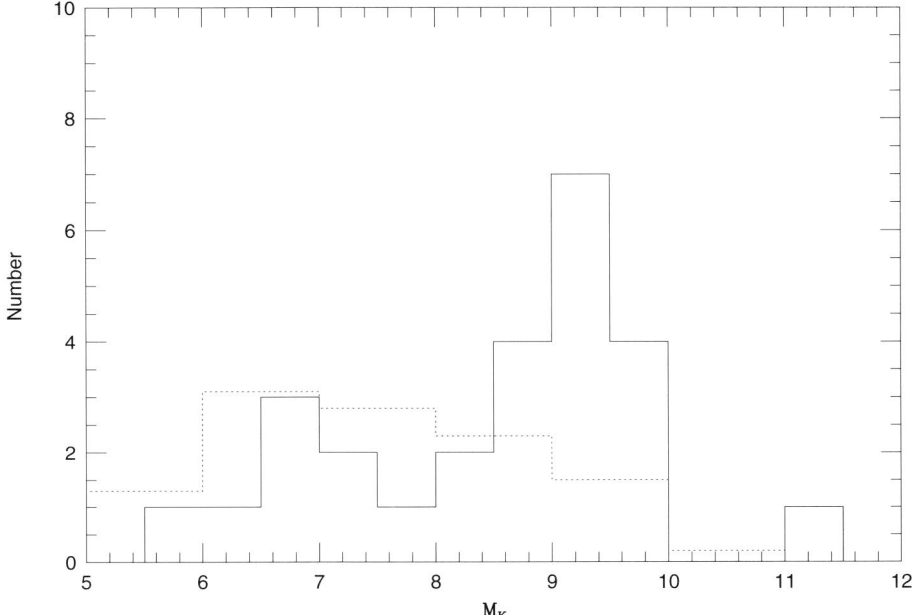

Figure 7.20. The K-band luminosity function derived from M dwarf companions of white dwarfs (solid line) [Z1]. The 8-parsec sample luminosity function, scaled to match at $M_K = 6.75$, is plotted as a dotted line.

pencil-beam photometric surveys, using colours to deduce luminosities, hence distances and densities. These two approaches give results that are in good agreement, once the appropriate colour–magnitude relationships are adopted for the photometric parallax analysis. In particular, the 'Wielen dip' at $M_V \sim +7$ has been identified in data for stars more than 1 kpc above the Galactic Plane. The luminosity function determined for the Pleiades cluster, the only well-studied open cluster which is sufficiently young to be dynamically intact, is also in reasonable agreement with the Solar Neighbourhood data. Taken together, these results provide a solid basis for assuming that the luminosity function derived from the nearby stars is representative of the 'mature' (ages of 0.5 Gyr or more) population of the Galactic disk.

The available data are combined to derive the current best estimate of $\Phi(M_V)$ (Table 7.3). At absolute magnitudes $M_V \leq 8$, the densities are computed from the data included in the third *Catalogue of Nearby Stars*, limiting the sample to $r < 22$ pc, $\delta > -33°$, and taking *Hipparcos* parallax measurements into account; for $8 \leq M_v \leq 15$, the statistics are from the Reid *et al.* [R11] spectroscopic analysis of CNS3 stars, adopting the distance limits listed in Table 7.2 but including new *Hipparcos* parallax data; and at fainter magnitudes, the densities are based directly on the 8-parsec sample.

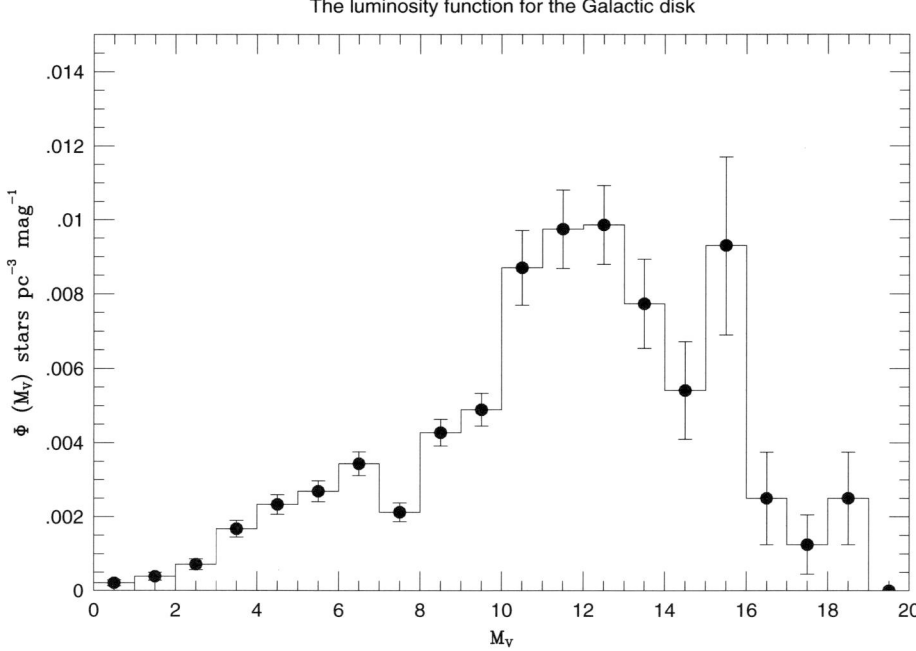

Figure 7.21. The stellar luminosity function, $\Phi(M_V)$.

Table 7.3. The stellar visual luminosity function.

M_V	N_{star}	$\Phi(M_V)$ $\times 10^{-3}$ pc^{-3}	M_V	N_{star}	$\Phi(M_V)$
0.5	7	0.21 ± 0.08	10.5	75	8.7 ± 1.0
1.5	13	0.39 ± 0.11	11.5	84	9.7 ± 1.1
2.5	24	0.72 ± 0.15	12.5	85	9.9 ± 1.1
3.5	56	1.7 ± 0.22	13.5	42	7.7 ± 1.2
4.5	78	2.3 ± 0.26	14.5	17	5.4 ± 1.3
5.5	90	2.7 ± 0.28	15.5	15	9.3 ± 2.4
6.5	115	3.4 ± 0.32	16.5	4	2.5 ± 1.25
7.5	71	2.1 ± 0.24	17.5	2	1.25 ± 0.9
8.5	143	4.3 ± 0.36	18.5	4	2.5 ± 1.25
9.5	123	4.9 ± 0.44			

7.9   REFERENCES

A1   van Altena, W. F., 1969, *AJ*, **74**, 2.
A2   Armandroff, T. E., 1983, *IAU Colloquium 76, The Nearby Stars and the Stellar Luminosity Function*, ed. A. G. Davis Philip and A. R. Upgren, p. 229.
B1   Bessell, M. S., 1986, *PASP*, **98**, 1303.
B2   van Biesbroeck, G., 1944, *AJ*, **51**, 61.
B3   van Biesbrock, G., 1961, *AJ*, **66**, 528.
B4   Boesgaard, A. M., 1990, *ApJ*, **351**, 467.
B5   Bok, B. J., Macrae, D. A., 1941, *Annals New York Acad. Sci.*, **42**, 219.
B6   Boss, L., 1908, *AJ*, **26**, 31.
B7   Bryja, C., Humphreys, R. M., Jones, T. J., 1994, *AJ*, **107**, 246.
B8   van Bueren, H. G., 1952, *Bull. Astr. Inst. Neth.*, **11**, 385.
C1   Chandrasekhar, S., 1942, *Principles of Stellar Dynamics*, Dover Publications Inc., New York.
C2   Comstock, G. C., 1910, *Astr. Nachr.*, **185**, 297.
D1   Dahn, C. C., Liebert, J., Harrington, R. S., 1986, *AJ*, **91**, 621.
E1   Eggen, O. J., 1950, *ApJ*, **112**, 144.
E2   Eggen, O. J., 1984, *AJ*, **89**, 830.
F1   Faber, S. M., Burstein, D., Tinsley, B. M., King, I. R., 1976, *AJ*, **81**, 45.
F2   Foster, D. C., Byrne, P. B., Hawley, S. L., Rolleston, W. R. J., 1997, *A&AS*, **126**, 81.
F3   Friel, E. D., 1995, *ARA&A*, **33**, 381.
F4   de la Fuente Marcos, R. 1997, *A&A*, **322**, 764.
G1   Gilmore, G. F., Reid, I. N., Hewett, P. C., 1986, *MNRAS*, **213**, 257.
G2   Giclas, H. L., Burnham, R., Thomas, N. C., 1972, The Lowell Observatory Proper Motion Survey, *Lowell Obs. Bull.* 158.
G3   Gliese, W., 1957, *Veröff. astr. Rechen Inst. Heidelberg*, **8**.
G4   Gliese, W., 1969, *Veröff. astr. Rechen Inst. Heidelberg*, **22**.
G5   Gliese, W., 1972, *QJRAS*, **13**, 138.
G6   Gliese, W., Jahreiss, H., 1979, *A&AS*, **38**, 423.
G7   Gliese, W., Jahreiss, H., 1991, *Preliminary Version of the Third Nearby Star Catalogue CNS3*, Strasburg.
G8   Gould, A., Bahcall, J. N., Flynn, C., 1996, *ApJ*, **465**, 759.
G9   Gould, A., Bahcall, J. N., Flynn, C., 1997, *ApJ*, **482**, 913.
G10 Gunn, J. E., Griffin, R. F., Griffin, R. E. M., Zimmerman, B. A., 1988, *AJ*, **96**, 198.
H1   Hambly, N., Hawkins, M. R. S., Jameson, R. F., 1993, *MNRAS*, **263**, 647.
H2   Hambly, N. C., Steele, I. A., Hawkins, M. R. S., Jameson, R. F., 1995, *MNRAS*, **273**, 505.
H3   Hanson, R. B., 1975, *AJ*, **80**, 379.
H4   Hanson, R. B., 1979, *MNRAS*, **186**, 875.
H5   Hanson, R. B., 1983, *IAU Coll. 76*, p. 51, ed. A. R. Upgren, L. Davis Press, Schenectady.
H6   Hanson, R. B., Vasilevskis, S., 1983, *AJ*, **88**, 844.

H7   Hawkins, M. R. S., Bessell, M. S., 1988, *MNRAS*, **234**, 177.

H8   Hawley, S. L., Gizis, J. E., Reid, I. N., 1996, *AJ*, **112**, 2799.

H9   Herbst, W., Dickman, R. L., 1983, in *Nearby Stars and the Stellar Luminosity Function*, p. 187, ed. A. R. Upgren and A. G. Davis Philip, L. Davis Press, Schenectady.

H10 Herschel, W., 1785, *Phil. Trans.*, **75**, 213.

H11 Hertzsprung, E., 1934, *Bull. Astr. Inst. Neth.*, **7**, 258.

J1   Jarrett, T. H., Dickman, R. L., Herbst, W., 1994, *ApJ*, **424**, 852.

J2   Jones, B. F., Stauffer, J. R. 1991, *AJ*, **102**, 1080.

J3   Jones, D. H. P., 1973, *MNRAS*, **161**, 19P.

K1   Kapteyn, J. C., 1902, *Publ. Astr. Lab. Gröningen*, No. 11.

K2   Kapteyn, J. C., 1909, *ApJ*, **29**, 46.

K3   Kapteyn, J. C., 1914, *ApJ*, **40**, 43.

K4   Kapteyn, J. C., van Rhijn, P. J., 1924, *ApJ*, **52**, 23.

K5   Kibblewhite, E. J., Bridgeland, M. T., Bunclark, P., Irwin, M. J., 1984, *Proceedings of the Astronomical Microdensitometry Conference*, ed. D. A. Klinglesmith, NASA Conf. Publ. 2317.

K6   King, I. R., 1977, *Highlights of Astronomy*, Vol. 4, ed. E. A. Müller, Reidel: Dordrecht.

K7   Kirkpatrick, J. D., McGraw, J. T., Hess, T. R., Liebert, J., McCarthy, D. W., 1994, *ApJS*, **94**, 749.

K8   Klare, G. Schaifers, K., 1966, *Astr. Nachr.*, **289**, 81.

K9   Klein-Wassink, W. J., 1927, *Publ. Kapteyn Astr. Lab.*, No. 41.

K10 Kroupa, P., 1995, *ApJ*, **453**, 358.

K11 Kuiper, G. P., 1942, *ApJ*, **95**, 201.

L1   van Leeuwen, F., 1982, PhD thesis, Leiden University.

L2   van Leeuwen, F., Hansen Ruiz, C. S., 1997. in *Hipparcos Venice '97*, eds B. Battrick, M. A. C. Perryman, ESA, p. 689.

L3   Leggett, S. K., Dahn, C. C., Harris, H., 1994, *AJ*, **108**, 944.

L4   Leggett, S. K., Hawkins, M. R. S., 1988, *MNRAS*, **234**, 1045.

L5   Leggett, S. K., Hawkins, M. R. S., 1989, *MNRAS*, **238**, 145.

L6   Lutz, T. E., Kelker, D. H., 1973, *PASP*, **85**, 573.

L7   Luyten, W. J., 1923, *Lick Obs. Bull.*, **11**, 1.

L8   Luyten, W. J., 1924, *ApJ*, **62**, 8.

L9   Luyten, W. J., 1939, *Publ. Obs. Minn.*, **2**, 121.

L10 Luyten, W. J., 1941, *Annals New York Acad. Sci.*, **42**, 201.

L11 Luyten, W. J., 1968, *MNRAS*, **139**, 221.

L12 Luyten, W. J., 1974, *Proper Motion Survey with the 48-inch Schmidt Telescope*, XLVI.

L13 Luyten, W. J., 1977, *Highlights in Astronomy*, Vol. 4, p. 89, ed. E. A. Müller, Reidel, Dordrecht.

L14 Luyten, W. J., Hill, G., Morris, S., 1981, *Proper Motion Survey with the 48-inch Schmidt Telescope*, LIX, University of Minnesota, Minneapolis.

M1   McCuskey, S. W., 1963, *Vistas in Astronomy*, **7**, 141.

M2   Mazzei, P., Pigatto, L., 1988, *A&A*, **193**, 148.

M3   van Maanen, A., 1937, *ApJ*, **85**, 26.

M4   Malmquist, K. G., 1936, *Stockholm Obs. Medd.*, Nr. 26.

M5   Martini, P.. Osmer, P. S., 1998, *AJ*, **116**, 2513.

M6   Mermilliod, J.-C., Robichon, N., Arenou, F., Lebreton, Y., 1997, in *Hipparcos Venice '97*, eds B. Battrick and M. A. C. Perryman ESA, 643.

M7   Meusinger, H., Schilbach, E., Souchay, J., 1996, *A&A*, **312**, 833.

M8   Mihalas, D., Binney, J., 1981, *Galactic Astronomy*, W. H. Freeman & Company, San Francisco.

M9   Murray, C.A. 1988, *IAU Symposium 133: Mapping the Sky*, p. 143, ed. S. Débarbat, J. A. Eddy, H. K. Eichhorn and A. R. Upgren, Reidel, Dordrecht.

M10 Murray, C. A., Sanduleak, N., 1972, *MNRAS*, **157**, 273.

O1   Oort, J. H., 1932, *Bull. Astr. Inst. Neth.*, **6**, 249.

O2   Oort, J. H., 1965, *Galactic Structure*, ed. M. Schmidt, University of Chicago Press.

O3   Oort, J. H., 1974, *Highlights of Astronomy*, Vol. 3, p. 317, ed G. Contopoulos, Reidel, Dordrecht.

P1   Pels, G., Oort, J. H., Pels-Kluyver, H. A., 1975, *A&A*, **43**, 423.

P2   Perryman, M. A. C., Brown, A. G. A., Lebreton, Y., Gómez, A., Turon, C., Cayrel de Strobel, G., Mermilliod, J.-C., Robichon, N., Kovalevsky, J., Crifo, F., 1998. *A&A*, **331**, 81.

P3   Pesch, P., 1972, *ApJ*, **177**, 519.

P4   Pesch, P., Sanduleak, N., 1978, *AJ*, **83**, 1090.

P5   Pesch, P., Dahn, C. C., 1982, *AJ*, **87**, 122.

P6   Pinsonneault, M. H., Stauffer, J., Soderblom, D. R., King, J. R., Hanson, R. B., 1998. *ApJ*, **504**, 170.

P7   Probst, R. G., 1983, *ApJ*, **274**, 237.

P8   Prosser, C. F., Stauffer, J. R., Kraft, R. P., 1991, *AJ*, **101**, 1361.

P9   Prosser, C. F., Randich, S., Stauffer, J. R., Schmitt, J. H. M. M., Simon, T., 1996, *AJ*, **112**, 1570.

R1   Raboud, D., Mermilliod, J.-C., 1998a, *A&A*, **329**, 101.

R2   Raboud, D., Mermilliod, J.-C., 1998a, *A&A*, **333**, 897.

R3   Reid, I. N., 1982, *MNRAS*, **201**, 51.

R4   Reid, I. N., 1984, *MNRAS*, **206**, 1.

R5   Reid, I. N., 1991, *AJ*, **102**, 1428.

R6   Reid, I. N., 1992, *MNRAS*, **257**, 257.

R7   Reid, I. N., 1993, *MNRAS*, **265**, 785.

R8   Reid, I. N, Gizis. J. E., 1997, *AJ*, **113**, 2246.

R9   Reid, I. N, Gizis. J. E., 1997, *AJ*, **114**, 1992.

R10 Reid, I. N., Gilmore, G. F., 1982, *MNRAS*, **201**, 73.

R11 Reid, I. N., Hawley, S. L., Gizis, J. E., 1995, *AJ*, **110**, 1838.

R12 Reid, I. N., Hawley, S. L., Mateo, M., 1995, *MNRAS*, **272**, 828.

R13 Reid, I. N., Hawley, S. L., 1999, *AJ*, **117**, 343.

R14 van Rhijn, P. J., 1916, *Publ. Kaptyen Astr. Lab. Gröningen*, No. 26.

R15 van Rhijn, P. J., 1925, *Publ. Kaptyen Astr. Lab. Gröningen*, No. 38.

R16 van Rhijn, P. J., 1936, *Publ. Kaptyen Astr. Lab. Gröningen*, No. 47.

R17 Rolleston, W. R. J, Byrne, P. B., 1997, *A&AS*, **126**, 357.
S1 Sanduleak, N., 1965, PhD thesis, Case Institute of Technology.
S2 Sanduleak, N., 1976, *AJ*, **81**, 350.
S3 Schmidt, M., 1974, *Highlights of Astronomy*, Vol. 3, p. 450, ed G. Contopoulos, Reidel, Dordrecht.
S4 Schmidt, M., 1975, *ApJ*, **202**, 22.
S5 Schmitt, J. H. M. M., Fleming, T. A., Giampapa, M. S., 1995, *ApJ*, **450**, 392.
S6 Seares, F. H. 1924, *ApJ*, **59**, 310.
S7 von Seeliger, H., 1898, *Abhandlungen der K. Bayer. Akademie d. Wiss., II. Kl.,* **19**, 564.
S8 Simons, D. A., Henry, T. J., Kirkpatrick, J. D., 1996, *AJ*, **112**, 2238.
S9 Smethells, W. G., 1974, PhD thesis, Case Western Reserve University.
S10 Spitzer, L., 1940, *MNRAS*, **100**, 396.
S11 Starikova, G. A., 1960, *Soviet Astr.*, **4**, 451.
S12 Stauffer, J. R., Klemola, A., Prosser, C., Probst, R., 1991, *AJ*, **101**, 980.
S13 Stauffer, J. R., Liebert, J., Giampapa, M., Macintosh, B., Reid, I. N., Hamilton, D., 1994, *AJ*, **108**, 160.
S14 Stobie, R. S., 1986, *Patt. Rec. L.*, **4**, 317.
S15 Stobie, R. S., Ishida, K., Peacock, J. A., 1989, *MNRAS*, **238**, 709.
T1 Terlevich, E., 1987, *MNRAS*, **224**, 193.
T2 Thé, P. S., Staller, R. F. A., 1974, *A&A*, **36**, 155.
T3 Tinney, C. G., 1993, *ApJ*, **414**, 279.
T4 Trumpler, R. J., 1920, *PASP*, **32**, 43.
T5 Trumpler, R. J., 1930, *Lick Obs. Bull.*, 420.
T6 Trumpler, R. J., Weaver, H. F., 1953, *Statistical Astronomy*, University of California Press.
U1 Upgren, A. R., Armandroff, T. E., 1981, *AJ*, **86**, 1898.
W1 Weidemann, V., Jordan, S., Iben, I., Casertano, S., 1992, *AJ*, **104**, 1876.
W2 Weistrop, D. W., 1972, *AJ*, **77**, 849.
W3 Weistrop, D. W., 1976, *ApJ*, **204**, 113.
W4 Weistrop, D. W., 1980, *AJ*, **85**, 738.
W5 Wielen, R., 1974, *Highlights of Astronomy*, Vol. 3, p. 395, ed. G. Contopoulos, Reidel, Dordrecht.
W6 Wielen, R., Jahreiss, H., Krüger, R., 1983, *IAU Colloquium 76, The Nearby Stars and the Stellar Luminosity Function*, ed. A. G. Davis Philip and A. R. Upgren, p. 163.
W7 Woolley, R. v.d.r., Epps, E. A., Penston, M. J., Pocock, S. B., 1970, *Royal Obs. Bull.*, No. 5.
Z1 Zuckerman, B, Becklin, E. E., 1992, *ApJ*, **386**, 260.
Z2 Zwart, S. F. P., Hut, P., Makino, J., McMillan, S. L. W., 1998, *A&A*, **337**, 363.

8

The mass function

8.1 INTRODUCTION

The number of stars per unit mass – the stellar mass function – describes how a molecular cloud redistributes its material to form stars. This parameter is also a necessary ingredient for determining the mass distribution and the total stellar mass in star clusters and external galaxies. It is therefore fundamental to both star-formation theory and Galactic structure.

Deriving $\Psi(M)$ is more demanding than determining $\Phi(M)$. Given a well-defined sample of stellar systems, individual distances and apparent brightnesses can be measured, and $\Phi(M_V)$ calculated directly. In contrast, masses are currently known for relatively few stars, limited almost exclusively to components of resolved or eclipsing binary systems. As a result, indirect techniques must be employed to estimate masses for the overwhelming majority of stars, even amongst those within 8 parsecs of the Sun.

The necessity of relying on indirect mass estimates would not constitute a complication in determining $\Psi(M)$ if there were a simple, single-valued relation between a directly measurable quantity and mass. Unfortunately, this is not the case. In main-sequence stars, luminosity depends strongly on mass, but is also dependent to a lesser extent on factors such as age and chemical composition. As already discussed in Chapter 7, luminosity evolution invalidates direct comparison between $\Phi(M_V)$ for nearby stars and results for young open clusters and star-forming regions, where stars are still in the process of contracting onto the main sequence. Conversely, the low metal abundance of elderly subdwarfs in the Galactic halo leads to their having luminosities and colours significantly different from those of solar-abundance dwarfs of the same mass.

Despite the observational and theoretical difficulties, the stellar mass function can be determined for stars in a range of environments. The opening sections of this chapter outline the terminology used in describing the mass function and the methods used to measure stellar masses. The latter sections summarise results derived for stars in the Galactic disk: the nearest stars; intermediate-aged clusters such as the Pleiades; and young associations. Metal-poor subdwarfs and the halo mass function are discussed in Chapter 11.

8.2   DEFINING THE MASS FUNCTION

8.2.1   Basic terminology

The following symbols are used to represent the stellar luminosity function and mass function:

$$\Phi(M) = \frac{dN}{dM}: \quad \text{the luminosity function in stars pc}^{-3}\,\text{mag}^{-1}$$

$$\Psi(M) = \frac{dN}{dM}: \quad \text{the mass function in linear units, stars pc}^{-3}\,M_\odot^{-1}$$

$$\xi(M) = \frac{dN}{d\log M}: \text{the mass function in logarithmic units, stars pc}^{-3}\log(M_\odot)^{-1}$$

A luminosity function is transformed to a mass function, by applying the appropriate mass–luminosity relationship (MLR), dL/dM,

$$\frac{dN}{dM} = \frac{dN}{dL}\frac{dL}{dM} \tag{8.1}$$

where the luminosity function is given in general by dN/dL. Strictly speaking, the term *mass–luminosity relationship* applies to the correlation between mass and bolometric luminosity. The expression, however, is often used to refer to the mass calibration for individual passbands, such as the mass–M_V relationship, just as *luminosity function* is used to describe the number of stars per unit M_V, M_I, M_K, in addition to per unit luminosity.

The linear and logarithmic mass functions are related as follows:

$$\Psi(M) = \frac{dN}{dM} = \frac{dN}{d(\log M)}\frac{d(\log M)}{dM}$$

$$= \log_{10}e\,\frac{dN}{d(\log M)}\frac{d(\ln M)}{dM}$$

$$= \frac{0.4343}{M}\,\xi(M)$$

Finally, the *star-forming mass distribution* is defined as

$$\Xi(M) = \frac{M \times \Psi(M)}{\int_{\min}^{\max}\Psi(M)\,dM} \tag{8.3}$$

This function describes the mass budget for star formation – the fraction of the total mass devoted to forming stars of a specific individual mass.

8.2.2   The Salpeter mass function

The first serious attempt to deduce the form of the initial mass function for disk stars was by Salpeter [S1]. Many of the basic concepts, including the parameterisation of the mass function as a power-law, were introduced in this classic paper, albeit using different terminology. Salpeter defined the *original mass function*, $\xi(M)$, as

$$\xi(M) = \frac{T_0}{d(\log M)} \frac{dN}{dt}$$

where dN is the number of stars in the mass range $d(\log M)$ created in time dt, and T_0 is the age of the Galaxy (which Salpeter took to be 5 Gyr). $\xi^i(M)$ (or $\Psi^i(M)$ in linear units) is now known as the *initial mass function*, or IMF.

Salpeter combined the van Rhijn [R7] and Luyten [L8] analyses to derive a present-day luminosity function, $\Phi(M_V)$. With no direct measurement of the Galactic star-formation history, Salpeter assumed that the present star formation rate was characteristic of times past, and used estimates of the main-sequence lifetimes of massive stars to allow for stars which had evolved off the main sequence. The observed mass function could then be transformed to the initial mass function (Figure 8.1), given reasonably well by the approximation

$$\xi^i(M) \sim 0.03 \left(\frac{M}{M_\odot} \right)^{-1.35} \text{stars pc}^{-3} [\log(M)]^{-1} \tag{8.5}$$

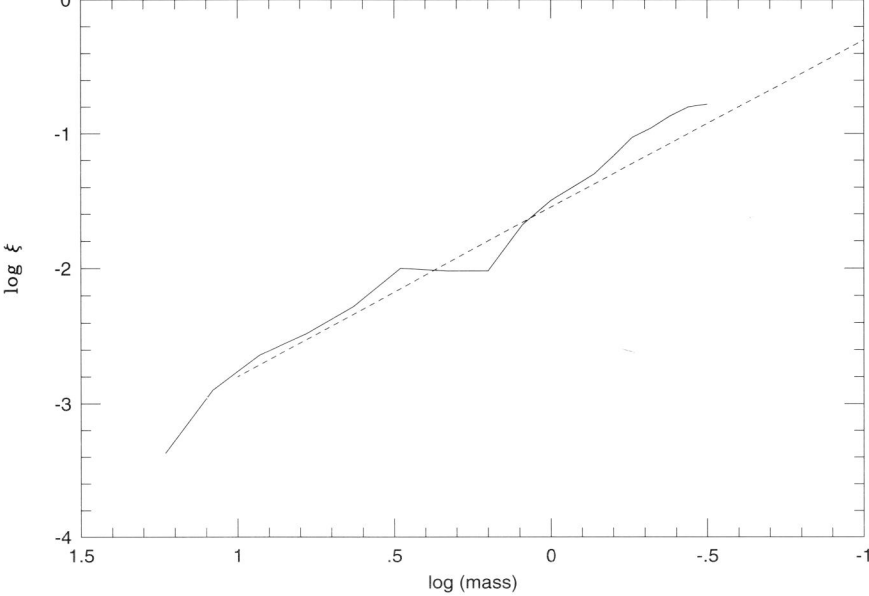

Figure 8.1. The Salpeter mass function. The solid line plots Salpeter's empirical results, and the dotted line is a power-law mass function, $x = -2.35$.

In linear units, this is

$$\Psi^i(M) \propto M^{-2.35} \text{ stars pc}^{-3} M_\odot^{-1} \tag{8.6}$$

Power-law mass functions are often written using the terminology $\xi(M) \propto M^{x+1}$ or $\Psi(M) \propto M^{-\alpha}$, where $\alpha = -x$. A power-law mass function with slope $\alpha = 2.35$ is known as a Salpeter function; a mass function with slope $\alpha > 2.35$ is described as being 'steeper' than the Salpeter function, while $\alpha < 2.35$ is 'flatter' than the Salpeter value.

The main developments in studying the mass function since Salpeter's initial work are surveyed in review articles by Scalo and others [M7], [S2], [R1], [S3]. The first of these papers [M7] introduced much of the terminology currently in use in the subject, particularly in the analysis of stars with masses exceeding $1\,M_\odot$.

8.2.3   The mass function for intermediate- and high-mass stars

While M dwarfs represent the main focus of this book, their properties must be placed in a more general context. This is particularly relevant in studying the mass function, where one of the main goals is assessment of the relative proportion of mass bound *in perpetuo* as low-mass stars, and the proportion recycled in the interstellar medium as massive stars evolve. Unlike M dwarfs, stars more massive than $\sim 1\,M_\odot$ have lifetimes shorter than the age of the Galactic disk, and the observed luminosity function includes only a fraction of the total number formed during the history of the Milky Way. The 'missing' earlier generations of stars have evolved to become white dwarfs, neutron stars and black holes. In addition, higher-mass main sequence stars have a lower velocity dispersion than M dwarfs (hence Parenago's discontinuity, Section 6.5.1), and are therefore confined more closely to the Galactic plane. Both effects, together with possible variations in the star-formation history of the Milky Way, must be taken into account when deriving an estimate of the initial mass function. Indeed, it is these corrections, rather than uncertainties in the mass–luminosity relationship, which are the greatest source of uncertainty in $\Psi(M)$ at supra-solar masses.

In observational terms, Miller and Scalo define the present-day mass function for main-sequence stars as

$$\xi_{ms}^{PD}(\log M) = \Phi(\mathrm{M}_V) \left| \frac{d\mathrm{M}_V}{d\log(M)} \right| 2H(\mathrm{M}_V) f_{ms}(\mathrm{M}_V) \tag{8.7}$$

where $d\mathrm{M}_V/d\log(M)$ is the slope of the $(\mathrm{M}_V, \text{mass})$ relationship; $H(\mathrm{M}_V)$ is the exponential scaleheight of the vertical density distribution of stars, absolute magnitude M_V; and $f_{ms}(\mathrm{M}_V)$ is the fraction of stars with absolute magnitude M_V which are on the main sequence. ξ_{ms}^{PD} can also be written as

$$\xi_{ms}^{PD}(\log M) = \int_{T_0-\tau}^{T_0} C(\log M, t)\, dt \tag{8.8}$$

for stars with main-sequence lifetimes, τ_{ms}, shorter than the age of the Galactic disk, T_0, or

$$\xi_{ms}^{PD}(\log M) = \int_0^{T_0} C(\log M, t)\, dt \tag{8.9}$$

for stars with $\tau_{ms} > T_0$. The latter expression holds for M dwarfs.

$C(\log M, t)$ is the creation function, the convolution of the stellar birthrate, $B(\log M, t)$, and the initial mass function, $\xi(\log M, t)$. If we assume that the latter two functions are separable – that is, that the initial mass function is time invariant – then the creation function can be written as

$$C(\log M, t) = \frac{\xi(\log M)}{T_0} \frac{B(t)}{\langle B \rangle} \tag{8.10}$$

where $\langle B \rangle$ is the time-averaged total birthrate,

$$\langle B \rangle = \frac{\int_0^{T_0} B(t)\, dt}{T_0} \tag{8.11}$$

Hence

$$C(\log M, t) = \frac{\xi(\log M)}{T_0} b(t) \tag{8.12}$$

where $b(t)$ is the relative birthrate, which expresses the star-formation rate at time t in terms of the average rate of the history of the Galactic disk. Combining these relationships, the present-day mass function can be written as

$$\xi_{ms}^{PD}(\log M) = \frac{\xi(\log M)}{T_0} \int_{T_0 - \tau_{ms}}^{T_0} b(t)\, dt \tag{8.13}$$

for stars with $\tau_{ms} < T_0$, and

$$\xi_{ms}^{PD}(\log M) = \frac{\xi(\log M)}{T_0} \int_0^{T_0} b(t)\, dt = \frac{\xi(\log M)}{T_0} \tag{8.14}$$

for stars more massive than $\sim 1\, M_\odot$ which have main-sequence lifetimes longer than the age of the disk.

The rapid variation in main-sequence lifetime with increasing mass means that results deduced for $\xi(\log M)$ at high masses are dependent on the functional form adopted for B(t). While the problem can be constrained by requiring that there are no discontinuities in the inferred $\xi(\log M)$, most studies of higher-mass stars tend to concentrate on young open clusters and associations, representing individual episodes of star formation of determinate age. However, few Galactic associations lie within the range of direct trigonometric parallax measurement, so there are uncertainties in the distance calibration. In addition, small number statistics and interstellar reddening contribute to the uncertainties in $\xi(M)$. Finally, spectroscopic observations are required to determine masses for O and B stars ($\geq 6\, M_\odot$). With temperatures exceeding 15,000 K, the UBVRI passbands lie on the long-wavelength

Rayleigh-Jeans tail, and show little variation with spectral type. Indeed, as the peak radiation shifts to shorter wavelengths, the hottest O stars become intrinsically fainter than B stars at visual wavelengths, so even M_V becomes a poor indicator of intrinsic luminosity, and hence of mass.

Observational studies show that the initial mass function has a steep slope at masses exceeding $\sim 2\,M_\odot$. There are significant variations in the results measured in different clusters, probably due in part to the difficulties outlined above. Scalo [S3] has assembled more than 50 separate measurements of the slope of the stellar mass function (fitted as a power law) at masses of $2\,M_\odot$ or above, and the average value is $\langle\alpha\rangle \sim 2.5 \pm 0.4$, close to Salpeter's original result. There is no obvious correlation with either metal abundance or environment; neither, it should be emphasised, is there strong evidence that the data require a power-law representation.

A full discussion of the complexities involved in determining $\Psi^i(M)$ at high masses is outside the scope of this book. For present purposes, we characterise the mass function for stars of $1\,M_\odot$ and above as a single power-law, slightly steeper than Salpeter at $\Psi(M) \propto M^{-2.5}$. Following the discussion above, we assume that the mass function derived for stars with $M < 1\,M_\odot$ represents the total space density of those stars, integrated over the star-formation history of the disk. Where it is necessary to extend analysis to include the full stellar mass range, we normalise the higher-mass function to match the observed space density of $1\,M_\odot$ stars.

8.3 MEASURING STELLAR MASSES

As noted in the introduction, direct mass measurement is possible for only a small number of individual stars. In the near future it may prove possible to use gravitational lensing techniques to determine stellar masses, but for the present, binary stars provide the only effective means of determining stellar masses. The stellar mass function cannot be derived directly from those few stars with known mass, but, if that sample is representative of the general population, those stars can be used to calibrate main-sequence mass–luminosity relations.

Mass determination for stars in a binary system requires measurement of seven orbital elements: the semi-major axis, a; the period, P; the time of periastron passage, T_0; the eccentricity, e; the longitude of periastron passage, ω; the angle of the ascending node, Ω; and i, the inclination of the orbital plane with respect to our line of sight. An orbital inclination of $90°$ corresponds to an edge-on (eclipsing) binary. These parameters can be determined using two observational techniques, direct imaging and radial velocity observations. Detailed descriptions of binary star orbit analysis may be found in [B4], [C1].

8.3.1 Astrometric binaries

If both stars in a binary system are resolved, then the relative orbit of the secondary about the primary can be measured directly. Binary star astrometry began with William Herschel, who in 1776 started recording visual measurements of the separation (ρ) and position angle (θ, measured from north through east) of double stars, with the assumption that the fainter star was more distant, and could therefore provide a reference for the determination of the parallax of the 'nearer' brighter star. Instead of measuring parallax, Herschel detected orbital motion in some systems, including α Geminorum (Castor) and γ Virginis.

Visual measurements, primarily using bifilar micrometers, continued throughout the nineteenth and even twentieth century, with extensive catalogues compiled by (amongst others) Wilhelm and Otto Struve at Dorpat and Pulkovo Observatories, Robert G. Aitken and William J. Hussey at Lick Observatory and S. W. Burnham at Yerkes and other observatories. While photography extended observations to fainter stars, and allowed for more accurate astrometry of well separated systems, visual observations still held the edge for resolved binaries with small ($<2''$) angular separations. Exposure times of even a few minutes produce images with an angular resolution equivalent to average seeing. In contrast, an experienced visual observer can take advantage of fleeting moments of superb seeing to measure a close double which would be unresolved on a photographic plate. Thus, double star observers such as George van Biesbroeck were making visual observations with large reflectors – such as the McDonald Observatory 82-inch – as late as the 1960s and early 1970s. In recent years, higher quantum-efficiency detectors, such as CCDs, and techniques such as speckle interferometry (Section 1.7.1) have replaced visual observations.

The relative orbit of a binary is an ellipse with the primary star at one focus. Since the orbital plane is generally inclined with respect to the plane of the sky, observations map the projected orbit, which is also an ellipse, but with the primary no longer at an apparent focus. Measuring the latter's displacement from the focus of the apparent orbit allows determination of the inclination. Once i has been determined, the eccentricity and semi-major axis derived from the apparent orbit can be corrected to their true values. Finally, knowledge of the distance of the system is essential when converting the angular measurement of the semi-major axis, α, to a linear measurement, a.

Given a and P, the total mass of the binary system follows from Kepler's third law:

$$(M_1 + M_2) = \frac{a^3}{P^2} \tag{8.15}$$

where mass is measured in solar masses, a is measured in astronomical units and P in years. The determination of individual masses demands measurement of the relative distance of each star from the barycentre of the system. This requires absolute positional astrometry – measurement of (α, δ) for each component, rather than of the relative orientation (θ) and separation (Δ) of the two components. The contribution of parallax and proper motion can then be computed and removed to recover the individual orbits. The semi-major axes are inversely proportional to the masses:

$$\frac{M_1}{M_2} = \frac{a_2}{a_1} = \frac{\alpha_2}{\alpha_1} \tag{8.16}$$

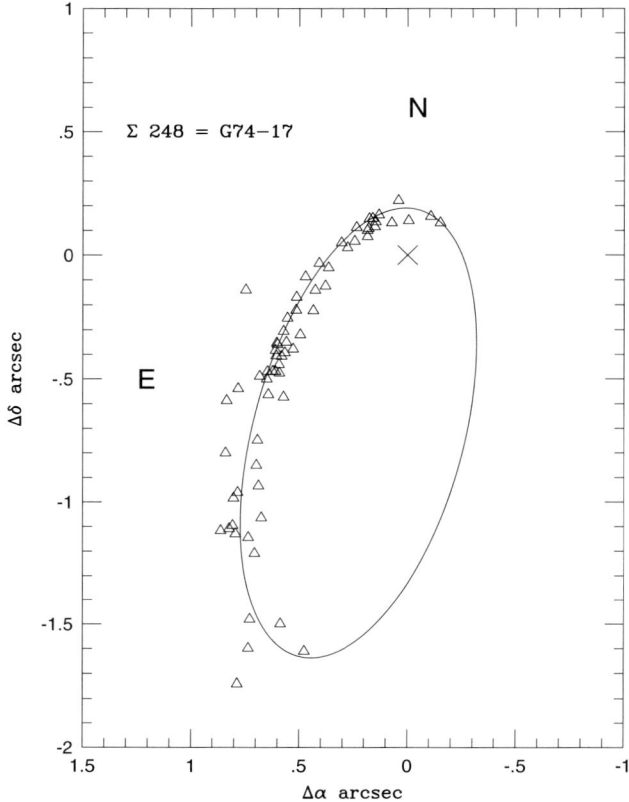

Figure 8.2. The relative orbit of the G/K dwarf binary Σ248. The cross marks the position of the primary and observations span over 160 years, from W. Struve's original measurements in 1830 (at lower left) to speckle imaging data from 1991. (Data and orbit from [T2].)

In many binaries the lower-luminosity component is either fainter than the magnitude limit or is unresolved from the primary star, and the existence of a companion is deduced only by the presence of systematic positional residuals when solving for parallax and proper motion. It was through applying this technique in 1840 that F. W. Bessel deduced that Sirius had an invisible companion of comparable mass (the white dwarf, Sirius B). Under such circumstances, absolute astrometric data can be used to determine the orbit described by the primary star about the barycentre. If the distance is known, and the orbital inclination estimated, then the semi-major axis, a_1, of the primary-star orbit can be calculated. Kepler's third law can be written in the form

$$\frac{M_2^3}{(M_1 + M_2)^2} = \frac{a_1^3}{P^2} \tag{8.17}$$

and the individual masses follow from

$$M_1 = \frac{1 - B}{B^3} \frac{a_1^3}{P^2}$$

$$M_2 = \frac{1}{B^2} \frac{a_1^3}{P^2} \tag{8.18}$$

where $B = M_2/(M_1 + M_2)$ is the scale factor.

These equations can be solved for (M_1, M_2) only if B is known, which requires at least one observation of *both* components. If absolute positions for both the primary and the barycentre are known, the separation between the two components determines the relative distance of each from the barycentre, and hence the mass ratio. In the absence of direct detection, the only option is the use of indirect methods (such as setting an upper limit to the luminosity of the companion) to constrain the mass of both components.

A potential complication is introduced because the photocentroid is measured in unresolved systems; the intensity-weighted centroid of the point-spread function due to both components. If the secondary makes a significant contribution to the total flux, then the measured centroid is offset from the actual position of the primary, and α_1 is underestimated. In the extreme case of equal components (such as Gl 866AB), the photocentroid shows no evidence for orbital motion. This bias can be compensated for if the relative luminosities of the two components are known; for example, modern high-resolution imaging techniques might be used to detect a low-luminosity companion in a star with a long history of photographic astrometry. However, it is often necessary to use observations acquired with a wide variety of photographic emulsions, electronic detectors and filters, each with their own particular spectral response. The derivation of an accurate orbit from such data is a process fraught with uncertainty.

Until recently, atmospheric seeing restricted astrometry to systems in which the components had separations of at least 1–2 arcsec. Even at distances of 5–10 parsecs, this demands orbits with semi-major axes exceeding 5 AU and consequent periods of decades, requiring long time-series of accurate astrometric observations. As recently as 1985, only a handful of M dwarfs could be characterised as having mass estimates of even moderate precision: only 10 stars had masses known with a formal uncertainty of less than 20% [M2]. With the development of infrared speckle interferometry, complemented by Fine Guidance Sensor (FGS) observations on the Hubble Space Telescope, it has become possible to obtain astrometric observations of systems with separations as low as 0″.1 and periods as little as 10 years [B3]. Higher precision mass measurements stem from more precise astrometry and more complete orbital coverage.

8.3.2   Spectroscopic binaries

The alternative to direct imaging is the determination of stellar orbits by monitoring radial velocities. If velocities for both stars are measurable – either through their being resolved or because they have similar luminosities and form a double-lined spectroscopic binary – then all orbital elements except inclination can be calculated. The mass ratio follows directly from the ratio between the two velocity amplitudes (K_1, K_2 – Figure 8.3), since

$$\frac{M_1}{M_2} = \frac{K_2}{K_1} \tag{8.19}$$

Integrating each velocity curve over the full cycle allows computation of the projection of each orbit onto the line of sight (perpendicular to the plane of the sky) and hence derive $a_1 \sin(i)$ and $a_2 \sin(i)$. If the system is eclipsing, then the inclination must be close to 90° and individual masses can be determined. For non-eclipsing systems, at least one direct image of both components is required to determine i. Marcy and Moore [M2] provide an excellent discussion on synthesising results from radial velocity observations and astrometry of the close binary Gl 623 AB. Lacking such data, lower limits on the individual masses are given by

$$(M_1 + M_2) \geq \frac{(a_1')^3}{P^2} \tag{8.20}$$

where $a_1' = a \sin(i)$, the value for the semi-major axis deduced from the velocity data.

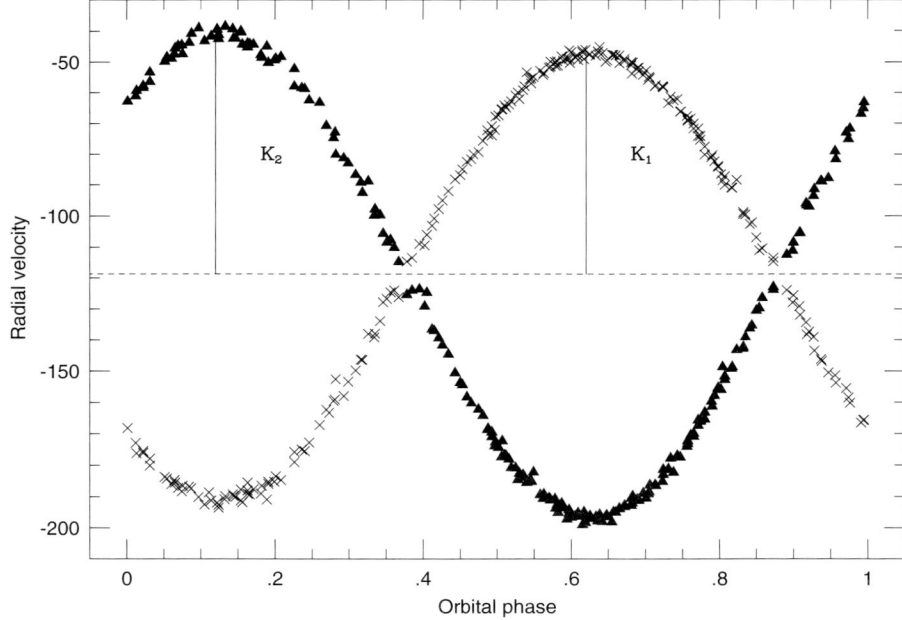

Figure 8.3. Radial velocity curves for the double-lined spectroscopic binary CM Draconis (from [M4]). The ratio of the semi-amplitude velocities of the two curves, K_1 and K_2, is inversely proportional to the mass ratop.

If the secondary component is faint, the binary is single-lined, allowing measurement only of the radial-velocity curve of the primary. In this case, the mass function can be determined:

$$\frac{M_2^3 \sin^3(i)}{(M_1 + M_2)^3} = \frac{(a_1')^3}{P^2} \tag{8.21}$$

The amplitude of the reflex motion, K_1, is given by

$$K_1 = \frac{M_2 \sin(i)}{M_1^{2/3} \sqrt{1 - e^2}} \left(\frac{2\pi G}{P} \right) \tag{8.22}$$

where i is the inclination of the pole of the orbit to the line of sight, e is the orbital eccentricity, and P is the period. Expressing the masses in solar units and the period in years, this becomes

$$K_1 = 29.79 P^{-1/3} \frac{M_2 \sin(i)}{M_1^{2/3} \sqrt{1 - e^2}} \, \text{km s}^{-1} \tag{8.23}$$

Again, such observations allow limits to be set on the mass of the secondary star, but do not permit direct mass-determination.

Only four M dwarf eclipsing binaries are known: YY Gem [L1], CM Dra [M4], the recently discovered, GJ 2069Aab [D3] and BW3 V38 [M1]. Of these, the last may be an interacting system, and it currently lacks radial-velocity data.

8.3.3   Mass determination from gravitational lensing

Light is deflected when it passes through a gravitational field. The classic example is the positional change of 1″.75 measured for stars in the vicinity of the Sun during the 1919 solar eclipse [D4]. A massive object can act as a lens, amplifying the total flux delivered to a given observer from a background source to an extent which depends on the angular separation of the sources (the impact parameter) and the mass of the object [R2]. Paczyński [P1] originally pointed out the potential of this (microlensing) technique for detecting compact objects which *might* be constituents in the invisible dark-matter Galactic corona. Several projects (MACHO, EROS, DUO, OGLE) are currently underway with the goal of using statistical techniques to estimate the mass distribution of these hypothetical objects. Microlensing also offers the possibility of direct mass-measurement of isolated stellar objects.

The basic geometry of gravitational microlensing is illustrated in Figure 8.4(a) (following [P2]). Consider a source, S, at distance D_S from an observer, O. The light from that source passes close to an object, L, mass M, at distance D_L. The object acts as a gravitational lens, deflecting light from S through angle α, so that the source appears to be at position S′. In fact, if L is effectively a point source, then S will produce two images (as with QSO 0957+561, [W3]). If

$$D = \frac{(D_S - D_L)/D_L}{D_S} \tag{8.24}$$

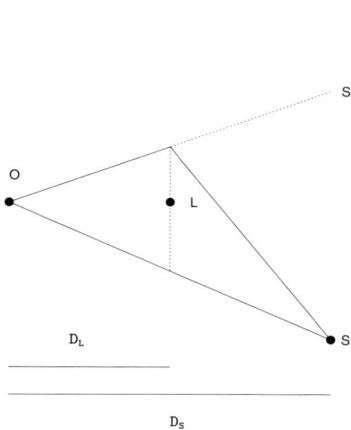
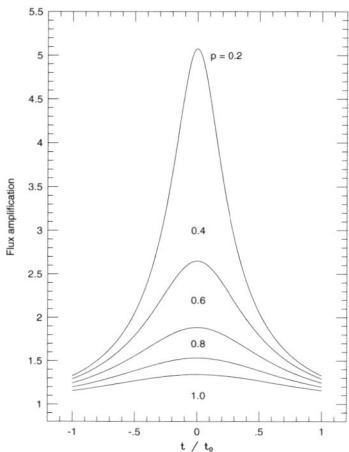

Figure 8.4. (a) Lensing geometry: O is the observer, S is the source, L is the lens and S' the observed position of the source from O. (b) Flux amplification for a range of impact parameters (see equation (8.31)).

is the effective lens distance,

$$R_g = \frac{2GM}{c^2} \qquad (8.25)$$

the gravitational radius of object L, and

$$R_E = \sqrt{2R_g D} \qquad (8.26)$$

the linear Einstein radius, then the positions of the images are offset by

$$R_{+,-} = 0.5 \left[R_S \pm \sqrt{(R_S^2 + 4R_E^2)} \right] \qquad (8.27)$$

If the impact parameter is small, these images are unresolved, and the total flux of the lensed object is amplified by a factor

$$A = \frac{u^2 + 2}{u\sqrt{(u^2 + 4)}}, \quad \text{where } u = \frac{R_S}{R_E} \qquad (8.28)$$

In the case of a Galactic source, the angular diameter of the Einstein ring (in mas) is given by

$$r_E = \frac{R_E}{D_L} = 0.902 \sqrt{\left(\frac{M}{M_\odot}\right)\left(\frac{10\,\text{kpc}}{D_L}\right)\left(1 - \frac{D_L}{D_S}\right)} \qquad (8.29)$$

Equation (8.28) shows that the source amplification is dependent on the impact parameter (that is, the minimum angular separation between L and S). In the case of a Galactic source, this parameter can be expected to vary on relatively rapid timescales due to differences in the proper motions of S and L. The characteristic time for this variation is given by the time to traverse one Einstein radius, or

$$t_0 = 0.214\,\mathrm{yr}\;\sqrt{\frac{M}{M_\odot}\frac{D_L}{10\,\mathrm{kpc}}\left(1-\frac{D_L}{D_S}\right)\frac{200\,\mathrm{km\,s}^{-1}}{V}}$$

(8.30)

The impact parameter is defined in terms of the Einstein radius,

$$p = \frac{r_{\min}}{r_E}$$

(8.31)

Figure 8.4(b) shows the predicted light amplification for unresolved sources. This variation is wavelength-independent, providing a key property used to identify lensing events in the vast databases accumulated by the various lensing surveys. The average mass of the lensing objects is

$$\langle M \rangle = \frac{c^2\,V_{\mathrm{rms}}^2}{GD_S}t_{0,av}^2$$

(8.32)

where $\log t_{0,av} = \langle \log t_0 \rangle$ and V_{rms} is the tangential velocity dispersion of the lenses. Current microlensing surveys are directed towards either the Magellanic Clouds or the Galactic Bulge (depending on the time of year) – targets which provide a large number of background sources, assumed to have known distance. The velocity dispersion of the lenses is usually taken as ~ 200–$300\,\mathrm{km\,s}^{-1}$, as might be expected for a non-rotating, pressure-supported population (Chapter 6). The most recent results suggest $\langle M \rangle \sim 0.5 \pm 0.3M_\odot$, although the nature of these Massive Compact Halo Objects (MACHOs) remains unclear [A2], [A3].

Equations (8.28) to (8.30) show that if both the lens and the source can be resolved and observed, then it is possible to estimate a mass for L. Alternatively, astrometric measurement of the curved paths described by the lensed source(s) as the lens traverses in front of the source position can also provide a means of estimating the mass of L if the lens itself is invisible. These measurements all require high-resolution interferometry and absolute astrometry, but may be possible for the forthcoming Space Interferometry Mission (SIM).

8.3.4 Summary

Mass estimates are currently available for stars in more than 100 systems, both eclipsing binaries (see the compilation by [A5]) and nearby astrometric binaries [H6], [H7]. The majority of known eclipsing systems are of spectral type G or earlier, whereas most astrometric analyses centre on K and M binaries. This dichotomy does not necessarily imply that eclipsing systems are more common amongst early-type stars: rather, it reflects the spectral type distribution of stars brighter than 10th magnitude, which over the last 150 years have been studied in more detail than have stars at fainter apparent magnitudes.

In general, orbital parameters are determined to higher precision for the eclipsing binaries than for the astrometric systems. The majority of the former have relatively short periods (days or hours) and large velocity amplitudes ($>50 \, \text{km s}^{-1}$, Figure 8.3). Radial velocities can be measured to a precision of better than $100 \, \text{m s}^{-1}$, while the short periods allow for complete phase coverage. In contrast, those in the sample of local astrometric binaries have typical separations of one to several arcsec and periods of decades to centuries. While speckle interferometry and HST FGS observations can measure the relative positions with a typical precision of $\sim 0\overset{''}{.}05$, the long periods demand integration of older, lower-precision data into the solution. Even the inclusion of the latter observations does not guarantee coverage of the full orbit (Figure 8.2). Thus, astrometric mass determinations are generally less accurate.

8.4   MASS–LUMINOSITY RELATIONSHIP

Stellar luminosity functions usually exhibit significant structure. Equation (8.1) shows that this structure need not be an intrinsic property of the underlying mass distribution, but can be introduced by features in the mass–luminosity conversion. As described in Chapter 3, changes in stellar structure – notably due to H_2 dissociation, the disappearance of the radiative core and the onset of degeneracy at low masses – lead to variations in the mass–temperature and/or mass–radius relationship, with consequent effects on the (M, L) relationship. Variations in both the overall energy distribution and the growth and decline of specific spectral features, especially molecular bands, have a strong influence on the absolute magnitude/mass relation for specific passbands. Accurate definition of the relevant mass-luminosity relationship is therefore vital in obtaining an accurate estimate of $\Psi(M)$ for a given population.

8.4.1   Empirical mass–luminosity calibration

Figure 8.5 plots mass–luminosity data for late-type dwarfs with masses determined to an accuracy of at least 20%. The observational data for these stars are listed in Table 8.1. Ideally, one would like to establish the mass–luminosity relationship directly from this diagram, and use that relationship to constrain structural parameters in stellar models. This ideal can be achieved for stars with masses above $1 \, M_\odot$, but the lower precision of mass estimates for subsolar-mass stars precludes strictly empirical analysis in that régime. As outlined in Section 8.3.4, this reflects the necessity of relying on data for astrometric binaries among the lower-mass stars.

Despite the significant uncertainties, Henry and McCarthy derived empirical mass–luminosity calibrations based on results for stars included in their initial analysis [H6]. Semi-empirical relationships have also been derived [K3], and are plotted in Figure 8.5 together with theoretical (mass, M_V) and (mass, M_K) isochrones [B1]. As noted in Section 8.3, we make the implicit assumption that the calibrating stars are a representative subset of the parent population. Any bias amongst the calibrators towards, for example, younger stars still in pre-main sequence contraction, or metal-poor stars, would lead to biased mass estimates for main-sequence stars in the field. It is unlikely that such is the case for the present sample, which includes only binaries with measureable orbits from the volume-limited 8-parsec sample. As described in Chapter 6, the dispersion in velocity effectively guarantees that these stars are representative of the disk as a whole. Spectroscopy of individual stars shows no evidence of bias in either age or abundance.

The three calibrations plotted in Figure 8.5 are in reasonable agreement in both M_V and M_K. Each incorporates points of inflection where the slope of the mean relationship changes rapidly. These are defined explicitly as the nodal points between the separate relationships contributing to the empirical [H6] and semi-empirical [K3] calibrations – at 0.5 and 0.18 $M_\odot$, and 0.6 and 0.33 $M_\odot$ respectively, with the latter relationship also flattening at masses below 0.2 $M_\odot$ – and implicitly within the theoretical calibration. Each node can be identified with a specific physical

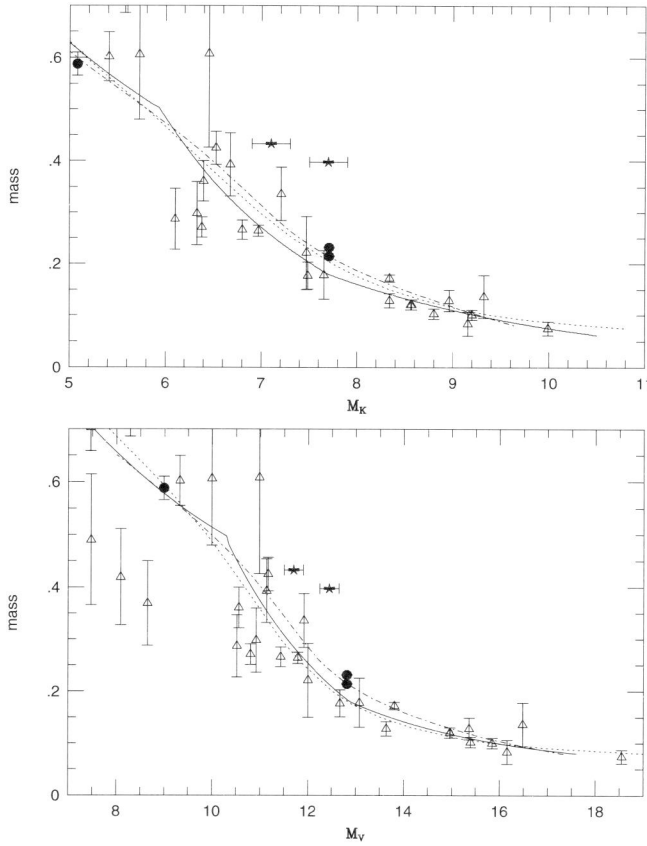

Figure 8.5. Empirical mass–luminosity data. Astrometric binaries are plotted as triangles; YY Gem and CM Dra are shown as solid dots; the five-point stars mark the two components of GJ 2069. The empirical relationships from [H6] are plotted as solid lines; the [K3] semi-empirical relationships are plotted as dot–dash lines; and the dotted lines show the 1-Gyr isochrones from the Lyon solar-abundance models [B1].

Table 8.1. Subsolar-mass stars with well-determined masses.

System	Spectral type	Mass	M_V	M_K	a	e	Reference
Gl 278 CD	M0	0.62 ± 0.03	9.0	5.1			YY Gem, 1
	M0	0.59 ± 0.02	9.0	5.1			
Gl 630.1AB	M4.5	0.2307 ± 0.0010	12.82	7.7	0.0175 ± 0.0001	0.005	CM Dra, 2
	M4.5	0.2136 ± 0.0010	12.8	7.75			
GJ 2069Aab	M3.5	0.4329 ± 0.0018	11.7	7.1	0.01737 ± 0.00003	0.0	3
	M3.5	0.3975 ± 0.0015	12.45	7.7	0.01891 ± 0.00003		
Gl 22AC	M2	0.361 ± 0.039	10.56	6.40	4.95 ± 0.16	0.05	4, 5
		0.128 ± 0.014	13.64	8.34			
Gl 65AB	M5.5	0.102 ± 0.010	15.40	8.80	5.12 ± 0.17	0.62	4, 5
	M5.5	0.100 ± 0.010	15.85	9.20			
Gl 67AB	G1.5	0.736 ± 0.231	4.45	3.04	7.73 ± 0.63	0.42	4
		0.221 ± 0.071	12.0	7.47			
Gl 166BC	DA	0.564 ± 0.019			333.52 ± 1.72	0.41	6
	M4.5	0.177 ± 0.026	12.68	7.49			
Gl 234AB	M4.5	0.177 ± 0.047	13.08	7.66	4.17 ± 0.36	0.40	4, 5
		0.083 ± 0.023	16.16	9.16			
Gl 340AB	K3	0.696 ± 0.101	6.64	4.21	11.34 ± 0.58	0.35	4
	K3	0.625 ± 0.090	6.86	4.29			
Gl 352AB	M3	0.287 ± 0.060	10.52	6.10	5.11 ± 0.51	0.05	4
		0.298 ± 0.062	10.92	6.33			
Gl 469AB	M3.5	0.24	11.63				5
		0.17	13.22				
Gl 473AB	M5	0.12	14.97	8.57			5
	M5	0.12	14.96	8.56			
Gl 508AB	M0.5	0.865 ± 0.179	8.29	5.58	14.22 ± 1.18	0.225	4
		0.606 ± 0.126	10.00	5.73			
Gl 570BC	M1	0.602 ± 0.047	9.33	5.41	0.821 ± 0.059	0.76	4
		0.425 ± 0.032	11.17	6.53			
Gl 623AB	M2.5	0.608 ± 0.182	10.98	6.45	2.06 ± 0.24	0.566	7
		0.136 ± 0.042	16.04	9.32			
Gl 661AB	M3.5	0.271 ± 0.020	10.80	6.38	4.45 ± 0.11	0.80	4
		0.266 ± 0.019	11.43	6.80			
Gl 677AB	K5	0.419 ± 0.092	8.09	4.66	12.45 ± 1.32	0.18	4
		0.369 ± 0.081	8.65	4.94			
Gl 702AB	K0	0.888 ± 0.056	5.67	3.87	22.83 ± 0.47	0.50	4
	K5	0.703 ± 0.045	7.46	4.12			
Gl 704AB	F7	0.741 ± 0.187	4.12	2.55	16.9 ± 1.13	0.74	4
	K5	0.490 ± 0.124	7.48	4.61			
Gl 725AB	M3.0	0.393 ± 0.061	11.14	6.68	48.5 ± 2.44	0.53	4
	M3.5	0.336 ± 0.052	11.92	7.21			
Gl 748AB	M3.5	0.26	11.30	6.7			5
		0.17	13.11				

Table 8.1 (*cont.*)

System	Spectral type	Mass	M_V	M_K	a	e	Reference
Gl 831ABC	M4.5	0.19	12.70	7.2			5
		0.12	14.80				
		0.11	15.4				
Gl 860AB	M3	0.264 ± 0.011	11.79	6.97	9.46 ± 0.13	0.41	4
	M4	0.172 ± 0.007	13.81	8.34			
Gl 1081AB	M3.5	0.25	11.49				5
		0.17	13.16				
GJ 1245AC	M5.5	0.128 ± 0.021	15.37	8.96	3.60 ± 0.20	0.32	4
	M5.5	0.074 ± 0.013	18.55	9.99			
GJ 2005AD	M6	0.10	16.26				5
		0.07	18.72				
G250−29AB	M3	0.28	11.04				5
		0.19	12.65				

[1] Lacy [L1].
[2] Metcalfe *et al.* [M4].
[3] Delfosse *et al.* [D3].
[4] Henry and McCarthy [H6].
[5] Henry *et al.* [H7].
[6] Reid [R4].
[7] Marcy and Moore [M2].

phenomenon: the influence of H_2 dissociation and the emergence of H^- as a substantive opacity source (0.5–0.6 $M_\odot$); the development of full convection ($\sim 0.3\,M_\odot$) and increasing degeneracy ($< 0.2\,M_\odot$).

The significance of non-linearities in the mass–luminosity relationship can best be assessed by comparing the derivative of the mass-luminosity relationship against the luminosity function in the appropriate passband [K5]. If dM/dL is large, then a small range in luminosity corresponds to a large range in mass; in contrast, a numerically small value for the derivative implies that a given interval in mass is distributed over a large range in luminosity. Hence, a monotonic mass function can be redistributed to produce a maximum in $\Phi(M)$ in the former case, and an extended minimum in the latter. Figure 8.6 compares the first derivatives of the [B1] and [K3] (M_V, mass) relationships against $\Phi(M_V)$ for nearby stars (from Chapter 7). The maximum in the latter function at $M_V \sim 12$ lies close (but not exactly coincident with) peaks in the former functions, while both the Wielen dip and the extended tail at low luminosities match minima in the dM/dM_V relations.

None of the mass–luminosity relationships discussed above provide a definitive calibration for the lowest-luminosity stars ($M_V > 15$, $M < 0.1\,M_\odot$) which lie close to the brown dwarf domain. As discussed in the following chapter, this has produced consequent uncertainty in the exact location of the hydrogen-burning limit in the observational plane. Indeed, well-known low-luminosity dwarfs, such as LHS 2065 and LHS 2397a, have been proposed as possible brown dwarfs. It now seems clear

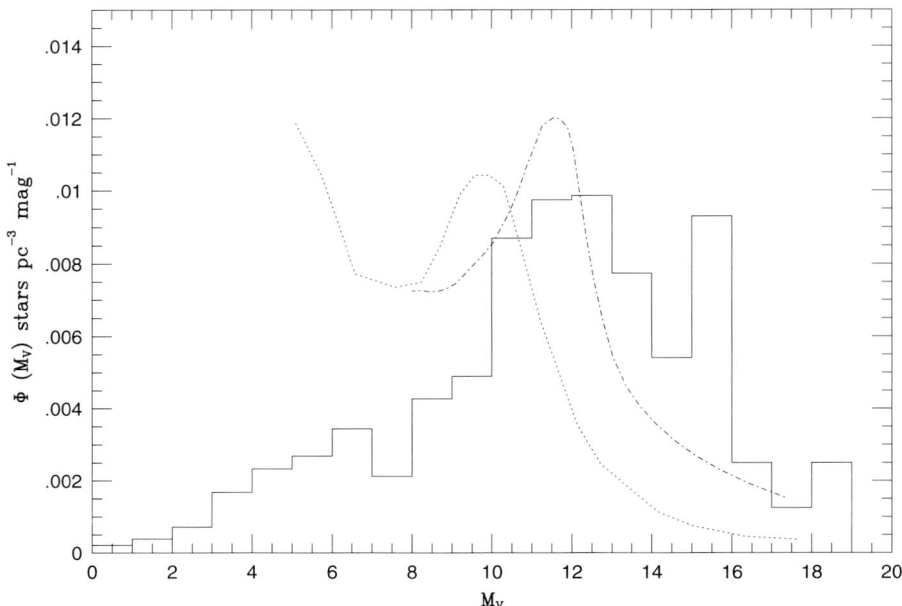

Figure 8.6. A comparison of the derivatives of the [B1] (dotted line) and [K3] (dash–dot) (mass, M_V) relations $(d(\text{mass})/dM_V)$, and $\Phi(M_V)$ for the nearby stars. The peak in the latter corresponds approximately with maxima in the former.

that these M dwarfs are fusion-powered, and that the stellar/substellar boundary lies at $M_V > 19$ and $M_K > 10.5$. Chapter 9 extends the discussion to these lower-mass objects.

8.4.2 Mass–luminosity relationship for pre-main sequence stars

There are currently no reliable empirical mass estimates for individual T Tauri stars to provide direct calibration of the mass–luminosity relationship during the pre-main-sequence phase. Binaries appear to be more common among some pre-main-sequence stars than amongst even G dwarfs in the solar neighbourhood, with multiplicity fractions rising to close to 100% in the Taurus cloud [G3], [G4]. There are, however, significant variations between star-forming regions, perhaps indicating that environmental factors influence binary formation. In any case, the nearest regions, in the Taurus cloud and Orion, lie at distances of 150–500 parsecs, which limits even speckle techniques to systems with minimum separations of $\sim 20\,\text{AU}$. Orbits can be determined for high-mass (shorter-period) systems, but to date only incomplete results are available [M3]. The relatively small number of spectroscopic binaries with preliminary orbital analyses are also drawn from high-mass T Tauri stars in even the nearby Taurus cloud.

Lacking empirical data, studies which aim at determining mass functions for young $(\tau < 15\,\text{Myr})$ clusters must rely on theoretical stellar models. There are, however, techniques which can use multiple pre-main sequence systems to test at least some aspects of these models, provided that the systems are spatially resolved.

Colour–magnitude data for intermediate-age and old open clusters, such as the Pleiades and Hyades, can be matched against theoretical predictions, since stars in these clusters are effectively coeval. Similar data for extremely young clusters show substantially more dispersion, even after allowing for effects due to differential reddening. This probably reflects the fact that star formation is not an instantaneous event, but persists for 2–3 Myr within a given region – a small fraction of the age of a cluster like the Pleiades, but a substantial fraction of the whole in an active star-forming region. The increased dispersion in the H–R diagram precludes a simple test of pre-main sequence models. However, components of a binary (or multiple) star system can be expected to form over a much shorter period – effectively, simultaneously. If temperatures and luminosities of each component can be determined, and if the distance is known with moderate accuracy, then all are expected to fall on the same isochrone.

The first studies of this type were undertaken by Hartigan *et al.* [H3], who analysed data for 39 multiple star systems, one of which is the interesting quadruple, GG Tau (recently reanalysed [W4]. The four T Tauri stars in that system form a double binary, with GG Tau Aa and Ab, separated by 0″.25, or 35 AU, lying 10″.1 from the 1″.48 (207 AU) separation, lower-luminosity pair, GG Tau Ba and Bb. All have radial velocities consistent with their forming a single system. The brighter pair have spectroscopic observations by the Faint Object Spectrograph on the HST, while the wider pair were observed using the Keck 10-m telescopes, allowing estimation of the effective temperatures; luminosities are known from infrared photometry. The former parameters are somewhat problematic for the lower-luminosity pair Ba and Bb, which have spectral types of M5 and M7. As described in Chapter 4 (and illustrated in Figure 2.8), the lower gravity of T Tauri stars leads to their having optical spectra resembling M giants. Mid-to-late-type M giants have higher temperatures, by 200–300 K, than M dwarfs of the same spectral type, and it remains unclear whether the dwarf or giant temperature scale is more appropriate for pre-main sequence stars.

Figure 8.7 compares empirically-derived parameters for the GG Tau quadruple against two sets of isochrones (from [B1]). The models differ in the value of the mixing-length factor (also denoted as α; see Chapter 4) adopted for stars more massive than $0.6 M_\odot$: $\alpha = 1$, the standard model; or $\alpha = 1.9$, favoured by more recent studies. The figure also shows the temperature range bracketed by the dwarf and giant scales for the components of the low-mass binary. In either case, all four stars can be matched against a single isochrone, with ages of 5 Myr ($\alpha = 1$) or 2 Myr ($\alpha = 1.9$, $M < 0.6 M_\odot$) by choosing appropriate intermediate values of T_{eff} for the low-mass pair. Both analyses imply a mass close to $0.05 M_\odot$ for the lowest-mass object in the system, significantly below the H-burning limit. Once a reliable calibration of the temperature scale for low-mass T Tauri dwarfs becomes available, this technique, applied to other systems, will provide more stringent tests of the accuracy of pre-main sequence models.

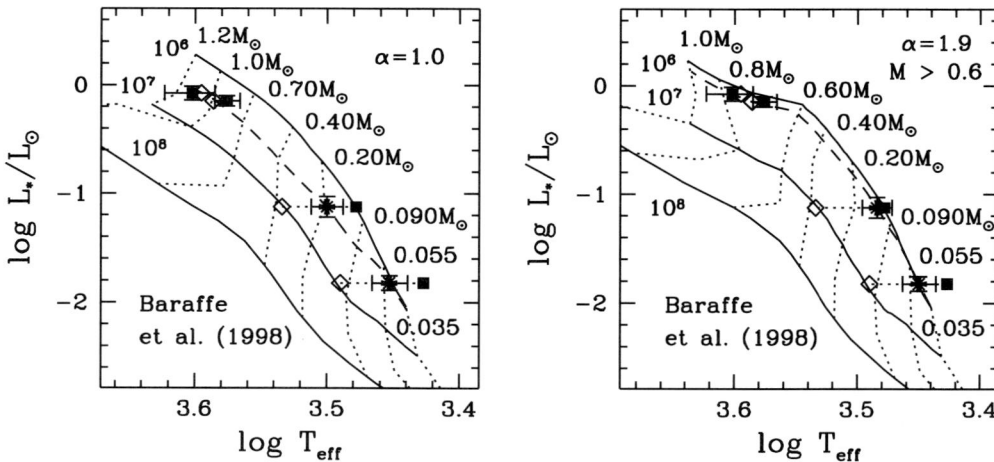

Figure 8.7. A comparison of the location of the four components of the pre-main sequence quadruple GG Tauri against theoretical isochrones computed by Baraffe *et al.* All four stars are expected to fall on the same isochrone. The calculations in the left panel are for a mixing length (α) of 1.0; the right panel shows models with $\alpha = 1.9$ for masses above 0.6 $M_\odot$. The solid square and open diamond mark the upper (giant) and lower (dwarf) temperature estimates for the low-mass pair; the asterisks mark the inferred temperatures based on extrapolating the isochrone which best matches the higher-mass pair. (From [W4], reproduced courtesy of the *Astrophysical Journal*.)

8.5   THE FIELD-STAR MASS FUNCTION

Investigations of $\Psi(M)$ in the general Galactic field can be placed under two headings: statistical determinations, which compute the luminosity function as an intermediate step; and more direct analyses, which use the mass–luminosity relationship to estimate individual stellar masses before combining the data to estimate $\Psi(M)$. Both approaches are liable to similar systematic uncertainties arising from uncertainties in the MLR. Rebinning luminosity function data, however, can smooth over important features, while the star-by-star approach offers the potential for detailed assessment. Regardless of the approach adopted, the most crucial step in the process lies in defining the properties of the sample being analysed. Most erroneous mass function determinations have their origin in a misunderstanding of the properties of the parent sample.

8.5.1   Statistical analyses of $\Psi(M)$

The majority of published studies derive $\Psi(M)$ from luminosity function data, usually $\Phi(M_V)$, either through direct application of the mass–luminosity relationship or through the inverse approach of adjusting $\Psi(M)$ (via the MLR) to match $\Phi(M_V)$. Most also reach erroneous conclusions regarding the form of $\Psi(M)$, identifying a maximum at a mass of $\sim 0.25\,M_\odot$ – a result which can be traced to various inaccuracies in the adopted luminosity functions.

In their influential paper, Miller and Scalo [M7] use the direct method to transform $\Phi(M_V)$ to $\xi(\log M)$, adopting a variety of star-formation histories and age estimates for the Galactic disk. Several functional forms were used to represent their results, with the most frequently cited being the log-normal form,

$$\xi(\log M) = C_0 \exp[-C_1(\log M - C_2)^2] \qquad (8.33)$$

where $C_0 = 106.0$, $C_1 = 1.09$ and $C_2 = -1.02$ for a constant stellar birth-rate in a 12-Gyr old Galactic disk. The Miller–Scalo function is a close match to a power-law with index $\alpha = 2.5$ for masses between 1 and $10\,M_\odot$. However, unlike a power-law representation, the log-normal form reaches a maximum density (here at $\sim 0.25\,M_\odot$) and declines thereafter. In averaging the available luminosity function data to derive the $\Phi(M_V)$ which underlies their mass function determination, Miller and Scalo follow Luyten in smoothing over the Wielen dip, and also underestimate the number density of stars at $M_V \sim 12$. The former error was corrected by Scalo [S2] in his reanalysis of the problem, but both he and Rana [R1] adopt (mass, M_V) relationships which fail to take into account the significant changes in slope below $0.6\,M_\odot$ [K4]. Their smooth relationships redistribute stars in mass to give the apparent maximum of $\Psi(M)$.

The photometric parallax studies of the luminosity function described in Chapter 7 use direct calibration to estimate $\Psi(M)$. The resulting mass functions are in general agreement in showing a peak in number density at $\sim 0.25\,M_\odot$, a subsequent decline, and the suggestion of a rising function near the hydrogen-burning limit [R3], [H4], [S7], [T1]. Unfortunately, the good agreement stems from incorporating the same systematic error – an absolute-magnitude calibration which smoothes over the 'step' in the main sequence at $(V–I) \sim 2.9$. As was demonstrated in Chapter 7 (Figure 7.11), using a linear $(M_V, V–I)$ relationship to estimate absolute magnitudes in that region leads to stars being misplaced within $\Phi(M_V)$, enhancing the peak at $M_V \sim 12$, and decreasing the apparent numbers of fainter stars. Thus, even though the (mass, M_V) relationship is more accurate than the Miller/Scalo relationship, the final result is similarly incorrect.

Kroupa *et al.* [K2], [K3] pioneered the use of the inverse approach, varying $\Psi(M)$ to match a fiducial luminosity function. Their later study takes as its reference a smoothed version of the [W5] nearby-star luminosity function, supplemented by the results for VLM dwarfs within 5.2 parsecs [D1]. Using this as a constraint, they adjust the form of the mass function to minimise χ^2 in the residuals between the required (mass, M_V) relationship and empirical data on binary star masses (from [P4]). Fitting a three-segment power-law, they derive

$$\alpha = 2.7, M > 1.0\,M_\odot; \quad \alpha = 2.2, 1.0 > M > 0.5\,M_\odot; \quad \alpha = 1.3, M < 0.5\,M_\odot \qquad (8.34)$$

The change in slope at $\sim 0.5\,M_\odot$ is almost certainly introduced by systematic incompleteness in the reference luminosity function. Adopting a similar approach,

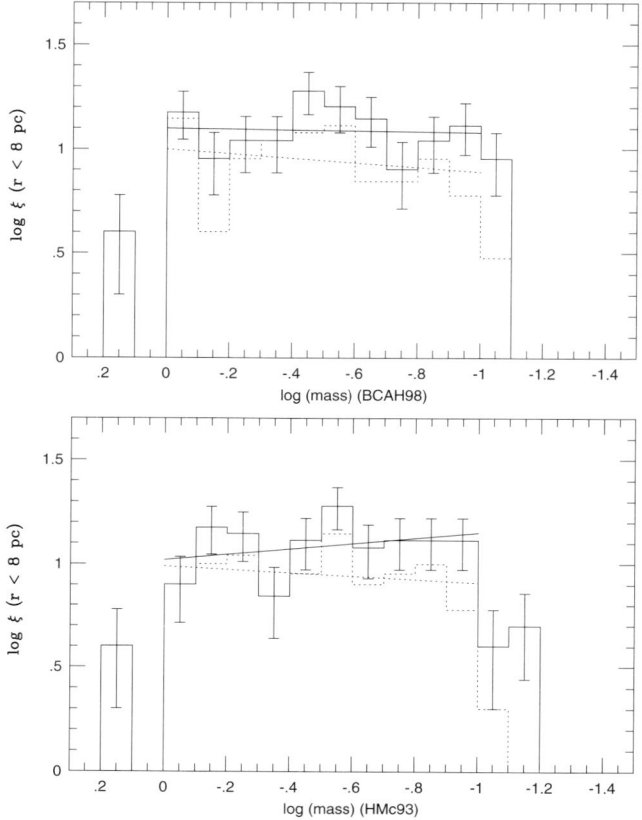

Figure 8.8. The nearby-star mass function. The Henry and McCarthy [H6] empirical (mass, M_K) relationship is adopted in determining $\Psi(M)$ in the lower panel; Baraffe *et al.* [B1] 1-Gyr models are used in the upper panel. In both cases the dotted lines plot the systemic mass function. The best-fit power-law relationships are given in the text.

a single power-law fit to the mass range 0.1–1 $M_\odot$ gives limits of $0.3 < \alpha < 0.9$, constrained mainly at the lowest masses [H5]. Reanalysis of the [K3] data resulted in a somewhat flatter relationship than the original study, with $0.66 < \alpha < 1.44$ for masses below $0.5\,M_\odot$ [K4].

8.5.2   The mass function for nearby stars

The most straightforward method of deriving $\Psi(M)$ is to define a volume-limited stellar sample and determine masses on a star-by-star basis. Figure 8.8 plots the results of this exercise as applied to the northern 8-parsec sample (excluding white dwarfs), where masses have been computed both using Henry and McCarthy's empirical (mass, M_K) relationship and the 1-Gyr (mass, M_K) isochrone from the

Lyon models. M_K is likely to be a more reliable mass indicator than passbands at shorter wavelengths, since the flux in the K band is less strongly dependent on abundance, and therefore provides a better tracer of the bolometric luminosity. In addition to $\Psi(M)$ defined by all main-sequence stars, we also plot $\Psi(M)$ for systems – single stars and primary stars in binaries or multiple systems.

There is reasonable agreement between the two mass functions plotted in Figure 8.8. Both are well matched by a linear relation (a power-law in $\Psi(M)$) over the 0.1–1.0 $M_\odot$ mass range. Indeed, the theoretical calibration suggests that this power-law can be extended to lower masses. The best-fit relationships plotted are

$$\log \xi(\log M) = -(0.13 \pm 0.14) \log M + (1.02 \pm 0.08) \tag{8.35}$$

for the empirical calibration, and

$$\log \xi(\log M) = (0.02 \pm 0.14) \log M + (1.10 \pm 0.08) \tag{8.36}$$

for the theoretical calibration. Both are in units of stars per $0.1 \log M$ for a volume equivalent to the northern 8-parsec sample (1,608 pc^3). These relationships correspond to linear mass functions $\Psi(M) \propto M^{-1.13}$ and $\propto M^{-0.98}$ respectively (that is, $\alpha \sim 1$), with uncertainties of ± 0.15 in the derived slopes.

8.5.3   The disk mass function

These results can be combined to determine a current best estimate for the form of the disk mass function. Eliminating photometric-parallax analyses as unreliable due to the calibration problems near $M_V \sim 12$, the remaining studies rest on nearby-star samples to establish their normalisations. Those studies represent the mass function as either a two-segment or three-segment power-law, with recent results favouring the simpler model: Figure 8.8 shows no evidence for a significant change in slope at $\sim 0.5 \, M_\odot$. The exact form of $\Psi(M)$ above 0.4 $M_\odot$ can be derived from stars in the Hipparcos catalogue once the necessary supplementary information (notably chemical composition and binarity) is available for the $\sim$4,000 G and K dwarfs identified within 30 parsecs of the Sun. For the present, there is no evidence for statistically significant deviations from a single power-law approximation over the 0.1–1 $M_\odot$ mass range.

The 8-parsec analysis gives the present-day mass function, representing a summation of the complete history of star formation for stars less massive than 1 $M_\odot$. As discussed in Chapter 6, there are currently only limited constraints on the variability of the stellar birthrate over the history of the disk; the available data are consistent with a constant star formation rate. Even if that assumption were invalid, the following relationship represent the average level of star formation over the history of the disk:

$$\Psi(M) = 0.035 \times M^{-2.5\pm0.3} \, T_D^{-1} \, \text{stars pc}^{-3} \, M_\odot^{-1} \, \text{Gyr}^{-1}, \quad M \geq 1.0 \, M_\odot \tag{8.37}$$

$$\Psi(M) = 0.035 \times M^{-1.05\pm0.15} \, T_D^{-1} \, \text{stars pc}^{-3} \, M_\odot^{-1} \, \text{Gyr}^{-1}, \quad M \leq 1.0 \, M_\odot \tag{8.38}$$

where T_D is the age of the disk in Gyrs, and the normalisation matches the local density of stars in the 8-parsec sample.

The significant steepening in $\Psi(M)$ at $1\,M_\odot$ is almost certainly due to stellar evolution rather than to an intrinsic property of star-formation. Stars with masses only slightly greater than $1\,M_\odot$ have main-sequence lifetimes which are shorter than the probable age of the disk, and the first generations have evolved off the main sequence. As discussed in the following section, results for the Pleiades cluster suggest that the $\alpha \sim 1$ region of the initial mass function may extend to $\sim 1.5\,M_\odot$.

Equations (8.37) and (8.38) should not be taken as an assertion that the mass function *is* a power-law. It is not yet clear whether there is any underlying physical significance to the functional form chosen, or whether this is only a convenient mathematical fitting function. We merely note that power-laws provide an adequate representation of the available data. These relationships do, however, provide a baseline for comparison both with data drawn from other environments within our Galaxy, and with results for other Local Group galaxies. Studies of individual star-forming regions, open clusters and associations are required to assess the extent of any departures from the mean.

8.6   MASS FUNCTIONS OF OPEN CLUSTERS

Reliable mass functions can obviously be derived only for open clusters with reliable luminosity functions. For detailed surveys of low-mass stars, the cluster must be near enough to allow observations to extend to close to the hydrogen-burning limit, and sufficiently young that mass segregation and evaporation have not modified the cluster mass function significantly from its original values. The only cluster which both meets these criteria and has received the necessary observational attention is the Pleiades, although both the Hyades and, to a lesser extent, Praesepe, offer some insight into the dynamical evolutionary processes.

8.6.1   The Pleiades and intermediate-age open clusters

With an age of ~ 125 Myr and a distance of only ~ 132 parsecs, the Pleiades is an ideal cluster for mass function analysis. Observations extend below the hydrogen-burning limit, and the cluster is sufficiently young that mass segregation appears to be limited to high mass (~ 2–$5\,M_\odot$) A and B stars, probably reflecting initial conditions rather than dynamical evolution (Section 7.6). The age of the cluster, however, is sufficient that stars with masses above $\sim 0.4\,M_\odot$ have reached their main-sequence configurations, and the field mass–luminosity relationship (Figure 8.5) should be appropriate for estimating masses for those stars. Figure 8.9 compares the (mass, M_V) relationship predicted by the Lyon models for ages of 10^8, 10^9 and 10^{10} years. While there are relatively slight differences between the 1 and 10 Gyr isochrones, 100-Myr-old $\sim 0.1\,M_\odot$ stars are expected to be overluminous by a factor of ~ 2.

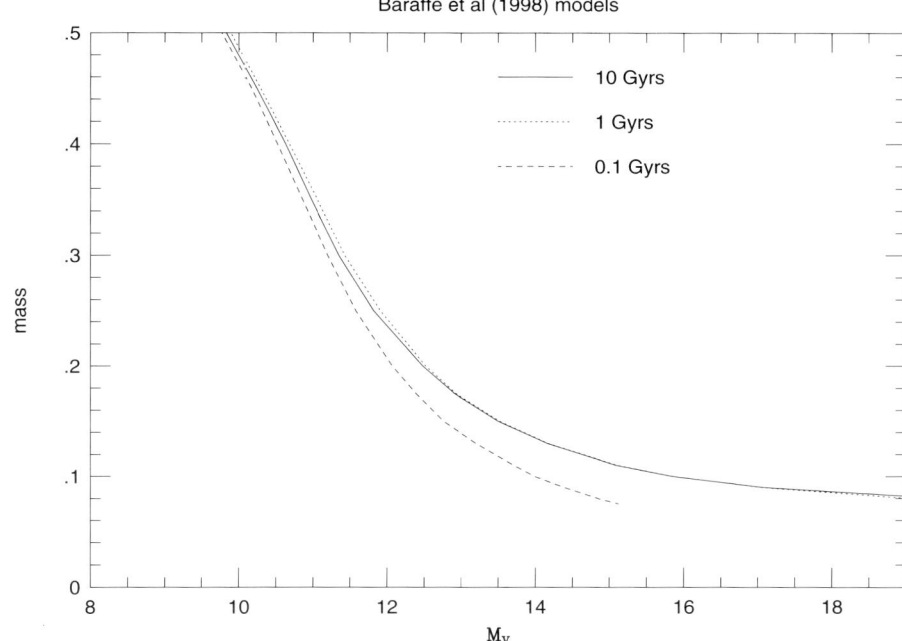

Figure 8.9. Theoretical (mass, M_V) relationships for ages of 0.1, 1 and 10 Gyr. The latter relationships are almost indistinguishable. (Models from [B1].)

There is no single, comprehensive survey of the Pleiades spanning the full mass range of cluster members. However, it is possible to combine results from the Meusinger *et al.* [M5] proper-motion survey – which is complete to $M_V \sim 12$ ($\sim 0.3\,M_\odot$) within its 16.5 square degree coverage – with proper-motion data for cluster K and M dwarfs [S6], [H1]. In their analysis, [M5] use eight different sets of main-sequence and pre-main sequence isochrones to convert their derived $\Phi(M_V)$ to a mass function. The individual results are in good agreement for masses above $0.6\,M_\odot$, and Figure 8.10 plots the weighted average.

The other proper-motion surveys aimed specifically at detecting lower-mass members, and reach luminosities close to the hydrogen-burning limit. All of the stars have VRI photometry, allowing reliable calculation of M_{bol}, and estimate masses [R5]. The resulting mass function is also plotted in Figure 8.10, scaled to match the Meusinger *et al.* results in the region of overlap.

When combined, these two analyses suggest that $\Psi(M)$ in the Pleiades is closely similar to the field-star function plotted in Figure 8.8. The number density distribution for high-mass stars (~ 1.2–$4\,M_\odot$) matches a power-law, $\alpha \sim 3$, while at lower masses the data are broadly consistent with $\Psi(M) \propto M^{-1.1\pm0.2}$. The turnover at $\sim 0.25\,M_\odot$ is clearly due to incompleteness, and the same probably holds for the lowest-mass binaries in the Reid $\Psi(M)$. As discussed in the following chapter, recent surveys of the Pleiades have extended coverage to substellar masses.

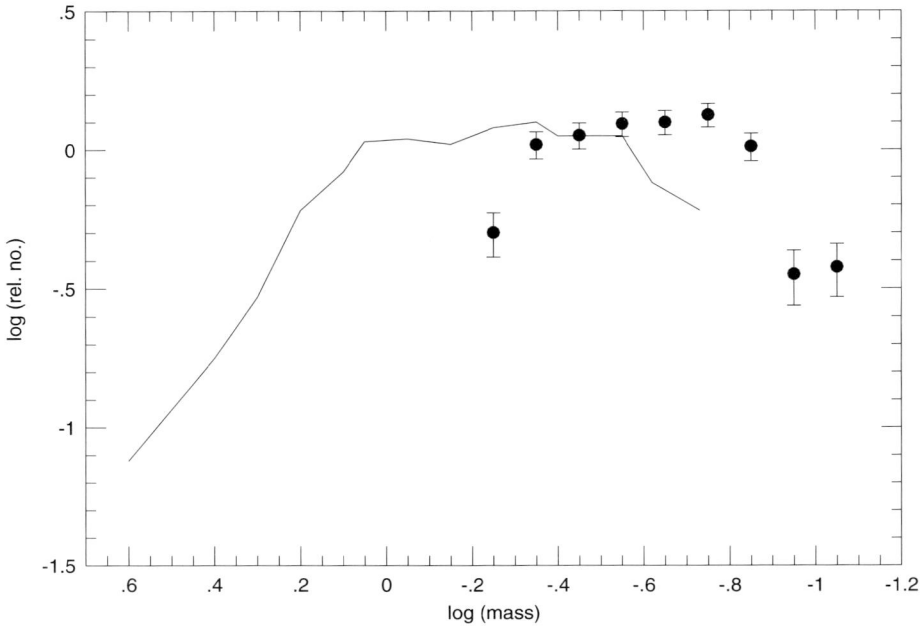

Figure 8.10. The Pleiades mass function, derived by combining the data for stars with $M_V < 12$ [M5] (solid line) with analysis of lower-luminosity K and M dwarfs [R5] (solid points).

There is evidence that some young open clusters have mass functions which, unlike the Pleiades, do not match $\Psi(M)$ for the field. In particular, Figure 7.18 indicates that $\Phi(M_V)$ peaks at $M_V \sim 6$ in both IC 2391 and IC 2602, implying a maximum at $\sim 0.4\, M_\odot$ in $\Psi(M)$. This suggests that there are environments where the star-formation process becomes less efficient in producing low-mass stars. Both of these clusters lie at distances of only ~ 160 parsecs from the Sun, but are in highly-crowded regions of the Galactic Plane, and the current results may reflect incomplete membership catalogues. More thorough surveys for faint cluster members are currently in progress.

8.6.2   The Hyades and older clusters

The mass segregation present in clusters older than a few hundred Myrs not only renders it difficult to derive an unbiased estimate of $\Psi^i(M)$, but also presents an obstacle to the determination of even the present-day mass function. As discussed in Chapter 7, few clusters have been surveyed in sufficient detail to allow unambiguous identification of complete samples of members at large distances from the cluster centre – radii where the contrast against the field is extremely low. Since those outlying members have a mass distribution skewed towards lower masses, the tendency is to underestimate the relative number of later-type stars in those clusters.

At the present time, the Hyades cluster is the only intermediate-age cluster that has been well-surveyed to large distances from the centre. Outlying low-mass members have been identified through both proper-motion and photometric surveys. The latter surveys, including an analysis based on near-infrared data from the 2MASS survey (see Chapter 9), are capable of detecting Hyades members with masses below the hydrogen-burning limit. None have been detected – the lowest-mass member known currently has M $\sim 0.083\,M_\odot$ (from [R6]). Figure 8.11 outlines the cluster mass function derived by combining all the available data. Fitting a single power-law to $\Psi(M)$ leads to an index of $\alpha \sim 0.5$ – flatter than the field-star function – and it is clear that the distribution drops even more sharply towards the low-mass limit. Reconstructing the initial mass distribution from these data is a task beyond current dynamical models.

In contrast, surveys of Praesepe – a cluster of similar age and abundance to the Hyades – suggest that the mass function is a closer match to the field. Hambly *et al.* [H2] derive a best-fit power-law index of $\alpha = 1.5$ based on their photographic survey, while Williams *et al.* [W6] find $\alpha = 1.34 \pm 0.25$ from deeper CCD data covering a smaller area of the cluster. Pinfield *et al.* [P1] have recently extended analysis to even fainter magnitudes and derive a similar result. None of these surveys is based on a

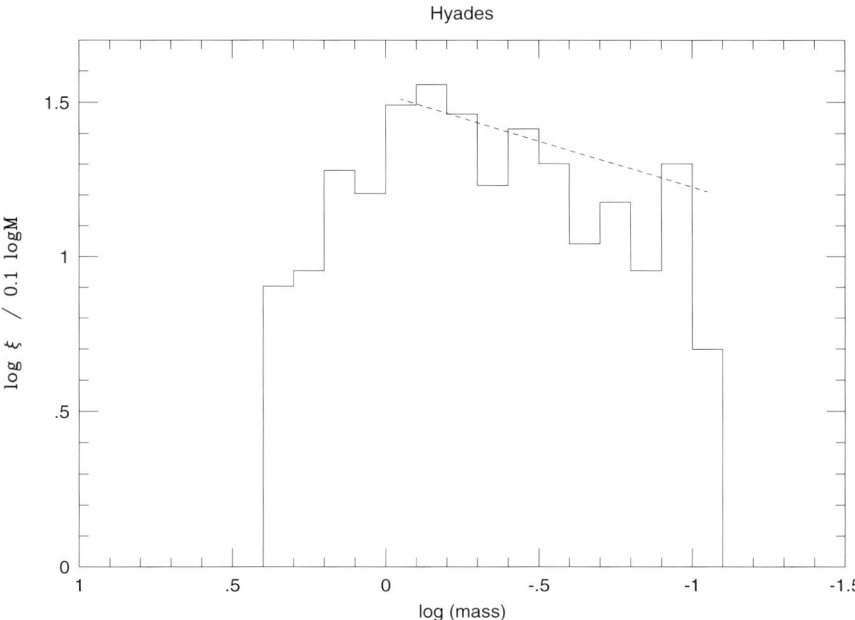

Figure 8.11. The Hyades mass function. Several surveys have been combined (see Chapter 7) as described in [R6].

spectroscopically-confirmed sample, and experience has shown that the addition of those data will reduce the number of cluster members. It is unlikely, however, that cluster membership will be depleted enough to match the Hyades observations. Thus, despite their similarities in age and abundance, there appear to be significant differences in the present-day mass functions of the Hyades and Praesepe clusters.

8.7 PROTOCLUSTERS AND STAR-FORMING REGIONS

Surveys of clusters younger than $\sim 20\,\mathrm{Myr}$ offer both advantages and disadvantages for studies of $\Psi(M)$ at subsolar masses (see [M6]). On the positive side, low-mass stars are at their most luminous during this phase of pre-main-sequence contraction (Chapter 3). A $0.3\,M_\odot$ star has a luminosity of $\sim 0.1\,L_\odot$ at age $10\,\mathrm{Myr}$, as compared with a luminosity of $\sim 0.013\,L_\odot$ at $100\,\mathrm{Myr}$, or $\sim 0.008\,L_\odot$ on the main sequence. These objects can therefore be identified at substantially greater distances from the Sun than is possible for their older counterparts in Pleiades-like clusters.

Young clusters offer the further advantage of being largely free of the effects of dynamical evolution. Some degree of mass segregation exists in even the youngest systems, with the highest-mass stars often more centrally concentrated than the average cluster member. Given typical crossing times of between a few $\times 10^5$–$10^6\,\mathrm{yr}$, this concentration is more likely to be a result of the conditions prevailing during the earliest stages of star formation rather than a consequence of dynamical relaxation. Solar-type and lower-mass stars appear to follow nearly identical radial density distributions (as in the Pleiades), thereby minimising potential biases due to incomplete areal coverage.

Youth, however, also carries its disadvantages. Since the cluster members are still in the pre-main sequence contraction stage, and no empirical mass estimates are yet available (Section 8.4.2), theoretical calculations offer the only method of estimating a mass–luminosity relationship. Applying these calibrations demands reliable age estimates, which are complicated by the fact that clusters do not form instantaneously: that is, cluster members span a range of ages. The resultant uncertainty in individual ages leads to corresponding uncertainty in mass, which is highest at these young ages when stars evolve so rapidly.

Further complications can arise from the presence of circumstellar disks in some systems. The material in those disks can dominate the energy distribution in the mid- and far-infrared, and, given sufficiently high temperatures, may even make a significant contribution to the observed flux at shorter wavelengths, leading to an overestimate of both the photospheric bolometric luminosity and the mass. Finally, and most importantly, protoclusters are still embedded within the remains of the parent molecular cloud. The highly-variable obscuration caused by dust within that cloud is a serious impediment to an accurate interpretation of observations, at least at optical wavelengths.

Dust obscuration amounts to a loss of a few magnitudes at visual wavelengths for a typical few-Myr-old star cluster (such as IC 348, [L2]), and can reach levels of $A_V \sim 20$–50 magnitudes, or more, in the denser regions of embedded protoclusters such as NGC 2024 (in Orion) or ρ Ophiuchi. The absorption is variable on scales of $\sim 1,000$–$10,000$ AU, or ~ 1–10 arcsec for the nearer clusters. Under such circumstances, optical surveys are capable of identifying only the most luminous and least obscured cluster members, and are therefore poorly suited to providing catalogues for statistical analysis. However, the scattering properties of interstellar dust lead to significantly less absorption at longer wavelengths. In particular, the absorption at $2.2\,\mu$m, A_K, is almost a factor of 10 less than that at $0.5\,\mu$m. Typical cloud temperatures are between 50 and 100 K, so dust *emission* peaks at $\lambda \sim 60\,\mu$m, but is negligible at near-infrared wavelengths.

Given these circumstances, star-formation regions have long been recognised as interesting targets for infrared observations. However, while even the earliest scans led to notable discoveries – such as massive protostars like the Becklin–Neugebauer object and the surrounding Kleinmann–Low nebula in Orion [B2], [K1] – initial surveys were limited to either bright sources or small solid angles, and often both. It is only with the development in the mid-1990s of large-format infrared arrays and high-sensitivity spectrographs that it has become possible to undertake studies capable of detecting protostars with masses below $0.1\,M_\odot$ over the entire area of major star-formation regions. As a result, the full potential of infrared studies remains to be realised. Nonetheless, preliminary results are intriguing.

In general, investigations of the stellar mass function in young, star-forming regions have followed two broadly complementary lines of attack: statistical analysis of deep starcounts, and more detailed source-by-source analyses.

8.7.1 Infrared imaging surveys

Near-infrared number–magnitude counts offer a straightforward means of probing the stellar content in obscured, star-forming regions. The luminosity function of cluster members can be determined statistically by comparing source counts centred on the cluster against counts made within nearby, off-cluster fields. If the cluster distance is known, the apparent luminosity function $\phi(m_K)$ can be transformed to $\Phi(M_K)$. Since this technique is based on direct imaging, it offers the possibility of obtaining a complete census of even the lowest-luminosity cluster members through a series of simple and efficient observations. With the current generation of infrared CCDs, data can be obtained covering an entire cluster in a matter of only a few nights on an intermediate-sized telescope.

There are, however, complications in deriving a mass function from the resultant K-band luminosity function:

- The (mass, M_K) relationship must be appropriate for the age of the cluster, τ_C, and therefore requires both an accurate estimate of τ_C and reliable pre-main-sequence mass–luminosity relationships.
- The cluster stars are unlikely to be exactly coeval. Rather than the cluster forming in a single burst at time $T = T_0 - \tau_C$, where T_0 is the present time, individual stars span a range of ages, $\tau_i = \tau_C \pm \Delta\tau$.

- While working in the near-infrared minimises the effects of obscuration, differential reddening (either foreground or within the cluster itself) is likely to be present at the 0.1–1.0 magnitude level in A_K. Moreover, there may be significant differences between the total obscuration in cluster and off-cluster fields: dust within the young cluster usually leads to higher reddening of background stars. This can produce systematic errors in $\phi(m_K)$.
- Emission from circumstellar disks can contribute significantly to the flux in the thermal infrared ($\lambda > 2\,\mu m$) in young protostars.
- Finally, source counts alone cannot distinguish single and multiple stars.

Many of these problems can be addressed: optical and near-infrared colours can be used to probe the extent of differential reddening, while excess radiation at longer wavelengths (above the predicted photospheric flux) can be used to assess possible contributions from hot circumstellar dust.

The usual technique is to compute the expected K-band luminosity function (KLF) based on an estimated initial mass function, age and star-formation history. The last two parameters can be determined to some extent from photometric and/or spectroscopic data, although a degree of guess-work is also often required. The predicted luminosity function is matched against the observed KLF (in the observational plane), and the input parameters adjusted until reasonable concordance is achieved. Initial analyses [Z2] were based on single-burst star-formation models, but more recent studies [L2], [L3] have adopted more complex (and more realistic) star-forming histories. In general, this approach is effective at ruling out inappropriate models, but only identifies consistent (rather than unique) solutions.

8.7.2   H–R diagram analyses

An alternative method of studying young clusters is to use spectroscopic and photometric observations to estimate bolometric luminosities and effective temperatures for each cluster member. Given these data, each star can be placed on the H–R diagram, making due allowance for foreground reddening and circumstellar dust emission, while eliminating foreground and background field stars. Comparison with pre-main sequence evolutionary tracks permits the estimation of masses and ages on a star-by-star basis (see Figure 8.7), whilst the mass function and star-formation history follow from summation of the individual results.

In principle, this approach offers higher precision than the statistical KLF analysis. There are, however, significant practical obstacles – notably in obtaining spectroscopic data of the requisite accuracy for the faintest, and most highly obscured, cluster members. Simultaneous observations of tens of candidate cluster members, using optical or near-infrared multi-object spectrographs can go some way towards addressing the latter problem, but photon scarcity limits the full-scale application of this method.

8.7.3   Two case studies: IC 348 and the Orion Nebula Cluster

To illustrate the relative merits of these analyses we consider their application to IC 348 and the Orion Nebula Cluster (the Trapezium), two well-studied, young star-forming regions. Observations of these clusters demonstrate the complementary nature of the techniques.

IC 348

This cluster lies at a distance of $\sim 320\,\mathrm{pc}$ $((m-M)_0 = 7.5$ magnitudes) in the Perseus Molecular Cloud. Originally catalogued by Dreyer, the cluster has 40–50 members detectable at optical wavelengths, and the brightest star (type B5) is surrounded by a prominent reflection nebula (Figure 8.12). Herbig's [H8] spectroscopic observations of strong $H\alpha$ emission in 16 stars within this region confirmed the presence of a young cluster. Infrared scans [S8] reveal a moderate to large population of optically-invisible sources, embedded in what is presumably the remnants of the parent molecular cloud [K6]. The general characteristics point to a young age for the system – initially estimated as between 5 and 20 Myrs [S8]. More recent studies [H9], [L6], [L7] find that most of the stars have ages between 1 and 5 Myr.

Lada and Lada [L2] surveyed the central 0.1 square degree of this cluster, reaching limiting magnitudes of 16.5, 15.5 and 14.5 in the J, H and K passbands. Of the >600 sources detected, most are concentrated in a relatively small region

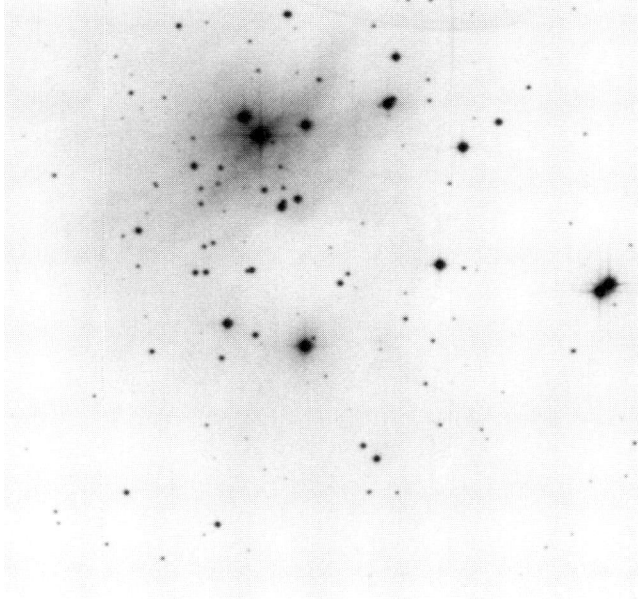

Figure 8.12. An R-band image of the cluster IC 348. The field is 10 arcmin square. (Courtesy of Palomar Observatory/STScI.)

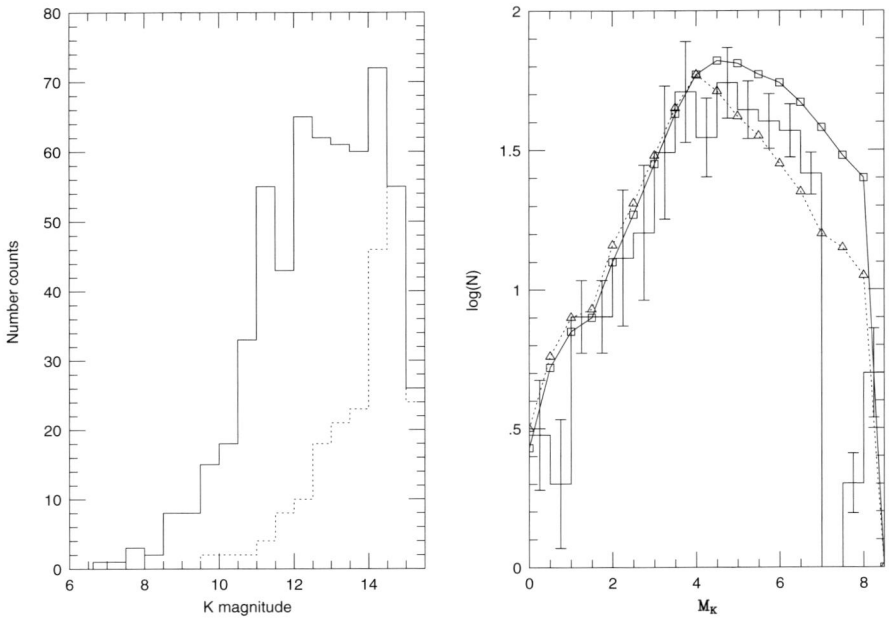

Figure 8.13. (a) K-band source counts towards IC 348. The dotted histogram shows the likely contribution from background/foreground sources [L2]; (b) the K-band luminosity function (histogram) matched against predictions based on a log-normal mass function (open triangles) and a truncated function (open squares – from [L3]).

south of o Persei, where the source density rises to $\sim$230 stars pc^{-2}, equivalent to a mean stellar separation of $\sim$0.25 pc. The contamination from background and foreground stars is estimated through observations of control fields, offset by 1° from the cluster centre. The resulting K-band luminosity function (Figure 8.13) has a broad peak at $M_K \sim 4.5$, or $K = 11.5$, well above the limiting magnitude of the survey. As in studies of other young clusters, the brighter source counts follow a near-power law distribution with magnitude, index 0.38 at K and 0.32 at J. This is much steeper than that measured for either intermediate-age open clusters or field stars at similar magnitudes and reflects a compression of the mass–luminosity relationship: a smaller interval in L spans a larger range in M.

Rather than attempting a direct transformation of $\Phi(M_K)$ to $\Psi(M)$, [L2] and [L3] invert the analysis, using computer simulations to predict $\Phi(M_K)$ given $\Psi(M)$ and a star-formation history. Starting with a Miller–Scalo log-normal mass function, pre-main sequence tracks from [D2] are used to estimate (L, T_{eff}) as a function of (mass, age). Transforming to M_K via temperature-dependent bolometric corrections, the resulting KLFs are compared to the observed luminosity function. The star-count data match models in which star formation has progressed at a relatively uniform rate over the last 5–10 Myr [L4]; single-burst (coeval) star-formation models produce luminosity functions which are either too narrow or fail to match the observed slope of the source counts at bright magnitudes. Recent spectroscopic analysis of cluster members [L6], [L7] has refined these time-scales: as indicated above, the majority of the cluster stars prove to be younger than ~ 3 Myr. Results favour a mass function matching the Miller–Scalo formulation to $\sim 0.3\, M_\odot$, but better represented by a power-law $0.3 < \alpha < 0.8$ at lower masses.

The Orion Nebula Cluster

This is the young star cluster surrounding θ^1 Ori – the four O-type stars of the Trapezium which ionise the Orion nebula. Lying at a distance of 470 ± 70 pc [G1], the Orion Nebula Cluster (ONC) is one of the most prominent of a series of active star-forming regions in the Orion OB association [G2]. Immediately behind the cluster lies the L1640 molecular cloud whose very substantial density leads to extinctions of $A_V > 80$ magnitudes, effectively eliminating any contribution by background stars to even near-infrared source counts. The ONC itself is ~ 3 parsecs in diameter, and lies primarily within the low gas-density region excavated by winds from the Trapezium stars (the overall morphology of the area is described by Zuckerman [Z3]). Proper-motion studies [J1] show that the overwhelming majority of stars within ~ 20 arcmin of the Trapezium are cluster members, simplifying source-count analysis. The star density rises to $\sim 20,000$ stars pc^{-3} in the central ~ 0.3 parsecs.

The Orion complex has been the subject of numerous spectroscopic and photometric investigations of both the stellar and gaseous content (see [G2] and [H10] for summaries), including recent high-resolution Hubble Space Telescope imaging [P5] which provides direct observations of the 'silhouettes' of circumstellar disks ('proplyds' – Figure 3.19, in colour section) around a number of cluster stars. The cluster was an early target of infrared observations [A4]; the relatively narrow KLF can be modeled as a single burst of star formation, aged 10^6 years, with a Miller–Scalo mass function [L2].

The most detailed study of the ONC, however, has been undertaken at optical wavelengths. Hillenbrand [H10] has compiled V–I photometry for 1,600 of the 3,500 sources identified within the central 5 pc of the cluster, as well as spectroscopy for more than 980 of the brighter stars. While the photometric sample includes only 40% of the complete population, the spatial distribution of the observed sources is similar to the complete sample, and there is no evidence for significant differences between the constituent stars in the two datasets. Thus, the optically-selected sample is probably a fair subset of the population in the ONC cluster.

Hillenbrand's estimate of the ONC mass function is derived using the star-by-star approach. Effective temperatures are estimated from the spectral types, while the photometry provides M$_{bol}$, allowing each star in the spectroscopic subsample to be placed on the H–R diagram. Comparison with theoretical evolutionary tracks (Figure 8.14) provides the age and mass of each star; and the mass function and star-formation history follow by combining those results. As in the star-count analyses, age and mass calibrations rest entirely upon the theoretical models, and choosing a different set of models as reference can lead to different conclusions.

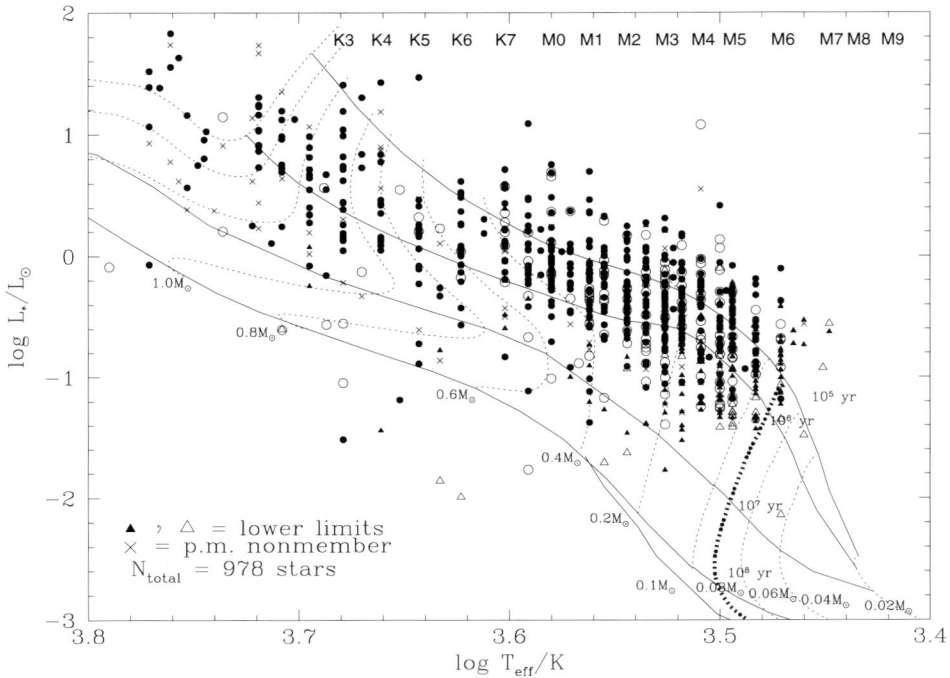

Figure 8.14. The H–R diagram for lower-luminosity stars in the Orion Nebula Cluster. (From [H10], courtesy of L. Hillenbrand and the *Astronomical Journal*.)

However, restricting the age analysis to the lower main sequence, the [D2] tracks indicate an average age of <1 Myr, with an overall spread in ages of ~ 2 Myr; the youngest stars lie nearer the cluster centre.

The derived stellar mass function (Figure 8.15) is reminiscent of the Miller–Scalo log-normal function, with the possible addition of a feature near $\sim 1\,M_\odot$: a dip in $\Psi(M)$, implying a relative scarcity of solar-like stars. Matching $\Psi(M)$ at higher masses with a power-law gives $\alpha \sim 2.5$ for $M > 0.25\,M_\odot$, consistent with the high-mass field function, but with no indication of the flattening below $1\,M_\odot$. The ONC function peaks at $\sim 0.25\,M_\odot$, with a subsequent sharp decline in number densities. While incompleteness might account at least partially for the scarcity of very low-mass stars (and brown dwarfs – see Chapter 9), the higher-luminosity stars should be fully represented in the optical sample.

Summary

These individual studies illustrate the relative merits of the two types of analysis. Infrared star-counts provide a broad overview and a statistically well-defined sample, but ambiguous results; the HR-diagram analysis permits more insight into details of the star-formation history and the exact form of the mass function. It is

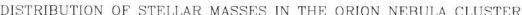

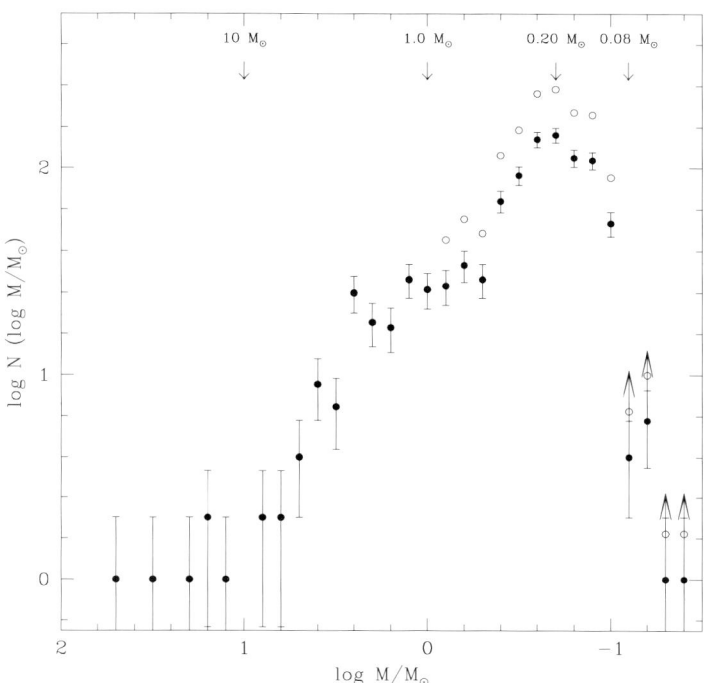

Figure 8.15. The mass function deduced by Hillenbrand for the Orion Nebula Cluster. Solid points mark the raw number counts; the data plotted as open circles include corrections for incompleteness. (From [H10], courtesy of L. Hillenbrand and the *Astronomical Journal*.)

notable that both IC 348 and the ONC appear to have mass functions which reach a maximum near $0.25 M_\odot$ – a result which suggests a preferred mass for star formation, and which is at odds with data for both the Solar Neighbourhood field stars and the Pleiades cluster. There also appears to be a relative scarcity of brown dwarfs in both young clusters (a matter further discussed in the following chapter).

The origin of the discrepancy between the form of $\Psi(M)$ in the field and in young clusters is not yet understood, but may be related to the relatively high densities and high star-formation rates present in both IC 348 and the ONC. Providing partial support for this hypothesis are results for the less active, lower total-mass ρ Ophiuchi cluster, where $\Psi(M)$ matches a power-law with $\alpha \sim 1$ from $4 M_\odot$ to beyond the hydrogen-burning limit [W7]. Lower-mass stars may also form with greater frequency (or high-mass stars with less frequency) in the more placid environment offered by dispersed star-forming regions such as the Auriga–Taurus association. Characterised by low star-formation rates, the latter regions extend over tens of parsecs, and are correspondingly more difficult (technically) to survey completely than are dense, easily identifiable clusters. All-sky near-infrared surveys currently under way (Chapter 9) will provide a means of identifying complete samples of protostars within these regions, and allow mass function determinations comparable with the IC 348 and ONC studies.

8.8   THEORETICAL INTERPRETATIONS OF THE IMF

One of the principal reasons for investigating the form of $\Psi^i(M)$, and its range of possible variations, is the fact that these results illuminate aspects of the star formation process. Considerable progress has been made in determining the different stages involved in the formation of an individual star from a cloud core (as outlined in Chapter 3). At the same time, several processes have been proposed to account for the mass spectrum of cloud cores, including coalescence of fragments (a bottom-up approach) [S4], [M8]; fragmentation and subsequent accretion (a top-down process) [A1], [Z1]; complex interaction of accretion and protostellar outflow [S5]; and fractal networks within clouds [L4], [E1]. All are qualitative to some degree, and none provide a fully acceptable explanation of why a molecular cloud should redistribute its mass to match the observed stellar mass function. A full discussion of these complex models is left to other authors.

The classical analysis of star formation began with the description of the Jeans mass, defined as

$$M_J = \left(\frac{5kT}{G\mu m_H} \right)^{3/2} \left(\frac{3}{4\pi\rho} \right)^{1/2} \tag{8.39}$$

where T and ρ are the temperature and density of the cloud, m_H is the mass of a hydrogen atom, and μ is the mean molecular weight. At this mass, gravitational forces exceed gas pressure, and the cloud collapses. If the potential energy released can be radiated away, then the Jeans mass can decrease, allowing fragmentation and further collapse. Eventually, the internal opacity becomes sufficient to allow hydrostatic equilibrium. This is opacity-limited fragmentation.

As with any fragmentation process, an increasing number of objects with decreasing mass can be expected, with the mass spectrum generally predicted to follow a power-law to a lower mass-limit of $\sim 0.01\,M_\odot$ [B5]. This is qualitatively similar to the observations, but provides no explanation as to why the power-law index should change substantially at some characteristic mass ($1.3\,M_\odot$ in the Pleiades). Moreover, Adams and Fatuzzo [A1] argue that clumpiness, turbulence and ordered motion due to stellar winds in *real* molecular clouds invalidate the general assumption of undisturbed collapse underlying this model. Indeed, they assert that 'the Jeans mass has virtually nothing to do with the masses of forming stars.'

Adams and Fatuzzo have proposed their own mechanism for organising star formation, in which the final mass spectrum is obtained almost by accident. They claim that the mass of a given protostellar core and the final mass of the stellar end-product depend on particular values of a moderate number of parameters, such as temperature, density, local turbulence, composition and so forth. If these parameters are largely independent, then applying the central limit theorem to their summed contribution predicts that the final distribution matches the log-normal Miller–Scalo mass function. Unfortunately, the latter function is no longer regarded as providing an adequate match to $\Psi^i(M)$ in the general field, although it may be more appropriate for individual star-forming regions.

8.9   THE STAR-FORMING MASS DISTRIBUTION AND THE LOCAL MASS DENSITY

Multiplying the initial mass function – the number of stars per unit mass – by mass, and normalising that function by the total mass of stars formed, gives $\Xi(M)$. As outlined in Section 8.2, this function describes the fraction of the total star-forming mass of a given molecular cloud which becomes bound up in stars of a given mass, providing a measurement of the relative efficiency of the star-formation process. In the case of the field-star IMF, $\Xi(M) \propto M^0$ for $0.1 < M/M_\odot < 1$; that is, mass is equally divided among stars in this mass range. Integrating $\bar{\Psi}(M)$, stars with masses between the hydrogen-burning limit and $1\,M_\odot$ account for $\sim 48\%$ of the total star-forming mass, while all except $\sim 1.5\%$ of the remaining mass is devoted to forming 1–$10\,M_\odot$ stars. (The probable contribution from substellar-mass brown dwarfs is discussed in the following chapter.)

Integrating $\bar{\Psi}(M)$ also allows an upper limit to be set on the contribution made by stars to the local mass density, ρ_0. Subsolar-mass stars contribute $0.036\,M_\odot\,\mathrm{pc}^{-3}$, with half of the mass density in M dwarfs. A straightforward integration of $\bar{\Psi}(M)$ for higher-mass (1–$100\,M_\odot$) stars would indicate that those stars make a similar total contribution. However, the calculation fails to take into account the fact that most of those stars have evolved off the main sequence, and have recycled much of their material in the interstellar medium. As a result, high-mass stars make a relatively small contribution to ρ_0 at any given time. In the local inter-arm region, 1–$10\,M_\odot$ stars contribute only $\sim 0.004\,M_\odot\,\mathrm{pc}^{-3}$ to the total mass density. Stellar remnants make a similarly small contribution, amounting to $\sim 0.005\,M_\odot\,\mathrm{pc}^{-3}$, giving a total stellar mass density of $0.045\,M_\odot\,\mathrm{pc}^{-3}$.

As discussed in Section 7.3, one of the main stimulants in the surge of interest in M dwarfs in the 1970s was the discrepancy between the Oort dynamical mass density and the summed contribution from known constituents of the Solar Neighbourhood: the hypothesis of 'missing mass' in the Galactic disk. How do matters stand at the turn of the century? The observed mass density remains largely unchanged: interstellar gas and dust contributes $\rho_{ism} \sim 0.03$ to $0.05\,M_\odot/\mathrm{pc}^3$ which, combined with the stellar mass density, gives $\rho_{obs} \sim 0.075$ to $0.095\,M_\odot/\mathrm{pc}^3$. Dynamical estimates, however, have changed. The most recent, based on Hipparcos data, derives a mass density of $\rho_{dyn} \sim 0.076 \pm 0.015\,M_\odot/\mathrm{pc}^3$ [C2], significantly lower than Oort's $0.15\,M_\odot/\mathrm{pc}^3$, and consistent with the observed value. Thus, there is no longer evidence for significant quantities of dark matter within the Galactic disk.

8.10   SUMMARY

We have reviewed the definition of the stellar mass function, $\Psi(M)$, and its importance in understanding star formation mechanisms and Galactic structure parameters, such as the local mass density. At present, masses can be derived directly only for stars in binary systems with well-determined orbits, although microlensing projects may offer additional data in the future. Those stars are used to calibrate empirical mass–luminosity relationships, which can then be used to derive the field star mass function from the observed luminosity function. The most recent studies indicate that the mass function for disk stars can be well represented by a power-law, $\Psi(M) \propto M^{-1}$ for masses between 0.1 and 1 $M_\odot$.

Theoretical mass–luminosity calibrations offer the only means of determining $\Psi(M)$ for young stars still in the process of contracting onto the main sequence. There are reasonable grounds for suspecting that variations in the form of $\Psi^i(M)$ occur from one star-forming region to another. The strongest support for this hypothesis comes from detailed observations of the Orion Nebula Cluster, which strongly favour a mass function both steeper than the field and which turns over at $\sim 0.25\,M_\odot$; and from data for the young clusters IC 2602 and IC 2391, which seem to have significantly fewer low-mass M dwarfs than do the Pleiades. Star formation theory, however, has not yet succeeded in identifying either the cause of these variations, or the physical mechanism underlying the entire process of transforming molecular gas into stars with masses distributed between 100 and $< 0.08\,M_\odot$.

8.11   REFERENCES

A1   Adams, F. C., Fatuzzo, M., 1996, *ApJ*, **464**, 256.
A2   Alcock, C. *et al.*, 1997, *ApJ*, **479**, 119.
A3   Alcock, C. *et al.*, 1997, *ApJ*, **486**, 697.
A4   Ali, B., Depoy, D., 1995, *AJ*, **109**, 1379.
A5   Andersen, J., 1991, *Astr. Ast. Rev.*, **3**, 91.
B1   Baraffe, I., Chabrier, H., Allard, F., Hauschildt, P. H., 1998, *A&A*, **337**, 403.
B2   Becklin, E. E., Neugebauer, G., 1967, *ApJ*, **147**, 799.
B3   Benedict, G. F., McArthur, N., Nelan, E. *et al.*, 1994, *PASP*, **106**, 327.
B4   Binnendijk, L., 1960, *Properties of Double Stars*, University of Pennsylvania Press, Philadelphia.
B5   Boss, A. P., 1991, *Nature*, **351**, 298.
C1   Couteau, P., 1981, *Observing Visual Double Stars*, trans. A. H. Batten, The MIT Press, Cambridge, Massachusetts.
C2   Crézé, M., Chereul, E., Bienayme, O., Pichon, C., 1998, *A&A*, **329**, 920.
D1   Dahn, C. C., Liebert, J., Harrington, R. S., 1986, *AJ*, **91**, 62.
D2   D'Antona, F., Mazzitelli, I., 1994, *ApJS*, **90**, 467.
D3   Delfosse, X., Forveille, T., Mayor, M., Burnet, M., Perrier, C., 1999, *A&A*, **341**, L63.
D4   Dyson, F. W., Eddington, A. S., Davidson, C. R., 1920, *Mem. RAS*, **62**, 291.
E1   Elmegreen, B. G., 1997, *ApJ*, **486**, 944.

G1   Genzel, R., Reid, M. J., Moran, J. M., Downes, D., 1981, *ApJ*, **224**, 884.

G2   Genzel, R., Stutski, J., 1989, *ARA&A*, **27**, 41.

G3   Ghez, A. M., Weinberger, A. J., Neugebauer, G., Matthews, K., McCarthy, D. W., 1995, *AJ*, **110**, 753.

G4   Ghez, A. M., McCarthy, D. W., Patience, J. L., Beck, T. L., 1997, *ApJ*, **481**, 378.

G5   Gizis, J. E., Reid, I. N., Monet, D. G., 1999, *AJ*, **118**, 997.

H1   Hambly, N. C., Hawkins, M. R. S., Jameson, R. F., 1993, *MNRAS*, **263**, 647.

H2   Hambly, N. C., Steele, I. A., Hawkins, M. R. S., Jameson, R. F., 1995, *MNRAS*, **273**, 505.

H3   Hartigan, P., Strom, K. M., Strom, S. E., 1994, *ApJ*, **427**, 961.

H4   Hawkins, M. R. S., Bessell, M. S., 1988, *MNRAS*, **234**, 77.

H5   Haywood, M., 1994, *A&A*, **282**, 444.

H6   Henry, T. J., McCarthy, D. W., 1993, *AJ*, **106**, 773.

H7   Henry, T. J., Franz, O. G., Wasserman, L. H., Benedict, G. F., Shelus, P. J., Ianna, P. A., Kirkpatrick, J. D., McCarthy, D. W., 1999, *ApJ*, **512**, 864.

H8   Herbig, G. H., 1954, *PASP*, **66**, 19.

H9   Herbig, G. H., 1998, *ApJ*, **497**, 736.

H10 Hillenbrand, L. A., 1997, *AJ*, **113**, 1733.

J1   Jones, B. T., Walker, M. F., 1988, *AJ*, **95**, 1755.

K1   Kleinmann, D. E., Low, F. J., 1967, *ApJ*, **149**, L1.

K2   Kroupa, P., Tout, C. A., Gilmore, G. F., 1990, *MNRAS*, **244**, 76.

K3   Kroupa, P., Tout, C. A., Gilmore, G. F., 1993, *MNRAS*, **262**, 545.

K4   Kroupa, P., 1995, *ApJ*, **453**, 358.

K5   Kroupa, P., Tout, C. A., 1997, *MNRAS*, **287**, 402.

K6   Kutner, M. L., Machnik, D. E., Tucker, K. D., Dickman, R. L., 1980, *ApJ*, **237**, 734.

L1   Lacy, C. H., 1979, *ApJ*, **228**, 817.

L2   Lada, E. A., Lada, C. J., 1995, *AJ*, **109**, 1682.

L3   Lada, E. A., Lada, C. J., Muench, A., 1998, in *The 38th Herstmonceux Conference*, ed. G. Gilmore, Cambridge University Press.

L4   Larson, R. B., 1992, *MNRAS*, **256**, 641.

L5   Leggett, S. K., Hawkins, M. R. S., 1989, *MNRAS*, **238**, 145.

L6   Luhman, K., Rieke, G. H., Lada, C. J., Lada, E. A., 1998, *ApJ*, **508**, 347.

L7   Luhman, K., 1999, *ApJ*, **522**.

L8   Luyten, W. J., 1941, *Annals New York Acad. Sci.*, **42**, 201.

M1   Maceroni, V., Rucinski, S. M., 1997, *PASP*, **109**, 782.

M2   Marcy, G. W., Moore, D., 1989, *ApJ*, **341**, 961.

M3   Mathieu, R. D., Ghez, A. M., Jensen, E. L. N., Simon, M., 1999, in *Protostars and Planets IV*, ed. V. Mannings, A. Boss and S. Russell, University of Arizona Press.

M4   Metcalfe, T. S., Mathieu, R.D., Latham, D.W., Torres, G. 1996, *ApJ*, **456**, 356.

M5   Meusinger, H.. Schilbach, E., Souchay, J. 1996, *A&A*, **312**, 833.

M6   Meyer, M. R., Adams, F. C., Hillenbrand, L. A., Carpenter, J. M., Larson,

R. B., 1998, in *Protostars and Planets IV*, ed. V. Mannings, A. Boss and S. Russell, University of Arizona Press.

M7   Miller, G. E., Scalo, J. M., 1979, *ApJS*, **41**, 513.
M8   Murray, S. D., Lin, D. C., 1996, *ApJ*, **467**, 728.
P1   Paczyński, B., 1986, *ApJ*, **304**, 1.
P2   Paczyński, B., 1996, *ARA&A*, **34**, 419.
P3   Pinfield, D. J., Hodgkin, S. T., Jameson, R. F., Cossburn, M. R., von Hippel, R., 1997, *MNRAS*, **287**, 180.
P4   Popper, D. M., 1980, *ARA&A*, **18**, 115.
P5   Prosser, C. F., Stauffer, J. R., Hartmann, L., Soderblom, D. R., Jones, B. F., Werner, M. W., McCaughrean, M. J., 1994, *ApJ*, **421**, 517.
R1   Rana, N. C., 1987, *A&A*, **184**, 104.
R2   Refsdal, S., 1964, *MNRAS*, **128**, 295.
R3   Reid, I. N., 1987, *MNRAS*, **225**, 873.
R4   Reid, I. N., 1996, *AJ*, **111**, 2000.
R5   Reid, I. N., 1998, in *The Stellar Initial Mass Function*, ASP Conf. Ser. vol. 142, ed. G. Gilmore and D. Howell, p. 121.
R6   Reid, I. N., Hawley, S. L., 1999, *AJ*, **117**, 343.
R7   van Rhijn, P. J., 1936, *Publ. Kapteyn Astr. Lab. Gröningen*, No. 47.
S1   Salpeter, E. E., 1956, *ApJ*, **121**, 161.
S2   Scalo, J. M., 1986, Fund. Cos. Phys., **11**, 1.
S3   Scalo, J. M., 1998, in *The 38th Herstmonceux Conference*, ed. G. Gilmore, Cambridge University Press.
S4   Scalo, J. M., Pumphrey, W. A., 1982, *ApJ*, **258**, L29.
S5   Silk, J., 1995, *ApJ*, **438**, L41.
S6   Stauffer, J. R., Klemola, A., Prosser, C., Probst, R., 1991, *AJ*, **101**, 980.
S7   Stobie, R. S., Ishida, K., Peacock, J. R., 1989, *MNRAS*, **238**, 709.
S8   Strom, S. E., Strom, K. M., Carrasco, L., 1974, *PASP*, **86**, 798.
T1   Tinney, C. G., 1993, *ApJ*, **414**, 279.
T2   Torres, G., 1995, *PASP*, **107**, 524.
W1   Walker, M. F., 1969, *ApJ*, **155**, 447.
W2   Warren, W. H., Hesser, J. E., 1977, *ApJS*, **34**, 207.
W3   Walsh, D., Carswell, R. F., Weymann, R. J., 1979, Nature, **279**, 381.
W4   White, R. J., Ghez, A. M., Reid, I. N., Schultz, G., 1999, *ApJ*, **520**, 811.
W5   Wielen, R., Jahreiss, H., Krüger, R., 1983, *IAU Colloquium 76, The Nearby Stars and the Stellar Luminosity Function*, ed. A. G. Davis Philip and A. R. Upgren, p. 163.
W6   Williams, D. M., Rieke, G. H., Stauffer, J. R., 1995, *ApJ*, **454**, 144.
W7   Williams, D. M., Comeron, F., Rieke, G. H., Rieke, M. J., 1995b, *ApJ*, **454**, 144.
Z1   Zinnecker, H., 1984, *MNRAS*, **210**, 43.
Z2   Zinnecker, H., McCaughrean, M. J., Wilking, B., 1993, in *Protostars and Planets III*, ed. by E. H. Levy and J. I. Lunine, University of Arizona Press, p. 429.
Z3   Zuckerman, B. 1973, *ApJ*, **181**, 863.

9

Brown dwarfs

9.1 INTRODUCTION

The existence of substellar-mass, star-like objects was first considered seriously by Kumar [K11] (Section 3.3), who outlined their essential properties: no central energy source due to hydrogen fusion, degeneracy and short luminous lifetimes. Like low-mass M dwarfs, these objects – renamed 'brown dwarfs' by Tarter [T1] – are ideal baryonic dark matter candidates, and have been the targets of a wide variety of surveys over the last three decades. As with any search for a newly-hypothesised object, the first task is to determine whether any exist. In answering this question, one can utilise all available techniques, without regard to the complications of biases and selection effects. However, once the issue of existence has been answered in the affirmative, the emphasis of observational programmes must change, moving from simple discovery searches to addressing specific issues through statistically well-controlled surveys. With the near-contemporaneous discoveries of Gl 229 B and the planet orbiting 51 Peg, both brown-dwarf and planet surveys have moved from the first to the second stage. Early results from the DENIS, 2MASS and SLOAN sky surveys are providing the data necessary for a transition from phenomenology to statistics.

This chapter summarises results from brown dwarf surveys obtained prior to mid-1999, and Chapter 10 considers planet detections. After outlining characteristics which serve as observational signatures of substellar-mass brown dwarfs, we review some of the early unsuccessful programmes before describing the first successes (not all immediately recognised as such). The field as a whole underwent a revolution during the three years following the discovery of Gl 229B, with the identification of isolated brown dwarfs in both the general field and young clusters, most recently including objects as cool as Gl 229B itself. This much-enlarged sample of substellar objects provides an empirical description of the changing properties of these objects with age and temperature, and those matters form the focus of the latter sections of the chapter.

9.2   HOW TO RECOGNISE A BROWN DWARF

Brown dwarfs are defined by their mass. Unfortunately, this parameter is rarely measurable in a direct fashion (Section 8.2). As yet, no brown-dwarf binary has been identified which can provide direct calibration of the substellar régime, although the Pleiad PPl 15 (Section 9.4.4) may eventually prove suitable. Lacking such data, we have to infer masses by comparing secondary indicators – luminosity and temperature – against theoretical predictions. The theoretical stellar models described in Chapter 3 indicate that objects with $L < 10^{-4} L_\odot$ (Figure 3.8) or $T_{eff} < 2,000\,K$ (Figure 3.9) are at or below the stellar mass limit. Classification is more ambiguous for younger, hotter brown dwarfs, since current empirical and theoretical temperature calibrations are of insufficient accuracy to distinguish stars from brown dwarfs between $\sim 10^{-2}$ and $10^{-4} L_\odot$ (Figure 3.12).

A technique which can be used to identify some lower-mass brown dwarfs is the lithium test – detection of the 6708 Å Li I absorption line in mid- to late-M dwarfs [M2]. The primordial lithium abundance is estimated as [Li] = 3.3 on a logarithmic scale where [H] = 12.0: that is, there are 5×10^8 hydrogen atoms for every lithium atom. Lithium absorption is detected easily in M-type T Tauri stars with ages of a few million years, but, as discussed in Section 3.3, primordial lithium is destroyed by reaction 5^{II} of the PPII chain. The reaction rate is temperature-dependent, requiring a minimum temperature of $\sim 3 \times 10^6\,K$. As a result, the rate of depletion varies with mass, and objects with masses below 0.055–0.065 $M_\odot$ (some uncertainty remains in the model calculations [B1], [B8], [B10]) retain lithium at the primordial abundance level throughout their lifetime. Figure 9.1 shows the lithium-depletion line as a function of age and temperature superimposed on evolutionary tracks from the Tucson models. A 0.075 $M_\odot$ dwarf, lying on the brown dwarf threshold, is predicted to deplete lithium in $\sim 1.5 \times 10^8$ years, by which time its effective temperature is $\sim 2,800\,K$, equivalent to a spectral type of $\sim$M6. Thus, these calculations imply that any dwarf with a spectral type later than M6 and detectable lithium absorption is a substellar-mass brown dwarf.

Both detailed model calculations and the definition of the lithium test are developments of the 1990s, postdating many surveys for substellar objects. Since it has been clear from the outset that brown dwarfs evolve rapidly to effective temperatures below 2,500 K, those earlier surveys used broadband photometry to search for objects with extremely red colours – either as extensions of photometric parallax surveys for low-mass stars in the field, or as targeted observations, searching for low-mass companions to known nearby stars. At these low temperatures, brown dwarfs emit most of their energy at wavelengths beyond 1 μm. Thus, while initial surveys were undertaken at red or far-red optical wavelengths (R–I), the emphasis shifted to the near-infrared (1–2.5 μm) once appropriate technology became available.

With the actual discovery of brown dwarfs, it became apparent that searching for objects which have red colours in *all* passbands is not an infallible discovery technique. In particular, methane absorption at temperatures below $\sim 1,200\,K$ leads to blue near-infrared colours – a characteristic predicted by Tsuji in 1964

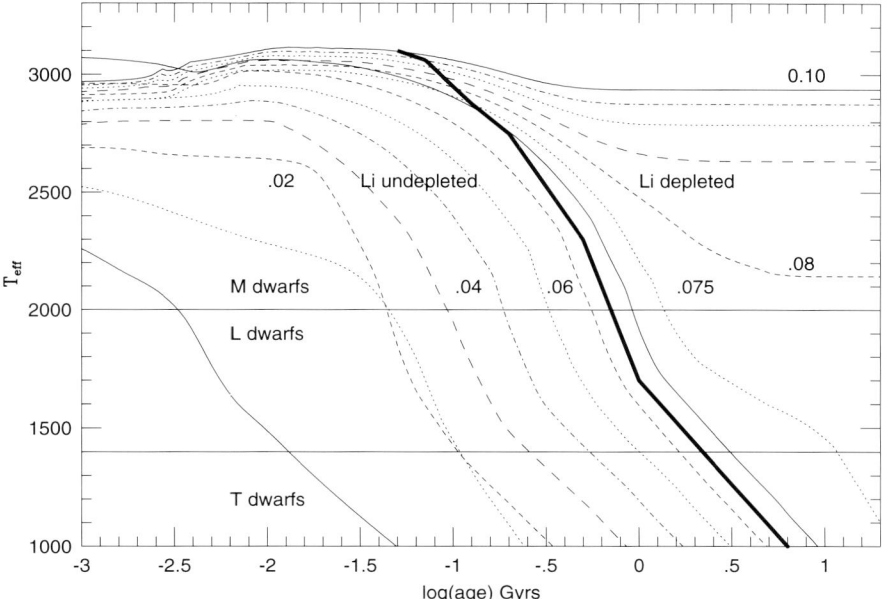

Figure 9.1. The lithium depletion line superimposed on evolutionary tracks for low-mass stars and brown dwarfs (labelled by mass in $M_\odot$) predicted using the Tucson models [B10]. Models to the right of the solid line have depleted lithium by over 90%. Brown dwarfs with masses below $\sim 0.065\,M_\odot$ are predicted to retain lithium at the primordial abundance throughout their lifetime.

[T14], but accorded little recognition until quite recently. However, there is a substantial parameter space between the location of the lowest-mass stars and the onset of methane formation, and most of the L dwarfs discussed in Section 2.2.3 have $(J–K_S)$ colours redder than 1.3 magnitudes. Thus, the initial fixation on searching for objects which are red at all wavelengths is less of a bias than occasionally claimed.

Finally, since brown dwarfs can occur as companions to more massive objects (and in brown dwarf/brown dwarf binaries), we must consider the distinction between brown dwarfs and planets. In principle we can discriminate between these alternative classifications based on the mode of formation: planets form in a disk around a more massive central object; brown dwarfs form as separate, accreting entities, like stars. In practice, we cannot reverse time and reconstruct the history of each low-mass companion, but must segregate based on present properties. There have been suggestions that the brown dwarf/planet boundary should be set at the mass limit for deuterium burning ($\sim 0.013\,M_\odot$, or 14 Jupiter masses, M_J, at solar abundance [B10]). This classification matches the fusion criterion which defines the stellar/brown dwarf boundary, and would apply to both single and multiple objects. Deuterium, however, is almost impossible to detect in objects at these low temperatures, so this creates another observationally imprecise division. In fact, as discussed further in Chapter 10, Nature may supply its own division. There are indications of a minimum in the mass distribution of companions at $10\text{--}30\,M_J$, suggestive of a break between the mass distributions of brown-dwarf and planetary companions. For the present, we acknowledge that a grey area exists in classifying the lower extreme of the brown dwarf mass distribution, but will consider all companions with masses exceeding $15\,M_J$ and all isolated T-type (Gl 229B-like) dwarfs as potential brown dwarfs.

9.3   FIRST STEPS: SOME INTERESTING MISTAKES

Text-books occasionally give the impression that science progresses in a smooth, orderly fashion, with one discovery leading naturally to the next. In fact, this is generally not the case: the path to greater knowledge follows a zig-zag route, sometimes with unforeseen tangential diversions and the occasional U-turn out of a blind alley. The route toward the discovery of brown dwarfs is strewn with the carcasses of hopeful monsters. It is instructive to review some of these discarded hypotheses.

Many late-type M dwarfs have been suggested as brown dwarf candidates at various times during the last two decades of the twentieth century, including VB 10, LHS 2924 and the unusual dMe dwarf, PC0025+0447 (Section 5.6.1). Indeed, Kirkpatrick *et al.* [K2] suggested that the spatial distribution of the latest-type M dwarfs indicated that those objects are closely confined to the Plane, as would be expected for a very young population (Section 6.5). That possibility has largely been eliminated by the subsequent photometric survey by Tinney [T4], [T9], whose sample of late-type M dwarfs is well distributed above and below the Plane. Moreover, analysis of the kinematics of very low-mass dwarfs shows that the velocity dispersions are higher than the $5\text{--}10\,\mathrm{km\,s^{-1}}$ that would be expected for a $\sim 10^8$-year-old population. Other mis-steps have had wider ramifications.

9.3.1   VB 8B: the disappearing brown dwarf

Although eventually proven spurious, the hypothetical low-mass companion to Gl 644C (van Biesbrock 8) had more impact on the field of brown dwarf surveys than many subsequent actual discoveries. The source was identified by McCarthy *et al.* [M1] from one-dimensional speckle scans of VB 8 (Section 1.7.1). The visibility function suggested the presence of a companion, three magnitudes fainter than the primary (K = 8.8 magnitudes), at a separation of 1.0 arcsec. Both this star and VB 10 had been previously suspected of having astrometric companions [H4], and the speckle data appeared to provide confirmation for VB 8. The low luminosity at $2.2\,\mu\mathrm{m}$ together with the red infrared colours ((H–K) ~ 1.3) suggested a brown dwarf companion, and VB 8B was accepted as the first clear case of a substellar-mass dwarf. Assuming an age of $\sim 3.5 \times 10^9$ years for the Gl 644/Gl 643 system, VB 8B was inferred to have a mass between ~ 0.065 and $0.085\,M_\odot$, placing it just below the hydrogen-burning limit.

This 'discovery' prompted the first conference devoted exclusively to brown dwarfs (the 1986 Astrophysics of Brown Dwarfs Workshop) stimulating a subsequent spate of brown dwarf surveys. VB 8B, however, did not long survive that meeting. Skrutskie *et al.* [S7] obtained direct K-band images of VB 8 using one of the first near-infrared cameras (equipped with a 32 × 32-element Rochester InSb array). Those data, taken at K- and L-band under excellent seeing conditions (a FWHM of 0″.64 in a 5-s exposure), showed no evidence of the expected companion, although simulations showed that such an object should have been detected easily even at separations as small as 0.6 arcsec. The astrometric data were inconsistent with a short period and hence a rapid change in the projected separation since the speckle observations.

The decisive non-observation of VB 8B was made by Perrier and Mariotti [P2], who obtained new K- and L-band speckle interferometric scans, and found no evidence for any companion brighter than K $\sim$ 14.5 magnitude at the expected position. The inferred mass ratios and standard orbital dynamics rule out the possibility of the 'companion' lying along the line of sight to VB 8. The only conclusion possible is that the original detection was an observational artefact, probably due to the chromatic effects of atmospheric refraction. McCarthy *et al.* compared their scans of VB 8 against data for a known point-source, but at significantly different altitudes, and therefore with different image profiles. VB 8B thus became one in a long line of brown dwarf candidates that have failed to survive rigorous scrutiny.

9.3.2   G29-38: dust or a brown dwarf?

G29-38 is a DA white dwarf which lies near the lower-temperature boundary of the ZZ Ceti pulsational-instability strip, $T_{eff} \sim 11,500\,\text{K}$. The star was identified as a white dwarf candidate in the Lowell proper-motion survey and confirmed as a DA (hydrogen atmosphere) white dwarf (EG 159) by Eggen and Greenstein [E1]. Also known as ZZ Psc, the star was included by Zuckerman and Becklin among the sample of white dwarfs surveyed at JHK (Section 7.7.2), and aperture photometry revealed the presence of a substantial infrared excess [Z4]. Rather than the expected (V–K) colour of −0.3, the observed colour is (V–K) = +0.3. Subtracting the white dwarf flux distribution from the original photometry, which extended to the 4.8 μm M-band, revealed a distribution peaking at $\sim 3.5\,\mu\text{m}$ (Figure 9.2).

The initial interpretation was that the infrared excess was due to a low-mass companion. Greenstein [G9] estimated a black-body temperature in the range 1,100–1,500 K, a luminosity of $\sim 4 \times 10^{-5}\, L_\odot$ and a mass of ~ 0.04–$0.08\, M_\odot$, with the white dwarf and the brown dwarf making equal contributions at 2.2 μm. Direct imaging shows G29-38 to be unresolved at the ~ 0.8 arcsec level [T10]. One-dimensional speckle scans suggested that there was a slight elongation in the north–south direction, consistent with two point-sources of equal luminosity (at 2.2 μm) separated by 0″.23 [H1], but recent speckle observations contradict that hypothesis [K9]. There is therefore no reliable evidence for a resolved companion.

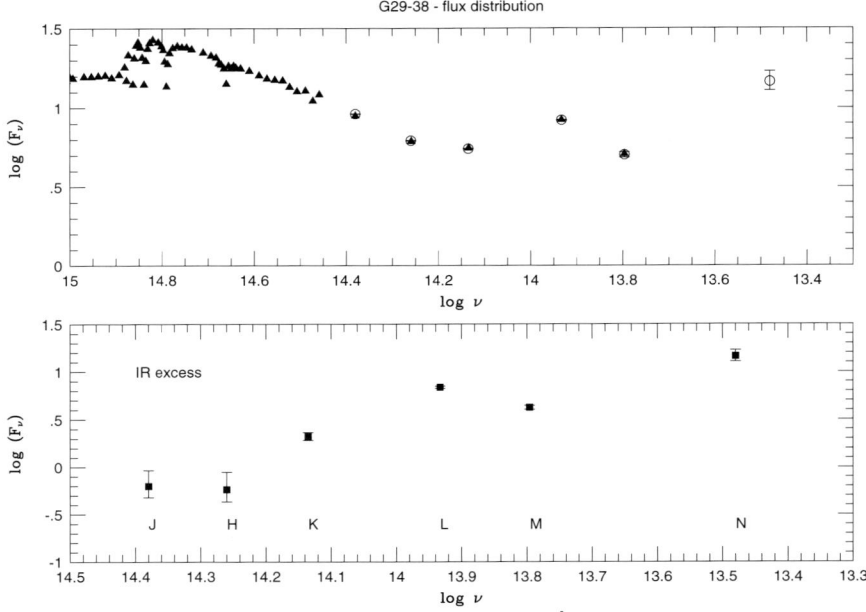

Figure 9.2. The flux distribution of G29-38. The features at optical wavelengths are the Balmer lines.

As more photometric data were acquired, inconsistencies arose between the observations and a model incorporating an unresolved brown dwarf companion. First, the implied radius of the companion is $\sim 0.15\,R_{\odot}$ [Z4] – approximately twice that predicted by models. Second, infrared spectroscopy [T11] fails to reveal any features in the K-band window (1.95–2.45 μm): either CO or CH_4 absorption is expected for a cool dwarf companion. Finally, the system is detected at 10 μm [T2], [T11] at a level well above that expected for even a 1,100 K companion. No single-temperature black-body can fit all of the photometric data, although the mid-infrared excess might be accommodated by adding dust to a low-temperature atmosphere.

G29-38s pulsations provide the most stringent test of the companion hypothesis. ZZ Ceti stars exhibit a complex range of flux variations due to non-radial pulsations with periods in the range of 200–1,000 seconds and amplitudes of up to 0.3 magnitudes. G29-38 is one of the largest-amplitude variables of this type. In these stars the variations are typically dominated by a single mode, but there are also numerous other lower-amplitude variations at different frequencies. In order to disentangle the complex interplay among these different modes, Nather et al. [N3] devised the concept of the Whole Earth Telescope (WET), which consists of a series of telescopes distributed around the world so that a given star can be monitored continuously for days. Continuous coverage allows unambiguous determination of the intrinsic mode of variation, free of the aliasing problems introduced by being unable to observe during the day at a single site. If the targeted ZZ Ceti star has a binary companion, the distance to Earth, and hence the light-travel time, changes with a period equal to that of the orbit, leading to long-term modulation of any stable intrinsic pulsational frequency.

In the analysis of the initial dataset [W4], one particular mode of variation (P = 615 seconds) appeared to show exactly the behaviour expected if the star were in a binary orbit. The residuals (the observed minus calculated time of maximum for that mode) implied that the orbital period was 90 days and that the companion had a mass of at least 0.3 $M_\odot$ – well above the brown dwarf limit. However, with such a massive companion, one would expect to see substantial radial velocity variations ($\sim\pm28\,\mathrm{km\,s^{-1}}$) synchronised with the modulation in the pulsation. These variations are not observed [G5]. Lower-amplitude ($\pm5\,\mathrm{km\,s^{-1}}$) variations may be present with a period of 11.2 months [B2], but non-radial pulsations can introduce apparent velocity shifts of this order [C1].

Patterson *et al.* [P1] monitored the star photometrically at both blue and infrared wavelengths, and find no evidence for long-period modulations of any of several stable modes visible at optical or infrared wavelengths, including the P = 615 seconds mode. Crucially, a number of the higher-frequency variations are detected at both optical and infrared wavelengths, showing that these variations are driven by the white dwarf. The data are consistent with their being temperature variations due to non-radial pulsations. Finally, Kleinmann *et al.* [K7] analyse WET data from a series of observing runs spaced over a period of five years. Several modes of variation – notably a 284-second periodicity – are evident throughout the full set of observations, and there is no evidence for orbital modulation. This effectively excludes the brown dwarf hypothesis.

The infrared excess, however, is clearly real, and has been verified by numerous observers, most recently using the ISO satellite. The only reasonable alternative to a discrete companion is circumstellar dust. This can account both for the presence of pulsations at infrared wavelengths, through reprocessing of light from the white dwarf, and for the fact that the infrared excess fails to match a single-temperature black-body curve. However, there are also a number of complications involved in this interpretation of the available data.

First, maintaining temperatures of $\sim$1,000–1,200 K requires that the dust particles are at radii of only $\sim$1 $R_\odot$ from the white dwarf. At these small radii, the Poynting–Robertson effect – a net drag exerted on small particles by the absorption and re-emission of radiation, leads to short orbital lifetimes. Indeed, it was on this basis that Zuckerman and Becklin originally rejected the dust-cloud hypothesis. The Poynting–Robertson depletion time-scale is given by

$$t_{PR} \sim 2 \times 10^4\,r\,\text{years} \tag{9.1}$$

where r is the radius of the particle in millimetres. Since we observe periodic infrared pulsations, the dust particles must be small enough to have thermal time-scales that allow them to respond to the rapid changes in the surface temperature of the white dwarf. This sets an upper limit of $\sim$100 mm on the particle radius [G4], implying a depletion time of less than 2 million years – more than 100 times shorter than the cooling time of the white dwarf. Thus, we are either lucky enough to witness a rare, short-lived phenomenon (unlikely – but consider comet Shoemaker–Levy 9), or the radiative grains are being replenished from an external source.

This raises the matter of the origin of the dust. G29-38 has a cooling time (as a white dwarf) of 1–2 Gyr, so one would expect any residual dust from the planetary nebula to have dispersed. Particulate matter at radii of a few $R_\odot$ in the main-sequence system should have been engulfed during the red-giant stages of evolution. The dust may originate from a disk of relatively large (>metres) fragments – perhaps the remnants of a disrupted asteroid – with collisions supplying the reservoir of small fragments that produce the infrared excess [G4]. Whatever the source, the detection of neutral and ionised absorption lines of calcium and magnesium in the visible spectrum [K9], [Z7] suggests that dust may be accreting onto the surface of the white dwarf. In summary, G29-38 is clearly an object of considerable astrophysical interest, but it is unlikely that the system contains a substellar-mass brown dwarf.

9.4   BROWN DWARFS REVEALED

Brown dwarfs can occur in isolation or as companions to other hydrogen-burning or non-stellar objects. Almost all surveys aim to discover brown dwarfs under one or other of these circumstances, but not both. Results from these two types of survey address different astrophysical issues. The space density of field brown dwarfs feeds directly into the determination of the initial mass function, and the total mass locked up in long-lived, star-like objects within the Galactic disk. Brown dwarf companions of main-sequence stars, however, are of little importance for global dynamics: even if every stellar system in the Solar Neighbourhood were to include a $0.07\,M_\odot$ brown dwarf, the total mass density would be increased by only $0.004\,M_\odot$, or $\sim 10\%$ of the *known* stellar mass density. Moreover, there is growing evidence that the mass distribution of companions is not identical to the field mass function [R5]. Thus, the relative frequency and the mass–ratio distribution of multiple systems are relevant to the understanding of binary-star formation, rather than for studies of the initial mass function and questions concerning Galactic structure.

JSearches for isolated brown dwarfs utilise different techniques than surveys for brown dwarfs as companions. We illustrate this by next describing the key observational discoveries which have verified the existence of brown dwarfs. It should be noted that brown dwarf candidates in several young (few Myr old) clusters were identified contemporaneously with the field discoveries, but ambiguities in interpreting evolutionary tracks led to the former discoveries having less impact.

9.4.1   GD 165B

The sensitivity of any search for VLM objects in binary systems is set by the contrast required to detect companions, either spatially against the point-spread function of the primary star, or photometrically against the flux distribution. As described in Section 7.7.2, white dwarfs are excellent targets in searches for low-mass companions since they have low luminosities and high effective temperatures. The presence of a cool companion therefore leads to unusual optical/IR colours.

Like G29-38, GD 165 was identified as a probable white dwarf based on the blue $(m_{pg}-m_v)$ colour measured in the course of the Lowell proper-motion survey. Subsequent observations confirmed it as type DA with a surface temperature of $\sim 13,400$ K, placing it near the hotter boundary of the ZZ Ceti instability strip [G8]. The star is also known as CX Boö, and the current best estimate of the trigonometric parallax is $0\overset{''}{.}0317 \pm 0.0025$ (USNO observations), corresponding to a distance of 31.5 parsecs. Recent observations have revealed that the white dwarf is a single-lined spectroscopic binary, with a cooler white dwarf companion, GD 165C, in a short-period orbit [S1].

GD 165 was included by Zuckerman and Becklin [Z6] as a target in their search for low-mass companions, and was among the first to be observed using an infrared array. Those observations revealed the presence of a faint companion at a separation of only 3.5 arcsec, and with an unusually red (J–K) colour of 1.6 magnitudes, indicating a temperature well below 2,500 K [Z5]. Subsequent astrometry confirmed that GD 165B shares the proper motion of the white dwarf, and is therefore unquestionably a physical companion, rather than a foreground or background star (or galaxy).

Based on the known distance to GD 165, the companion has $M_K = 11.66$ and $M_{bol} \sim 14.6$, or $L \sim 10^{-4}L_\odot$ – close to the substellar régime. Black-body fitting to the flux distribution suggested a temperature close to 1,800 K. Infrared spectroscopy showed that the object has extremely strong H_2O absorption bands [J1], but is not qualitatively different from late M dwarfs (Figure 9.3). The first optical spectrum [K1] revealed markedly different spectral properties than are observed in late-type M dwarfs such as LHS 2924 (spectral type M9). Instead of strong TiO and VO molecular absorption, the spectrum is relatively smooth. However, the proximity to GD 165A rendered the observations difficult, and hence perhaps less reliable, as well as raising the possibility of atmospheric 'pollution' of the companion during the asymptotic giant branch phase.

It is now clear that GD 165B was the first L dwarf to be discovered. More recent observations using the Keck telescopes provide a better definition of the optical spectrum, indicating a spectral type of L4 [K5]. The best-fit temperature derived by matching those data against the latest model atmosphere calculations is $1,900 \pm 100$ K. The mass depends on the inferred age of the system, which in turn demands an estimate of the mass of the GD 165A progenitor. The latter can be estimated from the present-day mass of the white dwarf, measured by fitting the Balmer line profiles to stellar models and deriving the gravity [B6]. The initial/final mass relationship suggests that low-mass white dwarfs are descended from low-mass progenitors [W2]. Current estimates for GD 165A indicate a mass of 0.56–0.65 $M_\odot$, suggesting a progenitor mass of 1.2–3 $M_\odot$, and main-sequence lifetimes of from 4.6–0.4 Gyr.[1] The white dwarf

---

[1] The estimate of the main-sequence lifetime is complicated by the presence of GD 165C, which is cooler, and therefore older, and may well have transferred mass to GD 165A. In this case, the progenitor may have had a lower mass and a longer lifetime.

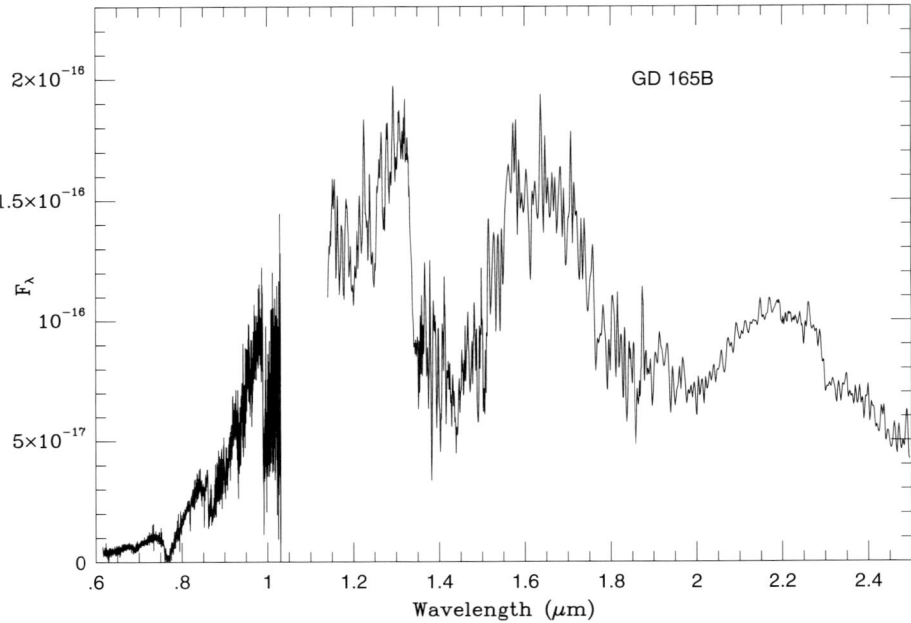

Figure 9.3. The optical and infrared spectrum of GD 165B, the prototype L dwarf. (Optical data from [K5]; infrared data from [J1].)

cooling time of ~ 0.6 Gyr must be added to this value. Hence, GD 165B probably has an age between 1 and 5 Gyr, implying a mass between 0.065 and 0.075 $M_\odot$.

9.4.2   HD 114762B

GD 165B was detected through direct imaging. As described in Section 2.7, an alternative technique better suited to identifying close companions is to search for reflex orbital motion in the primary star. In most cases, it is possible to set only a lower limit to the mass of the companion, since the orbital inclination, i, is usually unknown. However, under certain circumstances the mass can be constrained within reasonable limits.

HD 114762 is a 7th magnitude F9 dwarf (V = 7.36, (B–V) = 0.56) lying at a relatively high Galactic latitude. The Hipparcos parallax (0.''0247 ± 0.0014) places it at a distance of 40.5 ± 2.5 parsecs. It is slightly metal-poor, with [Fe/H] = −0.67 and a relatively low gravity, $\log g$ = 4.17 [G7], suggesting that the star is both old and has evolved slightly from the main sequence. Age estimates range from ~ 5 to 10 Gyr [H2], [H7]. The star was chosen as a possible addition to an improved set of radial-velocity standards, but observations over a period of 10 years showed unambiguous velocity variations of ±0.6 km s^{-1} with a period of 84 days [L1], [C2]. The non-symmetric nature of the velocity curve shows that the orbit has a significant eccentricity, e = 0.38 (Figure 9.4). The spectral type implies a mass of 1.17 $M_\odot$,

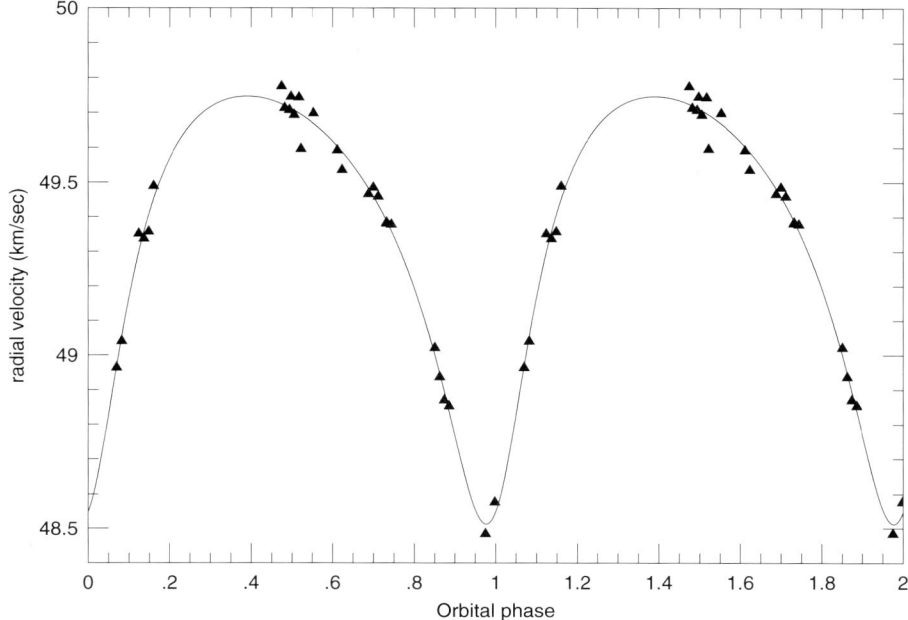

Figure 9.4. Radial velocity curve of HD 114762. The observations are taken from [C2] and have an accuracy of 0.034 km s⁻¹.

giving a projected semi-major axis of $a\sin(i) = 0.3\,\mathrm{AU}$ and a companion mass of $M_2\sin(i) = 0.011\,M_\odot$.

Clearly, the crucial parameter in determining whether this system harbours a brown dwarf, or even a high-mass planet, is the orbital inclination: the companion has a mass exceeding the hydrogen-burning limit only if $i < 8\overset{\circ}{.}5$. Photometric monitoring shows no evidence of eclipses [R10], [H7], setting an upper limit of $i < 89°$ for a Jovian-diameter companion. Line profile analysis offers a possible means of setting stronger constraints by determining the rotational velocity of the primary, $v\sin(i_{eq})$, and comparing that value with the average for similar F stars. This assumes that the companion has an orbit which is coplanar with the stellar equatorial plane, which is reasonable given the small separation [H2]. Coplanarity would be consistent with formation in a disk (that is, a planet), although the net angular momentum of the parent molecular cloud core might align rotation and orbital axis in a protobinary system. Analyses of high-resolution spectra [C2], [H2] indicate that the profiles are dominated by macroturbulence (random motions in the stellar atmosphere), with $v\sin(i_{eq}) \approx 0.8\,\mathrm{km\,s^{-1}}$. Given the probable age (τ) and the measured chromospheric activity, the intrinsic rotational velocity is estimated as 2.3 ($\tau = 8\,\mathrm{Gyr}$) to $3.0\,\mathrm{km\,s^{-1}}$ ($\tau = 5\,\mathrm{Gyr}$), implying $i \sim 15$–20°, with a 1σ uncertainty of $\pm 10°$.

These results emphasise the inherent ambiguity of this method for individual objects. If the binary orbit is coplanar with stellar rotation, then the most likely value for M_2 is $\sim$0.04–0.05 $M_\odot$. However, the uncertainties in i allow M_2 to enter the stellar régime. If such is the case, then the orbital semi-major axis of 0.35 AU. (similar to the Mercury–Sun distance) corresponds to an angular separation of $\sim$9 mas. Interferometric observations may eventually settle the issue of whether HD 114762B is a star, a brown dwarf, or a planet.

9.4.3   Gl 229B

The identification of Gl 229B marked a pivotal point in brown dwarf research. Like GD 165B, this low-luminosity companion was discovered through direct imaging – in this case using a coronagraph to occult most of the radiation from the primary star. The characteristics of Gl 229B place it unequivocally below the hydrogen-burning limit, and render it the prototype for spectral class T.

Gl 229 is an early-type M-dwarf, $M_V \sim 9.4$, lying at a distance of only 5.7 parsecs (see Appendix). As one of the nearest M dwarfs, the star was included in Henry's near-infrared speckle survey [H8], and those observations showed no evidence for any companions within $\pm$20 AU ($\pm$2 arcsec). The star was also observed as part of a nearby-star companion search by Nakajima and collaborators using the Johns Hopkins coronagraph on the Palomar 60-inch in October 1994 [N1]. This system uses an apodising mask and occulting disk to obscure the central 3 arcsec and cut down scattered light, while tip–tilt correction (using the reflected image of the primary) provides image sharpening over the full 60×60-arcsec field of view.

The Gl 229 I-band images (Figure 9.5, see colour section) reveals the presence of a faint companion, 7.8 arcsec distant (that is, outside Henry's survey limits) and some 13 magnitudes fainter than the primary star in the I-band ($M_I \sim 20.3$) [N2]. Subsequent measurements in October 1995 confirmed that the companion shares the proper motion of the primary ($\sim$1 arcsec yr^{-1}), while K-band observations with the Hale 200-inch showed that the companion has extremely red optical/infrared colours, $(I–K) = 6.2$ magnitude [M9]. Calculating the bolometric luminosity from the available broadband photometry leads to a value of $L \sim 6.4 \pm 0.6 \times 10^{-6} L_\odot$, locating the object firmly in the brown dwarf régime.

Initially, the most surprising aspect of these observations were the infrared colours, $(J–H) \sim 0.0$, $(J–K) \sim -0.1$ – strikingly different from late M dwarfs, and even GD 165B. Low-resolution spectroscopy [O1] supplied the answer to this puzzle, showing that the 1–2.5 μm spectrum exhibits strong molecular bands cutting into the longer-wavelength sections of both the H and K atmospheric windows (Figure 9.6). These are due to methane, the presence of which had been predicted for objects with temperatures below 1,200 K ([T14]; Chapter 4). Thus both the low luminosity and the cool effective temperature provide unambiguous evidence that Gl 229B has a mass insufficient to initiate hydrogen core-burning.

Detailed comparison with theoretical models [M5] indicates that Gl 229B has an effective temperature of 960 ± 70 K and a surface gravity of less than 2,200 m s^{-2} (the solar gravity is 3×10^4 m s^{-2}). The continued presence of CO, as evidenced by the detection of the 4.7 μm (1–0) band [N6], is unexpected, since that molecule should have an abundance ratio of 10^{-4} with respect to CH_4 at 950 K [F2]. Similarly, the

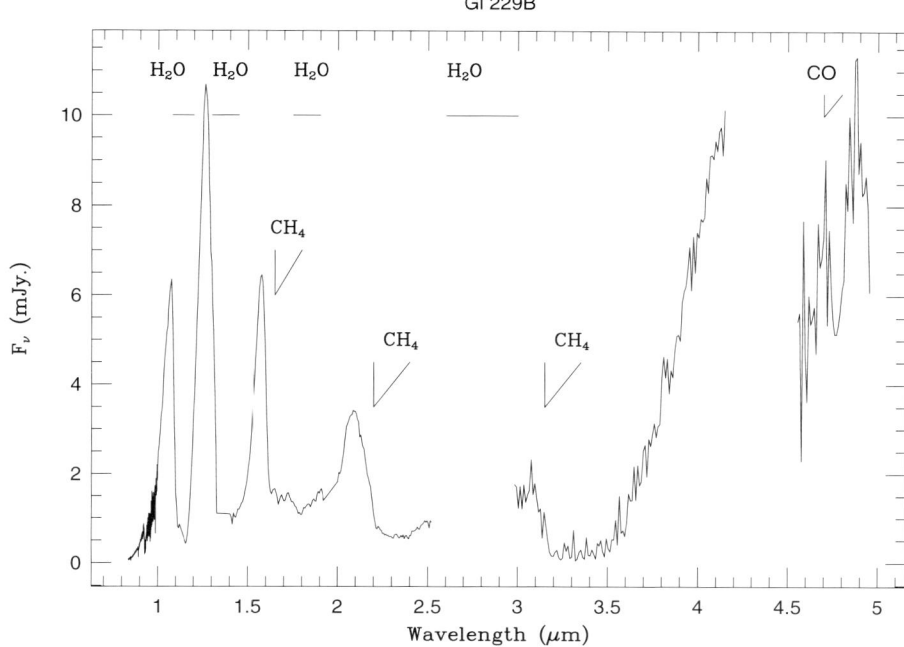

Figure 9.6. The spectral energy distribution of Gl 229B from 0.8 to 5 μm. The terrestial atmosphere is opaque due to H_2O absorption between 2.52 and 3 μm, and 4.15 and 4.5 μm. (Data from [O3], courtesy of B. Oppenheimer.)

detection of Cs atomic lines in the optical spectrum contradicts predictions that the species should be locked up in molecular CsCl. Both observations can be explained if material from deeper, hotter layers is convectively mixed in the photosphere [O3]. Alternatively, the atmospheric structure may (indeed should) be extremely non-grey, allowing one to see to different depths, and different temperatures, at different wavelengths. The concept of a single effective temperature may be irrelevant in these complex objects.

The depth of the H_2O features in the near-infrared suggests that dust is essentially absent – either because it is concentrated in clouds with only a small covering factor [T15] or because it has dropped below the level of the photosphere [O3]. As with GD 165B, the mass depends on the assumed age which, based on the absence of chromospheric activity in Gl 229A, is at least 1 Gyr, and may be as old as 5 Gyr. Under these circumstances, Gl 229B has a mass of between 30 and 60 Jovian masses.

9.4.4   PPl 15

Surveys for brown dwarf members of open clusters have the advantage that, with age, abundance and distance as known factors, observations can be targeted to cover a restricted range of luminosity and colour. Moreover, young clusters offer the best

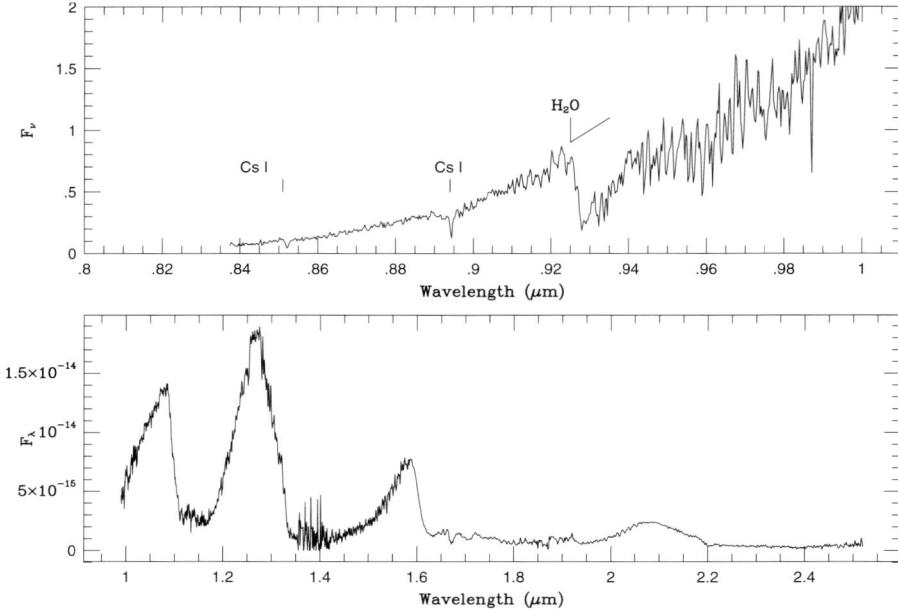

Figure 9.7. Intermediate-resolution optical and near-infrared spectra of Gl 229B. (The former data are from Keck observations by Oppenheimer *et al.* [O2]; the latter are UKIRT data from Geballe *et al.* [G1].)

chances of detecting at least higher-mass brown dwarfs, since those objects have temperatures comparable to M dwarfs and luminosities exceeding $10^{-2} L_\odot$. Given its proximity and age of $\sim 125\,\mathrm{Myr}$, the Pleiades has been a prime target of such surveys.

None of the large-area photometric surveys described in Section 7.6.3 reach luminosities below the hydrogen-burning limit. The first observations to extend to such faint magnitudes were obtained by Stauffer *et al.* [S9], [S10], using a CCD camera on the Hale 200-inch telescope. These data achieve limiting magnitudes of $V \sim 23$ and $I \sim 21$ over 0.4 square degrees in the central regions of the cluster, and 18 sources with colours and magnitudes consistent with the extrapolation of the Pleiades main-sequence were identified. Six of the latter lie close to the theoretically-expected hydrogen-burning limit. Follow-up low-resolution spectroscopy [H3], [S11] confirmed most as late-type (Sp $\geq$ M5) dwarfs with radial velocities consistent with cluster membership.

High-resolution spectroscopy provided evidence that the faintest of the PPl objects, PPl 15 ($I = 17.8$, $(V–I) = 4.6$), is substellar in mass. Spectra obtained with the HIRES spectrograph on the Keck telescope, detect the 6708 Å Li I line at moderate strength (equivalent width 0.5 Å) [B3]. The line strength indicates that lithium has been depleted to 1% of the primordial value, implying that the object is a high-mass brown dwarf. Subsequent observations have revealed the source as a double-lined spectroscopic binary, with a period of 6 days and an orbital eccentricity

0.5. The two components, each with $M_I \sim 11.5$, are estimated to have masses of 0.065 and 0.059 $M_\odot$ [B4].

Follow-up observations of fainter photometric brown dwarf candidates – notably by the Tenerife group (Rebolo, Martín, Zapatero-Osorio and collaborators) – have resulted in the identification of a number of other brown dwarfs at lower luminosities. These sources have been used to estimate the underlying mass function. In addition, the location of the boundary in the colour–magnitude diagram between dwarfs which have completely depleted lithium, and objects with only partial depletion, provides a method of estimating the age of the cluster (see Section 9.7.1).

9.4.5   Brown dwarfs in the field

Isolated brown dwarfs have now also been identified in the general field. While the majority were uncovered in the course of the near-infrared sky surveys discussed in the following section, the initial discoveries came from a variety of sources. The first field M dwarf with lithium absorption was discovered in the course of spectroscopic follow-up observations of a $B_J RI$ photographic dataset [T3]. Known as 296 A (it lies within UK Schmidt survey field 296), the colours are relatively unremarkable ((I–K) ~ 2.8), with a spectral type of M6 and Li I 6708 Å absorption with an equivalent width of 0.5 ± 0.1 Å. Based on an estimated effective temperature of $2,800 \pm 200$ K, the object is likely to have a mass in the range 0.05–0.095 $M_\odot$, and an age of between 20 and 200 Myr. Despite its youth, 296 A is not associated with any obvious young cluster or association. It remains unclear whether it formed in isolation, or whether it was ejected from a cluster through dynamical interactions. A number of similar objects have been detected along the line of sight towards the Hyades and Pleiades (as described in Section 9.7.4).

Brown dwarfs have also been identified serendipitously in proper-motion surveys. The isolated L dwarf, Kelu 1, was discovered in the course of a photographic survey aimed at finding low-luminosity white dwarfs [R11]. Rather than the featureless spectrum of a cool DC white dwarf, follow-up observations revealed absorption similar to that observed in GD 165B. Since Kelu 1 is isolated, and three magnitudes brighter than the latter object, observations extend to 4000 Å and reveal broad Na D lines, weak Hα emission and atomic lithium 6708 Å in addition to the hydride bands in the far red. Current trigonometric parallax measurements indicate a distance of 16 parsecs, implying $M_I \sim 15.8$ and $M_K \sim 10.8$; Kelu 1 now serves as the primary standard for spectral class L2 (Figure 2.4).

Even Luyten's catalogues are providing examples of brown dwarfs. LP 944–20 was rediscovered independently by Irwin, McMahon and Hazard in their photometric survey for high-redshift QSO candidates, and re-named BRI 0337–3535; low resolution spectroscopy indicates a spectral type of M9.5 [K4]. The trigonometric parallax is $\pi = 0''.201$, giving a distance of only 5 parsecs [T7] and $M_K = 11.10$, $M_{bol} = 14.32$ – significantly fainter than LHS 2065, LHS 2924 and LHS 2397a. Intermediate-resolution spectroscopy by Tinney [T8] reveals another crucial difference: LP 944-20 has lithium absorption with an equivalent width of 0.53 Å. Like PPl

15, the abundance is depleted to $<1\%$ of the primordial value, leading to an age estimate of 0.5–1 Gyr and a mass between 0.06 and 0.07 $M_\odot$.

9.5   ISOLATED BROWN DWARFS I: FIRST-GENERATION SURVEYS

Brown dwarfs are cool objects. A 10^8-year old, 0.06 $M_\odot$ brown dwarf has a surface temperature of $\sim 2,700$ K (spectral type M5), while a 10^{10}-year old 10 M_J object has cooled to ~ 200 K. The peak in the emergent energy distribution of an F_λ black-body, temperature T_{BB}, is given by Wien's displacement law,

$$\lambda_{max} = \frac{2898}{T_{BB}} \text{ μm} \tag{9.2}$$

Even allowing for the presence of extensive molecular absorption, it is clear that surveys in the near- (1–5 μm) and mid-infrared (5–15 μm) offer the best prospect of finding examples of the field brown dwarf population.

Cool brown dwarfs, typified by Gl 229B, are most luminous in the mid-infrared, and surveys at those wavelengths are best suited to the detection of older and/or lower-mass brown dwarfs. Ground-based observations longward of 3 μm are hampered by the poor transmission and high emissivity of the Earth's atmosphere. As a result, all wide-field mid-IR surveys (as opposed to pointed observations of previously-known objects) have been, or are being, carried out from space. These observations have the advantage of a much lower sky-background, although dust in the zodiacal light is a problem near the plane of the ecliptic, but the disadvantage that satellite observatories must be lifted into orbit (which imposes weight, and hence size, restriction), and must work reliably in a remote environment.

Near-infrared wavelengths, on the other hand, are readily accessible from the ground. While the JHKL passbands are not optimum for identifying cool, methane-rich brown dwarfs, the much larger telescopes available at ground-based observatories, and the less stringent constraints on instrumentation, permit observations to achieve sensitivities similar to the space-based surveys. Segregating *bona fide* brown dwarfs from the plethora of other Galactic and extragalactic sources demands both accurate near-infrared photometry and supplementary observations at shorter wavelengths. It is only recently that technology has become capable of meeting these requirements, but the first near-infrared sky surveys – none of which included brown dwarf discovery as a primary goal – set important constraints on the probable space density of these objects.

9.5.1   The Two Micron Sky Survey

Early infrared observations were limited by the availability of only single-element detectors. Under such circumstances, aperture photometry is the sole technique available for detecting sources and determining magnitudes. The observations consist of measuring the total flux with the target centred in the aperture, and determining the contribution from the sky background through measurements at an offset, blank sky position. The near-infrared background is both bright and rapidly variable in the near-infrared – particularly longward of 2 µm, where thermal emission from both atmosphere and telescope is significant (Table 1.1), necessitating cycling between on-source and off-source observations every few seconds for faint targets. Given a typical circular aperture of 10-arcsec diameter for single-channel photometry, the effective sky brightness is K ∼ 8.2 magnitudes. If the overall observation efficiency is 30%, allowing for losses in the telescope and camera optics and the quantum efficiency of the detector, a 1-m telescope detects ∼ 300,000 photons s^{-1} at this magnitude. Applying the signal-to-noise formula given in Chapter 1, a 10-second exposure gives 10% photometry at K = 13.5 – a respectively faint magnitude. Unfortunately, with a 10-arcsec aperture, approximately 7×10^9 such observations (and ∼ 2,200 years integration time) are required to cover the entire sky.

Working under these restrictions, the first near-infrared sky survey, the Two Micron Sky Survey (TMSS) [N4], aimed merely at detecting the infrared equivalent of naked-eye stars. The observations were made using a custom-built 62-inch f/1 aluminised plastic mirror, with two detectors: a set of eight PbS cells, mounted in four pairs along the north–south line, with 2.0–2.5 µm filters; and an adjacent Si detector with an I-band filter. Each PbS detector covered 10×10 arcmin, so the array of detectors spanned 40 arcmin in declination, while the Si detector had a 20-arcmin field of view. Observations were made by scanning the telescope in RA at 15 or 30 times the sidereal rate, chopping back and forth between the detectors by nodding the primary mirror. The limiting magnitudes for detection were I ∼ 9 and K ∼ 5, but the final catalogue was limited to 3rd magnitude at K. In total, the TMSS includes only 5,612 sources, none of which (obviously) are brown dwarfs; indeed, the only M-type stars are giants. However, this survey provided the first census of the infrared sky.

9.5.2   Mid-infrared surveys: AFGL and IRAS

The mid-infrared equivalent of the TMSS is the Air Force Geophysics Laboratory (AFGL) four-colour survey, covering 90% of the sky at wavelengths of 4.2, 11.0, 19.8 and 27.4 µm [P3]. These observations were made using 16.5-cm telescopes, lifted to the upper boundaries of the atmosphere in a series of nine rocket launches. A total of 2,363 sources were detected to a flux limit of ∼ 50 Jy at 4.2 µm. As with the TMSS, these objects are primarily of high intrinsic luminosity, and include most of the major star-forming regions.

The Infrared Astronomical Satellite (IRAS), launched in 1983 as a joint venture by the USA, the UK and the Netherlands, was directed more towards the average objects in the Galaxy. The main mission project consisted of an all-sky survey in four mid-infrared passbands, centred at 12, 25, 60 and 100 µm, with the detectors scanning the sky through apertures matched to the appropriate resolution for the given wavelength. Details are given in Table 9.1. Individual sources were identified and extracted from the continuous scans, and positions, flux densities and crude morphological parameters were determined. Cross-referencing scans in the four passbands permitted measurement of colours, and hence an estimate of the effective temperature of each source.

Table 9.1. The Infrared Astronomical Satellite (IRAS).

Purpose	All-sky survey at 12, 25, 60 and 100 μm.
Launched	January 1983 into 900-km 103-min orbit, $i = 99°$.
Ceased operations	November 1983; surveyed 96% of the sky.
Telescope	Two-mirror Ritchey–Chrétien, primary diameter 60 cm; cooled to between 2 and 5 K by liquid helium.
Instrumentation	62 Ge : Ga detectors over 30′-wide focal plane.
Operational mode	Great circle scans. The satellite orbit was oriented perpendicular to the Earth–Sun radius vector, with the telescope pointing radially outward from the Earth. The satellite completed one great circle scan through both ecliptic poles every 103 minutes.
Positional accuracy	~20 arcsec at 12 μm after full analysis; diffraction-limited at 26 μm.
Major catalogues	*Point Source Catalogue* (1984) 250,000 sources to ~0.5 Jy at 12, 25, 60 μm; 1.5 Jy at 100 μm *Extended Source Catalogue* *Faint Source Catalogue* (1992), $b > 10°$ 173,000 sources to ~0.2 Jy at 12, 25 μm.

The IRAS 12 μm observations are best suited to detection of brown dwarfs. However, the sensitivity is limited by both the short integration time and the small aperture of the telescope. As a result, even young, higher-mass brown dwarfs are detectable to distances of only ~1 parsec, while 10^{10}-year-old brown dwarfs could be identified only if they lay within 0.01 parsec (2,000 AU!) of the Sun. With such a tiny sampling volume, the substellar mass function would need to rise with a slope close to the Salpeter index before even a single brown dwarf detection would be expected. Even the *Faint Source Catalogue* – derived from extensive reprocessing of the survey scans covering latitudes above 10° and with a detection limit of 0.17 Jy at 12 μm – is capable of detecting, at most, four brown dwarfs over the full sky.

These pessimistic predictions have been upheld by analysis of the catalogues. There is a total of 250,000 sources in the PSC, and 173,000 sources in the FSC. Almost all of these can be eliminated by cross-referencing against the POSSI or UKST photographic sky surveys, and none of the remaining sources have proven to be brown dwarfs when subjected to more detailed scrutiny. With so few brown dwarfs predicted under the most optimistic circumstances, however, the absence of detections sets only weak constraints on the local space density.

9.6   ISOLATED BROWN DWARFS II: SECOND-GENERATION SURVEYS

The development of array detectors has permitted deep, wide-field surveys at near-infrared wavelengths. Arrays offer two advantages: first, larger areal coverage in a single exposure, and with higher spatial resolution than was possible with aperture photometry; second, simultaneous measurement of the sky background, and therefore more accurate sky subtraction. Two separate, but similar projects are currently taking advantage of these advances to survey to flux levels more than 10,000 times fainter than achieved by the TMSS, extending the source catalogues from 5,600 to $>10^8$ objects.

At optical wavelengths, all of the currently available large-scale surveys rely on photographic plate material – the Palomar Oschin Schmidt POSS I and POSS II surveys in the north, and the UK and ESO Schmidt surveys in the south. Optical CCDs are more sensitive than photographic plates by factors of ten or more, but until recently were limited to relatively small size. Now that $2,048 \times 2,048$ format chips are readily available, wide-angle surveys extending to much fainter magnitudes than was possible with IIIa photographic emulsion may be undertaken.

9.6.1 The Deep Near Infrared Survey (DENIS)

The Deep Near Infrared Survey is being developed by a European/Brazilian consortium, headed by a group from the Observatoire de Paris [E2], [C5]. The basic operational parameters are given in Table 9.2. The survey covers wavelengths between 0.8 and 2.5 μm; Figure 9.8 illustrates an advantage offered by the choice of passbands. Not only can one identify extremely cool GD 165B-like objects, which are red in both infrared and optical/infrared colours, but 'methane dwarfs' like Gl 229B stand out through having red (I–J) colours and blue (J–K) colours. A potential problem lies in the bright limiting magnitude in the I-band: late-type M dwarfs with moderately red (J–K) colours and K magnitudes fainter than 13 may not be detected at I, requiring further observations to distinguish these sources from cool candidate brown dwarfs.

DENIS data covering 1% of the sky were analysed in a 'brown dwarf mini-survey' as a preliminary search for unusually cool objects [D4]. None of the sources

Table 9.2. The Deep Near Infrared Survey (DENIS).

Purpose	Near-infrared survey of the southern sky, $-88° < \delta < +2°$.
Telescope	ESO 1.0 m, La Silla, Chile.
First light	January 1996.
Projected duration	3–5 years.
Instrumentation	2×256-square HgCdTer (NICMOS3) arrays at J, K_S; one 1,024-square optical CCD at I-band; dichroics allow simultaneous observations.
Plate-scales	1 arcsec pix^{-1} at I; 3 arcsec pix^{-1} at J, K_S.
Operational modes	Step and stare along strips in declinations.
Limiting magnitude	S/N = 10 at I ~ 17.5, J ~ 16, K_S ~ 13.
Projected catalogues	Point source and extended source.

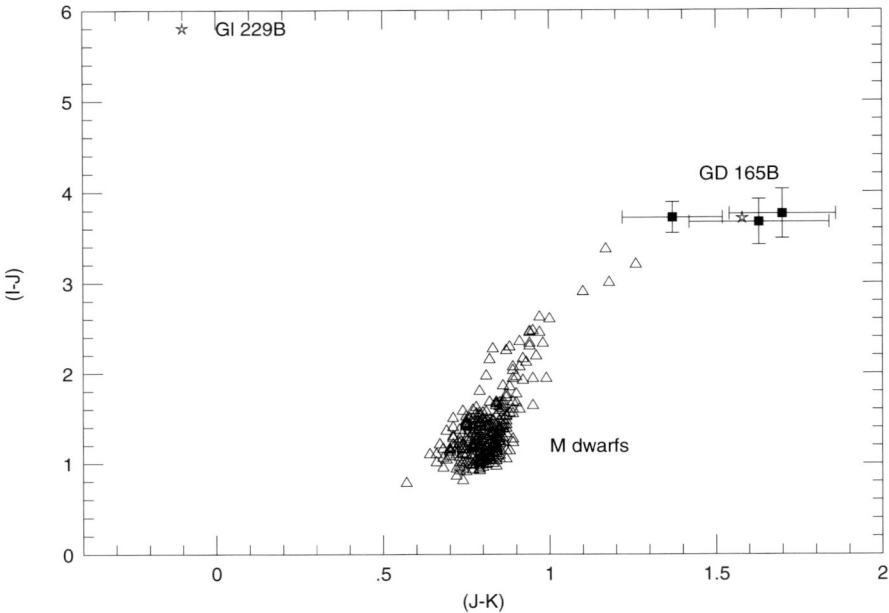

Figure 9.8. The ((I–J), (J–K)) two-colour diagram for late-type stars. The positions occupied by GD 165B and Gl 229B are marked by stars, while the three brown dwarf candidates from the DENIS mini-survey are identified as solid squares.

detected prove to have colours matching Gl 229B, but three objects lie in the vicinity of GD 165B (Figure 9.8). Follow-up optical spectroscopy shows that all three are L dwarfs, with weak or no TiO and VO absorption, strong alkali lines and metal hydride absorption bands (see Section 4.7) [T6]. One of the three – DENIS-P 1228.2-1547 – has a strong (2.3 Å equivalent width) lithium absorption line at 6708 Å, confirming at least that object as substellar in mass. None show evidence of methane absorption at near-infrared wavelengths [D4], [T12]. Under the newly defined classification system illustrated in Figure 2.5, their spectral types are L3, L5 and L7.

9.6.2   The Two Micron All Sky Survey (2MASS)

The Two Micron All Sky Survey [S8] is similar in concept to DENIS. However, unlike DENIS, the 2MASS survey includes only near-infrared observations, in the J, H and K_S passbands (Table 9.3). 2MASS data can be matched (astrometrically) against scans of the POSS I, POSS II and UK Schmidt surveys compiled by the US Naval Observatory, Flagstaff [M14]. Those photographic observations provide optical/infrared colours (or lower limits) for candidate late-M and L-type dwarfs. Indeed, the initial sample of 2MASS brown dwarf candidates was selected using three criteria: $(J–K_S) > 1.3$; $K_S < 14.5$; and no optical counterpart visible on the

Table 9.3. The Two Micron All-Sky Survey (2MASS).

Purpose	All-sky near-infrared survey.
Telescopes	1.3-m telescope at Mt. Hamilton, Arizona, USA.
	1.3-m telescope at Cerro Pachón, Chile.
First light	June 1997 (north); March 1998 (south).
Projected duration	4 years.
Instrumentation	3×256-square NICMOS3 arrays at J, H and K_S; dichroics allow simultaneous observations;
Plate-scale	2.0 arcsec pix^{-1}.
Operational mode	Great circle scans along $6°$ arcs.
Limiting magnitudes	$S/N = 10$ at $J \sim 16.5$, $H \sim 15.8$, $K_S \sim 14.5$.
Projected catalogues	Point source, extended source, image atlas.

red-sensitive POSS I E plates (Figure 9.9). The last criterion ensures an $(R–K_S)$ colour redder than 5.5 magnitudes, corresponding to a spectral type later than M6.

2MASS has two significant advantages over DENIS: a higher sensitivity limit; and, for the northern survey, access to follow-up observations using the Keck 10-m telescopes and their associated instrumentation, notably the Low Resolution

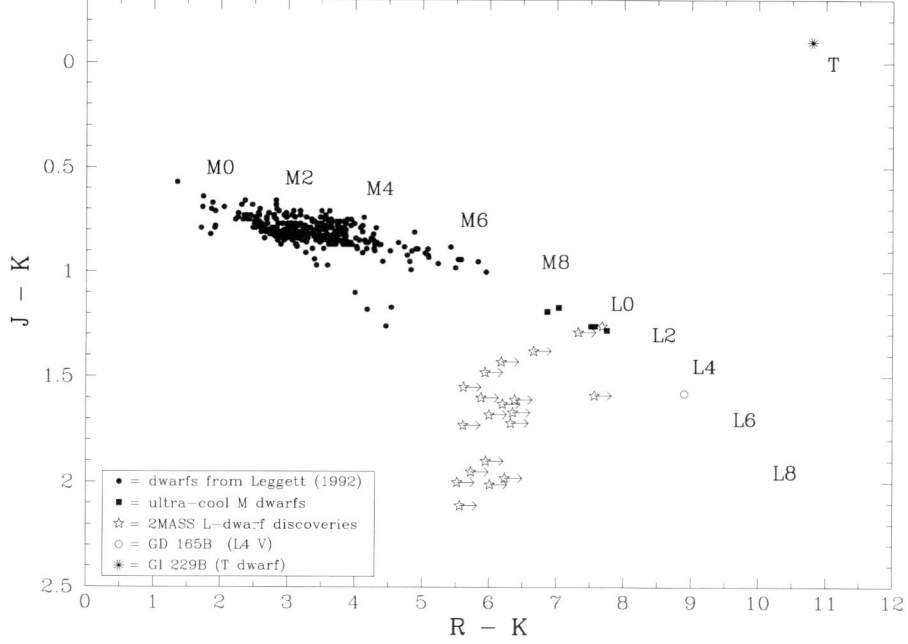

Figure 9.9. Selection criteria for the initial 2MASS L dwarf sample: sources were targeted for follow-up spectroscopy based on red (J–K) colours and non-detection on red POSS I plates. (From [K6], courtesy of the *Astrophysical Journal*.)

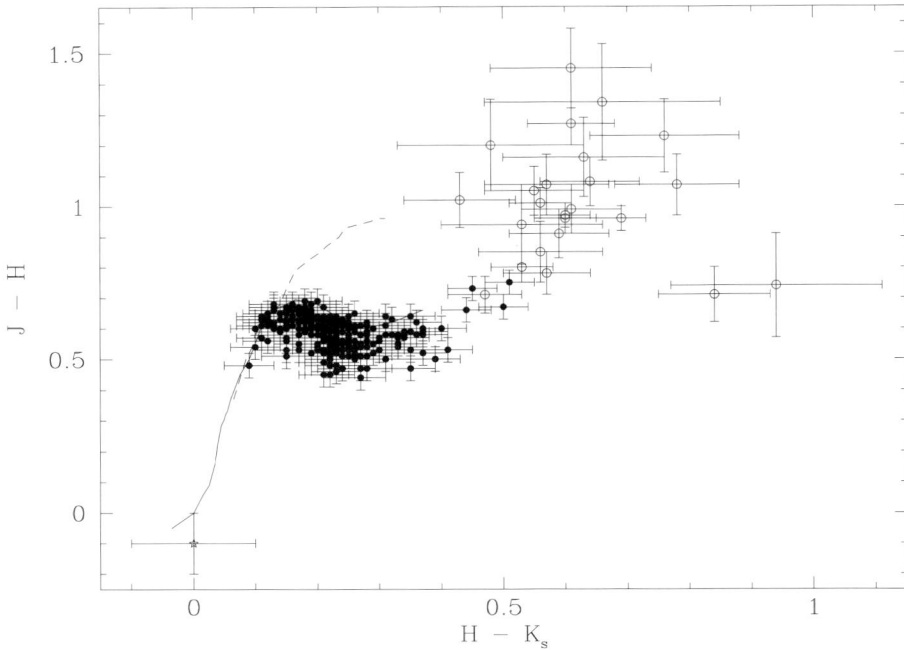

Figure 9.10. The ((J–H), (H–K)) two-colour diagram for M dwarfs (solid points) and L dwarfs (open squares). (From [K6], courtesy of the *Astronomical Journal*.)

Imaging Spectrograph (LRIS) [O1] and the HIRES echelle spectrograph [V2]. The initial analysis of 2MASS data [K6] is based on scans of an area of 370 square degrees, comparable in size to the DENIS mini-survey, and has resulted in the identification of 20 L dwarfs, only one of which meets the DENIS detection criterion of $K_S < 13.5$. These objects form the basis of the definition of spectral class L illustrated in Figure 2.4. Subsequent observations of other candidates have raised the total number of 2MASS-identified L dwarfs to more than 85 sources.

Section 4.7 outlined how a comparison between the molecular species present in the L dwarf spectra and theoretical predictions suggests that these objects span an effective temperature range of $\sim 2{,}000$–$\sim 1{,}200\,K$, with the lower boundary set by the onset of methane absorption. The latter value is uncertain by at least $\pm 150\,K$. Dust-free theoretical models of brown dwarfs in this temperature range predict (J–K) colours scarcely redder than those of late-type M dwarfs, with rapid blueward evolution. These predictions are clearly at odds with the observations (Figure 9.10). Indeed, the near-infrared colours are sufficiently distinctive to allow late-type L dwarfs to be identified from (J–K) alone. The resolution of this discrepancy lies in incorporating the effects of dust formation within the atmosphere at temperatures below $\sim 2{,}600\,K$, or spectral types later than $\sim M7$. TiO is expected to condense first, as perovskite ($CaTiO_3$) or calcium titanate, followed by VO, the metal hydrides and, at much lower temperatures, the alkali metals, Na, Cs, Rb, K and Li, as chlorides or hydroxides. Extensive theoretical work is currently underway by a number of groups (notably Fegley and Lodders, Burrows and collaborators, and Allard and collaborators) investigating the complex evolution of the atmospheres of these cool dwarfs. It is clear that atmospheric dynamics (cloud formation) are becoming important: dusty atmospheres which retain the dust at the height at which it forms become increasingly redder with decreasing temperatures, exceeding the colours of the reddest dwarfs by a substantial margin. The redward evolution is probably truncated as the dust 'rains out' to depths below the 'photosphere', although the latter term should be used with caution. L and T dwarf atmospheres are extremely non-grey, with the $\tau = 1$ level varying significantly in physical depth as a function of wavelength. In many ways, modelling these atmospheres is more akin to planetary science than to stellar astrophysics.

One of the consequences of dust formation is high atmospheric transparency at optical wavelengths as the main molecular opacity sources (TiO, VO) are removed to the solid phase. As a consequence, the $\tau = 1$ level lies deep within the atmosphere, and the alkali atomic lines become prominent due to the high column density, high gas-pressure and increased contrast relative to the continuum. These lines increase in strength with decreasing temperature until the parent elements form molecular species. In the case of the two most abundant elements, sodium ($[Na] = 6.31$) and potassium ($[K] = 5.13$), the lowest-excitation (resonance) lines achieve white dwarf proportions in the coolest dwarfs. The Na D lines form a $1500\,\text{\AA}$-wide bowl-like depression by spectral type L5 (Figure 9.11), while the K I $7665/7699\,\text{\AA}$ doublet broadens to a width of $500\,\text{\AA}$ by type L8, and exceeds $1000\,\text{\AA}$ amongst the cooler T dwarfs (see further below). Both of these *atomic* lines become sufficiently strong that they affect the flux emitted in the *broadband* V and I passbands respectively.

There are unequivocal brown dwarfs amongst the field L dwarfs. Approximately one-third of the 2MASS L dwarfs have detectable lithium absorption (Figure 9.12), with equivalent widths ranging from $\sim 1.5\,\text{\AA}$ to almost $15\,\text{\AA}$ among the later types. Observations of the extended 2MASS sample of L dwarfs suggests that the relative fraction with lithium increases with later spectral types. This is not unexpected, since the relative number of very low-mass stars and higher-mass brown dwarfs should decrease towards lower temperatures. Hα emission is still present in almost 25% of the sample, but is confined mainly to earlier-type L dwarfs (spectral types $<$L4). As discussed in Sections 5.6 and 5.7, the underlying origin of the residual chromospheric activity in these dwarfs remains a subject of investigation.

At some critical temperature, T_{meth}, the thermochemical balance for carbon shifts from CO to CH$_4$ as the dominant molecule. At that point, methane absorption will start to affect the flux emitted in the H and K bands, and the dwarf must evolve blueward in the JHK diagram away from the L dwarfs to a point near Gl 229B. To date, infrared spectroscopy has not revealed any indication of CH$_4$ absorption in the $2.2\,\mu$m overtone band in any of the field L dwarfs – even the latest known (spectral type L8) (Figure 9.13). If one considers the relationship between colour and spectral type (Figure 9.14), the effects of methane absorption in T dwarfs is striking, but there is little evidence for a blueward trend in colour among the latest L dwarfs. There may be a plateau in the (J–K$_S$) colour due to H$_2$ absorption in the K-band in late L

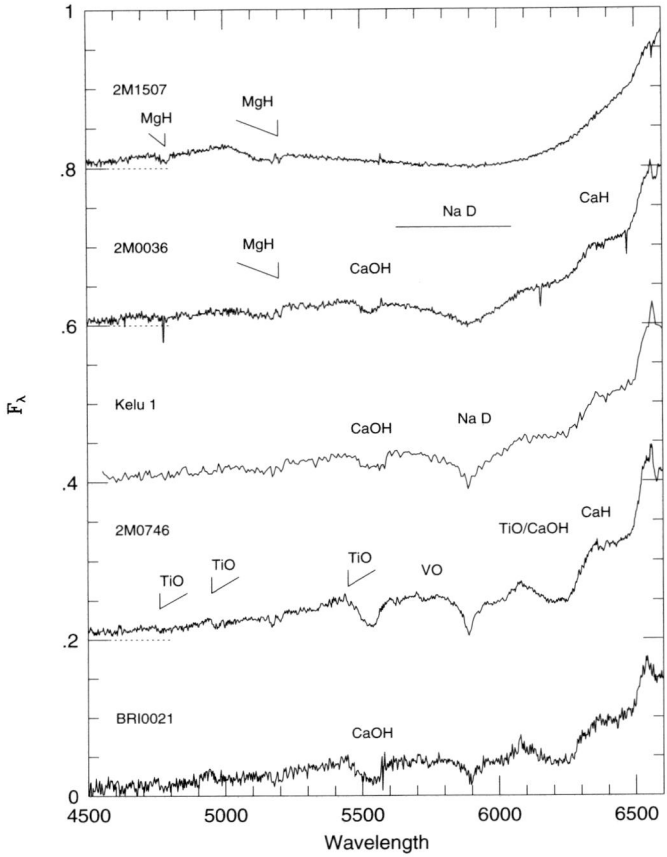

Figure 9.11. Optical blue–green spectra of late M and L dwarfs. Note the behaviour of the Na D lines in the latter types. (From [R8], courtesy of the *Astronomical Journal*.)

dwarfs. Theoretical calculations place T_{meth} in the broad range of 1,100–1,500 K [B11], [L3], depending on the appropriate pressure, but suggest a relatively rapid transition in an homogeneous atmosphere. The latter condition may not hold in these cool brown dwarfs.

The colour–magnitude diagram outlined by L dwarfs with known trigonometric parallaxes (the latter determined primarily by the US Naval Observatory, [D1]) suggests that the L8 spectral class is only a few hundred degrees hotter than the measured temperature of Gl 229B ($\sim$970 K). This implies that the L/T transition is rapid. The lowest luminosity L dwarf, 2MASSJ1523 – a wide companion of the nearby star Gl 584 – is only $\sim$0.4 magnitudes brighter at M_J than Gl 229B. Current estimates indicate that late L dwarfs and T dwarfs have similar bolometric corrections in the J-band, with $BC_J = 1.5$–2 magnitudes, implying that the difference

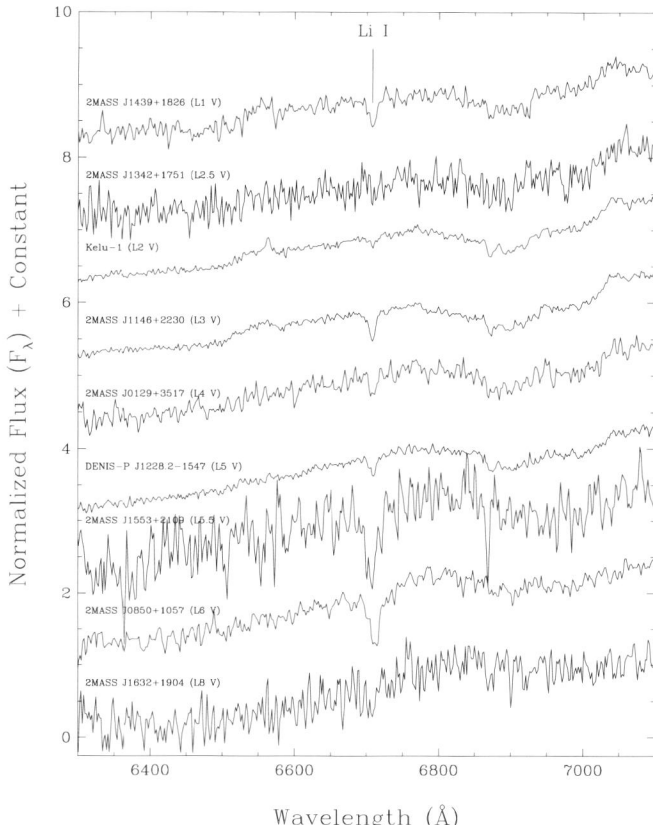

Figure 9.12. The 6300–7100 Å region of the spectra of several L dwarfs with Li 6708 Å absorption. Kelu 1 and 2MASSJ1146 also exhibit Hα in emission. (From [K6], courtesy of the *Astrophysical Journal*.)

in bolometric magnitude between L8s and Gl 229B is no more than ~ 1 magnitude. Since degeneracy fixes brown dwarf radii at a nearly constant value, it can be assumed that

$$\frac{L_1}{L_2} = \frac{T_1^4}{T_2^4} \qquad (9.3)$$

Hence, $\Delta M_{bol} = 1$ magnitude implies $\Delta T (\text{LB-Gl 229B}) \sim 15\%$ (~ 250 K), or $T_{\text{eff}}(\text{L8}) \sim 1{,}200$ K. Clearly, extensive theoretical and observational studies are required to verify this preliminary calculation.

Gl 229B-like objects have been referred to as 'methane dwarfs'. This is misleading, since methane is also expected to be present in the outer atmospheres of cooler L dwarfs. Indeed, CH_4 has a fundamental absorption band at 3.3 μm which is stronger than the overtone feature, and is predicted to be detectable in late-type L dwarfs,

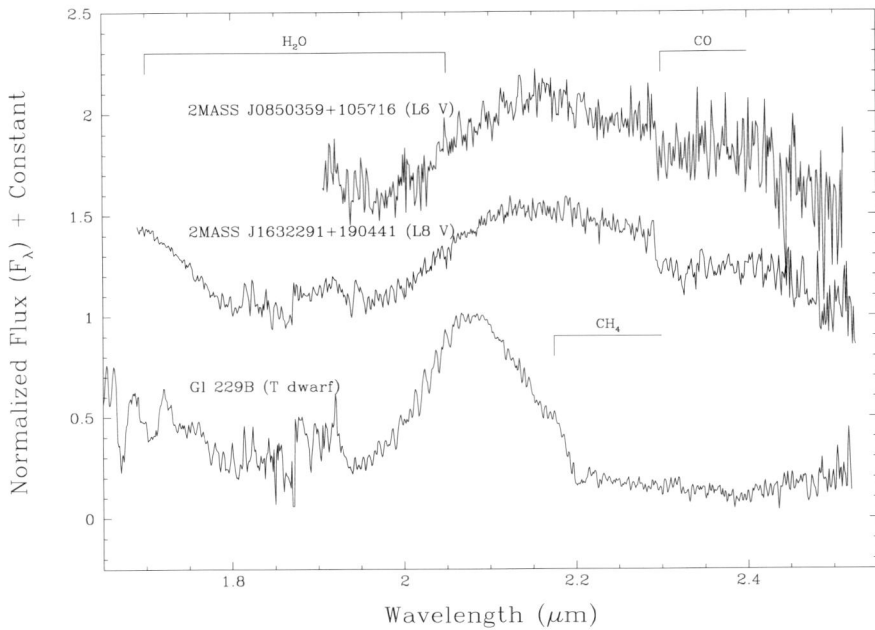

Figure 9.13. Infrared spectra of L dwarfs. The depression at ~2.12 μm in these spectra is due to H_2 absorption, but there is no evidence for the CH_4 evident in Gl 229B. (Spectra obtained by S. K. Leggett on UKIRT; from [K6], courtesy of the *Astrophysical Journal*.)

before the 2.2 μm feature becomes prominent. However, observations to date, which include dwarfs of spectral type L7, have not yet succeeded in identifying the feature. It is possible that photodissociation (due to background Galactic radiation) may inhibit CH_4 formation in the low density outer atmosphere.

Field dwarfs with properties similar to those of GL 229B are starting to be uncovered by deep, wide-angle photometric surveys. The first of these objects were identified in the Sloan Digital Sky Survey (see below). Shortly afterwards, four further examples were identified from an exhaustive search of the initial 2MASS scans of the northern sky [B12]. A similar number have since been discovered in the southern sky, including Gl 570D, almost 1 magnitude fainter than Gl 229B [B13]. The four northern 2MASS T dwarfs form a complete sample to a J magnitude limit of ~16 within an area of ~1,800 square degrees. The overall scarcity of T dwarfs folds into estimates of the substellar mass function, as described in Section 9.9.

9.6.3   The Sloan Digital Sky Survey (SDSS)

The Sloan Digital Sky Survey [G10] is a combined imaging and spectroscopic survey which will cover π steradians of the sky at high Galactic latitudes (Table 9.4). The

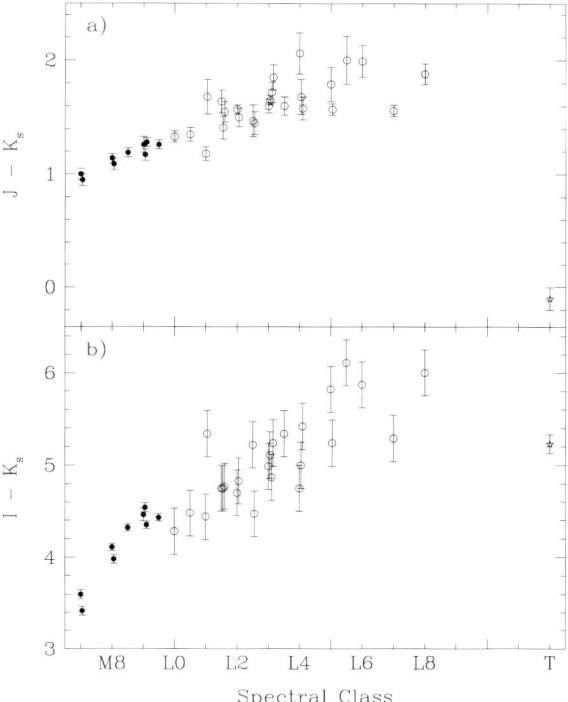

Figure 9.14. (I–K) and (J–K) colours as a function of spectral type for 2MASS L dwarfs. (From [K6], courtesy of the *Astrophysical Journal*.)

main scientific goals lie in the field of extragalactic astronomy, but the photometric data, taken in the u', g', r', i', and z' passbands,[2] extend to extremely faint magnitudes. Although primarily an optical survey, SDSS is capable of detecting DENIS/2MASS L dwarfs in at least the i and z passbands, and analysis of the initial scans resulted in the identification of several such objects. More significantly, these data produced the first example of an isolated T dwarf – a field counterpart to Gl 229B.

The first SDSS T dwarf (1624+0029) is distinguished by its having extremely red colours: $i \sim 23.8$, $(i - z) \sim 3.5$. In comparison, late-type M dwarfs, such as LHS 2924, have $(i - z) \sim 1.5$, while late-type L dwarfs have $(i - z) \sim 2.5$. Follow-up near-infrared photometry of $1624 + 0029$ reveals colours similar to those of Gl 229B – (J–H) ~ -0.3, (J–K) ~ -0.3, (I–K) ~ 7.3 – and near-infrared spectra confirm the presence of CH_4 absorption [S13]. The optical spectrum shows Cs I 8521 Å and 8943 Å atomic lines, together with strong H_2O absorption (Figure 9.16). Moreover,

---

[2] The filters are based on the modified Gunn system, with $r'(\lambda_{eff}) \sim 6230$ Å, $i'(\lambda_{eff}) \sim 7620$ Å, $z'(\lambda_{eff}) \sim 9130$ Å [F4]. The r' and i' magnitudes can be approximately transformed to the Cousins systems by: $R_C \sim r' - 0.11(r' - i') - 0.1$, $I_C \sim i' + 0.4$.

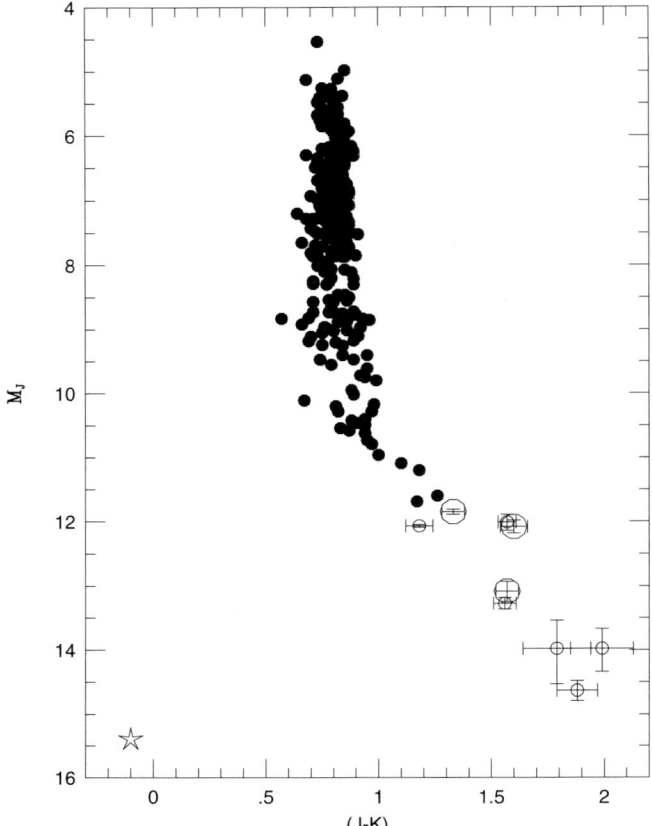

Figure 9.15. The $(M_J, (J-K_S))$ colour–magnitude diagram defined by M dwarfs (solid symbols) and L dwarfs with measured trigonometric parallaxes (open circles). Gl 229B is identified by a star.

the detection of flux at 7000 Å suggests that the highly-broadened potassium doublet, rather than atmospheric dust, is responsible for the steep spectral slope between 8000 and 8500 Å. This additional absorption is probably responsible for much of the ~ 1.5 magnitude offset in the (I–J) colour between Gl 229B and the latest L dwarfs (Figure 9.17).

SDSS 1624+0029 was only identified in late April 1999, and more extensive observations will obviously be obtained, but with a J magnitude of 15.5, the object probably lies within 10 parsecs of the Sun. A second SDSS T dwarf with similar properties was identified in early May 1999. Given that both are selected from a magnitude-limited sample, statistically these are likely to represent the most luminous, and therefore the hottest, T dwarfs. The similarity between these objects and Gl 229B ($T_{eff} \sim 970$ K) is therefore very interesting.

SDSS is in its earliest phase, but this result shows considerable promise for brown dwarf detection. The total area covered by the preliminary scans amounts to about

Table 9.4. The Sloan Digital Sky Survey (SDSS).

Purpose	Deep optical imaging/spectroscopic survey over π steradians at high Galactic latitude to study cosmological large-scale structure.
Telescope	2.5-m at Apache Point, New Mexico.
First light	June 1998.
Projected durations	2000–2005.
Instrumentation	30 2,048-square photometric CCDs; 24 astrometric CCDs; two 600-fibre multi-object fibre spectrographs.
Operational mode	Great circle scanning for imaging.
Limiting magnitudes	S/N = 10 at $r' \sim 23.1$, $i' \sim 22.5$, $z' \sim 20.8$.
Projected catalogues	Point-source and extended-source; image database; redshifts for $\sim 10^6$ galaxies; spectroscopic catalogue.

400 square degrees – approximately 1% of the sky. The colour-magnitude diagram plotted in Figure 9.17 emphasises that combining the deep SDSS optical data with the 2MASS infrared scans will provide an extremely effective method of identifying cool brown dwarfs in the field.

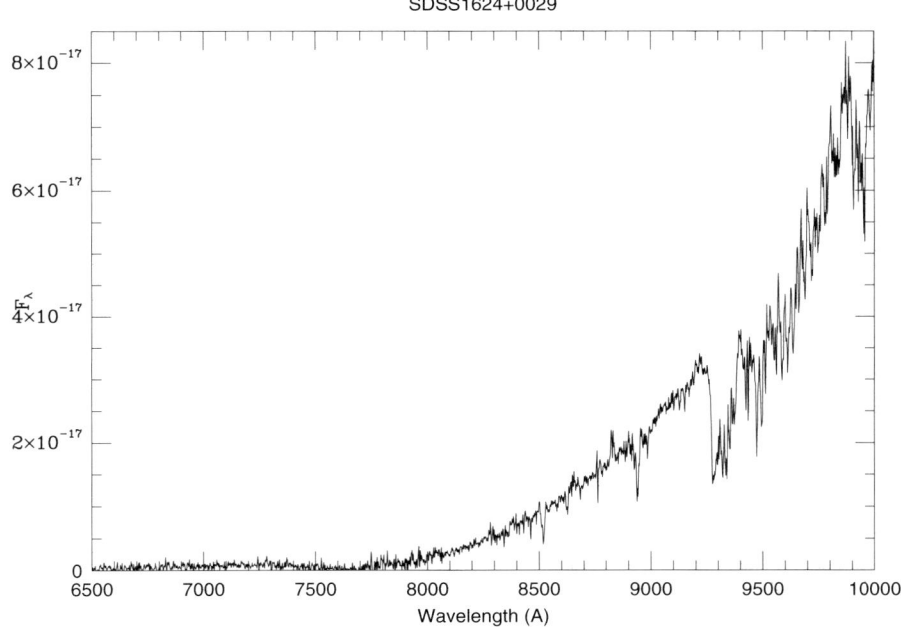

Figure 9.16. The far-red optical spectrum of the first field T dwarf, SDSS 1624+0029. (Keck LRIS spectrum obtained by the 2MASS consortium.)

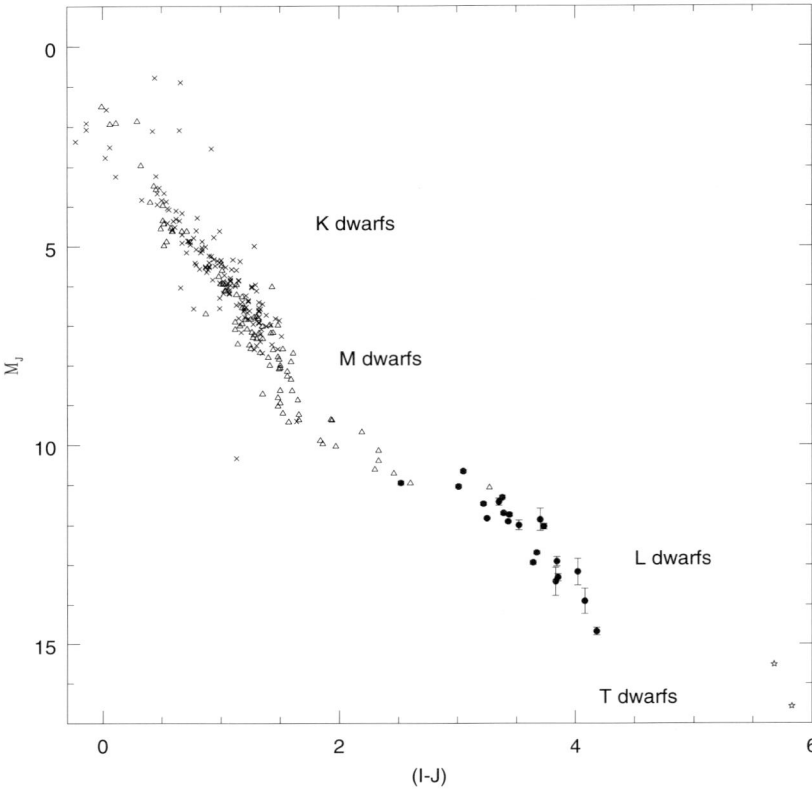

Figure 9.17. The $(M_J, (I–J))$ colour–magnitude diagram for M, L and T dwarfs with well-determined parallaxes.

9.7   BROWN DWARFS IN CLUSTERS

Surveys for brown dwarfs in clusters face the same problems encountered by the luminosity function studies described in Chapter 7: segregating cluster members from the dominant background population of field stars in the few nearby inter-mediate-age systems where brown dwarfs are observationally accessible; and addressing the variable absorption in young clusters, such as ρ Ophiuchi and the Orion Nebula Cluster. In both situations there is the advantage of searching for brown dwarfs which are still at M dwarf-like temperatures and luminosities, but mass determinations must rely entirely on theoretical models. As with surveys of the general field, the availability of infrared arrays has had a considerable impact, although wide-field I-band CCD (and photographic) imaging remains an effective tool for identifying intermediate-age brown dwarfs.

9.7.1   The Pleiades

The proximity and relative youth of the Pleiades make it the target of choice for brown dwarf surveys. Following the initial deep CCD surveys of Stauffer *et al.* [S9], the cluster has been the target of a systematic photometric study undertaken by a team of astronomers at the Instituto de Astrofísica de Canarias, Tenerife. Their first survey covered only 0.15 square degrees to R $\sim$ 20 and I $\sim$ 19 [Z1] and identified nine candidate low-mass members, seven of which prove to be background and foreground M dwarfs [M6]. Two objects, however – identified as Teide 1 and Calar 3 (named after Canarian volcanos) – have strong lithium absorption and are confirmed as Pleiad brown dwarfs [R2], [R3].

More recent surveys [B9], [Z3] have extended coverage to magnitudes as faint as I $\sim$ 22 within the central square degree. Almost 50 brown dwarf candidates have been identified, mainly through their location in the (I, (I–Z)) colour–magnitude diagram (where Z has $\lambda_{eff} \sim 9500$ Å). Not all are confirmed as Pleiads, but spectroscopy of several sources between 20th and 21st magnitude at I has resulted in the identification of the first Pleiad L dwarf [M8]. The faintest objects are expected to have masses of only $\sim 0.03 \, M_\odot$ if they are actually cluster members.

The detection of lithium in low-mass Pleiades dwarfs confirms them as brown dwarfs: the distribution of these objects in the colour–magnitude diagram provides a means of estimating the age of the cluster. Current theoretical models – both numerical [D1], [B1], [B10] and analytical [B8] – predict the rate of lithium depletion with time, and hence predict the evolution of the location of the lithium-depletion boundary as a function of mass. Extensive observations of Pleiads places this boundary at I $\sim$ 17.8 (Figure 9.18), corresponding to an age of ~ 125 Myr [S12]. This is almost twice the 70 Myr which had been found in many studies of the higher-mass cluster stars, but is consistent with main-sequence fitting analyses based on models which incorporate modest amounts of convective overshoot (such as [S2], [V1]). Thus, brown dwarfs serve as surprisingly effective chronometers.

9.7.2   Other open clusters

No other cluster has been subjected to such intense scrutiny as has the Pleiades, but individual low-mass dwarfs at or near the hydrogen-burning limit have been identified in at least two clusters.

α Persei cluster

AP J0323+4853 is the lowest-luminosity member so far identified in the 50–100-Myr-old α Persei cluster. Discovered by the Tenerife group [R1] through a deep (R, (R–I)) survey of the central regions of the cluster, the star has strong Hα emission (EW = 60 Å), a spectral type of M6 and a radial velocity consistent with cluster membership [Z1]. The luminosity is $\log L/L_\odot = -2.46$ or $M_{bol} = 10.8$. Uncertainties in both the effective temperatures and the cluster age prevent a definitive mass estimate. Adopting $T_{eff} \sim 3040 \pm 100$ K, a comparison with theoretical isochrones constrains the mass to lie between 0.05 and 0.1 $M_\odot$. However, there is no evidence

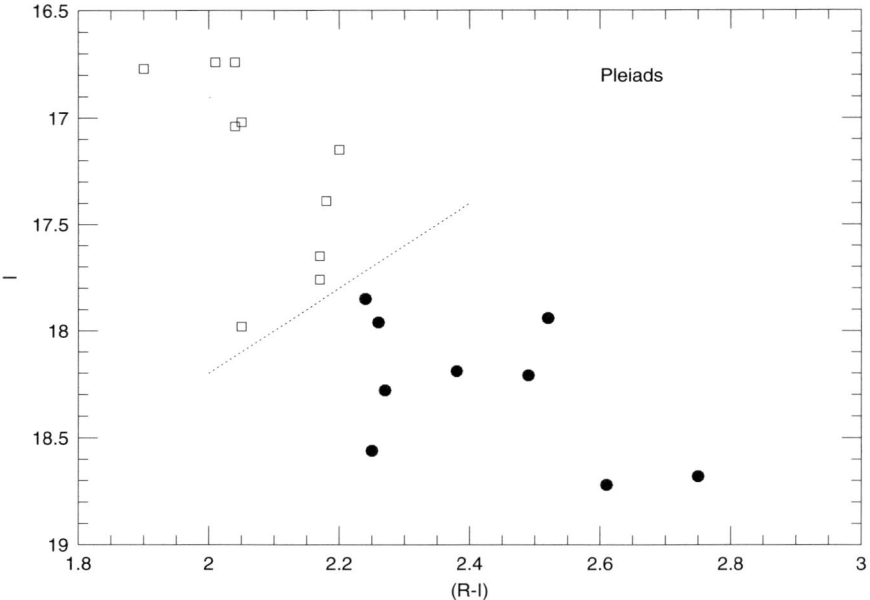

Figure 9.18. The lithium depletion boundary in the Pleiades. Solid points are dwarfs with lithium; open circles mark cluster members with no detected lithium. (Data from [S12].)

for lithium 6708 Å absorption, indicating that the bulk of the primordial lithium has been destroyed through nucleosynthesis. Based on lithium depletion calculations, this sets a lower limit to the mass of 0.08 $M_\odot$ if α Persei has an age of 100 Myr (current estimates favour 85 Myr). Thus, indications are that AP J0323+4853 lies close to, but above, the hydrogen-burning limit.

Praesepe

With an age of 600–800 Myr and a distance of 177 parsecs, Praesepe is a more difficult target for brown dwarf surveys. To date, deep photometric surveys have produced only one strong candidate: RPr1, I $\sim$ 21.0 and (I–K) $\sim$ 4.6, with a spectral type of M8.5 [M3]. Cluster membership is not yet confirmed, but if it is a member, the bolometric luminosity of $\log L/L_\odot = -3.6$ implies a mass of $0.075 \pm 0.01\, M_\odot$. Again, this places the dwarf close to the hydrogen-burning boundary. This dearth of spectroscopically or astrometrically confirmed members should be contrasted with the large number of unconfirmed candidates from purely photometric surveys in clusters.

9.7.3   Young embedded protoclusters

Chapter 8 outlined the problems associated with identifying substellar-mass objects in star clusters with ages of less than a few million years. The striking advantage offered by such locations is that even 0.01 $M_\odot$ brown dwarfs have luminosities of at least $10^{-2} L_\odot$, making them accessible to observational study. However, variations in the line-of-sight reddening from star to star, infrared excess radiation (and further obscuration) from residual circumstellar material, and a dispersion in age of 10^5 years or more within a given cluster, all serve to complicate the interpretation of the observed luminosities. Accurate masses can be derived only by determining photo-spheric temperatures, and by matching the position of each star against theoretical isochrones. Thus, the derived $\Psi(M)$ is model-dependent, although the relative cluster-to-cluster uncertainties can be minimised by calibrating against a self-consistent set of models.

Candidate brown dwarfs have been identified in several young star clusters – indeed, one could present a good case that the first brown dwarfs were detected through this type of observation (in, for example, ρ Ophiuchi [C3]). However, in most clusters the detection was a statistical excess in source counts, rather than observations of individual objects, while uncertainties in the reliability of theoretical models cloud the mass calibration.

Amongst the relatively small number of individual brown dwarf candidates, one of the strongest is $162349.8-242601$ in the 1 Myr-old ρ Ophiuchi cluster [L5]. This object is the least-obscured of the four brown dwarf candidates identified by Rieke and Rieke [R9], and has a spectral type of M8.5, with molecular and atomic line-strengths that indicate a lower gravity than expected for a dwarf on the main sequence. Given a distance of 160 parsecs, the luminosity is $\log L/L_\odot = -2.58$, while the effective temperature is $\sim 2{,}600$ K. Matching these parameters against models leads to a mass estimate of $0.035 \pm 0.025\ M_\odot$. These are infrared observations of a number of other low-mass sources in the same cluster [C4].

9.7.4   Not quite the Hyades

As the nearest substantial cluster, the Hyades should be as useful for brown dwarf surveys as the Pleiades, despite having an age of 625 Myr. However, as discussed in Section 8.6.2, the extensive surveys undertaken to date show that the cluster has very few low-mass members – either through dynamical depletion or through their never having formed. Nonetheless, deep photometric surveys in this region have succeeded in identifying brown dwarfs, but none are members of the cluster.

The most sensitive search for low-mass Hyads [L2] identified 12 potential members within the central 25 square degrees of the cluster. Follow-up spectroscopic observations by Reid and Hawley [R6] show that only one object, (LH 0418+13), is a possible cluster member, with a mass of $\sim 0.083\ M_\odot$, and six of the remaining stars are background M dwarfs. Five objects, however, have unusual spectra: weak Na I 8192/8196 Å, barely detectable K I 7665/7699 Å, strong Hα emission and weak CaH absorption (Figure 9.19). These are characteristic of low surface gravity and the most reasonable interpretation is that the objects are low-mass T Tauri stars. Indeed, trigonometric parallax measurements [H5] of three dwarfs place them at distances between 150 and 250 parsecs – comparable with the distance to the nearby Taurus–Auriga star-forming region. Two have lithium absorption, confirming their

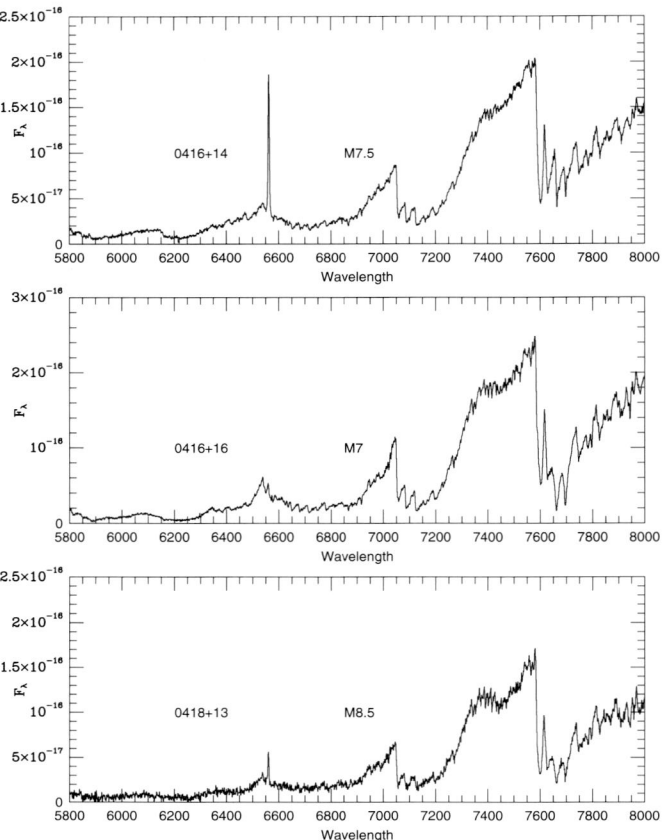

Figure 9.19. Low-resolution spectra of three of the Leggett/Hawkins candidate Hyads. LH 0418+13 is a possible cluster member; LH 0416+16 is a background M dwarf; LH 0416+14 is a background T Tauri star with a mass close to the hydrogen-burning limit. (From [R6], courtesy of the *Astronomical Journal*.)

substellar nature. As Figure 9.20 shows, the derived luminosities and effective temperatures indicate masses between ~ 0.08 and $0.05\,M_\odot$.

These young brown dwarfs are not particularly rare: Oppenheimer *et al.* [O2] found a similar late-type T Tauri star in the direction of the Pleiades, while Gizis *et al.* [G3] have identified several other such dwarfs in 2MASS data covering the Hyades. The main conundrum centres on their environment. Typical age estimates are 10 to 20 Myr, while the dwarfs lie up to 50 parsecs from the nearest star-forming region. A similar phenomenon is evident among the near-solar-mass T Tauri stars detected by ROSAT [N5]. This displacement requires a velocity difference of 5–10 km s^{-1} with respect to the parent cluster, and it remains unclear how such substantial differences can be achieved for so many stars. Further study of these sources may provide interesting information on the initial disruption of open clusters.

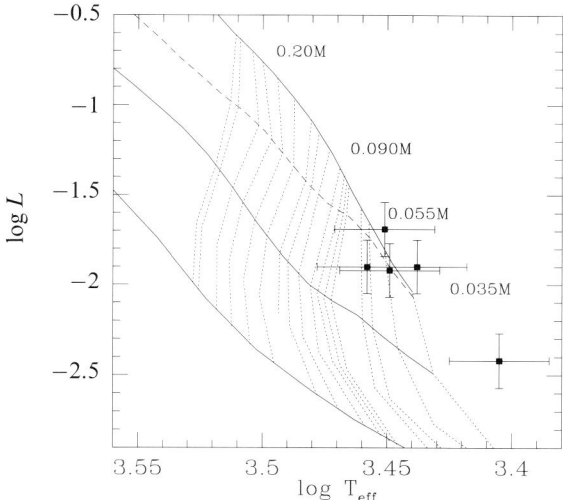

Figure 9.20. An H–R diagram showing the location of the five T Tauri dwarfs behind the Hyades, identified in [R6]. The theoretical tracks are from [B1].

9.8 BROWN DWARFS AS COMPANIONS

> Nerissa: When the Moon shines, we do not see the candle
> Portia: So doth the greater glory dim the less ...
> Shakespeare, The Merchant of Venice

As with surveys for stellar binaries, both direct imaging and radial velocity observations can be used to search for brown dwarf companions of nearby stars. The latter technique offers potentially the most effective method of detecting brown dwarfs which have cooled to temperatures of only a few hundred degrees. The first brown dwarfs to be identified are all companions, and the nearest stars have been subjected to intense scrutiny since those initial discoveries. There is no guarantee, however, that the mass function of companions, $\Psi_C(M)$, is identical to $\Psi(M)$ for isolated dwarfs, and the later surveys have yielded few additional candidates.

9.8.1 Brown dwarfs as companions of white dwarfs

Cool companions to white dwarfs produce excess flux at infrared wavelengths. The first searches based on this method were carried out using aperture photometry. Subsequent surveys [K10], [S4], (see also Section 7.7.2) revealed no additional candidate brown dwarf companions. The most extensive survey includes K-band aperture photometry and direct images of almost 150 white dwarfs [Z6]. While 18 of these stars have M-dwarf companions with masses in the range 0.3–$0.15\,M_\odot$, GD 165B remains the only object found which lies at (or below) the hydrogen-burning limit (Section 9.4.1).

9.8.2   Brown dwarfs as companions of main sequence stars: imaging surveys

Photographic proper-motion surveys provided the first opportunity to search for low-luminosity companions with separations exceeding 3 arcsec, and Luyten's Palomar surveys resulted in the detection of numerous wide, common proper-motion pairs. The most successful study in terms of detecting low-luminosity brown dwarf candidates was van Biesbroeck's [B7] search for distant companions of nearby stars, which yielded some 20 systems including VB 8 (Gl 644C – spectral type M7) and VB 10 (Gl 752B – spectral type M8), the archetypal ultracool M dwarfs. At smaller separations, low-luminosity M dwarfs offer the best targets in searching for even lower-luminosity brown dwarfs (*vide* Portia's remark). Recent studies have concentrated on infrared observations [S6], [H8], and have failed to detect any companions. For the 8-parsec sample, these data have been supplemented with deep, wide-field optical images [S5] and, in many cases, coronagraphic observations [O3]. No new brown dwarfs have been discovered. A possible detection of a low-luminosity ($M_R \sim 20$) companion to Proxima Centauri, using the HST Faint Object spectrograph as a camera with Proxima placed behind an occulting bar, has recently been reported [S3]. Radial velocity observations, however, rule out the presence of any companion more massive than $1.1–0.22 M_J$ in the appropriate period range, suggesting that the 'detection' is spurious [K12].

Brown dwarfs have been found as companions of stars beyond the 8-parsec distance limit. An $\sim$L2 companion some 300 AU distant from the M dwarf Gl 96-3 was discovered in an imaging survey of chromospherically active stars within 40 parsecs of the Sun [R4]. The underlying premise of such surveys is that active stars are relatively young; hence brown dwarfs companions are correspondingly luminous and easier to detect. Given an age of 20–100 Myr for Gl 96-3, the companion has a mass between 15 and $50 M_J$. Two of the 2MASS L dwarfs are companions of nearby G dwarfs, Gl 417 and Gl 584. Like Gl 96-3, Gl 417 is chromospherically active, suggesting an age of ~ 50 Myr and a relatively low mass for the L5 companion (which has strong Li absorption). Gl 584 – known to be a binary G dwarf – is not active, but the companion has an extremely late spectral type with some evidence of variable bandstrengths. Finally, the lowest-luminosity T dwarf currently known is Gl 570D, a member of one of the multiple systems in the 8-parsec sample [B13]. In all of these cases, the brown dwarf companion lies several hundred AU from the primary star.

Low-luminosity stars in clusters have also been surveyed for binary companions. The most sensitive observations of late-type Hyades M dwarfs are with the HST Planetary Camera [G2], [R5]. The I-band (F870LP filter) Planetary Camera observations are diffraction limited, with a resolution of 0.1 arcsec, and include 53 cluster members with $M_V > 11.9$ (spectral types later than M3). Several new stellar binaries were identified, but no substellar-mass companions, although the survey was capable of detecting $0.05 M_\odot$ Hyades-age brown dwarfs.

These non-detections amongst the nearby stars and the Hyades can be used to constrain the mass function for companions, $\Psi_C(M)$. If $\Psi_C(M) \propto M^{-1}$, then one expects equal numbers of companions with $0.25 \geq M/M_\odot > 0.1$ and 0.10

$\geq M/M_\odot > 0.05$. Between 10% and 20% of M dwarfs in both the field and the Hyades have companions in the former mass range; only Gl 229 has a known brown dwarf companion. Even allowing for nearby brown dwarfs which have cooled below the detection limit of current surveys, this suggests that brown dwarf secondaries are relatively rare.

9.8.3   Brown dwarfs as companions of main sequence stars: radial-velocity surveys

Radial-velocity surveys are most successful at detecting companions lying at small separations from the primary star. Given sufficient signal, conventional methods as simple as averaging visual measurements of the positions of narrow atomic lines in the stellar spectrum can yield velocities accurate to better than $1\,\mathrm{km\,s^{-1}}$, and such methods have been used during most of the twentieth century. The last decade, however, has seen the development of more specialised techniques designed to improve the velocity resolution by more than two orders of magnitude, and radial velocities with an accuracy of $<5\,\mathrm{m\,s^{-1}}$ are within the reach of current programmes. These techniques are aimed mainly at the identification of planetary-mass companions around nearby stars (discussed in the next chapter). Here we consider the limits that these programmes have set on the frequency of higher-mass ($M > 0.01\,M_\odot$) brown dwarf companions of main-sequence stars.

Since $v\sin(i)$ is measured, radial-velocity variations offer only a statistical tool for investigating $\Psi_C(M)$. If there is no direct measurement of the linear separation, one is limited to determining $M_2\sin(i)$, and employing models to constrain the true mass distribution of companions based on the observed distribution, $f(M_2\sin(i))$. These techniques were pioneered by Trimble [T13], and have been developed further by a number of groups [M4], [M10], [M11], [M13], [W1]. Given an estimate of the probable distribution of semi-major axes, typically taken from observations of stellar binaries, a functional form for $\Psi_C(M)$ is adopted. Assuming random orbital inclinations, one can predict $f_{\mathrm{th}}(M_2\sin(i))$ and compare that against $f_{\mathrm{obs}}(M_2\sin(i))$. Models can be tailored to match both the accuracy and the duration of specific surveys [M4]. The crucial point is that projection effects allow a companion-star mass function truncated at the hydrogen-burning limit to result in a certain fraction of systems with $M_2\sin(i) < 0.08\,M_\odot$. Thus, the observed fraction of apparently substellar-mass companions is an upper limit to the actual number of such systems.

Few brown dwarf candidates have been identified spectroscopically. As described in more detail in the following chapter, the majority of stars targeted by radial-velocity programmes are G and early K dwarfs, but observations extend to stars as luminous as early F-type main-sequence stars, and include a few relatively bright M dwarfs. Surveys can be divided into high-precision ($<50\,\mathrm{m\,s^{-1}}$) and moderate precision ($\sim0.5\,\mathrm{km\,s^{-1}}$) observations. No brown dwarf candidates have been identified in high-precision surveys, which currently include ~ 150 nearby stars, although several have planetary-mass companions (Chapter 10). The relatively small number of candidate brown dwarfs (Table 9.5) are all drawn from the 640 stars surveyed at moderate precision by Mayor *et al.* [M11]. As [M4] demonstrate, those

Table 9.5. Candidate brown dwarf companions.

Star	M_V	(B–V)	$m \sin(i)$ ($M_\odot$)	P (days)	e
HD 110833	6.15	0.94	0.016	270	0.69
BD −4° 782	7.83	1.22	0.020	241	0.28
HD 112758	5.95	0.78	0.033	103	0.16
HD 98230	5.45	0.45	0.035	4	0.00
HD 18445	5.79	0.96	0.037	555	0.54
HD 29587	6.37	0.64	0.038	1,158	0
HD 140913	4.68	0.54	0.044	148	0.61
HD 283750	7.16	1.08	0.048	2	0.02
HD 89707	4.49	0.55	0.052	198	0.95
HD 217580	6.32	0.95	0.057	454	0.52

detections are consistent with the expected number of stellar companions in high-inclination orbits if $\Psi_C(M) \propto M^{-0.8}$ with a cut-off at $\sim 0.1 M_\odot$. Indeed, *Hipparcos* astrometry reveals significant reflex orbital motion in most of the primaries listed in Table 9.5, indicating that those systems harbour low-mass stars rather than brown dwarfs.

The conclusion drawn from these surveys is that brown dwarfs are rare as companions to main-sequence stars – a scarcity which has been termed the 'brown dwarf desert' by Marcy. There is as yet no theoretical explanation for this result, but it might be hypothesised that potential low-mass companions are easily disrupted during the early stages of protostar evolution. This problem remains a focus of considerable theoretical and observational effort.

9.9   THE SUBSTELLAR MASS FUNCTION

The availability of deep photometric surveys of several clusters – notably the Pleiades – and the initial results for field brown dwarfs permits the first serious investigation of the form of $\Psi(M)$ at masses below the hydrogen-burning limit. As noted above, the scarcity of brown dwarfs as companions of main-sequence stars suggests strongly that $\Psi_C(M)$ differs significantly from the mass function appropriate for isolated objects.

9.9.1   Brown dwarfs in clusters

As with the stellar mass function, the fact that age is a known quantity allows masses to be derived for each known (or suspected) member of a given cluster, using the appropriate theoretical models. In an intermediate-age cluster such as the Pleiades, the dispersion in individual ages is small enough compared to the average age that a single-age mass–luminosity relationship can be utilised. The main uncertainties rest with the relatively small number of confirmed substellar cluster members. Nonetheless, there is reasonable agreement amongst the several studies, favouring a mass function consistent with a power-law $\Psi(M) \propto M^{-\alpha}$ with $\alpha = 1 \pm 0.5$ [F3], [B9], [M8]. This is compatible with extrapolation of the stellar mass function (Figure 8.8).

In younger clusters, the significant dispersion in individual ages relative to the mean age of the system dictates the use of full evolutionary tracks for mass determinations of both stellar and substellar objects. As discussed in Section 8.7, some of the most intensely studied clusters – such as IC 348 and the ONC – appear to have mass functions which reach a maximum at $\sim 0.25\,M_\odot$ [H6]. Both clusters have brown dwarf constituents, but deducing the exact form of $\Psi(M)$ at the lowest masses requires further observational work, as those sources are the faintest in the cluster, and are therefore the most difficult targets for detailed observations. Other studies have concentrated on smaller star-forming regions, and tend to derive mass functions which are increasing across the hydrogen-burning limit. Results include $\Psi(M) \propto M^{-1.2\pm0.1}$ for ρ Ophiuchi (91 sources) [C3]; and $\Psi(M) \propto M^{-1}$ in L1495E [L4].

9.9.2   The substellar mass function in the field

Analysis of the field L dwarfs is more complicated, since, regardless of mass, brown dwarfs evolve through a very limited region of the H–R diagram. Thus, even though luminosities and temperatures can be determined for individual objects, mass estimates require knowledge of the age. The only method of tackling this problem is through statistical analysis: assuming an age for the Galactic disk, a birth-rate, $B(t)$, and an initial mass function, one constructs a model of the expected distribution of luminosities and temperatures for objects in the immediate vicinity of the Sun. Given estimates of the temperatures at the M/L and L/T spectral type transitions, and of the near-infrared bolometric corrections, the predicted (apparent magnitude, spectral type) distribution can be compared to observational results.

The mass-dependent rate of brown dwarf evolution means that current analyses cannot constrain $\Psi(M)$ at the lowest masses. Very low-mass ($< 0.01\,M_\odot$) brown dwarfs are extremely unlikely to be detected locally, since they cool to temperatures of 1,000 K, and luminosities of $L < 10^{-5}\,L_\odot$, in only 10^7 years. As there are no active star-forming regions within 100 parsecs of the Sun, any 'planetary-mass' brown dwarfs in the Solar Neighbourhood are likely to have temperatures of only a few hundred degrees, luminosities of less than $10^{-6.5}\,L_\odot$, and $M_K > 19$. Moreover, this type of analysis is vulnerable to the stochastic nature of star-formation. Both the Lyon [B1] and Tucson [B10] models predict that brown dwarfs with masses of $0.07\,M_\odot$ enter the L dwarf temperature régime (at $\sim 2,000$ K) at ages of 1 Gyr and require more than 2 Gyr to reach the L/T boundary, while 0.01–0.02 $M_\odot$ objects are L dwarfs for only $\sim 10^7$ years at ages of less than 10^8 years (see Figure 9.1). Variations in the star-formation rate will affect the number of detectable sources as a function of mass, and hence the form of the inferred mass function.

Investigations of the star-formation history of the disk (Section 6.5.5) suggest that there have been no major variations over the last few Gyrs. However, since even $0.07\,M_\odot$ brown dwarfs cool to effectively undetectable levels after $\sim 5\,\mathrm{Gyr}$, the distribution of brown dwarfs is predicted to be dominated by old, sub-1,000 K (undetectable) objects. Thus, the total mass density inferred from matching the observations depends almost linearly on the age assumed for the Galactic disk. 2MASS, SDSS and DENIS detect only the tip of a very large iceberg.

Given both the uncertainties in the modelling and the relatively sparse observational dataset, current analyses can set only broad constraints on the form of the mass function at substellar masses. The absence of strong absorption bands at $1.25\,\mu\mathrm{m}$ in either L dwarfs or T dwarfs renders the J passband the optimum choice for these simulations. Reid *et al.* [R1] discuss results from 2MASS and DENIS, which indicate a surface density of approximately one L dwarf brighter than $J = 16$ magnitude per 25 square degrees, or $\sim 1,600$ over the whole sky. Combining the 2MASS and SDSS detections produces six T dwarfs within an area of $\sim 2,400$ square degrees to the same magnitude limit, or ~ 100 over the whole sky. Matched against these observations are predictions from a model which adopts an age of 10 Gyr for the Galactic disk, a constant stellar birthrate, a temperature of 2,000 K for the transition from spectral type M to L, and a lower mass cut-off of $0.01\,M_\odot$ in the initial mass function. In practice, the rapid evolution of low-mass brown dwarfs means that the results are insensitive to the form of $\Psi(M)$ below $0.025\,M_\odot$.

Representing $\Psi(M)$ as a power law, if $\alpha = 0$ then the combined number of L and T dwarfs with $J < 16$ in an all-sky survey is predicted to be $N_{LT} \sim 100$; if $\alpha = 1$, $N_{LT} \sim 900$; and if $\alpha = 2$, $N_{LT} \sim 5,800$. Matched against the observations, these simulations favour a power-law index $1 < \alpha < 2$. The relative number of L and T dwarfs offers some constraints on T_{meth}, the L/T transition temperature: the higher T_{meth}, the more T dwarfs predicted, as illustrated in Figure 9.21. With an observed number ratio $N_T/N_L \sim 5\%$, current results favour $T_{meth} < 1,200\,\mathrm{K}$, consistent with the L8/Gl229B relative-luminosity argument outlined in Section 6.2.

The lithium test provides a consistency check on these results. The steeper the mass function, the higher the fraction of low-mass brown dwarfs amongst both L dwarfs and ultracool M dwarfs. The detection of lithium in the latter objects is difficult, since TiO absorption is still present and the Li 6708 Å line is relatively weak: LP 944-20 was observed several times at moderate and low resolution before Tinney's higher-resolution spectroscopy succeeded in resolving the 0.5 Å equivalent width lithium line. At present, few late M dwarfs have been observed at the requisite resolution and to the necessary signal-to-noise, and the proportion with lithium absorption remains undefined. However, the observed frequency of lithium detection in L dwarfs is $\sim 25\%$. Simulations based on $\Psi(M) \propto M^{-1}$ predict $\sim 30\%$ of L dwarfs should have masses below $0.06\,M_\odot$, while the fraction is closer to two-thirds for an $\alpha = 2$ mass function. Thus, these preliminary results favour an index close to $\alpha = 1$ for a power-law functional form. More definitive results should be forthcoming with the completion of both SDSS and 2MASS surveys.

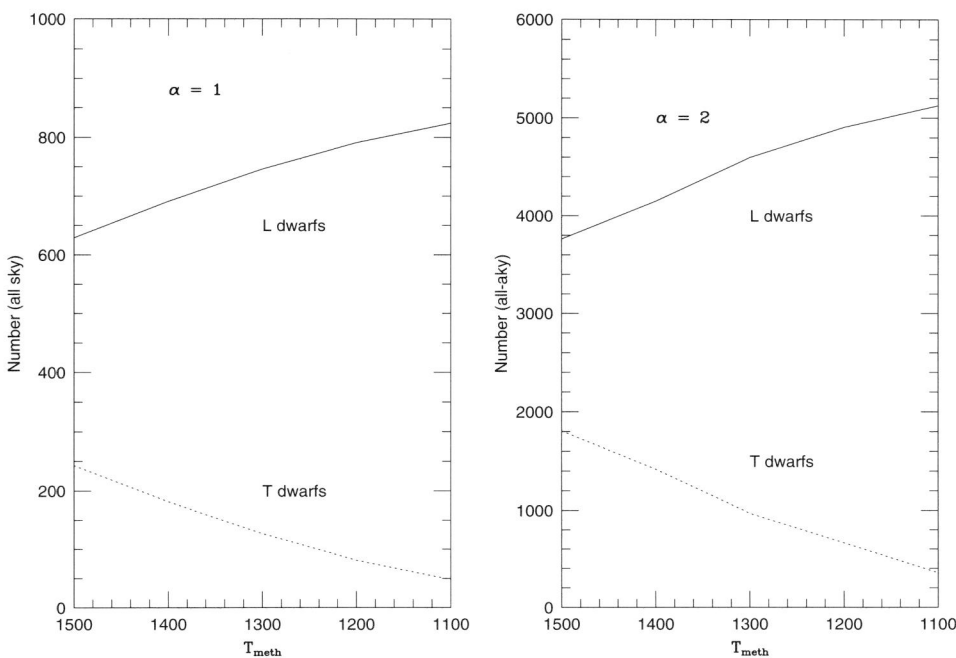

Figure 9.21. The predicted number of L and T dwarf detections as a function of the methane transition temperature for an all-sky survey to J = 16. The *left* panel plots predictions for $\alpha = 1$; the *right*, for $\alpha = 2$. In both cases, an upper temperature limit of 2,000 K is assumed for spectral type L0. The observed values derived in [R1] are 1,600 L dwarfs and 100 T dwarfs.

9.9.3 Summary

What are the implications of these results? Again, we must emphasise that while the power-law mass function model provides a useful means of analysing the observations, there is no unambiguous evidence that $\Psi(M)$ *is* a power-law. Nonetheless, the favoured indices are similar to those derived from fitting $\Psi(M)$ to the nearby-star data (Figure 8.8), suggesting a degree of continuity across the hydrogen-burning limit. Extrapolating the resulting mass function to 0.01 $M_\odot$ leads to the expectation of numerous cool, low-luminosity brown dwarfs within the Solar Neighbourhood, as illustrated in Figure 9.22 for $\alpha = 1$ and $\alpha = 2$ power-law mass functions. Quantifying these simulations, consider the case where $\Psi(M) \propto M^{-1.3}$ over the 0.01–0.075 $M_\odot$ mass range. This model predicts that brown dwarfs outnumber main-sequence stars by a factor of almost 2:1 in the solar neighbourhood. The average distance between individual systems is only 1.3 parsecs. However, the total mass density increases by only 0.005 $M_\odot$ pc^{-3}, or 15% of the contribution from hydrogen-burning stars. Thus, current results suggest that it is unlikely that brown dwarfs are a major constituent of Galactic dark matter.

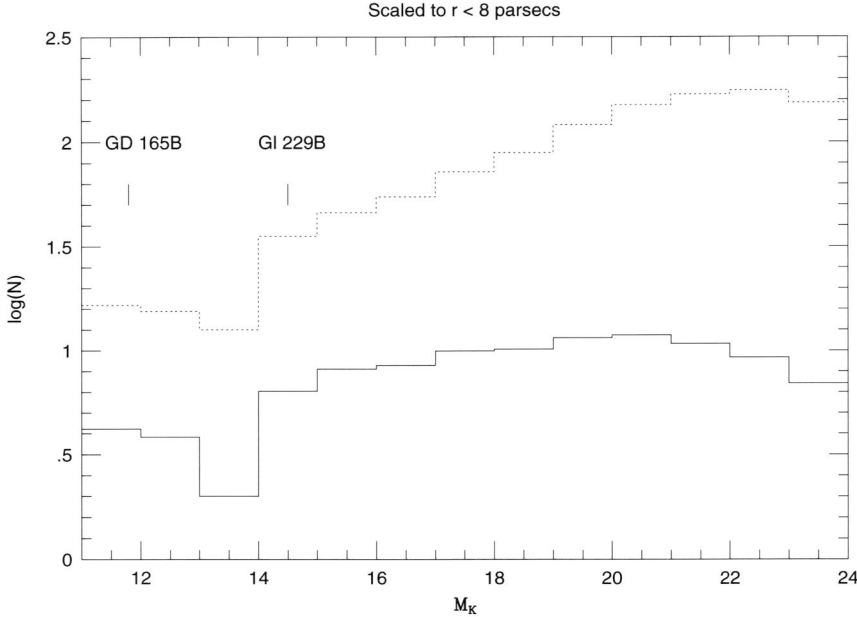

Figure 9.22. The predicted K-band luminosity function for brown dwarfs within 8 pc of the Sun. The solid line plots the expected distribution for $\Psi(M) \propto M^{-1}$; the dotted line shows results for $\Psi(M) \propto M^{-2}$. Both cases assume a uniform birth-rate and a 10-Gyr age for the Galactic disk.

9.10   FUTURE SPACE AND NEAR-SPACE MISSIONS: SOFIA, SIRTF AND NGST

The detection of members of the hypothesised local population of extremely cool brown dwarfs is very unlikely, given presently available observational resources. However, these circumstances should change in the early years of the twenty-first century. Figure 9.6 shows that the Gl 229B flux distribution peaks near 5 μm, and similar circumstances hold for cooler T dwarfs (Figure 9.23). Three mid-infrared projects which will come to fruition in the near future offer the possibility of finding and studying these objects.

SOFIA

SOFIA [E3], [B5] is the planned successor to the NASA Kuiper Airborne Observatory – a Boeing 747 modified to carry a 2.5-m telescope to an operating altitude of 41,000–46,000 ft (14 km). At that altitude, absorption due to atmospheric water vapour is reduced to almost negligible proportions over much of the spectrum,

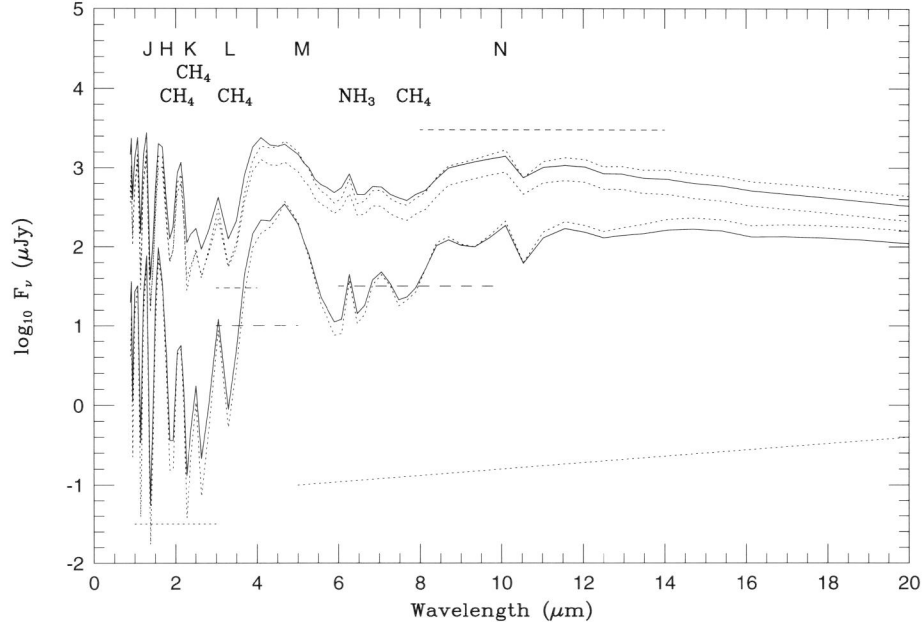

Figure 9.23. The predicted energy distribution of a 40 M_J brown dwarf at ages of 1 and 10 Gyr (solid line) and a 10 M_J brown dwarf at ages 0.1, 1 and 10 Gyr (dotted line). The temperatures are 1,090, 540, 1,000, 850 and 450 K respectively, and the flux distributions are scaled to a distance of 10 parsecs. The detection limits for SOFIA (short-dashed line), SIRTF (long-dash) and NGST (dotted) are shown. (Model calculations by A. Burrows.)

allowing access in particular to the 15–200 μm regions which are almost totally opaque, even from Mauna Kea (14,000 ft (4 km) altitude). The infrared sensitivity limits (3σ, one hour) expected for the instruments planned for SOFIA range from ~0.03 mJy at 3 μm to 3 mJy at 10 μm, and 10–20 mJy longward of 30 μm. The system should be capable of observing younger brown dwarfs within 10 parsecs (Figure 9.23). The proposed instrumentation is not well suited to large-scale surveys, and SOFIA is likely to be used for follow-up observations of previously identified targets, rather than as a tool for identifying cool brown dwarfs.

SIRTF

SIRTF, the Space Infrared Telescope Facility [W3], [F1] was envisioned originally as the infrared equivalent of the HST as part of NASA's series of Great Observatories. SIRTF has been scaled down to some extent since its conception, but still retains an effective complement of instruments (summarised in Table 9.6). The Infrared Array Camera – covering the 3–10 μm region with a field of view of 5 × 5 arcmin – is the most effective for searching for nearby, cool brown dwarfs. The expected sensitivity

Table 9.6. The Space Infrared Telescope Facility (SIRTF).

Purpose	Pointed photometric, imaging and spectroscopic observations from 3 to 180 μm.
Launch date	12 December 2001 into a solar orbit, trailing the Earth ~1 AU from Sun, ~0.32 AU from Earth after ~30 months.
Lifetime	~30 months.
Telescope	Two-mirror Ritchey–Chrétien, primary mirror diameter 85 cm.
Instrumentation	Infrared Array Camera (IRAC) – imaging at 3.5, 4.5, 6.3 and 8 μm; dichroic provides simultaneous 3.5/6.3 and 4.5/8.0 imaging of adjacent 5′.1 × 5′.1 fields, with 1.2 arcsec pixels; InSb detectors at 3.5 and 4.5 μm; Si : Ga at 6.3 and 8.0 μm.
	Multiband Imaging Photometer (MIPS) – 20–80 μm imaging: 128 × 128 Si : As array at 24 μm; 32 × 32 Ge : Ga array at 70 μm; 2 × 20 Ge : Ga array at 160 μm; low-resolution spectrometer.
	Infrared Spectrometer (IRS) – spectroscopy over 5 to 40 μm; low resolution ($R = 50$) from 5 to 40 μm; modest resolution ($R = 600$) from 10 to 38 μm; 128 × 128 Si : As and Si : Sb array detectors.
Operational mode	Pointed observations of individual fields and targets.

should allow observations to reach flux densities of ~1 μJy (3σ) in integration times of one hour. Such long integrations are out of the question for any large-scale survey, but integrations of 30–60 seconds should be capable of detecting flux densities as low as 10 μJy, bringing even old brown dwarfs at 10 parsecs within reach. Planning for SIRTF mission operations is currently at an early stage, but it is likely that there will be at least one major project devoted to searching for extremely cool brown dwarfs.

NGST

The Next Generation Space Telescope lies almost a decade in the future, and current plans are correspondingly sketchy. Building on the successes of the HST, the broad aim is to place a 4- to 8-m telescope in a trailing-Earth orbit; that is, orbiting the Sun in close proximity to the Earth, rather than orbiting the Earth itself. This configuration avoids problems such as terrestial eclipses during each orbit, and interference from the South Atlantic Anomaly (but renders servicing missions impossible with current technology). The telescope will be optimised for infrared observation, with minimal coverage of the 0.6–1 μm range. The instrumentation is expected to be capable of detecting 0.01 μJy sources at 1–3 μm in broadband imaging (10σ, 100 seconds), while the corresponding sensitivities are ~0.1–1 μJy between 5 and 30 μm. The overall field of view of any instrument is likely to be very small, but the NGST should be capable of obtaining spectroscopy of the majority of the brown dwarfs detected by SIRTF.

9.11 SUMMARY

The main conclusions of this chapter can be summarised as follows:

- The existence of brown dwarfs has now been amply confirmed by direct observations.
- The best current determinations of the average form of the mass function below the hydrogen-burning limit are consistent with a power-law with $1 < \alpha < 2$, with the index probably closer to 1. However, $\Psi(M)$ may well vary in form in different star-forming clusters.
- Wide-angle near-infrared surveys (DENIS, 2MASS) and deep optical surveys (SDSS) are finding numerous examples of brown dwarfs in the general field.
- Brown dwarfs are rarely found as companions to main-sequence (or post main-sequence) stars, providing further evidence that $\Psi_C(M)$ is not identical with the mass function derived for isolated stars and primaries in multiple systems.

9.12 REFERENCES

B1   Baraffe, I., Chabrier, G., Allard, F., Hauschildt, P. H. 1998, *A&A*, **337**, 403.

B2   Barnbaum, C., Zuckerman, B. 1992, *ApJ*, **396**, L31.

B3   Basri, G., Marcy, G. W., Graham, J. R. 1996, *ApJ*, **458**, 600.

B4   Basri, G., Martín, E. L. 1998, Brown dwarfs and extrasolar planets, ASP Conf. Ser. 134, ed. R. Rebolo, E. L. Martín, M. R. Zapatero-Osorio, p. 284.

B5   Becklin, E. E. 1998, Proc. SPIE, **3356**, 492.

B6   Bergeron, P., Wesemael, F., Lamontagne, R., Fontaine, G. Saffer, R., Allard, N. F. 1995, *ApJ*, **449**, 258.

B7   van Biesbroeck, G. 1961, *AJ*, **66**, 528.

B8   Bildsten, L., Brown, E. F., Matzner, C. D., Ushomirsky, G., 1997, *ApJ*, **482**, 442.

B9   Bouvier, J., Stauffer, J. R., Martín, E. L., Navascues, D. B. Y., Wallace, B., Bejar, V. J. S., 1998, *A&A*, **336**, 490.

B10 Burrows, A., Marley, M., Hubbard, W. B., Lunine, J. I., Guillot, T., Saumon, D., 1997, *ApJ*, **491**, 856.

B11 Burrows, A, Sharp, C. M., 1999, *ApJ*, **512**, 843.

B12 Burgasser, A. J., Kirkpatrick, J. D., Brown, M. E., Reid, I. N *et al.*, 1999, *ApJ*, **524**, L.

B13 Burgasser, A. J., Kirkpatrick, J. D., Cutri, R. M., Skrutskie, M. F. *et al.*, 1999, *ApJ*, 524.

C1   Clemens, J. C., van Kerkwijk, M. H., Wu, Y., 2000, *MNRAS*, **312**, in press.

C2   Cochran, W. D., Hatzes, A. P., Hancock, T. J., 1991 *ApJ*, **380**, L35.

C3   Cómeron, F., Rieke, G. H., Burrows, A., Rieke, M. J., 1993, *ApJ*, **416**, 185.

C4   Cómeron, F., Rieke, G. H., Claes, P., Torra, J., Laureus, R. J., 1998, *A&A*, **335**, 522.

C5   Copet, E. *et al.*, 1997, *Ap. Sp. Sci.*, **217**, 25.

D1 Dahn, C. C., Harris, H. L. *et al.*, 2000, *From Cool Stars to Giant Planets*, ed. C. Griffiths and M. Marley, ASP Conf. Ser., in press.

D2 D'Antona, F., Mazzitelli, I., 1986, *A&A*, **162**, 80.

D3 D'Antona, F., Mazzitelli, I., 1994, *ApJS*, **90**, 469.

D4 Delfosse, X. *et al.*, 1997, *A&A*, **327**, L25.

D5 Duquennoy, A., Mayor, M., 1991, *A&A*, **248**, 485.

E1 Eggen, O. J., Greenstein, J. L., 1965 *ApJ*, **141**, 83.

E2 Epchtein, N. *et al.*, 1994 in Science with Astronomical Near-Infrared Surveys, *Ap. Sp. Sci.*, **217**, 3.

E3 Erickson, E. F., 1992, *Sp. Sci. Rev.*, **61**, 61.

F1 Fanson, J. *et al.*, 1998, *Proc SPIE*, **3356**, 478.

F2 Fegley, B., Lodders, K., 1996, *ApJ*, **472**, L37.

F3 Festin, L., 1998, *A&A*, **333**, 497.

F4 Fukugita, M., Ichikawa, T., Gunn, J. E. *et al.*, 1996, *AJ*, **111**, 1748.

G1 Geballe, T. *et al.*, 1996, *ApJ*, **467**, L101.

G2 Gizis, J. E., Reid, I. N., 1996, *AJ*, **110**, 1248.

G3 Gizis, J. E., Reid, I. N., Monet, D. G., Kirkpatrick, J. D., 1999, *AJ*, 118.

G4 Graham, J. R., Matthews, K., Neugebauer, G., Soifer, B. T., 1990, *ApJ*, **357**, 216.

G5 Graham, J. R., McCarthy, J. K., Reid, I. N., Rich, R. M., 1990, *ApJ*, **357**, L21.

G6 Graham, J. R., Matthews, K., Greenstein, J. L., Neugebauer, G., Tinney, C. G., Persson, S. E., 1992, *AJ*, **104**, 2016.

G7 Gratton, R. G., Carretta, E., Castelli, F., 1997, *A&A*, **314**, 191.

G8 Greenstein, J. L., 1969, *ApJ*, **158**, 281.

G9 Greenstein, J. L., 1988, *AJ*, **95**, 1494.

G10 Gunn, J. E., Carr, M., Rockosi, C., Sekiguchi, M., *et al.*, 1998, *AJ*, **116**, 3040.

H1 Haas, M. , Leinert, Ch. 1990, *A&A*, **230**, 87.

H2 Hale, A., 1995, *PASP*, **107**, 22.

H3 Hamilton, D., Stauffer, J. R., 1993, *AJ*, **107**, 1855.

H4 Harrington, R. S., Kallarakal, V. V., Dahn, C. C., 1983, *AJ*, **88**, 1038.

H5 Harris, H. L. *et al.*, 1999, *AJ*, **117**, 339.

H6 Hillenbrand, L., 1997, *AJ*, **113**, 1733.

H7 Henry, G. W., Baliunas, S. L., Donahue, R. A., Soon, W. H., Saar, S. H., 1997, *ApJ*, **474**, 503.

H8 Henry, T. J., 1994, PhD thesis, University of Arizona.

J1 Jones, H. R. A., Longmore, A. J., Jameson, R. F., Mountain, C. M., 1994, *MNRAS*, **267**, 413.

K1 Kirkpatrick, J. D., Henry, T. J., Liebert, J., 1993, *ApJ*, **406**, 701.

K2 Kirkpatrick, J. D., McGraw, J. T., Hess, T. R., Liebert, J., McCarthy, D. W., 1994, *ApJS*, **94**, 749.

K3 Kirkpatrick, J. D., Henry, T. J., Simons, D. A., 1995, *AJ*, **109**, 797.

K4 Kirkpatick, J. D., Henry, T. J., Irwin, M. J., 1997, *AJ*, **113**, 1421.

K5 Kirkpatrick, J. D., Allard, F., Bida, T., Zuckerman, B., Becklin, E. E., Chabrier, G., Baraffe, I., 1999, *ApJ*, **519**, 834.

K6   Kirkpatrick, J. D., Reid, I. N., Liebert, J., Cutri, R., Nelson, B., Beichman, C. A., Dahn, C. C., Monet, D. G., Skrutskie, M. F., Gizis, J. E., 1999, *ApJ*, **519**, 802.

K7   Kleinmann, S. J. *et al.*, 1994, *ApJ*, **436**, 875.

K8   Koester, D., Provencal, J., Shipman, H. L., 1997, *A&A*, **320**, L57.

K9   Kuchner, M. J., Koresko, C. D, Brown, M. E., 1998, *ApJ*, **508**, L81.

K10 Kumar, C. K., 1985, *PASP*, **97**, 294.

K11 Kumar, S. S., 1963, *ApJ*, **137**, 1121.

K12 Kürster, M., Hatzes, A. P., Cochran, W. D., Dobereinerm S., Dennerl, K., Endl, M., 1999, *A&A*, **344**, L5.

L1   Latham, D. W., Mazeh, T., Stefanik, R. P., Mayor, M., Burki, G., 1989, *Nature*, **339**, 38.

L2   Leggett, S. K., Hawkins, M. R. S., 1989, *MNRAS*, **238**, 145.

L3   Lodders, K., 1999, *ApJ*, **519**, 793.

L4   Luhman, K. L., Rieke, G. H., 1997, *ApJ*, **497**, 354.

L5   Luhman, K. L., Liebert, J., Rieke, G. H., 1997, *ApJ*, **489**, L165.

M1   McCarthy, D. W., Probst, R. G., Low, F. J., 1985, *ApJ*, **290**, L9.

M2   Magazzu, A., Martín, E. L., Rebolo, R., 1993, *ApJ*, **404**, L17.

M3   Magazzu, A., Rebolo, R., Zapatero-Osorio, M. R., Martín, E. L., Hodgkin, S. T., 1998, *ApJ*, **497**, L47.

M4   Marcy, G. W., Butler, R. P., 1994, in *The Bottom of the Main Sequence and Beyond*, ed. C. G. Tinney, Springer-Verlag, Berlin, p. 98.

M5   Marley, M. S., Saumon, D., Guillot, T., Freedman, R. S., Hubbard, W. B., Burrows, A., Luninie, J. I., 1996, *Science* **272**, 1919.

M6   Martín, E. L., Rebolo, R., Zapatero-Osorio, M. R., 1996, *ApJ*, **469**, 705.

M7   Martín, E. L., Basri, G., Delfosse, X., Forveille, T., 1997, *A&A*, **327**, L29.

M8   Martín, E. L., Basri, G., Gallegos, J. E., Rebolo, R., Zapatero-Osorio, M. R., Bejar, V. J. S., 1998, *ApJ*, **499**, L61.

M9   Matthews, K., Nakajima, T., Kulkarni, S., Oppenheimer, B. R., 1996, *AJ*, **112**, 1678.

M10 Mayor, M., Duquennoy, A., Halbwachs, J.-L., Mermilliod, J.-C., 1992, in *IAU Colloquium 135*, ASP Conf. ser. vol. 32, p. 73, ed. H. McAlister and W. Hartkopf.

M11 Mayor, M., Queloz, D., Udry, M., Halbwachs, J.-L., 1997, in *Bioastronomy 96*, IAU Colloquium 161.

M12 Mazeh, T., Goldberg, D., 1992, *ApJ*, **394**, 592.

M13 Mazeh, T., Goldberg, D., Duquennoy, A., Mayor, M., 1992, *ApJ*, **401**, 265.

M14 Monet, D.G. *et al.*, 1998, *USNO A2.0 Catalogue*.

N1   Nakajima, T., Durrance, S. T., Golimowski, D. A., Kulkarni, S. R., 1994, *ApJ*, **428**, 797.

N2   Nakajima, T., Oppenheimer, B. R., Kulkarni, S. R., Golimowski, D. A., Matthews, K., Durrance, S. T., 1995, *Nature*, **378**, 463.

N3   Nather, R. E., Winget, D. E., Clemens, J. C., Hansen, C. J., Hines, B. P., 1990, *ApJ*, **361**, 909.

N4 Neugebauer, G., Leighton, R. B., 1969, *The Two Micron Sky Survey*, NASA SP-3047.

N5 Neuhauser, R., Torres, G., Sterzik, M. F., Randich, S., 1997, *A&A*, **325**, 647.

N6 Noll, K. S., Geballe, T. R., Marley, M., 1997, *ApJ*, **489**, L87.

O1 Oke, J. B. *et al.*, 1995, *PASP*, **107**, 375.

O2 Oppenheimer, B. R., Kulkarni, S. R., Matthews, K., Nakajima, T., 1995, *Science*, **270**, 1478.

O3 Oppenheimer, B. R., Basri, G., Nakajima, T., Kulkarni, S. R. 1998, *AJ*, **113**, 296.

O4 Oppenheimer, B. R., Kulkarni, S. R., Matthews, K., van Kerkwijk, M. H.. 1999, *ApJ*, **502**, 932.

P1 Patterson, J., Zuckerman, B., Becklin, E. E., Tholen, D. J, Hawarden, T. 1989, *ApJ*, **374**, 330.

P2 Perrier, C., Mariotti, J.-M., 1987, *ApJ*, **317**, L27.

P3 Price, S. D., Walker, R. G. 1976, *The AFGL Four Color Infrared Sky Survey*. AFGL-CR-83-0161, Air Force Systems Command, USAF.

R1 Rebolo, R., Martín, E. L., Maguzzu, A. 1992, *ApJ*, **389**, L83.

R2 Rebolo, R., Osorio, M. R. Z., Martín, E. L. 1995, *Nature*, **377**, 129.

R3 Rebolo,R., Martín, E. L., Basri, G., Marcy, G., Zapatero-Osorio, M. R. 1996, *ApJ*, **469**, L53.

R4 Rebolo, R., Zapatero Osorio, M. R., Madruga, S., Bejar, V. J. S., Arribas, S., Licandro, J. 1998, *Science*, **282**, 1309.

R5 Reid, I. N., Gizis, J. E. 1997, *AJ*, **114**, 1992.

R6 Reid, I. N., Hawley, S. L., 1999, *AJ*, **117**, 343.

R7 Reid, I. N., Kirkpatrick, J. D., Liebert, J., Burrows, A., Gizis, J. E., Burgasser, A., Dahn, C. C., Monet, D., Cutri, R., Beichman, C. A., Skrutskie, M. 1999, *ApJ*, **521**, 613.

R8 Reid, I. N., Kirkpatrick, J. D., Gizis, J. E., Dahn, C. C., Monet, D. G., Williams, R. J., Liebert, J., Burgasser, A. J. 2000, *AJ*, **119**, in press.

R9 Rieke, G. H., Rieke, M. J., 1990, *ApJ*, **362**, L21.

R10 Robinson, E. L., Cochran, A. L., Cochran, W. D., Shafter, A. W., Zhang, E., 1990, *AJ*, **99**, 672.

R11 Ruiz, M. T., Leggett, S. K., Allard, F., 1997, *ApJ*, **491**, L107.

S1 Saffer, R. A., Livio, M., Yungelson, L. R., 1998, *ApJ*, **502**, 394.

S2 Schaerer, D., Charbonnel, C., Meynet, G., Maeder, A., Schaller, G., 1993, *A&AS*, **102**, 339.

S3 Schulz, A. B. *et al.* 1998, *AJ*, **115**, 345.

S4 Shipman, H. L., 1987, in *Astrophysics of Brown Dwarfs*, ed. M. C. Kafatos, R. S. Harrington and S. P. Maran, Cambridge University Press, p. 71.

S5 Simons, D. A., Henry, T. J., Kirkpatrick, J. D., 1996, *AJ*, **112**, 2238.

S6 Skrutskie, M. F., Forrest, W. J., Shure, M. A., 1987 in *Astrophysics of Brown Dwarfs*, ed. M. C. Kafatos, R. S. Harrington and S. P. Maran, Cambridge University Press, p. 82.

S7 Skrutskie, M. F., Forrest, W. J., Shure, M. A., 1987, *ApJ*, **317**, L55.

S8   Skrutskie, M. F. *et al.*, 1997, in *The Impact of Large-Scale Near-IR Sky Surveys*, ed. F. Garson *et al.*, Kluwer, The Netherlands, p. 25.

S9   Stauffer, J., Hamilton, D., Probst, R., Rieke, G., Mateo, M., 1989, *ApJ*, **344**, L21.

S10   Stauffer, J., Hamilton, D., Probst, R. G., 1994, *AJ*, **108**, 155.

S11   Stauffer, J., Liebert, J., Giampapa, M., Macintosh, B., Reid, I. N, Hamilton, D., 1994, *AJ*, **108**, 160.

S12   Stauffer, J., Schultz, G., Kirkpatrick, J. D., 1998, *ApJ*, **499**, L199.

S13   Strauss, M. *et al.*, 1999, *ApJ*, **524**.

T1   Tarter, J. C., 1974, PhD thesis, University of California at Berkeley.

T2   Telesco, C. M., Joy, M., Sisk, C., 1990, *ApJ*, **358**, L17.

T3   Thackrah, A., Jones, H., Hawkins, M. R. S., 1997, *MNRAS*, **284**, 507.

T4   Tinney, C. G., 1993, *ApJ*, **414**, 279.

T5   Tinney, C. G., Reid, I. N., Gizis, J. E., Mould, J. R., 1995, *AJ*, **110**, 3014.

T6   Tinney, C. G., Delfosse, X., Forveille, T., 1997, *ApJ*, **490**, L95.

T7   Tinney, C. G., 1996, *MNRAS*, **281**, 644.

T8   Tinney, C. G., 1998, *MNRAS*, **296**, L42.

T9   Tinney, C. G., Reid, I. N., 1998, *MNRAS*, **301**, 1031.

T10   Tokunaga, A. T. *et al.*, 1988, *ApJ*, **332**, L71.

T11   Tokunaga, A. T., Becklin, E. E., Zuckerman, B., 1990, *ApJ*, **358**, L21.

T12   Tokunaga, A. T., Kobayashi, N., 1999, *AJ*, **117**, 1010.

T13   Trimble, V., 1990, *MNRAS*, **242**, 79.

T14   Tsuji, T., 1964, *Ann. Tokyo Obs.*, ser. II, **9**, 1.

T15   Tsuji, T., Ohnaka, K., Aoki, W., Nakajima, T., 1996, *A&A*, **308**, L29.

V1   Ventura, P. Zeppert, A., Mazzitelli, I., D'Antona, F., 1998, *A&A*, **331**, 1011.

V2   Vogt *et al.*, 1994, S.P.I.E., **2198**, 362.

W1   Walker, G., Bohlender, D. A., Walker, A. R., Irwin, A. W., Yang, S. L. S., Larson, A., 1992, *ApJ*, **396**, L91.

W2   Weidemann, V., 1990, *ARA&A*, **28**, 103.

W3   Werner, M. W., 1998, *Science with the NGST*, ASP Conf. Ser. No. **133**, 53.

W4   Winget, D. E. *et al.*, 1990, *ApJ*, **357**, 630.

Z1   Zapatero-Osorio, M. R., Rebolo, R., Martín, E. L., García López, R. J., 1996, *A&A*, **305**, 519.

Z2   Zapatero-Osorio, M. R., Rebolo, R., Martín, E. L., 1997, *A&A*, **317**, 164.

Z3   Zapatero-Osorio, M. R., Rebolo, R., Magazzu, A., Martín, E. L., Steele, I. A., Jameson, R. F., 1999, *A&AS*, **134**, 537.

Z4   Zuckerman, B., Becklin, E. E., 1987, *Nature*, **330**, 138.

Z5   Zuckerman, B., Becklin, E. E., 1988, *Nature*, **336**, 656.

Z6   Zuckerman, B., Becklin, E. E., 1992, *ApJ*, **386**, 260.

Z7   Zuckerman, B., Reid, I. N., 1999, *ApJ*, **505**, L143.

10

Extrasolar planets

'Innumerable Suns exist; innumerable Earths revolve around those
Suns in a manner similar to the way the seven planets revolve around
our Sun.'

Giordano Bruno, *De l'Infinito, Universo e Mundi*

10.1 INTRODUCTION

Many astronomical investigations are focused on addressing specific questions,
providing incremental additions to the growing body of astrophysical knowledge.
There are some issues, however, which have much wider ramifications. For example,
observations of the binary pulsar PSR1913+16 provide stringent tests of general
relativity [H7], [T1], while interest in the existence of brown dwarfs was stimulated by
their possible contribution to dark matter and galaxy formation. No study resonates
as strongly as the search for planetary systems amongst the stars in our Galaxy. This
fascination, both lay and scientific, stems from the obvious link with questions
concerning the existence of extraterrestial life and mankind's place in the Universe.

The latter decades of the twentieth century have witnessed the instigation of
several programmes designed to search for evidence for extraterrestial life – notably
SETI, the Search for Extraterrestial Intelligence [L5]. Underpinning such projects,
which seek civilisations similar to our own, is an equation devised by Frank Drake,
one of SETI's pioneers:

$$N = R_* f_p n_e f_l f_i f_c L \tag{10.1}$$

where N is the number of civilisations capable of long-distance communication in
the Galaxy; R_* is the formation rate of suitable stars; f_p is the fraction of those stars
with planets; n_e is the number of terrestrial planets per system; f_l is the fraction of the
latter where life is found; f_i is the fraction of life-bearing terrestrial planets where
intelligence develops; f_c is the fraction where technological civilisations are established; and L is the average lifetime of those civilisations.

Almost all of the factors in this equation are accessible only through indirect analysis, either purely theoretical (sometimes speculative) or laboratory-based simulations. Hence, it is instructive to recast the equation in a manner which highlights the fundamental philosophical issues:

$$N_l = N_* P_p P_l \qquad\qquad (10.2)$$

where N_l is the number of life-bearing planets other than Earth at time T; N_* is the number of stars at time T, given by the convolution of $\Psi(M)$ and the star-formation rate, with due allowance made for the stellar deathrate; P_p is the average probability of a star having a planetary system; and P_l is the average probability of a planet harbouring life at time T. P_l thus collects together in a single parameter such diverse environmental factors as planetary mass, orbital dimensions, binarity, age and stellar temperature, all of which may contribute to defining the likelihood of abiogenesis and the subsequent length of existence of any life-forms. The likely range of parameter space which can accommodate habitable planets has been discussed at length [K4], [K8].

Philosophically, there are only two interesting solutions to the second equation: $P_l = 0$, $N_l = 0$, and Earth is a unique system; or $P_l > 0$, $N_l =$ many. At present, P_l remains a matter of theoretical conjecture. P_p, on the other hand, *is* accessible to empirical determination, and the last decade of the twentieth century has seen the first independently-verified detections of planetary-mass companions to a variety of primary stars [M8], [M2]. This chapter summarises the highlights of these recent results. Further details can be found in Croswell's [C8] excellent popular review of this subject, and in technical review articles [K6], [M5].

10.2   PLANET FORMATION

10.2.1   Classical scenarios

Until very recently, theoretical mechanisms for forming planetary systems were constrained by observations of one system – our own. The inherent pitfalls in dealing with what Wetherill [W7] terms 'statistics of one' were well appreciated by planetary scientists and cosmogonists. Nonetheless, models for the formation of our Solar System came to represent the standard paradigm for planetary formation.

The main characteristics of the Solar System are: first, a significant change in the mean composition with increasing distance from the Sun, with the terrestial planets and asteroids at radii of less than 4 AU, gas giants at intermediate distances and icy planetesimals at radii beyond 30 AU; second, nearly coplanar and low-eccentricity orbits for eight planets and many lower-mass objects; third, angular momentum vectors in both orbital motion and rotation which are well aligned with the direction of solar rotation. Taken together, these properties strongly suggest an origin within a disk formed by the collapse of the pre-solar nebula. This concept of Solar System formation, first suggested by Kant [K3] and further elaborated by Laplace [L1], held sway during the nineteenth century, but was supplanted in the early years of the twentieth century by the near-collision theory originally proposed by Buffon [B12], and revived by Chamberlin [C3] and Moultin [M12], and later by Jeans [J2] and Jeffreys [J3]. The near-collision hypothesis envisaged a close encounter with a passing star leading to a long spindle of material being drawn from the Sun by tidal forces, with the planets condensing from that spindle.

The near-collision model had (to some) the philosophical attraction of requiring a very rare, perhaps unique, event. In contrast, the nebular hypothesis renders planet formation part of the natural sequence in star formation. The latter hypothesis was revived by von Weizsäcker [W5], who postulated the formation of cellular vortices due to instabilities within the protoplanetary disk – partly as a means of accounting for the numerical progression of planetary semi-major axes, known as Bode's law. The origin of modern theories, however, can be traced to Kuiper [K9], who not only suggested that the protoplanetary nebula was significantly more massive than the present-day sum of planetary masses, perhaps exceeding $0.1\,M_\odot$, but also proposed that the gas giants are the result of gaseous accretion on solid protoplanetary cores. Safronov [S2] and Hayashi *et al.* [H2] further extended this concept of building planets through accretion of planetesimals within a rotating disk of gas and dust. The identification in the 1980s of such disks associated with young T Tauri stars provides strong support for this conceptual model.

Current Solar System formation models are well summarised by [L12], [P2], [K6]. The original solar nebula is envisaged as having a mass of at least $\sim 0.02\,M_\odot$, based on adding cosmic proportions of hydrogen and helium to the current 'metallic' planetary masses, with a diameter of $\sim 100\,\mathrm{AU}$. These estimates are generally consistent with the masses and radii inferred from millimetre observations of T Tauri disks ([B12]; Section 3.6.2). HST near-infrared imaging of a number of those disks clearly shows that dust is present along the equatorial plane (Figure 3.20, colour section), and that dust provides vital building blocks for planet formation.

Classical formation scenarios envisaged the planets forming by progressive accretion within a relatively quiescent solar nebula. Refractory silicate grains are expected to form as the temperature drops below $1,700\,\mathrm{K}$, following a well-ordered 'condensation sequence'. The grains then aggregate to form 1–$10\,\mathrm{km}$ planetesimals, then $\sim 10^{-4}\,M_J$ planetary 'embryos' and, finally, terrestrial-mass planets [W6]. Recent, more detailed observations of T Tauri stars (Section 3.6.2) show that circumstellar disks are turbulent, and grain formation is likely to progress in a less ordered, more stochastic fashion. Processes such as collisional adhesion [W3], [C9] are likely to play a vital role in the formation of planetesimals which form the seeds for future planets.

The traditional method of forming gas giants is accretion onto a 'super-embryo', formed through the merging of 10–20 rocky embryos. The ambient temperature in the disk is expected to drop to $< 100\,\mathrm{K}$ at radii exceeding $4\,\mathrm{AU}$ for solar-type stars, allowing ices (mainly H_2O, but also CO_2, CO, NH_3, CH_4 and N_2) to condense, and proto-gas giants can accrete larger cores and massive envelopes [M11]. Envelope accretion, however, is predicted to require as much as 10^7 years [P2] – a time-scale which conflicts with the observed lifetime of optically thick disks in cTTs ($< 5\,\mathrm{Myr}$; Section 3.6.2). This is less of a problem for terrestrial planet formation, since the constituent planetesimals are expected to form on a more rapid time-scale. The apparent difficulty in forming massive planets led to the suggestion that such systems might be the exception, rather than the rule [W7]. The recent spate of planetary detections, however, shows that such systems are *not* rare, indicating that an alternative formation scenario may be required.

Boss [B8] revived the gravitational instability hypothesis originally proposed by Kuiper. The main difficulty faced by that mechanism lay in forming the ~ 0.02–$0.03\,M_J$ ice and rock cores required by contemporary models of Jupiter and Saturn. Recent theoretical calculations suggest core masses lower by a factor of three or more [G13], which fall within the range accommodated by models in which giant gaseous protoplanets form through the nebular disk breaking up under its own gravity. Boss's calculations suggest that Jovian-mass (and larger) objects can form in a matter of only a few thousand years – well within the projected lifetime of circumstellar disks.

Wetherill has pointed out that the existence of Jovian-mass planets at distances of several AU from the primary star may have a strong influence on the evolution of a solar system. Jupiter (and to a lesser extent, Saturn) acts as a guardian of the inner Solar System, ejecting cometary-mass objects to large radii, and therefore reducing the potential for catastrophic collisions (such as the impact of comet Shoemaker–Levy 9 on Jupiter) with the terrestrial planets. Uranus and Neptune perform similar tasks in the outer Solar System, marshalling comets into the Kuiper Belt. Thus, the development of complex life might be inhibited in a system lacking a Jupiter-like planet at the requisite radius, since the higher rate of cometary and asteroid impacts could disrupt an ecosystem on relatively short ($\sim 10^4\,\mathrm{yr}$) time-scales.

10.2.2   Definitions: brown dwarfs versus planets

As discussed in Chapter 9, it seems likely that the brown dwarf mass spectrum extends to masses below $\sim 0.01\,M_\odot$ ($11\,M_J$), reaching what might be considered the planetary régime. How, then, does one distinguish between a low-mass brown dwarf and a high-mass planetary companion? The pivotal distinction between the two rests with the mode of formation: brown dwarfs form by accretion within a giant molecular cloud in the same manner as hydrogen-burning stars; planets, by definition, form within the circumstellar disk of a protostellar nebula. This difference leads to a clear expectation of differences in chemical composition: observations of Jupiter and Saturn suggest abundances of $Z \sim 0.02$–0.06 and ~ 0.04–0.12 respectively, in at least the outer envelope, as compared with $Z \sim 0.02$ in the Sun. Unfortunately, for the foreseeable future there is little chance of direct detection of planetary-mass companions, let alone observations with sufficient detail to measure chemical abundances.

In most cases, the available observations allow only the measurement of a limited subset of orbital parameters, with the inclination usually indeterminate. Several authors, [C8], [B1], [B9], have suggested that the two types of companion can be distinguished on the basis of the orbital eccentricity, limiting 'planets' to objects in near-circular orbits. This definition, however, may well be biased by the 'statistics of one' argument; our Solar System may be atypical, and interactions between massive embryos have a reasonable probability of scattering objects into eccentric orbits [L11].

Given the observational limitations, it may be many years before we are able to do more than differentiate statistically between the two possible classifications.[1] Clearly, with only a measurement of $m\sin(i)$, any individual low-mass companion might be a brown dwarf (or even a low-mass star) in a low-inclination orbit. However, barring a cosmic conspiracy, the distribution of orbital inclination should be random, allowing a statistical estimate of the mass distribution. If the companions are predominantly brown dwarfs, a distribution which shows some continuity from the mass-function of companions above the hydrogen-burning limit might be expected; a distribution confined to near-Jovian masses, on the other hand, suggests a distinct, planetary origin. Current results (see Section 10.4.2) favour the latter.

10.3 SEARCHES FOR EXTRASOLAR PLANETS

Almost all of the techniques used to search for planetary-mass companions to main-sequence and evolved stars are also applied in searching for stellar or brown dwarf binaries, but since planetary targets have substantially lower masses and luminosities, the technical requirements are correspondingly more stringent. Most observations are capable of determining only a mass ratio with respect to the primary star, either m/M_* or, more commonly, $[m\sin(i)]/M_*$, where i is the orbital inclination. Since the distances to all of the target stars are well known, M_* can be estimated either using the standard main-sequence mass–luminosity relationship or, since some of the primaries have evolved off the main sequence, by comparing the position on the H–R diagram against theoretical isochrones (see [F2], [G9]).

10.3.1 Astrometric surveys

Planetary-mass companions are capable of introducing perturbations of no more than several mas in the motions of even the nearest stars to the Sun. As an example, Figure 10.1 shows the residuals introduced into the Sun's motion by the planets in the Solar System, scaling these motions to a viewpoint 10 parsecs distant along the line of the ecliptic poles. The semi-amplitude of the perturbation due to a single planet in a circular orbit is given by [G4]:

$$A = \frac{m}{d}\left(\frac{P}{M_* + m}\right)^{2/3}$$

$$= \frac{m}{M_J}\frac{5}{d}\left(\frac{P}{M_*}\right)^{2/3} \tag{10.3}$$

---

[1] The authors note that such predictions tend to be invalidated. We will happily accept the error if such proves to be the case for this prediction.

where m is the planetary mass, M_J is the mass of Jupiter, M_* is the primary-star mass in solar units; d is the distance in parsecs, and P is the period in years. The effective mass detection limit, m_d, for a given series of observations, is

$$m_d \propto \frac{\sigma d}{\sqrt{n}} \left(\frac{M_*}{P} \right)^{2/3} \tag{10.4}$$

where σ is the measurement uncertainty.

These equations show that for a planet of given mass, the astrometric signature and, in principle, the likelihood of detection, increase with decreasing mass of the primary star and increasing semi-major axis (longer period). Orbital inclination, and hence mass, can be determined directly from the astrometric orbit. In practice, however, planets with $A > 1$ mas have periods of many decades, which complicates the logistics of organising an observing programme. Additional obstacles to accurate interpretation are introduced if there is more than one massive planet, although Figure 10.1 shows that Jupiter dominates within the Solar System.

Despite these technical difficulties, astrometric searches for unseen companions have been undertaken for well over a century. There have been notable successes – such as Bessel's detections of the white dwarf companions of Sirius and Procyon – but there have also been many false alarms. The first purported detections of planetary-mass companions date from the 1940s, when Reuyl and Holmberg [R2] announced the identification of a $\sim$17-yr period, $\sim$0.01 $M_\odot$ companion of 70 Ophiuchi (Gl 702), while Strand [S6] identified a 4.8-yr period, 0.008 $M_\odot$ companion of one of the components of 61 Cygni (Gl 820). Similar claims were later made for, amongst others, AD Leo (Gl 388) [R3], Lalande 21185 (Gl 411) [L9], Stein 2051A (Gl 169.1A) [S7] and G96-45 (GJ 1081) [B3]. All of these analyses were based on measurements of several hundred photographic plates, spanning several decades, with typical deduced periods of 5–20 years and semi-major axes of

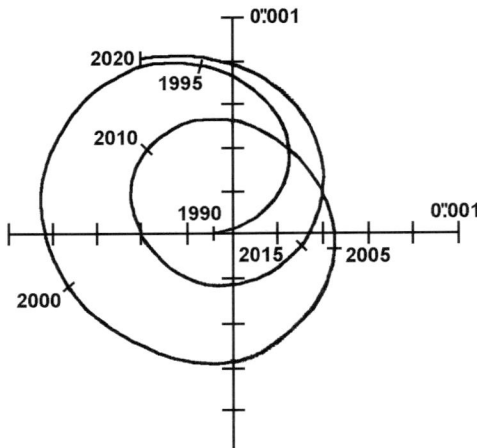

Figure 10.1. The astrometric displacement of the Sun due to Jupiter, as seen from a distance of 10 parsecs. (Courtesy of the Space Interferometer Mission).

10–20 mas. None of the candidates have been confirmed by subsequent and more accurate observations [L10], [H4].

The classic case is Barnard's Star (Gl 699), for which van de Kamp [K1], [K2] accumulated measurements of some 4,000 plates taken with the Sproul 24-inch refractor between 1916 and 1974. Analysing those data, van de Kamp identified astrometric perturbations which he ascribed to two Jovian-mass planets with orbital periods of 11.5 and 20–25 years respectively. Unfortunately, observations with the Allegheny 30-inch Thaw refractor – both photographic [G2], and with the photo-electric Multichannel Astrometric Photometer (MAP) – failed to confirm the predicted motions. Similarly, observations made at Leander McCormick Observa-tory by Fredericks and by Harrington, with the US Naval Observatory 61-inch, failed to detect significant perturbations; while Hershey [H5] identified similar 'wobbles' in astrometric analyses of Sproul photographic plates of other (more distant) stars. The balance of the data suggests that the original 'detections' stem from systematic positional errors, rather than from true perturbations. The most recent observations limit possible planetary companions of Barnard's Star to masses significantly below that of Jupiter [G6].

Current astrometric searches for planetary perturbations are based either on direct CCD observations [P3] or Ronchi-grating photometry [G5], with both techniques capable of approaching 0.1 mas. At the time of writing, only one unconfirmed astrometrically-detected planetary system remains. Gatewood [G7] (reversing an earlier result of his own [G3]) has claimed detection of perturbations in the secular acceleration of Lalande 21185. Secular acceleration, S_a, is a parameter which is measurable for only the nearest stars. It is the change in proper motion arising from the changing distance of a given star:

$$S_a = \frac{-2.05 \times 10^{-6}}{d} \mu V \text{ arcsec yr}^{-2} \tag{10.5}$$

with d in parsecs and V in km s^{-1}. In the case of Lalande 21185, the perturbations are ascribed to the presence of two planets, with masses of 10 and 20 M_J. No corresponding perturbations have been detected in the proper motion perpendicular to the direction of secular acceleration, nor have radial-velocity variations been observed. This implies that the planets have extremely elongated orbits, well-aligned with the star's radial velocity vector – which seems rather implausible. Further observations are required to confirm this possible detection.

10.3.2   Photometric detection and direct imaging

Direct imaging

The simplest form of photometric detection of a planet would be a direct image of a nearby planetary system. Currently available techniques, however, are not capable of direct detection of Jupiter-like gas giants around even the nearest stars. The flux contrast between Jupiter and the Sun is $\sim 10^{-9}$ (22 magnitudes) at optical and near-infrared wavelengths, so any planet can be expected to have a brightness well below the wings of the point-spread function of the primary star. Since the main source of radiation from the planet is reflected light at these wavelengths, there is no advantage gained in seeking planets around lower-luminosity primary stars. At mid-infrared wavelengths Jupiter is self-luminous, and the contrast drops to $\sim 10^{-4}$ at 20–100 μm. Unfortunately, current detectors lack both the sensitivity and spatial resolution to render surveys feasible at these wavelengths.

One method of enhancing the chances of direct observation is to search for planets around young stellar systems. As with brown dwarfs, theoretical models predict that Jovian planets are substantially more luminous during the initial stages of collapse.[2] Future developments in observational techniques may permit more detailed searches for protoplanets in young stellar clusters.

Eclipses and transits

An alternative to direct imaging is to search for planetary eclipses of the primary star. Jupiter has 1% of the surface area of the Sun; a corresponding dip in the total flux of a solar-type star can be expected during the transit of a gas giant. If the primary is a mid-type M dwarf, the eclipse can be as deep as 10%. In order for these circumstances to pertain, the orbital inclination must satisfy the condition

$$\tan(i) > \frac{a}{R_*} \tag{10.6}$$

where a is the orbital semi-major axis and R_* the radius of the star. The duration of the eclipse also depends on a, and concerns about instrumental stability make it more difficult to search for long-duration, long-period eclipses. Consequently, this technique is best suited to detecting planets in short-period orbits.

Defining $\tan(i_*) = a/R_*$, the probability that an orbit lies within the range $i_* < i < 90°$ is given by $\cos(i_*)$. Following Marcy and Butler [M5], if we consider the case of a Jupiter-sized planet orbiting a solar-type star with $a = 0.1$ AU, eclipses occur for $i > 87°.3$, corresponding to 4.7% of all systems for a random distribution of inclinations. Current results suggest that $\sim 4\%$ of G dwarfs have planets with $a < 0.1$ AU, implying that $\sim 0.2\%$ should experience eclipses. If the primary is an M2–4 dwarf, $R_* \sim 0.3\,R_\odot$, the corresponding values are $i > 89°.25$ and $P \sim 1.3\%$. Note that, given the similarity in radii, this technique cannot distinguish between a gas-giant planet, brown dwarf and low-mass stellar companion.

Known eclipsing binaries are good targets for this type of photometric monitoring, since angular momentum considerations suggest that it is likely that any planet will reside in the same orbital plane as the stellar orbit ($i = 90°$). CM Dra, the lowest-mass eclipsing binary currently known, is an obvious target [D1], and Guinan and collaborators have carried out an extensive photometric campaign from which they

---

[2] Terebey *et al.* [T3] have identified a candidate protoplanet based on 2 μm HST NICMOS images of the binary protostar TMR-1 in Taurus. Prompted by the presence of a luminous trail resembling a severed umbilical chord, they suggest that a nearby red object may be a young gas-giant ejected from the TMR-1 system. Follow-up infrared spectroscopy does not appear to confirm this hypothesis.

claim evidence for the presence of two low-mass companions [G14]. One is detected through observation of 0.08-magnitude depth eclipses, and is suspected to have a period of 70.3 ± 3 days and a mass of $64 \pm 4\,M_J$; the second appears to be more massive, with $M \sim 0.06\,M_\odot$ and a 39.6-day orbit. The latter manifests its presence through systematic residuals between the predicted and observed times of eclipse for the stellar binary. Neither can be regarded as fully confirmed at present, and further observations are underway.[3]

Protoplanetary and post-planetary disks

While direct imaging of planets lies beyond the bounds of current technology, potential planetary environments have been identified. As described in Section 3.6, molecular gas has been detected in association with many young T Tauri stars, and there is now overwhelming evidence that the gas settles into rationally supported circumstellar disks, which are obvious candidate sites for planetary formation. Millimetre observations are capable of tracing disk evolution until typical ages of 3–5 Myr, at which point the disk is sufficiently depleted of gas that it becomes optically thin and falls below current detection thresholds. It is only recently that it has become possible to trace the subsequent evolution through far-infrared observations of thermal radiation from the surviving dusty disk.

HR 4796 (Figure 10.2) is a nearby 10^7-year old A0 star which may be a member of the TW Hydrae association – a recently discovered loose cluster of 10–15 stars at a distance of 50–70 pc from the Sun [L13]. HR 4796 itself lies 67 ± 3 pc distant. Jura [J4] originally pointed out that IRAS observations reveal a substantial far-infrared excess, and suggested that this star has residual dust. Mid-infrared observations [K7], [J1] have succeeded in resolving the excess as a dusty disk at an inclination of $72°$ and extending at least 100 AU from the central star. Significantly, the 20 μm flux distribution does not increase monotonically with decreasing radius, but is best fit by a disk with substantially reduced densities at radii of less than 55 ± 15 AU (that is, a central hole). This model has been confirmed by near-infrared observations with the NICMOS camera on the HST, and a second system with similar characteristics, HD 141569 (spectral type A0), has been identified (Figure 10.3). This strongly suggests that relatively massive planetesimals have already formed and are transferring material from the inner to the outer disk. Residual emission at smaller radii, detectable at 12 and 20 μm, suggests the presence of hotter (200–300 K) dust. The outer disk lies at radii comparable to the Kuiper Belt in the Solar System, while the inner particles have properties reminiscent of zodiacal dust.

β Pictoris debris disks (Figure 10.4, colour section) represent a later stage in disk evolution, corresponding to ages of $\sim 10^8$ years. The spatial distribution of the debris is well matched with that inferred for the zodiacal disk in our own Solar System [A3],

---

[3] Subsequent observations have failed to confirm planetary eclipses in the CM Dra system (Doyle *et al.*, 2000, *ApJ*). However, planetary transits in the HD 209458 system were detected in September 1999. The depth of eclipse, $\sim 1.3\%$, implies a diameter somewhat larger than Jupiter (Charbonneau *et al.*, 2000, *ApJ Letters*).

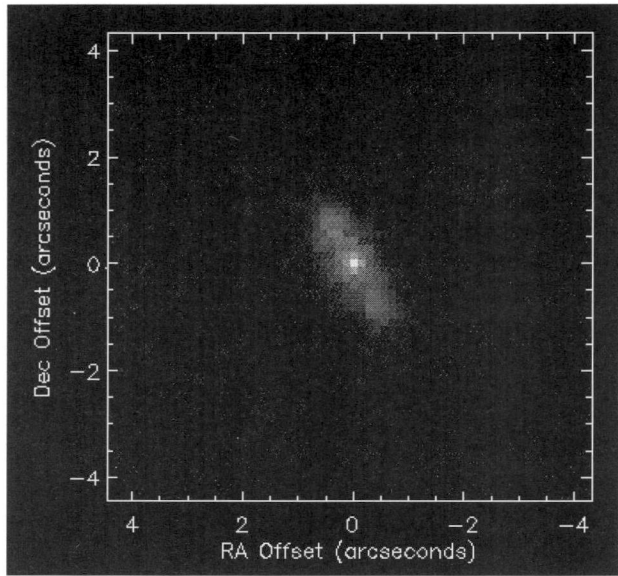

Figure 10.2. The HR 4796 dust disk. The image is oriented with north at the top and east on the left. (From MIRLIN observations, courtesy of D. Koerner.)

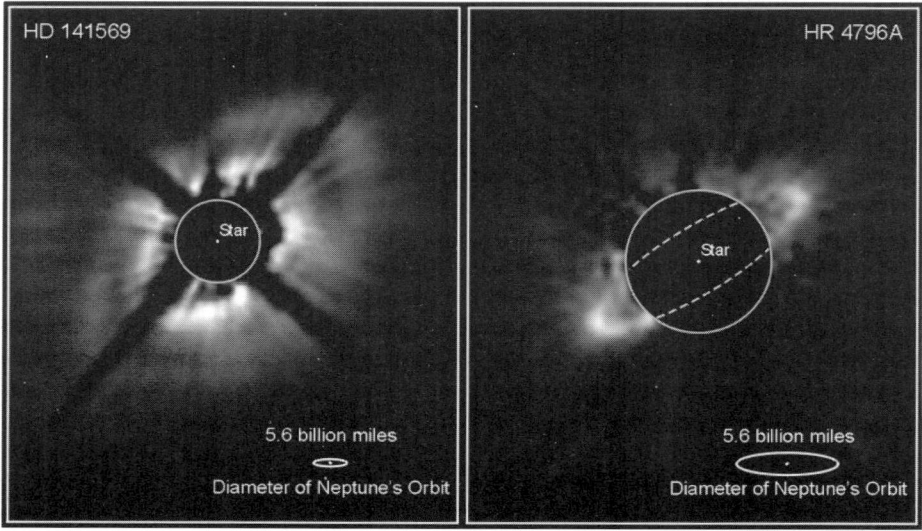

Figure 10.3. HST near-infrared observations of dusty disks in HD 141569 and HR 4796. The HR 4796 image is oriented with north on the left and east down (rotated $-90°$ with respect to Figure 10.2). (Courtesy of STScI.)

with a radius of $\sim 400\,\mathrm{AU}$ and a nearly exponential density profile perpendicular to the rotational plane. The infrared flux distribution implies the presence of grains spanning a wide range of sizes, from $< 1\,\mu\mathrm{m}$ to at least several millimetres [T2], [Z1]. The Poynting–Robertson effect and radiation pressure should lead to the smaller grains having orbital lifetimes of only $\sim 10^5$ years [A2], which, given the age of the primary star, suggests that particles must be replenished, perhaps by collisions between larger planetesimals [W4].

Detection of the $10\,\mu\mathrm{m}$ emission feature shows that silicate grains are present. However, the high albedo (0.5 ± 0.2) in the outer regions of the disk indicates the presence of more reflective material, probably 'dirty ices', as found in Solar System comets. Finally, and most significantly, there is strong evidence for a lower grain density in the inner portion of the disk $(r < 30\,\mathrm{AU})$, and the massive planetesimals represent the most likely mechanism of driving this inner-disk clearing. Indeed, rapid changes in the Ca II H and K absorption line profiles have been attributed to planetesimals passing in front of the stellar disk. It is likely that any major planets in this system are undergoing the sustained bombardment from planetesimals and cometesimals that appears to have characterised the early stages of evolution of the Solar System.

10.3.3   Detection using Doppler techniques

Radial-velocity surveys have proven to be the most successful method of identifying candidate planetary systems. Between 1994 and 1998, over a dozen stars with super-Jovian companions have been detected. However, the first planetary-mass companions, and the only terrestrial-mass objects found to date, were detected using a different type of Doppler observation: pulsar timing.

Pulsar planets

Pulsars are rapidly-rotating neutron stars, the remnants of Type II (massive star) supernovae. Initially rotating with periods of less than a millisecond, pulsars gradually spin down with time. During the early phases, abrupt changes can occur in the pulse period, prompted by changes in the internal structure, but such irregularities have not been detected in older systems, with periods of several milliseconds or more. The latter sources serve as remarkably accurate celestial clocks, in which binary motion reveals itself as systematic periodic residuals in the pulse arrival.

The first 'detections' of pulsar planets all proved to be spurious, with the most plausible (PSR B1829−10) [B1] arising from data reduction errors: adopting the wrong position for the pulsar and failing to allow for the eccentricity of the Earth's orbit led to incomplete correction for the Earth's orbital motion [L14] (see also [C8]). Almost simultaneously with the retraction of that 'detection', Wolszscan and Frail [W9] announced the discovery of two very low-mass companions to PSR B1257+12. Subsequent observations have not only confirmed their existence, but also detected the signature of their mutual gravitational interaction and identified a third

Table 10.1. The pulsar planets.

Planet	Period	a	$M \sin(i)$	e
a	25.34	0.19	0.015	0.0
b	66.54	0.36	3.4	0.018
c	98.22	0.47	2.8	0.026

planetesimal. All three are on short-period, near-circular orbits, and have masses closer to those of the terrestrial planets than to the gas giants (Table 10.1).

The origin of these low-mass planetesimals remains unclear. Given the relatively short lifetime of the supernova progenitor, the standard model of formation within the protostellar disk may not apply. Moreover, one would expect the substantial mass-loss during the supernova event to have severely disrupted any planetary system. Thus, it is possible that these terrestrial-mass objects originate in a 'born-again' circumstellar disk, formed either by material accreted from a now-defunct companion, or during the merger of two white dwarfs, which gave rise to the parent neutron star. If either of these scenarios holds, the pulsar planets may have little to do with conventional planetary systems.

Radial velocity surveys

Solar-type stars constitute the majority of the targets of high-precision radial velocity programmes. In part this reflects the underlying, often unstated, aim of finding planetary systems similar to our own, but technical considerations also favour observations of these stars. The characteristic velocity signatures of planetary companions are not easy to detect.

The reflex radial velocity induced by a companion of mass m is given by

$$K = \left(\frac{2\pi G}{P} \right)^{1/3} \frac{m \sin(i)}{(M_* + m)^{2/3}} (1 - e^2)^{-1/2}$$

$$= \left(\frac{Ga}{(M_* + m)} \right)^{1/2} (1 - e^2)^{-1/2} m \sin(i) \tag{10.7}$$

where K is the semi-amplitude of the velocity variation, and the other symbols have their usual meaning. In most cases, $M_* \gg m$, so $(M_* + m) \approx M_*$. Considering our solar system, Jupiter alone produces a near-sinusoidal variation of $\pm 12.5\,\mathrm{m\,s^{-1}}$ in the radial velocity of the Sun as measured from a vantage point in the plane of the ecliptic. Hence, an observational precision of better than $4\,\mathrm{m\,s^{-1}}$, maintained over a period of 5–10 years, is required for a 3σ detection of Jupiter-like planets in other solar-type stars.

Acquiring velocity measurements of such high precision demands high signal-to-noise spectra which, even with 10-m telescopes, limit observations to bright ($V < 11$th magnitude) stars. This favours either main-sequence stars earlier than spectral-type K or giant stars. Evolution beyond the main sequence might disrupt or

destroy existing planets, and many giant stars pulsate, complicating detection of
orbital motion. Substantial numbers of relatively narrow lines in the stellar spectrum
are also required to allow accurate velocity determination. This consideration
excludes stars hotter than $\sim 8,000\,$K (A-type or earlier), which have both a
scarcity of metallic absorption features and tend to have rapid rotation, with, as a
consequence, broad lines and poorly-determined line centroids. Hence, F, G and K
dwarfs are selected naturally as the prime targets for planet searches, with a few
bright M dwarfs also within reach of high-precision measurement. There are two
further major barriers to achieving the requisite long-term accuracy of a 3–$5\,\mathrm{m\,s}^{-1}$:
instrumental stability, and the intrinsic variability of stellar atmospheres.

Section 1.6 outlined the main technical characteristics of spectrographs. The usual
mode of observation is to intersperse programme-star exposures with spectra of
emission-line calibration lamps – usually combinations of He, Ne, Ar, Hg and Th.
While a suitable approach for most purposes, systematic errors in wavelength
calibration at the $\sim 100\,\mathrm{m\,s}^{-1}$ level can arise from a number of sources, including
time-dependent physical flexure of the spectrograph; thermal expansion of optical
and mechanical elements; and differences in the optical beam due to the way that the
star (a point-source) and the arc-lamp (diffuse source) illuminate the slit.

Most potential mechanical distortions can be eliminated by placing the spectrograph in a separate, temperature-controlled room, rather than on the telescope itself.
Indeed, the latter option is often not possible for spectrographs designed for 8-m
class telescopes, such as the HIRES (HIgh Resolution Echelle Spectrograph) [V1]
which has room-sized dimensions. Optical distortions can be taken into account if
the wavelength calibration is determined simultaneously with the programme-star
observations. Several techniques have been used: adapting a method used in
photographic spectroscopy, Mayor *et al.* [M8] use optical fibres to feed light from
both the target star and a calibration lamp to adjacent positions on the slit, allowing
simultaneous observation. The use of fibres for both star and calibration source
ensures similar illumination of the slit, and radial velocities can be measured to an
accuracy of $\epsilon_V \sim 4\,\mathrm{m\,s}^{-1}$. Similar precision can be attained if the wavelength
calibration is imprinted directly on the stellar spectrum. The simplest approach is
to take advantage of the natural grid of absorption lines due to H_2O, O_2 and OH
provided by the Earth's atmosphere, but variations in temperature and pressure limit
velocity uncertainties to $\epsilon_V \sim 20\,\mathrm{m\,s}^{-1}$. More stable calibrations result from introducing a gas-filled cell, usually either hydrogen fluoride [C1], [C2] or iodine [M1] into
the spectrograph beam immediately below the slit. This superimposes a fine grid of
narrow absorption lines onto the stellar spectrum, allowing the radial velocity to be
determined to an accuracy of $\epsilon_V \sim 3\,\mathrm{m\,s}^{-1}$ [B13].

The ultimate limit on velocity accuracy is set by the intrinsic stability of the
photosphere of the target star. Saar *et al.* [S1] have completed a detailed series of
observations of F, G and K main-sequence dwarfs, and find that the rms dispersion
of residuals about the mean velocity is well-correlated with the measured rotation
period, $\sigma_V \propto P^{-1.1}$. Stars with rotation periods of 10 days have $\sigma_V \sim 10\,\mathrm{m\,s}^{-1}$, rising
to $\sigma_V \sim 40\,\mathrm{m\,s}^{-1}$ at $P \sim 3$ days. These velocities are generally ascribed to starspots,
plages and other symptoms of stellar activity, which introduce changes in the stellar

Table 10.2. Radial-velocity planet searches.

	N_*	ΔT (yr)	σ_V (m s^{-1})	Status	N_{det}
Cochran and Hatzes [C4]	33	10	~10	Continuing	
Latham et al. [L2]	600		300	Complete	
Mayor et al. [M9]	140	3	~5	Continuing	2
Marcy and Butler [M5]	107	11	10–3	Continuing	
Noyes et al. [N1]	100	2	~3	Continuing	
Walker et al. [W1]	21	12	13	Complete	

line-profile by changing the flux distribution over the stellar surface. Similarly, convective motions in giant stars can lead to velocity 'jitter' of 0.5 km s^{-1} or more. Clearly, this added noise can easily hide the systematic velocity variations due to planetary-mass companions. Radial velocity surveys attempt to minimise these problems by selecting chromospherically inactive (1–2 Gyr or older) stars. In addition, known stellar binaries with separation of ~10 AU or less have so far received little attention.

Table 10.2 summarises the major radial-velocity-based planet searches undertaken during the last decade, several of which are still in progress (see also [M5]). The first solid planetary detection to emerge from these programmes[4] was Mayor and Queloz' [M8] identification of the low-mass companion to 51 Peg, which confounded all expectations by exhibiting a velocity semi-amplitude of more than 50 m s^{-1} and a period of only 4.23 days. The inferred mass of the companion, $m \sin(i)$, is only 0.47 M_J, while the projected semi-major axis is only 0.05 AU – one-sixth the radius of Mercury's orbit. Both Hipparcos astrometry and subsequent interferometric observations [B7] effectively rule out a stellar companion in a low inclination (near face-on) orbit. The surface temperature on a planetary companion at the inferred distance from the G5 primary is estimated to be ~1,300 K, and a gas giant is predicted to have 2–3 times the radius of Jupiter due to the high level of irradiation.

In the three years following that seminal result, more than a dozen planetary companions have been discovered. In most cases, the data are not sufficiently extensive to allow detection of more than a single massive perturber. Subsequent observations have not confirmed the initial suggestion [M8] that velocity residuals from their 51 Peg data, after subtracting the 4.23-day period, indicate a longer-period variation. Marcy and Butler [M5], however, have more convincing data for an 8-year variation in 55 Cnc, while 11 years of observations have culminated in the identification of three planets in the υ And system [B16] (Table 10.4). This last result shows that multi-planet systems, albeit somewhat unlike our own Solar system, exist

---

[4] Previous observations by Walker et al. [W1] attained the necessary precision for many of the subsequent detections. Unfortunately, their small sample did not include any of the (later) confirmed planetary systems.

elsewhere in the Galaxy – a result which is not unexpected, but which nonetheless carries significant implications.

The initial impression might be that the flood of detections reflects the culmination of decade-long observing programmes, but only two systems amongst those listed in Tables 10.3 and 10.4 have periods longer than 2 years. Moreover, the initial planetary detection was based almost entirely on observations with the Elodie echelle spectrograph at Haute-Provence, while many subsequent discoveries have been made using data obtained with the Hamilton echelle at Lick Observatory. Thus, contrary to appearances, the technological hardware for detecting extrasolar planets has been available for almost 15 years. The main breakthrough came with the availability of greater computing power and more sophisticated analysis programmes, combined with the knowledge that gas giants actually exist in short-period orbits.

With these discoveries, NASA efforts aimed at searching for extrasolar planetary systems have gained substantial momentum, leading to the development of the ORIGINS programme, underlying much of NASA's long-term scientific strategy in the early years of the twenty-first century. Planet searches are a key element in that programme, and significant resources are being devoted in several areas, including ground-based radial-velocity surveys. With improved reduction techniques and, in particular, the added light grasp of the new generation of 8–10 m telescopes, typical observational uncertainties of current observations are $\epsilon_V \sim 3\,\mathrm{m\,s^{-1}}$, making Jovian analogues accessible in the Solar Neighbourhood.

10.4 INTERPRETING THE RESULTS

10.4.1 Alternative explanations: planets or pulsations?

Stellar radial velocity variations can be produced by a variety of mechanisms other than binary motion, including magnetic activity, convective motion and stellar pulsation. All of these possibilities were considered, and rejected, by Mayor and Queloz in their first analysis of data for 51 Peg. In most cases the crucial factor is the highly periodic nature of the observed variations, maintained over many cycles (>300 for 51 Peg). More stochastic processes – such as starspot formation in active, rapidly rotating stars (such as HD 195019, [F1]) – can mimic this behaviour for a limited number of cycles, but usually fall out of phase after ~ 50 cycles. In the case of 51 Peg itself, the star is mildly active chromospherically, with a measured rotation period of ~ 37 days, significantly longer than the proposed planetary period. Chromospherically-induced velocity variations are also usually accompanied by photometric variability on a matched time-scale and a high level of Ca II H and K emission (Section 5.4.1).

Stellar pulsations can, in principle, give rise to long-term, stable variations. Gray [G10], and Gray and Hatzes [G12], identified apparent profile variations in the Fe I 6253 Å line in 51 Peg which are not predicted by the binary model, but would be consistent with stellar oscillations. There are, however, severe problems with this mechanism as an explanation for the overall stellar behaviour.

Stellar pulsations occur in two modes: p-modes, or pressure (acoustic) waves; and g-modes, or gravity (buoyancy) waves [G8]. Purely radial pulsations can be ruled out, since these would lead to variations of as much as 10–15% in the stellar radius (and concomitant effective temperature changes) which would be detected as periodic photometric variations. All of the G-stars with potential planetary-mass companions with periods of less than 10 days have been monitored extensively, and photometric variations of more than 0.0002 magnitudes can be ruled out [H3]. Thus, any pulsations must be non-radial, with no significant radius or temperature changes.

There are also significant problems involved in matching the observed periodicity. The frequency of harmonic oscillations for solar-type stars corresponds to periods of only 3–8 minutes for p-mode waves (as in the 5-minute solar oscillations, [L4]), and only ∼165 minutes for g-modes. Matching variations against the observed periods requires the dominant mode of oscillation to be a high overtone; the 115th overtone in g-mode oscillations is required to match the 51 Peg 4.23-day cycle [W8]. That oscillation should be accompanied by oscillations at other frequencies, but the latter are not detected in Fourier analysis of the 51 Peg velocity variations.

Finally, additional high-resolution spectroscopy has failed to confirm the original hypothetical profile variations [B11], [G11], [M5]. Stellar oscillations offer no prospect whatsoever of reproducing periodic variations of 1 year or more in G-type dwarfs; they create more problems than they solve for short-period variations, and the original supporting evidence has evaporated. Thus, there is currently no reasonable alternative to low-mass companions to explain the observed cyclical velocity variations in the stars listed in Table 10.3.

10.4.2   The mass distribution: planets or brown dwarfs?

Tables 10.3 and 10.4 list relevant parameters for the 16 systems with potential planetary-mass objects ($M \leq 10\,M_J$) discovered to date (June 1999). The low mass companion of HD 114762 [L2] is included, although, as discussed in Chapter 9, it probably has a mass falling within the brown dwarf régime. An astrometric orbit determination – and therefore a direct measurement of i – has not been determined for any of these stars, but almost all of them were observed by Hipparcos. With the exception of HD 114762, the latter data show no evidence for significant astrometric residuals, and therefore rule out high-mass brown dwarfs or low-mass stars in near face-on orbits.

Lacking direct measurement of i, the only method available of estimating the orbital inclination is by measuring the rotational velocity for the star, comparing that with the rotational period (if that can be determined using cyclical variations in chromospheric activity), and assuming that the companions lie within the stellar equatorial plane (effectively assuming formation within the protostellar disk). Since the majority of these stars are inactive, the expected rotation velocities are only 2–3 km s^{-1}, and it is seldom possible to set strong constraints on i. In the case of 51 Peg, $\sin(i) \sim 0.4$ [M8].

Table 10.3. The primary stars.

Name	Gl/GJ	Hipparcos	M_V	(B–V)	π (mas)	Mass ($M_\odot$)	[Fe/H]
HD 75289		43177	4.04	0.62	34.6	1.25	0.29
51 Peg	882	113357	4.59	0.74	65.1	1.05	0.20
HD 187123		97336	4.53	0.72	20.9	1.0	0.16
v And	61	7513	3.51	0.59	74.3	1.31	0.17
55 (ρ^1) Cnc	324A	43587	5.55	1.01	79.8	0.7	0.29
ρ CrB	606.2	78459	4.26	0.67	57.4	0.96	−0.29
HD 217107		113421	4.78	0.85	50.7	0.96	0.29
HD 210277	848.4	109378	4.99	0.86	47.0		
16 Cyg B	765.1B		4.37	0.66	35	0.97	0.06
Ross 780	876	113020	11.80	1.60	212.7	0.32	
47 UMa	407	53721	4.36	0.69	71.0	1.03	0.01
14 Her	614	79248	6.43	1.02	55.1	0.8	0.15
HD 195019		100970	4.12	0.72	26.8	0.98	0.0
HD 13445	86	10138	5.98	0.77	91.6	0.79	
τ Boö	527A	67275	3.57	0.53	64.1	1.36	0.34
HD 168443		89844	4.07	0.82	26.4		
70 Vir	512.1	65721	3.76	0.80	55.2	1.10	−0.03
HD 114762		64426	4.32	0.56	24.7	0.82	−0.60

Column 2 lists the designations from the *CNS3 Nearby-star Catalogue*, while column 3 gives the number in the Hipparcos catalogue. All of the photometry and parallax measurements are taken from the latter catalogue, except for 61 Cyg B, where the data are from the *CNS3 Nearby-star Catalogue*. Masses and abundance estimates are primarily from Gonzalez [G9].

The majority of the primaries are near-solar mass G stars, either on the main-sequence or subgiant branch, and the expected perturbations in the plane of the sky for planetary-mass companions are less than 0.1 mas in amplitude – well below the limits accessible with current technology. On the other hand, Gl 876, the lowest-mass planetary-system primary yet found, could exhibit reflex motions as high as ±1.6 mas, which should be measurable with the HST Fine Guidance Sensors [B4].

One of the main questions raised by these discoveries is whether the apparently low-mass companions represent the tail of the binary brown dwarf mass-distribution, or whether they are a distinctly different type of object (such as planets). As noted above, the continuity of the stellar-companion mass distribution, $\Psi_C(m)$, from above the hydrogen-burning limit to $\sim 0.001 \, M_\odot$, offers the best means of discriminating between these alternatives. Pending direct measurements of i, we are forced to rely on statistical reconstructions based on $\Psi_C(m\sin(i))$. Two approaches can be adopted: estimating the most likely intrinsic distribution given the available data, or estimating the likelihood of deriving the observed distribution given a particular parameterisation of $\Psi_C(m)$.

Adopting the former approach, it is reasonable to assume that the systems in Table 10.3 are drawn from a sample of planetary systems with orbital inclinations

Table 10.4. The companions.

Name	$m \sin(i)$ (M_J)	P (days)	a (AU)	e	Reference
HD 75289	0.42	3.51	0.046	0.054	14
51 Peg	0.47	4.229	0.05	0.0	1, 2
HD 187123	0.52	3.097	0.042	0.02	3
υ And	0.71	4.617	0.059	0.03	4
	2.11	241.2	0.83	0.18	16
	4.61	1,266.6	2.50	0.41	16
55 Cnc	0.84	14.65	0.11	0.05	4
	~5	~2,700	>4		
ρ CrB	1.1	39.65	0.23	0.028	5
HD 217107	1.28	7.11	0.04	0.14	6
HD 210277	1.37	437	1.15	0.45	7
16 Cyg B	1.5	804	0.6–0.7	0.67	8
Ross 780	2.1	60.85	0.21	0.27	9
47 UMa	2.8	1,099	2.11	0.03	10
14 Her	3.3	1,619	2.5	0.354	11
HD 195019	3.43	18.3	0.14	0.05	5
Gl 86	3.6	15.83	0.11	0.05	15
τ Boö	3.87	3.313	0.046	0.018	4
HD 168443	4.1	70	0.2	High	7
70 Vir	6.6	116.6	0.43	0.4	12
HD 114762	11	84.05	0.3	0.25	13

References: 1. Mayor and Queloz, 1995; 2. Marcy *et al.*, 1996; 3. Butler *et al.*, 1998; 4. Butler *et al.*, 1997; 5. Noyes *et al.*, 1997; 6. Fischer *et al.*, 1999; 7. Marcy *et al.*, 1999; 8. Cochran *et al.*, 1996; 9. Marcy *et al.*, 1998; 10. Butler and Marcy, 1996; 11. Mayor *et al.*, 1998; 12. Marcy and Butler, 1996; 13. Latham *et al.*, 1989; 14. Mayor *et al.*, 1999; 15. Queloz *et al.*, 1999; 16. Butler *et al.*, 1999.

which are distributed in a random fashion with respect to our line of sight. In this case,

$$\langle m \rangle = \frac{4}{\pi} m \sin(i) \tag{10.8}$$

where $\langle m \rangle$ is the expectation value of the mass.

Figure 10.5 plots the inferred mass-distribution of G-dwarf companions, including higher-mass objects from the lower-precision surveys [M9], [L2]. The data show a strong peak at masses below $10\,M_J$, with 18 of the 29 companions lying at what might be considered 'planetary' masses. Binning the data logarithmically places four companions in the mass range $0.1–1\,M_J$; 14 at masses of $1–10\,M_J$; and 11 with masses of $10–80\,M_J$. These numbers are consistent with a mass function $\Psi_C(M) \propto M^{-1}$ to $\sim M_J$. However, this simple calculation ignores the relative discovery probabilities: almost all of the companions with $M_J > 10$ are drawn from moderate-precision ($300\,\mathrm{m\,s^{-1}}$) observations of 640 stars, while the lowest-mass companions are from a sample of around 150 stars with high-precision

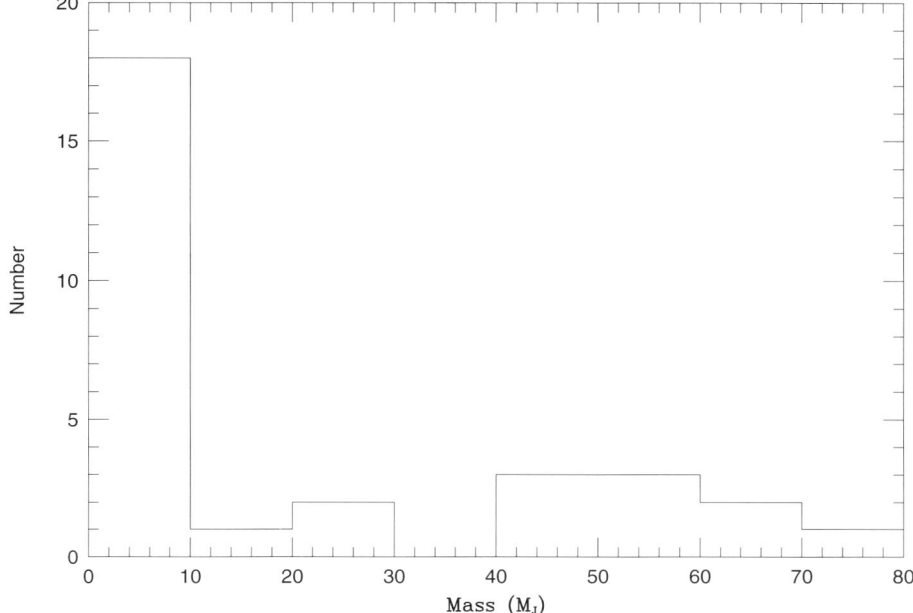

Figure 10.5. The most probable mass-distribution of low-mass radial velocity companions.

data. Hipparcos observations exist for almost all of the primary stars. Systems with companion mass $m\sin(i) > 10\,M_J$ show significant astrometric perturbations, implying near face-on orbits and a companion mass *above* the hydrogen-burning limit. Final results are not yet available, but it is likely that most of the companions plotted with $10 < m/M_J < 80$ in Figure 10.5 actually have masses exceeding $0.1 M_\odot$, preserving what Marcy has termed the 'brown dwarf desert' among G-dwarf companions.

A simple comparison of the relative numbers of binary companions in the different mass intervals reinforces this result. Consider parameterising the mass function of companions as $\Psi_C(M) \propto M^{-\alpha}$, as in the overall stellar mass function. Given this assumption, we can predict the expected relative numbers of companions in the mass ranges 0.001–0.01, 0.01–0.1 and 0.1–1 $M_\odot$ as $(0.1:1:10)$ for $\alpha = 0$; $(1:1:1)$ for $\alpha = 1$; and $(100:10:1)$ for $\alpha = 2$. Most known stellar binaries are excluded *a priori* from the radial velocity samples, but statistics from G dwarf surveys indicate a multiplicity fraction of $\sim 50\%$ [D2]. There should be little significant bias against late M dwarf or lower-mass companions amongst the radial-velocity samples, and results suggest that $\sim 7\%$ of G dwarfs have planetary-mass ($<0.01\,M_\odot$) companions, while no more than 1% of the companions fall in the brown dwarf/very low-mass star régime. Thus, the relative numbers of companions in the three mass ranges can be inferred as $(7:1:50)$ – a discontinuous distribution, which suggests separate origins and formation mechanisms for stellar or brown dwarf companions and the recently-discovered, planetary-mass objects.

10.4.3   Formation conundra

Jovian planets in Mercurian orbits

The existence of giant planets at distances of 0.1 AU or less from the parent star is clearly at odds with the standard formation theory. Temperatures at these radii in the protoplanetary disk exceed $\sim 1,000$ K, precluding the existence of any icy condensates. Refractory grains can exist, but there are two major problems with forming massive, rocky planets *in situ*: first, tidal forces, while insufficient in strength to disrupt a fully-formed Jovian-mass planet, are capable of preventing aggregation of smaller planetesimals; second, the heavy-element content in a 'rocky Jupiter' represents a significant fraction of the total mass of those elements in the parent star.

If giant planets cannot form at such small orbital radii, the obvious solution is to form them at conventional distances and identify a mechanism which allows the orbit to evolve inward over the stellar lifetime. Perhaps surprisingly, such mechanisms have been known for more than a decade. Indeed, their efficiency is so high that discussion has centred on means by which Jupiter has maintained its position at the present orbital radius [W2], [L6], [L8].

The two most successful mechanisms for planetary migration are based on angular momentum coupling between the planet and the protoplanetary disk, with the planet losing momentum and spiralling inwards. First, Lindblad resonances can form within the disk [A3] and interact with the planetary cores; second, massive planetesimals ($m > 0.1\,M_J$) sweep out gaps in the disk, and the net torque leads to inward migration [T4]. In both cases, theoretical simulations predict timescales of only ~ 3–5×10^5 yr for a Jovian-mass planet formed at 5 AU to accrete onto the central star. This is only $\sim 10\%$ of the expected lifetime of the parent disk. Lin [L6] suggests that a series of such planets form over the disk lifetime and march inward in succession, with the surviving planets being those which form just before the disk dissipates. As a variant of this process, inward migration may cease at small radii either due to tidal interactions, transferring momentum from the star's rotation to orbital motion, or because the star has cleared a low-density hole in the central regions of the disk.

There are, however, problems with mechanisms acting over such rapid time-scales. The accretion time of $\sim 3 \times 10^5$ yr is substantially shorter than the most optimistic estimates of the formation time of giant planets (Section 10.2.1). Moreover, *all* of the disk material takes part in the infall, so one expects the $\sim 0.01\,M_\odot$ within ~ 5 AU in the minimum-mass solar nebula to accrete onto the central star over 300,000 years – $dM/dt \sim 3 \times 10^{-8} M_\odot\,\text{yr}^{-1}$. Observations of T Tauri stars [H1] indicate accretion rates at or above this level in young ($< 10^6$ year-old) T Tauri stars, with a near power-law decline in dM/dt with age, $\dot{M} \propto \tau^{-2}$. Maintaining the inner disk over its few $\times 10^6$-yr lifetime therefore demands a substantial reservoir of material (0.05–0.10 $M_\odot$) at larger radii.

Orbital migration scenarios generally predict near-circular planetary orbits. In addition, at radii below ~ 0.15 AU, tidal interactions can lead to circularisation of the orbit, and synchronisation of the planetary rotational and sidereal periods, over moderately long timescales – 5 Gyr or less for 51 Peg, for example [M4].[5] In contrast to these expectations, several systems have notably eccentric orbits. This circumstance might result from gravitational scattering among a number of giant planets, as mutual perturbations cause the orbits to overlap [R1], possibly culminating in planet/planet mergers [L7]. If so, other giant planets in longer-period orbits should be present in the higher eccentricity systems, although as yet there have been no unequivocal detections. Other possible mechanisms for inducing eccentric orbits involve inhomogeneous structure and gravitational perturbations within the proto-planetary disk, or interactions with (as-yet undiscovered) stellar companions.

In summary, none of the proposed theoretical models for the dynamical evolution of protoplanetary disks provides a full explanation for the observed characteristics of 51 Peg-like planets. However, initial efforts have identified a number of promising avenues for further exploration through more detailed theoretical and observational studies.

Possible chemical signatures

Among the parameters listed in Table 10.3 are spectroscopic metallicity estimates. Almost all of the 13 stars with such measurements are at least as metal-rich as the Sun, with seven having $[m/H] > 0.15$. Figure 10.6 shows that the position of the stars on the H–R diagram tends to confirm this impression, with most of the FGK primaries lying along the upper edge of the main-sequence outlined by stars in the solar neighbourhood. The M dwarf Gl 876 is not included in this diagram, but lies toward the upper edge of the main-sequence in the $(M_V, (V–I))$ plane. Of the remaining stars, HD 195019, HD 168443 and 70 Vir are solar-abundance subgiants, with only HD 114762 ($[m/H] \sim -0.6$) and 14 Her ($[m/H] \sim 0.15$) lying toward the lower edge of the main sequence. As noted above, HD 114762 probably harbours a brown dwarf rather than a planet.

Figure 10.6 also plots the location of all of the stars being observed as part of the Keck planet-search programmes. These stars, and the stars on the longer-running Lick and Haute-Provence programmes, span main sequence spectral types later than F. While observations continue for all of these programmes, there is no obvious selection effect which should bias the initial discoveries towards metal-rich stars. Thus, the observed distribution might indicate a physical correlation.

Laughlin and Adams [L3] raise the alternative possibility that the higher abundances reflect the presence of planets – at least in the recent past. They hypothesise that Lin's sequence of formation and accretion of gas giants might lead to enhanced abundances in the outer photosphere. This mechanism is viable only for stars with shallow surface convection zones, where the accreted material can significantly alter the average metal content; that is, the metallic mass in the planet is comparable with the metallic mass in the convection zone. In dwarfs later than spectral type G1 (that is, most of the current sample) the convective zone extends to

---

[5] In the case of τ Boö, the companion is sufficiently massive to have spun up the stellar rotation period to match the planetary orbit.

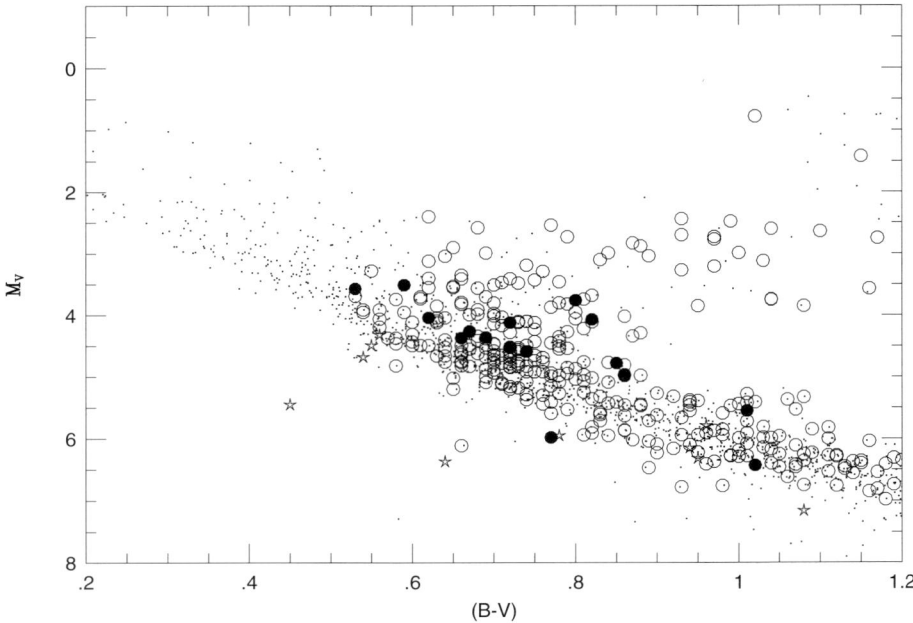

Figure 10.6. H–R diagram showing the local stars; open circles mark targets of the Keck planet-search surveys [M5]; solid circles mark stars with detected planetary-mass companions (Table 10.3), and five-point stars identify systems with candidate brown dwarf companions [L2], [M11].

much larger depths in the star, and the engorgement of several Jovian-mass planets has little effect on the mean metal abundance.

Planet formation depends on the presence of grains to form the dirty-ice planetesimals which collect to form the cores of giant planets, so a correlation with stellar abundance is not unreasonable. However, setting aside HD 114762 and 14 Her, the current distribution of primaries suggests a triggering mechanism which is intriguingly finely-tuned to near-solar abundances. Observations are currently underway of main-sequence stars in the $[m/H] \sim 0.15$ Hyades cluster, and the proportion of planetary systems in that sample should prove particularly interesting.

Three of the stars listed in Table 10.3 are components of wide ($> 1,000$ AU) binary systems. Of the three, two (55 Cnc and τ Boö) have M dwarf companions, but 16 Cygni A (the companion of the 16 Cygni B planetary system) is a G2 dwarf which shows no evidence of radial-velocity variations. The two G dwarfs in the 16 Cygni system have essentially identical heavy-element abundances, with one exception: while the lithium abundance in 16 Cygni A is [Li] = 3.51, 16 Cygni B has [Li] = 2.72 [K5] (note that the Sun has [Li] = 3.31). As with brown dwarfs, this implies that the convection zone in 16 Cygni B extends to higher temperatures (lower depth), leading to more rapid lithium destruction. However, the additional convective mixing must be limited, since beryllium – which is destroyed at $T \sim 3.5 \times 10^6$ K rather than 2.5×10^6 K – has identical abundance in both stars [G1].

Both theoretical [P1] and observational [S5] studies suggest that differences in lithium abundance can be driven by variations in stellar rotation, with deeper mixing occurring in slow rotators [M7]. Cochran *et al.* [C5] have proposed that observed differences between the 16 Cygni components may stem from differences in disk mass during the pre-main sequence phase, with more efficient braking of 16 Cygni B by magnetic coupling to a massive protoplanetary disk. While suggestive, this cannot provide a general rule for planetary formation, since the Sun, almost identical in temperature to both components, has a lithium abundance – and presumably an angular momentum history – closer to 16 Cygni A than to 16 Cygni B, the planetary primary.

10.5   THE FUTURE

The next decade should witness substantial progress in the characterisation of extrasolar planetary systems. Radial-velocity surveys currently underway have achieved the accuracy of $\epsilon_V = 3\,\mathrm{m\,s}^{-1}$ required to detect Jovian planets in Jovian orbits. The next step is to extend observations over the several-decade period necessary to separate individual signatures in multi-planet systems. With five to ten 8-m class telescopes approaching completion by the end of the twentieth-century, acquisition of the requisite data is only a matter of time.

The new frontier for planetary discoveries rests with optical and near-infrared interferometry. Major new facilities are being developed both on the ground – notably at the Keck Observatory and on Mount Wilson by the Georgia State University Center for High Angular Resolution Astronomy (CHARA) – and for future space missions – the Space Interferometer Mission (SIM) and the Terrestrial Planet Finder (TPF). These offer the prospect of attaining astrometric measurements at the microarcsecond level – possibly sufficient to detect perturbations due to terrestrial planets orbiting the nearest stars. Not surprisingly, significant technical difficulties must be overcome to achieve this goal.

Figure 10.7 presents a schematic of an interferometer: two telescopes, A_1 and A_2, separated by baseline B. Radio interferometry allows measurement of absolute positions over the entire sky, with the reference baseline set by the Earth's rotational motion. These techniques cannot be transferred directly to optical and near-infrared wavelengths, since ground-based observations must be made through the atmosphere, while defining a reliable absolute reference frame is problematic for space-based interferometry. However, under both circumstances it is possible to obtain extremely precise *relative* astrometry, measuring the position of the target (*1 in Figure 10.7) relative to a local reference (*2), provided that observations are limited to small angular scales.

Long experience with ground-based astrometry has shown that the accuracy with which one can measure an angular separation decreases with increasing separation. This arises because atmospheric turbulence is increasingly correlated on smaller angular scales. The measurement uncertainties can be estimated by modelling the atmosphere as a plane-parallel layer, height h, with a Kolmogorov turbulence

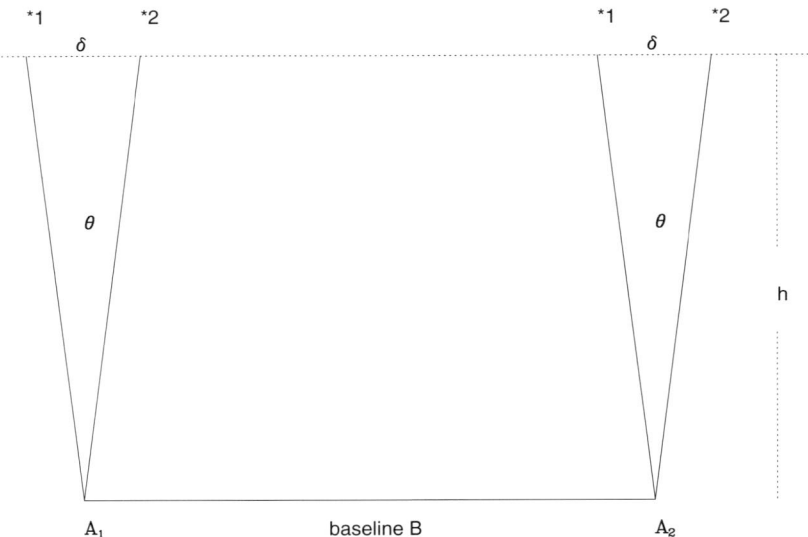

Figure 10.7. Schematic of a long-baseline interferometer.

spectrum [S3]. If we consider two stars separated by an angular distance θ, then in conventional astrometry the uncertainty in the measurement of the separation $\epsilon_\theta \propto \theta^{1/3}$. In contrast, if the stars are observed using an interferometer with baseline B, and if the separation of the two stellar light-paths at the top of the atmosphere, $\delta = \theta h$, is less than B (Figure 10.7), then $\epsilon_\theta \propto B^{-2/3}\theta$: that is, the inherent measurement accuracy decreases linearly with decreasing separation. The expected variation in ϵ_θ with θ for different baselines is plotted in Figure 10.8 (following [S3]), with the predicted accuracy exceeding 10 µarcsec for 100-m Keck-like baselines. Preliminary results support these predictions [C6].

Observations of the target star and the reference star must be undertaken simultaneously. The most effective method of satisfying this requirement is by using a dual-beam interferometer, in which the two stars are isolated at each aperture (A_1 and A_2 in Figure 10.7) and fed into separate beam-combiners. The angular separation is derived from the difference in path-length, measured by delay lines, required to give constructive interference and fringes. Note that (technical considerations aside) the astrometric accuracy cannot be improved to arbitrarily high levels by extending the interferometer baseline indefinitely, since at some point one begins to resolve stars and reduce fringe visibility. Solar-type stars have an angular diameter of ~ 0.5 mas at 10 parsecs, limiting optical interferometers to $B < 250$ m, or $B < 1{,}000$ m in the near-infrared (2.2 µm).

The most significant constraint on ground-based measurement is set by the atmosphere, which limits the field of view available for reference stars – radius ~ 20 arcsec for 10 microarcsecond astrometry at $B = 100$ m. As a result, a significant number of potential targets – particularly those at high galactic latitude, are not

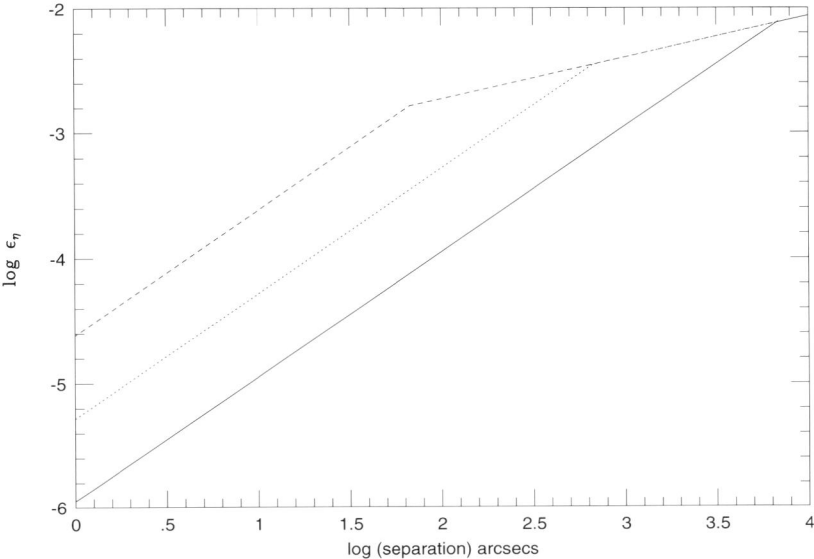

Figure 10.8. The predicted astrometric accuracy, ϵ_θ (arcsec) as a function of angular separation for baselines of 1 m (dashed line), 10 m (dotted line) and 100 m (solid line).

well-suited for ground-based interferometry – even for the Keck telescopes, with which the limiting magnitude is projected to reach $K = 21.8$ magnitudes. Moving above the atmosphere allows both access to reference stars at larger angular separations (degrees), and a potential astrometric precision of 1 microarcsecond [C7], [H6]. The most ambitious projects, all scheduled for launch between 2005 and 2010, aim to use interferometers for nulling; that is, adjusting the path-lengths from the individual apertures to produce destructive interference. The light from the on-axis stellar target is cancelled, allowing the detection of orbiting planets [B10], [A1]. 'Hot Jupiters', such as 51 Peg, are only ~ 5 magnitudes fainter than the primary star at 2.2 μm, and should be within the grasp of such systems. The detection of terrestrial planets requires nulling with an efficiency of 1 part in 10^6, which in turn demands maintaining the baselines of the spaceborne interferometer with nanometre accuracy – a very considerable technical challenge.

10.6 SUMMARY

The intensive radial velocity surveys undertaken over the past decade have provided irrefutable evidence of the existence of extrasolar planets. With more than a dozen such systems currently identified, the mass distribution suggests strongly that these very-low mass objects are not an extension of the mass distribution of stellar/brown dwarf companions. The obvious inference is formation as planets within the circumstellar disk of the parent protostar. Most of the currently-known planetary systems have G dwarf primaries, but to a large extent this reflects the distribution of spectral types targeted for observation. The discovery of a planet around Gl 876 – one of the few M dwarfs with high-accuracy radial velocity measurements – is particularly significant. Many important questions remain to be settled concerning the actual formation process, but the overall implication is that several percent of solar-type stars in the Galactic disk have planetary systems.

10.7   REFERENCES

A1   Angel, J. R. P., Woolf, N. J., 1997, *ApJ*, **475**, 373.
A2   Artymowicz, P., 1988, *ApJ*, **335**, L79.
A3   Artymowicz, P., 1993, *ApJ*, **419**, 166.
B1   Bailes, M., Lyne, A.G., Shemar, S. L., 1991, *Nature*, **352**, 311.
B2   Beckwith, S. V. W., Sargent, A. I., Chini, R. S., Güsten, R., 1990, *AJ*, **99**, 924.
B3   Behall, A. L., Harrington, R. S., 1976, *PASP*, **88**, 204.
B4   Benedict, G. F. *et al.*, 1992, *PASP*, **104**, 958.
B5   Black, D. C., 1995, *ARA&A*, **33**, 359.
B6   Black, D. C., 1997, *ApJ*, **490**, L171.
B7   Boden, A. F. *et al.*, 1998, *ApJ*, **504**, L39.
B8   Boss, A. P., 1997, *Science*, **276**, 1836.
B9   Boss, A. P., 1998, *Nature*, **393**, 141.
B10 Bracewell, R. N., McPhie, R. H., 1979, *Icarus*, **38**, 136.
B11 Brown, T. M., Kotak, R., Horner, S., Kennelly, E. J. *et al.*, 1998, *ApJS*, **117**, 563.
B12 Buffon, G. L. L., 1745, *De la formation des planétes*, Paris.
B13 Butler, R. P., Marcy, G. W., 1996, *ApJ*, **464**, L15.
B14 Butler, R. P., Marcy, G. W., Williams, E., McCarthy, C., Vogt, S. S., 1996, *PASP*, **108**, 500.
B15 Butler, R. P., Marcy, G. W., Vogt, S. S., Apps, K., 1998, *PASP*, **110**, 1389.
B16 Butler, R. P., Marcy, G., Fischer, D., Brown, T., Contos, A., Korzennik, S., Nisenson, P., Noyes, R., 1999, *ApJ*, **524**.
C1   Campbell, B., Walker, G. A. H., 1979, *PASP*, **91**, 540.
C2   Campbell, B., Walker, G. A. H., Yang, S., 1988, *ApJ*, **331**, 902.
C3   Chamberlin, T. C., 1901, *ApJ*, **14**, 17.
C4   Cochran, W. D., Hatzes, A. P., 1994, *Ap. Sp. Sci.*, **212**, 281.
C5   Cochran, W. D., Hatzes, A. P., Butler, R. P., Marcy, G. W., 1997, *ApJ*, **483**, 457.
C6   Colavita, M. M., 1994, *A&A*, **283**, 1027.
C7   Colavita, M. M., Shao, M., Rayman, M. D., 1993, *Appl. Opt.*, **32**, 1789.
C8   Croswell, K., 1998, *Planet Quest*, Harcourt Brace.
C9   Cuzzi, J. N., Dobrovolskis, A. R., Champney, J. M., 1993, *Icarus*, **106**, 102.
D1   Deeg, H. J., Doyle, L. R., Kozhevnikovm V. P., Martín, E. L. *et al.*, 1998, *A&A*, **338**, 479.

D2 Duquennoy, A., Mayor, M., 1991, *A&A*, **248**, 485.
F1 Fischer, D. A., Marcy, G. W., Butler, R. P., Vogt, S. S., Apps, K., 1999, *PASP*, **111**, 50.
F2 Fuhrmann, K., Pfeiffer, M. J., Bernkopf, J., 1997, *A&A*, **326**, 1081.
G1 García Lopez, R. J., Pérez de Taoro, M. R., 1997, *A&A*, **334**, 599.
G2 Gatewood, G, Eichorn, H., 1973, *AJ*, **78**, 769.
G3 Gatewood, G., 1974, *AJ*, **79**, 52.
G4 Gatewood, G., 1976, *Icarus*, **27**, 1.
G5 Gatewood, G. 1987, *AJ*, **94**, 213.
G6 Gatewood, G. 1995, *Ap. Sp. Sci.*, **223**, 91.
G7 Gatewood, G. 1997, *BAAS*, **28**, 885.
G8 Gautschy, A., Saio, H. 1995, *ARA&A*, **33**, 75.
G9 Gonzalez, G., 1998, *A&A*, **334**, 221.
G10 Gray, D., 1997, *Nature*, **385**, 795.
G11 Gray, D., 1998, *Nature*, **391**, 153.
G12 Gray, D., Hatzes, A., 1907, *ApJ*, **490**, 412.
G13 Guillot, T., Chabrier, G., Morel, P., Gautier, D., 1994, *Icarus*, **112**, 354.
G14 Guinan, E. F., Bradstreet, D. H., Ribas, I., Wolf, M., McCook, G. P., 1998, *BAAS*, **32**, 800.
H1 Hartmann, L., Calvert, N., Gullbring, E., D'Alessio, P., 1998, *ApJ*, **495**, 385.
H2 Hayashi, C., Nakazawa, K., Adachi, I., 1977, *PASJ*, **29**, 163.
H3 Henry, G.W., Baliunas, S. L., Donahue, R. A., Soon, W. H., Saar, S. H., 1997, *ApJ*, **474**, 503.
H4 Henry, T. J., 1994, PhD thesis, University of Arizona.
H5 Hershey, J. L., 1973, *AJ*, **78**, 769.
H6 Hog, E., Fabricius, C., Makarov, V. V., 1997, *Exp. Ast.*, **7**, 101.
H7 Hulse, R. A., Taylor, J. H., 1975, *ApJ*, **195**, L51.
J1 Jayawardhana, R., Fisher, S., Hartmann, L., Telesco, C., Pina, R., Fazio, G., 1998, *ApJ*, **503**, L79.
J2 Jeans, J. H., 1928, *Astronomy and Cosmogony*, London: Cambridge University Press.
J3 Jeffreys, H., 1929, *MNRAS*, **89**, 636.
J4 Jura, M., 1991, *ApJ*, **383**, L79.
J5 Jura, M., Malkan, M., White, R., Telesco, C., Pina, R., Fisher, R. W., 1998, *ApJ*, **505**, 897.
K1 van de Kamp, P., 1969, *AJ*, **74**, 241.
K2 van de Kamp, P., 1977, *Vistas in Astr.*, **20**, 501.
K3 Kant, I., 1755. *Allgemeine Naturgeschichte und Theorie des Himmels*, J. F. Petersen, Konisburg and Leipzeig.
K4 Kastings, J. F., Whitmire, D. P., Reynolds, R. T., 1993, *Icarus*, **101**, 108.
K5 King, J. R., Deliyannis, C.P., Hiltgen, D. D., Stephens, A., Cunha, K., Boesgaard, A. M., 1997, *AJ*, **113**, 1871.
K6 Koerner, D. W., 1997, *Origins Life Evol. Bio.*, **27**, 157.
K7 Koerner, D. W., Ressler, M. E., Werner, M. W., Backman, D. E., 1998, *ApJ*, **503**, L83.

K8   Koerner, D., LeVay, S., 1999, *Here Be Dragons*, Cambridge University Press.
K9   Kuiper, G., 1951, *Proc. Nat. Acad. Sci.*, **37**, 1.
L1   LaPlace, P. S., 1796, *Exposition du Système du Monde*, Circle-Sociale.
L2   Latham, D. W. *et al.*, 1989, *Nature*, **339**, 38.
L3   Laughlin, G., Adams, F. C., 1997, *ApJ*, **491**, L51.
L4   Leighton, R. B., Noyes, R. W., Simons, G. W., 1962, *ApJ*, **135**, 474.
L5   Lepage, A. J., MacRobert, A. M., 1998, *Sky & Telescope*, **96**, 44.
L6   Lin, D. N. C., 1986, in *The Solar System: Observations and Interpretations*, ed. M. G. Kivelson, Prentice Hall, p. 68.
L7   Lin, D. N. C., Ida, S. 1997, *ApJ*, **477**, 781.
L8   Lin, D. N. C., Papaloizou, J. C. B. 1986, *ApJ*, **309**, 846.
L9   Lippincott, S. L., 1960, *AJ*, **65**, 445.
L10   Lippincott, S. L., 1978, *Sp. Sci. Rev.*, **22**, 153.
L11   Lissauer, J. J., 1993, *ARA&A*, **31**, 129.
L12   Lissauer, J. J., 1995, *Icarus*, **114**, 217.
L13   Lowrance, P. J. *et al.*, 1999, *ApJ*, **512**, L69.
L14   Lyne, A. G., Bailes, M., 1992, *Nature*, **355**, 213.
M1   Marcy, G. W. Butler, R. P., 1992, *PASP*, **104**, 270.
M2   Marcy, G. W., Butler, R. P., 1996, *ApJ*, **464**, L15.
M3   Marcy, G., Butler, R. P., Williams, E., Bildsten, L., Graham, J., 1996, *ApJ*, **481**, 926.
M4   Marcy, G., Butler, R. P., Vogt, S., Fischer, D., Lissauer, J., 1998, *ApJ*, **505**, L147.
M5   Marcy, G., Butler, P., 1998, *ARA&A*, **36**, 57.
M6   Marcy, G., Butler, R. P., Vogt, S., Fischer, D., Liu, M., 1999, *ApJ*, **520**, 239.
M7   Martín, E. L., Claret, A., 1996, *A&A*, **306**, 408.
M8   Mayor, M., Queloz, D., 1995, *Nature*, **378**, 355.
M9   Mayor, M., Queloz, D., Udry, S., Halbwachs, J-L., 1997, in *Astronomical and Biochemical Origins and the Search for Life in the Universe*, ed. C. Cosmovici, S. Bowyer, D. Werthimer, IAU Coll. 161, 313.
M10   Mayor, M. *et al.*, 1998, in *Protostars and Planets IV*, ed. V. Manning, A. Boss and S. Russell, University of Arizona Press.
M11   Mizuno, H., 1980, *Prog. Theor. Phys.*, **64**, 544.
M12   Moultin, F. R., 1905, *ApJ*, **22**, 165.
N1   Noyes, R. W. *et al.*, 1997, *ApJ*, **483**, L111.
P1   Pinsonneault, M. H., Kawaler, S. D., Demarque, P., 1990, *ApJS*, **74**, 501.
P2   Pollack, J. B., Hubickyj, P., Bodenheimer, P., Lissauer, J. J., Podolak, M., Greenzweig, Y., 1996, *Icarus*, **124**, 62.
P3   Pravdo, S. H., Shaklan, S. B., 1996, *ApJ*, **465**, 264.
R1   Rasio, F. A., Ford, E. B., 1996, *Science*, **274**, 954.
R2   Reuyl, D., Holmberg, E., 1943, *ApJ*, **97**, 41.
R3   Reuyl, D., 1943, *ApJ*, **97**, 186.
S1   Saar, W. H., Butler, R. P., Marcy, G. W., 1998, *ApJ*, **498**, L153.
S2   Safronov, V. S., 1969, *Evolution of the Protoplanetary Cloud and Formation of Earth and Planets*, Moscow, Nauka Press; also NASA TTF-677, 1972.

S3   Shao, M., Colavita, M. M., 1992, *A&A*, **262**, 353.

S4   Smith, B. A., Terrile, R. J., 1984, *Science*, **226**, 4681.

S5   Soderblom, D. R. *et al.*, 1993, *AJ*, **106**, 1059.

S6   Strand, K. Aa., 1943, *Proc. Amer. Phil. Soc.*, **86**, 364.

S7   Strand, K. Aa., 1977, *AJ*, **82**, 745.

T1   Taylor, J. H., Weisberg, J. M., 1989, *ApJ*, **345**, 434.

T2   Telesco, C. M., Becklin, E. E., Wolstencraft, R. D., Decher, R., 1988, *Nature*, **335**, 51.

T3   Terebey, S., van Buren, D., Padgett, D.L., Hancock, T., Brundage, M., 1998, *ApJ*, **507**, L71.

T4   Trilling, D., Benz, W., Guillot, T., Lunine, J. I., Hubbard, W. B., Burrows, A., 1998, *ApJ*, **500**, 428.

V1   Vogt, S. S., Allen, S., Bigelow, B. *et al.*, 1994, *S.P.I.E.*, **2198**, 362.

W1   Walker, G. A. H., Walker, A. T., Irwin, A. W., Larson, A. M., Yang, S. L., Richardson, D. C., 1995, *Icarus*, **116**, 359.

W2   Ward, W. R., 1981, *Icarus*, **47**, 234.

W3   Weidenschilling, S. J., Spaute, D., Davis, D. R., Marzari, F., Ohtsuki, K., 1997, *Icarus*, **128**, 429.

W4   Weissman, P. R., 1984, *Science*, **224**, 987.

W5   von Weizsäcker, C. F., 1944, *Z. Astrophys.*, **22**, 319.

W6   Wetherill, G. W., 1990, *Ann. Rev. Earth Planet.*, **18**, 20.

W7   Wetherill, G. W., 1994, *Astrophys. Sp. Sci.*, **212**, 25.

W8   Willems, B., Van Hoolst, T., Smeyers, P., Waelkens, C., 1997, *A&A*, **326**, L37.

W9   Wolszczan, A., Frail, D. A., 1992, *Nature*, **355**, 145.

Z1   Zuckerman, B., Becklin, E. E., 1993, *ApJ*, **414**, 793.

11

M dwarfs in the Galactic halo

11.1 INTRODUCTION

Approximately 99.7% of the stars in the immediate vicinity of the Sun are members of the Galactic disk. The remaining stars belong to the Galactic halo – the fossil remnants of the first extensive burst of star formation in the history of the Galaxy (see Chapter 6). A clear distinction should be drawn between the stellar halo population – made of baryonic material and having a total mass of only $\sim 10^9 \, M_\odot$ – and the dark-matter halo, which is believed to be the dominant contributor to the Galactic potential. The dark-matter halo is held responsible for the relatively flat Galactic rotation curve, but its constituents have not yet been identified; those are the targets of gravitational lensing surveys. The present chapter concentrates on the stellar halo, and illustrates how observations of the lower-mass halo subdwarfs provide insight into the structure of the oldest stellar population so far identified in the Galaxy. The nature of the dark-matter halo is a subject in itself, and recent investigations are summarised elsewhere [T1].

Low-mass halo subdwarfs are best identified either in the immediate vicinity of the Sun – where they have relatively bright apparent magnitudes and can therefore be studied in detail, but are outnumbered substantially by dwarfs in the Galactic disk – or by probing the lowest-mass members of globular clusters, where subdwarfs dominate the starcounts, but have faint apparent magnitudes ($V > 22$). The high source density, and consequent image crowding, also complicates ground-based observations of globulars, and, as with open cluster studies (Chapter 7), dynamical evolution, mass segregation and tidal stripping remain important concerns in statistical analyses. The availability of the fully operational Hubble Space Telescope has gone a long way toward resolving these problems. In combination, observations of local stars and deep surveys of globular clusters provide complementary means of studying stars near the hydrogen-burning limit in the stellar halo.

11.2   HALO SUBDWARFS IN THE FIELD

As members of an old, and presumably kinematically well-mixed population, the nearby subdwarfs should provide a representative cross-section of the field halo. Studying the statistical properties of these stars, notably the luminosity and mass functions, demands the identification of a well-defined sample. Low-resolution objective-prism surveys and narrowband photometric surveys [B3] provide a means of searching for metal-poor F and G subdwarfs, looking for stars with weak Mgb or Ca II H and K absorption lines. Similar surveys could be undertaken for later-type stars – for example, searching for stars with strong hydride features – but to date most studies have concentrated on sifting through proper-motion catalogues to derive a suitable sample.

11.2.1   Photometric properties of M subdwarfs

Individual examples of nearby, late-type subluminous stars have been known since the discovery of Kapteyn's Star in the late nineteenth century. Most well-known subdwarfs are high-velocity stars, identified through their having large proper motions, and accurate trigonometric parallax measurements have been obtained for many of them. Given the variety of original sources, and consequent hetero-geneous selection criteria, these subdwarfs scarcely constitute a sample suitable for statistical analysis. However, they do provide an indication of the distribution in the H–R diagram.

Figure 11.1 plots the $(M_V, (V-I))$ and $(M_V, (B-V))$ H–R diagrams for late-type subdwarfs with parallax measurements. Photometric data for nearby stars are also plotted, providing a reference sequence for the Galactic disk. As with the FGK subdwarfs, the most subluminous stars, lying up to 3 magnitudes below the disk sequence, are also the most metal-poor (see Chapter 4). The subdwarf sequence has less dispersion in colour in the $(M_V, (B-V))$ diagram than in $(V-I)$, and crosses the disk sequence at $M_V \sim 11$. Later-type subdwarfs have redder $(B-V)$ colours than disk dwarfs of the same absolute magnitude, leading to an offset of 0.1–0.2 magnitudes between disk and halo stars in the $(B-V)/(V-I)$ two-colour diagram (see Figure 2.19).

A similar colour offset is evident between disk and halo in the $(J-H)/(H-K)$ diagram (Figure 2.20). As described in Chapter 4, metal-poor subdwarfs have increased opacity between 1 and 2 μm due to the pressure-induced dipole of H_2. This reduces the flux emitted at those wavelengths, particularly in the H-band (1.6 μm), leading to subdwarfs having bluer $(J-H)$ colours than disk dwarfs with the same $(H-K)$ colours. However, as with BVI colours, the offset amounts to less than 0.1 magnitude for even the most extreme subdwarfs, while uncertainties in the near-infrared colours are typically ±0.03 magnitude. Given these circumstances, broadband colours alone cannot provide reliable disk/halo discrimination, and offer only crude abundance estimates.

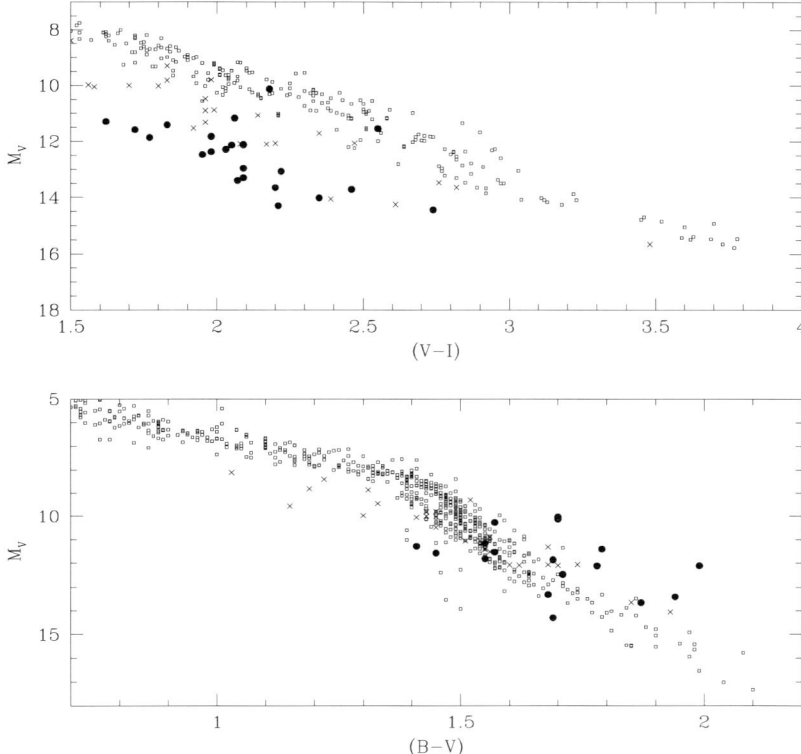

Figure 11.1. The $(M_V, (B–V))$ and $(M_V, (V–I))$ H–R diagrams for late-type subdwarfs. Disk dwarfs are plotted as open squares, sdM dwarfs (Section 11.3) are crosses and esdM dwarfs are filled circles. (From [G1], courtesy of J. Gizis and the *Astronomical Journal*.)

11.2.2   Searching for local subdwarfs

Proper-motion surveys are biased towards stars with high tangential velocities. These biases can be modelled if the underlying velocity distribution is known, as is the case for the local disk and halo populations. Indeed, the high-velocity bias is a distinct advantage in halo subdwarf identification, since the higher average velocity of halo stars amplifies the number of subdwarfs relative to disk dwarfs in a proper-motion selected sample. This is particularly important, given the extremely low proportion (1 in 300) of halo stars in a *volume*-limited sample.

The relative enhancement of halo stars within a proper-motion survey can be estimated as follows: consider a survey limited to stars with $\mu > \mu_{\lim}$. A star with transverse velocity, V_T, is just detected at a distance given by

$$d_{\lim} = \frac{V_T}{\kappa \mu_{\lim}} \qquad (11.1)$$

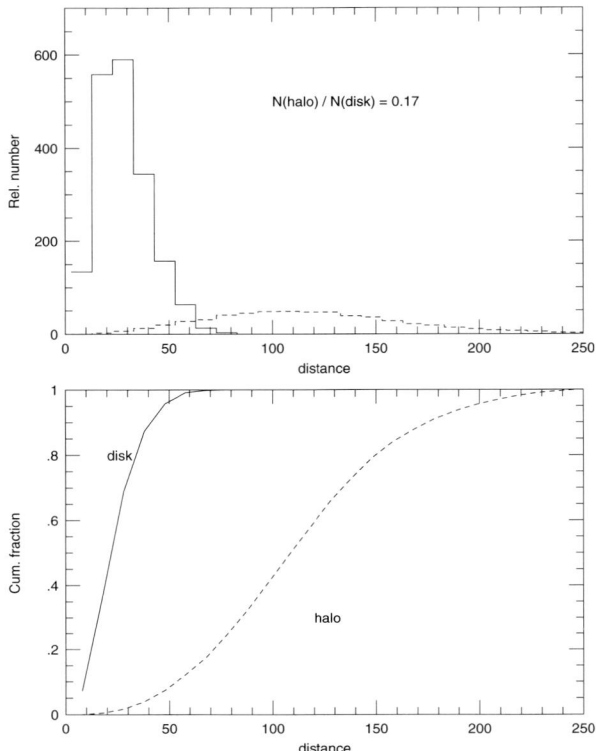

Figure 11.2. The expected distance distribution of a sample of stars selected by proper motion. Disk stars are indicated with solid lines, halo stars with dashed lines. Distances are given in parsecs.

where $\kappa = 4.74$. A stellar population with solar motion (U, V, W) and velocity dispersions $(\sigma_U, \sigma_V, \sigma_W)$ has a characteristic transverse velocity, $\langle V_T \rangle$. Thus, a characteristic distance can be defined for the detection of members of that population in the survey:

$$d_C = \frac{\langle V_T \rangle}{\kappa \mu_{\lim}} \qquad (11.2)$$

The distribution with distance of stars selected by proper motion depends on the form of the velocity distribution. Given a Schwarzschild velocity ellipsoid (that is, Gaussian velocity distributions in each co-ordinate), 25% of the population members with $d < d_C$ are predicted to have $\mu > \mu_{\lim}$, and these stars represent 80% of the contribution made by that population to the proper-motion survey (Figure 11.2). Hence, if a sample includes N_{tot} stars from a given stellar population, with known kinematics which provide a reliable estimate of d_C, then, in the absence of any other selection effects, the space density of that population can be estimated from

$$\rho = \frac{N_{\text{tot}} \times 0.8}{0.25 \times d_C^3} \tag{11.3}$$

Since all stars drawn from a given population have similar kinematics, intrinsically fainter stars on average have fainter apparent magnitudes than the more luminous members of that population.

Assume a mix of two stellar populations with different kinematics – the disk and the halo. Those two populations have average tangential velocities and characteristic distances of $(\langle V_T \rangle_d, d_d)$ and $(\langle V_T \rangle_h, d_h)$ respectively. The relative number of stars that each population contributes to a proper-motion survey is given by

$$\frac{N_h}{N_d} = \frac{\rho_h d_h^3}{\rho_d d_d^3} \tag{11.4}$$

where ρ_d and ρ_h are the local space densities of the two populations. Since $d_C \propto \langle V_T \rangle$, this ratio can be written as

$$\frac{N_h}{N_d} = \frac{\rho_h \langle V_T \rangle_h^3}{\rho_d \langle V_T \rangle_d^3} = \frac{\rho_h}{\rho_d} \times A_h \tag{11.5}$$

where A_h is the halo amplification factor.

The specific case of interest is the local mixture of halo and disk stars. The kinematic data listed in Table 6.3 show that disk dwarfs have an average tangential velocity of only $\sim 39\,\text{km}\,\text{s}^{-1}$. In contrast, the various representations of the kinematics of the halo predict average tangential velocities of 175–215 km s^{-1}. This implies an amplification factor, $A_h \sim 100$, for halo subdwarfs in a proper-motion limited sample. Thus, although the *number* density ratio of disk to halo stars is $\sim 300 : 1$, the higher tangential motions of the halo population can reduce this ratio to values as low as $3:1$ in proper-motion catalogues.

The full amplification factor, however, does not apply under all circumstances. The calculation above assumes that the stellar sample is defined only based on the proper motion; that is, the sample is proper-motion limited. Other factors can impose a lower effective distance limit; for example, the limiting apparent magnitude of the survey may set the detection limit, leading to a flux-limited sample. In this case, the effective amplification factor for a high-velocity population is reduced.

Consider the case of a sample drawn from the Luyten Half-Second (LHS) catalogue ($\mu_{\text{lim}} = 0.5\,\text{arcsec}\,\text{yr}^{-1}$). Luyten compiled this catalogue by matching the POSS I survey plates against second-epoch plate material, taken with the Palomar Schmidt, which has an effective limiting magnitude of $m_r \sim 19$. Given the average tangential velocities cited above, the characteristic distances for detection of disk and halo stars are ~ 32 and $\sim 140\,\text{pc}$ respectively. These distances correspond to distance moduli of 2.5 and 5.8 magnitudes respectively. As a result, the effective sampling volume for disk dwarfs with $M_r > 16.5$ ($M_V > 18$) and halo subdwarfs with $M_r > 13.2$ ($M_V > 14.5$), is set by the apparent magnitude limit, rather than by the proper-motion limit. The lowest luminosity stars are therefore drawn from a smaller volume, and each star carries more weight in determining both $\Phi_{halo}(M_V)$ and $\Psi_{halo}(M)$.

11.2.3   The reduced proper-motion diagram

Proper-motion data alone provide only the first cut in identifying nearby halo stars. Even with the optimum amplification factor, subdwarfs are still outnumbered by disk dwarfs by a factor of at least $4:1$. Further observations are required to distinguish disk and halo stars. Moreover, the subdwarfs themselves span a substantial range in abundance and consequently – as Figure 11.1 emphasises – a range of colour–magnitude (and mass–luminosity) relationships. A luminosity function based on *all* subdwarfs within a given volume combines stars drawn from different abundance-dependent mass intervals in a given interval, ΔM_V. It would be ideal to segregate the subdwarfs by abundance, and compare mass functions for intermediate and extreme metal-poor stars; but this demands a means of both differentiating halo subdwarfs from disk dwarfs, and determining approximate metallicities.

Photometric data offers some assistance in segregating disk and halo stars within a proper-motion selected sample. Reduced proper-motion diagrams (Section 7.2.3) are effective if data are available for the right passbands. The disk main sequence, halo subdwarfs and white dwarf sequences converge in the $(H_V, (B–V))$ and $(H_r, (m_{pg}–m_r))$ planes as the colour saturates at low luminosities. As a result, these diagrams are of only limited usefulness in identifying low-temperature subdwarfs. However, the sequences are well-separated in $(H_V, (V–I))$ or $(H_R, (R–I))$, allowing identification and elimination of the majority of late-type disk dwarfs from proper-motion selected samples. Figure 11.3 shows an example, plotting (V–I) data for stars from the LHS Catalogue.

11.3   CLASSIFYING LATE-TYPE SUBDWARFS

Section 4.6.4 describes how a reduction in metal abundance affects the emergent spectrum in low-mass dwarfs with effective temperatures below 4,000 K. In essence, TiO absorption decreases in strength with decreasing $[m/H]$, but the hydride bands are largely unaffected. The relative strength of those bands (TiO versus hydride), measured using either narrowband photometry or directly from spectra, provides the most effective technique for verifying the identify of late-type subdwarfs and quantifying their probable abundance. Magnesium hydride is the strongest feature present at optical wavelengths in the G and early-K halo subdwarfs [A1], [C6], while CaH is the most prominent absorber in cooler subdwarfs.

Narrowband indices can be designed to measure the flux within and adjacent to individual spectral features. The flux ratio between the on-band and sideband

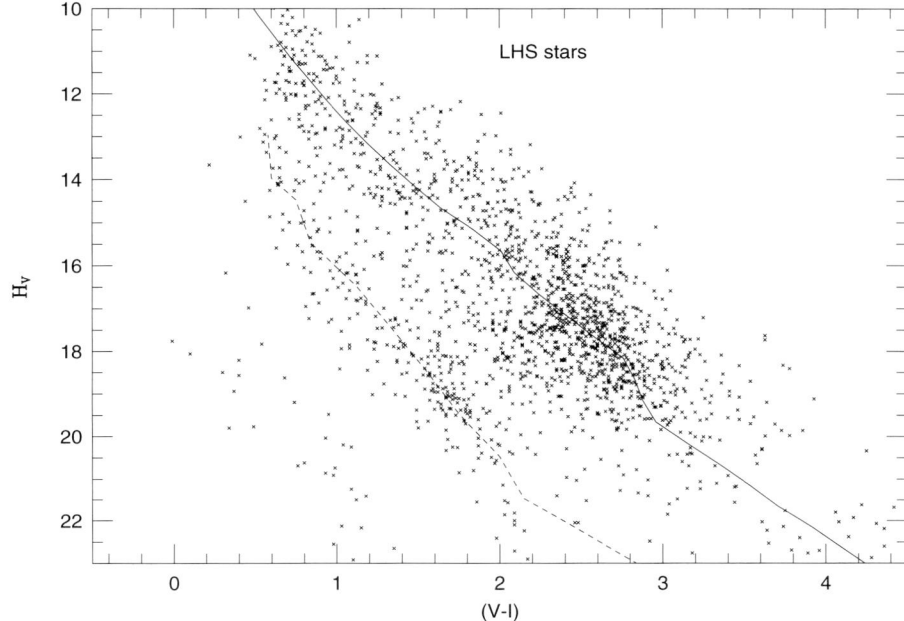

Figure 11.3. The $(H_V, (V–I))$ reduced proper motion diagram for stars in *Luyten's LHS Catalogue*. The solid line and dashed line mark the expected location of the disk and halo main-sequences at their respective mean tangential velocities. White dwarf stars contribute to the sequence at blue colours and high H_V.

('pseudocontinuum') measurements provides an estimate of the strength of the feature. As described in Section 2.2.6, the CaH and MgH bands have long been known as dwarf/giant discriminators [O1], and photometric indices designed to measure CaH strength (primarily the 6880 Å band) were originally devised to take advantage of the gravity sensitivity [J1]. As already discussed briefly in Section 2.4.1, hydride bandstrength can also be used to identify metal-poor late-type dwarfs. Following Mould's [M3] prediction that the CaH : TiO bandstrength ratio should also be abundance-sensitive, narrowband photometry was used to identify a number of late-type dwarfs with disk-like motions, but subnormal TiO strength [M4]. The objects were classified as 'metal-weak old disk stars'. Similar techniques were later used to classify some 50 high proper-motion stars as disk, intermediate (I) or halo (H) [H1].

Spectrophotometry permits a more finely-tuned measurement of variations in TiO and hydride bandstrength. The first extensive spectroscopic observations of M-type subdwarfs were undertaken by Bessell [B4], whose spectra confirmed the dominance of CaH in the coolest subdwarfs. These qualitative results have been quantified by Gizis [G1], who uses the narrowband indices listed in Table 11.1 (see also Figure 2.18) to measure the relative strengths of CaH and TiO features in the 6200–7500 Å

Table 11.1. Molecular bandstrength indicators.

Band	Sideband 1	On-band	Sideband 2
TiO 1	6703–6708	6718–6723	
TiO 2	7043–7046	7058–7061	
TiO 3	7079–7084	7092–7097	
TiO 4	7115–7120	7130–7135	
TiO 5	7042–7046	7126–7135	
CaH 1	6345–6355	6380–6390	6410–6420
CaH 2	7042–7046	6814–6846	
CaH 3	7042–7046	6960–6990	

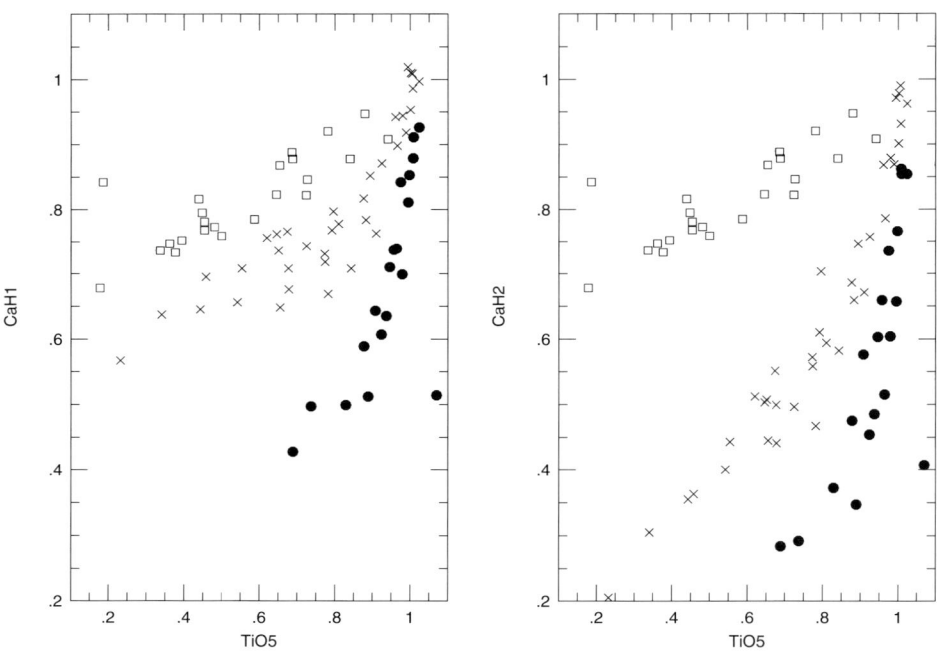

Figure 11.4. The narrowband CaH/TiO classification system. Open squares are disk dwarfs, sdM subdwarfs are marked as crosses, and esdM subdwarfs are solid points. (From [G1], courtesy of J. Gizis and the *Astronomical Journal*.)

region. Figure 11.4 plots these measurements for the parallax subdwarfs plotted in Figure 11.1. The stars have been grouped into three categories: disk-like; intermediate-bandstrength subdwarfs, class sdM; and extreme subdwarfs, class esdM.[1] The latter two categories correspond approximately to class (I) and class

---

[1] The majority of the [M4] metal-weak disk stars – including the well-known nearby high-velocity star, Barnard's Star (Gl 699) – are classified as disk-like.

(H) stars in the [H1] nomenclature. A comparison of Figures 11.1 and 11.4 shows that the esdM subdwarfs are also the most subluminous stars in the colour–magnitude distributions.

Gizis uses these narrowband spectrophotometric indices to define a spectral classification system for M subdwarfs which is comparable in utility to the system defined by Kirkpatrick *et al.* [K3] for near-solar abundance disk dwarfs. The strong dependence on metal abundance exhibited by the TiO bandheads renders those features less useful for unambiguous classification purposes, and Gizis ties his spectral-type scale to the abundance-insensitive CaH bandstrength. The latter offers a more direct correlation with effective temperature. The resultant sdM and esdM spectral sequences are plotted in Figures 11.5 and 11.6.

The relative distribution of the sdM and esdM stars in the colour–magnitude plane indicates that the esdM subdwarfs have lower metal abundance than the sdM subdwarfs. As described in Chapter 4, atmospheres computed for low-temperature metal-poor dwarfs [A3] allow the determination of preliminary quantitative estimates. These comparisons suggest an average metal abundance of $\langle [m/H] \rangle \sim -2$ for the esdM dwarfs, and $\langle [m/H] \rangle \sim -1.3$ for the sdMs. These estimates are supported both by spectroscopy of three sdM companions of FG subdwarfs of known abundance [G2], and by matching the observed colour–magnitude distribution against theoretical isochrones (Figure 3.15).

Gizis' classification therefore provides an effective method of dividing the local halo stars into moderately metal-poor and extremely metal-poor systems. The fact that mildly metal-poor stars, such as Barnard's Star, are excluded from even the sdM category demonstrates that contamination from disk dwarfs is limited to negligible proportions. This approach therefore opens the way for a measurement of statistical properties of the halo – notably the form of the luminosity function – as a function of abundance.

11.4   THE FIELD SUBDWARF LUMINOSITY FUNCTION

All except the most recent studies of the halo luminosity function predate quantitative abundance analysis of late-type subdwarfs, and therefore combine all stars within a single function, $\Phi_{halo}(M_V)$. Most analyses are based on proper-motion selected samples which, while biased towards including halo stars, still consist predominantly of disk dwarfs. The usual method of removing the disk contaminants is to determine transverse velocities – using either trigonometric or photometric parallaxes – and to include only stars with velocities exceeding some threshold, typically $V_T > 200$–$250 \, \mathrm{km \, s^{-1}}$. This criterion also restricts the sample to high-velocity halo stars, and directly computed space densities underestimate the true $\Phi_{halo}(M_V)$. However, Monte Carlo simulations can be used to compute the expected transverse velocity distribution for a given set of kinematics, $(U, V, W; \sigma_U, \sigma_V, \sigma_W)$, and hence determine the fraction of stars, f_c, with transverse velocities $V_T > V_{lim}$ [R5], [B1]. The results of recent analyses of halo kinematics are given in Table 6.2, and Figure 11.7 plots the expected tangential velocity distribution for two models.

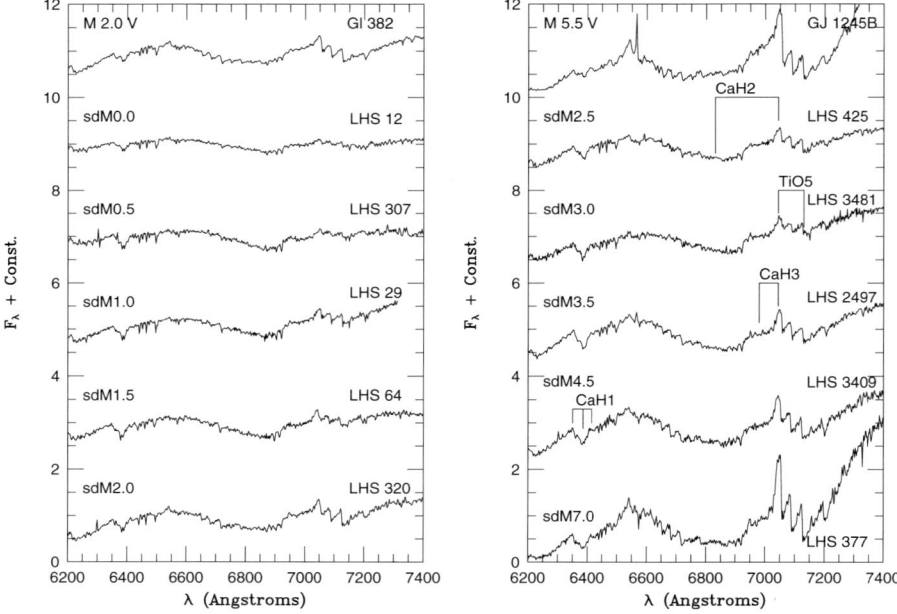

Figure 11.5. The sdM spectroscopic sequence [G1]. A disk dwarf is shown for comparison at the top of each panel. (From [GA], courtesy of J. Gizis and the *Astronomical Journal*.)

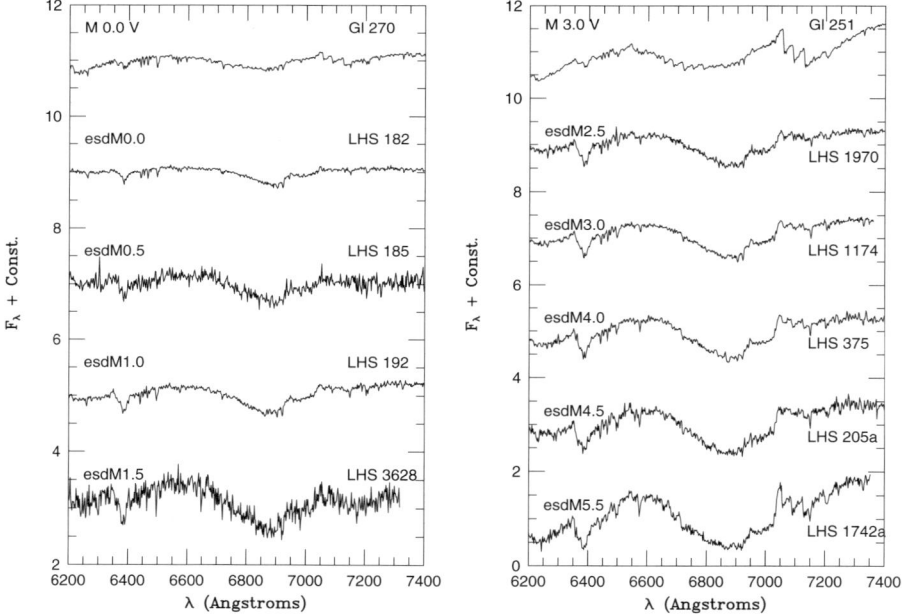

Figure 11.6. The esdM spectroscopic sequence. A disk dwarf is shown for comparison at the top of each panel. (From [G1], courtesy of J. Gizis and the *Astronomical Journal*.)

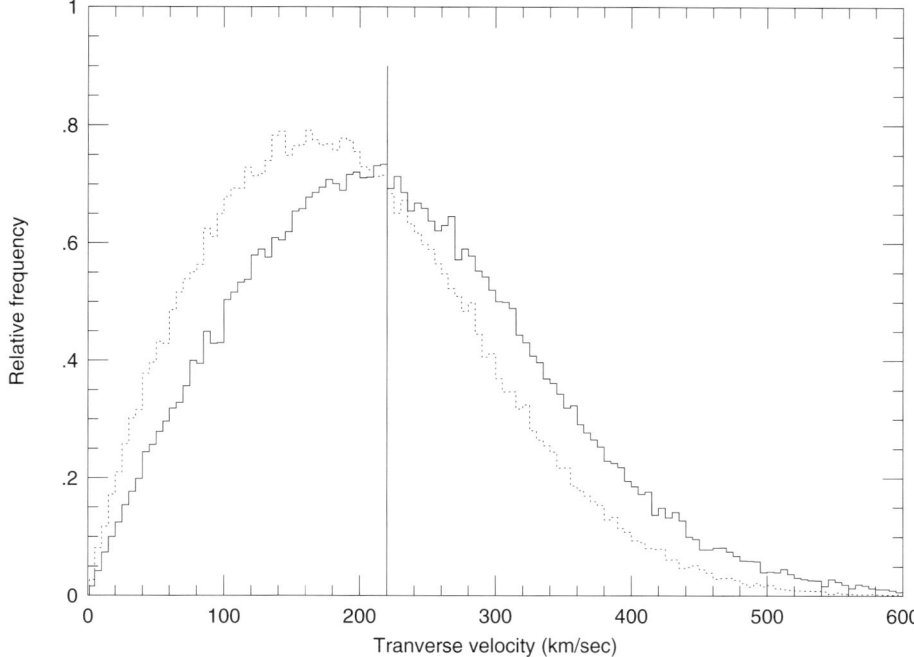

Figure 11.7. Tangential velocity distributions predicted by Monte Carlo simulations for kinematic models based on either Layden *et al.* RR Lyrae analysis [L1] (solid line), or Norris' [N1] observations of local subdwarfs (dotted line). In the former case, 51% of the stars have $V_T < 220\,\mathrm{km\,s^{-1}}$, while the lower kinematics derived in the latter analysis place $\sim 63\%$ of the sample below that velocity cut-off.

Appropriate correction factors, $\xi_c = 1/f_c$, typically ranging from 2 to 4, are deduced from these distributions and applied to the observed number-densities to estimate $\Phi_{\mathrm{halo}}(\mathrm{M}_V)$.

Deep star-count data can also be used to constrain the form of the luminosity function of field subdwarfs. Transforming these measurements to an estimate of the absolute space density of nearby subdwarfs is more complicated, since it requires accurate knowledge of the halo density distribution. This section summarises results deduced using these different techniques, and combines the most reliable data to obtain a best estimate of $\Phi_{\mathrm{halo}}(\mathrm{M}_V)$ for field subdwarfs.

11.4.1   Schmidt's derivation of the halo luminosity function

The first observational study of the halo luminosity function was carried out by Schmidt [S1]. His sample consisted of 126 stars, drawn primarily from the Lowell proper-motion survey, with $m_{pg} < 15.95$ and $\mu \geq 1\rlap{.}''295\,\mathrm{yr}^{-1}$. 121 of these stars had measured trigonometric parallaxes (in 1975), and distances to the remaining five stars were estimated. Based on the V/V_{max} statistic (Section 7.5.1), this sample has a uniform distribution, indicating that it is likely to be complete.[2] Schmidt eliminated disk dwarfs by restricting analysis to stars with $V_T > 250\,km\,s^{-1}$, the median velocity of a stellar population with kinematics matching those derived by Oort [O2] for metal-poor RR Lyrae stars (Table 6.2). The corresponding correction factor is $\xi_c \sim 2$.

Applying the cut-off in tangential velocity reduces the final sample to only 17 subdwarfs with $+5 < M_V < 12$, and one white dwarf. Schmidt derived space densities from these stars using the $1/V_{max}$ method: as in determining V/V_{max}, the maximum distance is calculated to which a given object can be moved and still be retained in the survey. The distance limit can be set by either the apparent magnitude limit or the proper-motion cut-off. Thus, the star G3-36 has $\mu = 2\rlap{.}''40\,yr^{-1}$, $m_{pg} = 13.4$, and $r = 26.3\,pc$. Moving it to $r = 48.7\,pc$ reduces μ below the survey limit of $1\rlap{.}''295\,yr^{-1}$, giving $V/V_{max} = (26.3/48.7)^3 = 0.157$. On the other hand, the star G75-47, at $\mu = 1\rlap{.}''42\,yr^{-1}$, $m_{pg} = 15.9$ and $r = 45.5\,pc$, has $r_{max} = 46.5\,pc$ set by the apparent magnitude limit of $m_{pg} < 15.95$.

The density contribution of each star is given by

$$\Phi_{tot} = \xi_c \times \sum \frac{1}{V_{max}} \tag{11.6}$$

and the total space density derived from Schmidt's halo star sample is 1.9×10^{-4} stars pc^{-3}. Two-fifths of the number density rests with the single white dwarf in the sample (G259-21, also known as Gl 699.1), so the space-density of halo subdwarfs in the range $5 < M_V < 12$ is 1.2×10^{-4} stars pc^{-3}; or a number ratio relative to disk dwarfs in the same absolute magnitude range of $\sim 1 : 350$.[3] Allowing for statistical limitations, there appear to be no radical differences between the shapes of the halo and disk luminosity functions (Figure 11.8).

11.4.2 $\Phi_{halo}(M_V)$ from proper-motion star samples

Schmidt's methods have been applied in several subsequent analyses which take advantage of larger, better calibrated proper-motion samples. Each adopts a different velocity criterion for segregating halo subdwarfs, different halo kinematics and, as a result, different values of ξ_C. However, with one exception the results are in reasonable agreement. The exception is Eggen's study [E1], where the halo selection criterion corresponds to $V_T > 85–130\,km\,s^{-1}$, leading to a resultant heavy contamination by moderate-velocity disk dwarfs and substantially higher inferred number densities. Setting aside those results, the other analyses can be combined to determine the local subdwarf luminosity function from $M_V = 5$ to near the hydrogen-burning limit.

---

[2] No stars matching Schmidt's criteria have been discovered in the intervening 20 or so years.
[3] Note that this comparison does not take account of the changing mass–M_V relationships followed by stars of varying abundance.

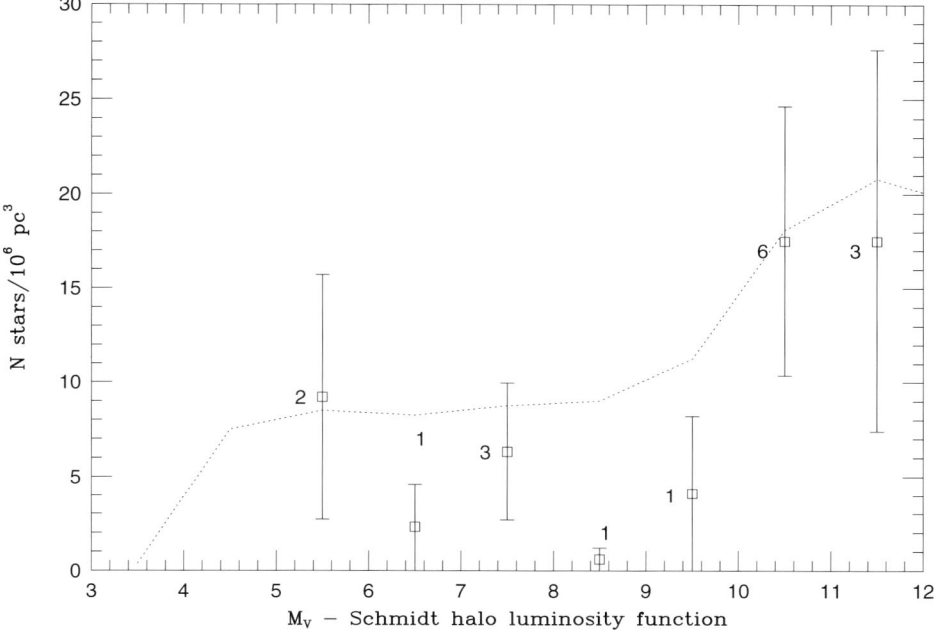

Figure 11.8. A comparison between Schmidt's halo luminosity function (transformed to M_V) and the nearby-star luminosity function scaled by a factor of 1/400.

Bahcall and Casertano [B1] provide the best statistics at bright magnitudes. They re-analyse Eggen's dataset of proper-motion stars, comprising stars with $V < 15$ and $\mu > 0''.7$ south of $\delta = +15°$, but with a tangential velocity cut-off of 220 km s^{-1}. This reduces the sample to only 94 stars and, since this is a southern sample derived partly from the Bruce proper motion survey ($m_{pg} < 15$), there may be incompleteness at fainter apparent (and absolute) magnitudes. Distances are based on photometric parallaxes (Eggen's M_I, (R–I) calibration) rather than trigonometric parallaxes. The adopted halo kinematics adopted by [B1] have relatively high net rotation, ($V = -154$ km s^{-1}; $\sigma_U = 140$, $\sigma_V = 100$, $\sigma_W = 75$ km s^{-1}), and imply a correction factor of $\xi_c = 3.0$ and a halo to disk density ratio of 1 : 300 for stars in the absolute-magnitude range $4 \leq M_V \leq 11$. Casertano *et al.* [C3] reconsider this analysis with revised halo kinematics, corresponding to a higher median transverse velocity and a lower correction factor of 2.4. Figure 11.9 plots the [B1] results with the [C3] scaling.

At fainter absolute magnitudes, the most detailed coverage is provided by Dahn *et al.* [D1] and Gizis and Reid [G3]. Both surveys are directed primarily towards low-luminosity subdwarfs approaching the hydrogen-burning limit. Both employ proper-motion selection as the initial criterion, and confirm spectroscopically that candidate subdwarfs are indeed metal-deficient. [D1] build on extensive observations of high motion stars in the LHS catalogue. Their luminosity function plotted in Figure 11.9

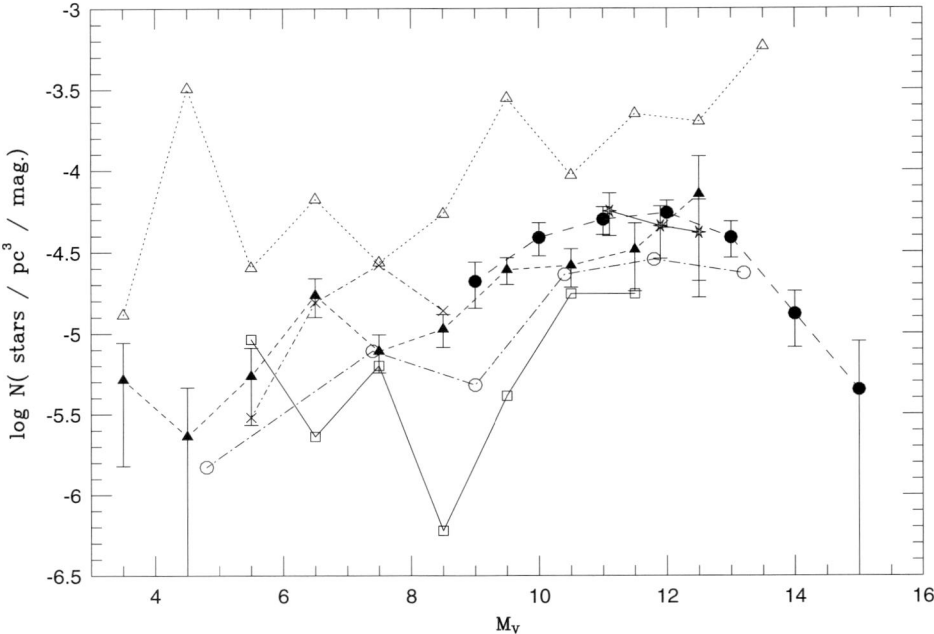

Figure 11.9. Several determinations of the halo luminosity function from proper motion studies: open squares [S1]; open triangles [E1]; crosses [R1]; open circles [D5]; solid triangles [B1], [C3]; six-point stars [G3]; and solid points [D1].

is based on 114 LHS stars with $0\rlap{.}''8 < \mu < 2\rlap{.}''5$, $11 < m_r < 18.1$, $\delta \geq -20°$ and $V_T > 220\,\text{km s}^{-1}$. 355 other stars meet the first three criteria, but have $V_T < 220\,\text{km s}^{-1}$. Space densities are calculated using the $1/V_{\max}$ method, with the correction factor of $\xi_c = 2.46$ based on the [M2] halo kinematics (Table 6.2). This survey provides the first statistically significant indication that the field subdwarf luminosity function flattens and turns over at $M_V \sim 13$.

The [G3] analysis is based on a more restricted stellar sample: proper-motion stars with $\mu > 0\rlap{.}''1\,\text{yr}^{-1}$ identified from scans of POSS I, POSS II and UK Schmidt R- and I-band plates [G1]. Surveying a total area of 300 square degrees, the sample extends to $R \sim 20$th magnitude – significantly fainter than any other study. Halo star candidates are selected from the $(H_R, (R-I))$ reduced proper-motion diagram, and confirmed through spectroscopic observations. The latter data also allow classification as either sdM or esdM, and calculation of the relative number-densities of those stars. The results encompass $(11 < M_V < 12)$ subdwarfs, and the number densities are in good agreement with [D1] (Figure 11.9). The relative numbers of esdM and sdM dwarfs are in the proportion 2:5, which is consistent with the relative numbers of metal-poor $([m/H] < -1.5)$ and intermediate abundance stars identified in spectroscopic studies of F and G subdwarfs [C1]. This suggests that there are no substantial changes in the overall shape of the luminosity function at these different abundances.

11.4.3 Φ_{halo} from star-count analyses

The alternative to local surveys is to use deep star-counts to study halo stars *in situ* at large distances above the Plane. The vertical density law deduced from Galactic structure analyses (Section 6.5.2) indicates that the halo component dominates the stellar distribution at heights of 5 kpc or more above the Plane. Selecting a sample at these distances avoids any significant problems due to contamination by disk dwarfs, but clearly also requires observations extending to faint apparent magnitudes – V > 23 and I > 20 for even the earlier-type M subdwarfs. These faint magnitudes make spectroscopic observations difficult and time-consuming. There is also the complication of transferring the observed number densities to the equivalent densities in the Plane.

The deepest star-count analysis combines V and I data from 53 HST Wide Field Camera images to derive $\Phi_{halo}(M_I)$ for the high halo ($|z| > \sim 6$ kpc) [G4]. The majority of these stars have I magnitudes fainter than 21st magnitude. None have spectroscopic observations, although their physical location in the Galaxy argues overwhelmingly that they are likely to be members of the halo. The general morphology of the derived luminosity function is consistent with local analyses. If one extrapolates the [G4] results to $z = 0$ using density laws derived from the star-count data, $\rho(r) \propto r^{-3.15\pm0.23}$, and an axial ratio $c/a = 0.82 \pm 0.13$, the implied space densities are a factor of two lower than those derived from local stars. The discrepancy probably rests with the composite nature of the halo: any flattened component (see [S2]; Section 6.3.2) makes a negligible contribution to far-halo star-counts, and therefore its contribution to the local halo cannot be predicted. Proper-motion surveys of local stars continue to provide the most effective method of probing Φ_{halo}.

11.4.4 The field-halo luminosity function

Figure 11.10 plots the composite halo luminosity function. We have combined the results from [B1] at bright magnitudes with [D1] at fainter magnitudes. Gizis and Reid's deeper survey predicts number densities in good agreement with the latter data at $M_V > 11$, and we therefore adopt the density normalisation of Dahn *et al.* (that is, their kinematic correction factors rather than the values suggested by Casertano *et al.*). Figure 11.9 shows that the two analyses are in reasonable agreement at $8 < M_V < 10$, but differ at fainter magnitudes. The [B1] deficit may reflect incompleteness in the parent proper-motion sample [B1], and we have not incorporated their results for $M_V \geq 9$. The inferred number density of halo subdwarfs in the solar neighbourhood with absolute magnitudes in the range $4 < M_V < 15$ is 3.02×10^{-4} stars pc^{-3} – a number ratio of 1:300 relative to the local density of disk dwarfs (Section 7.8.1).

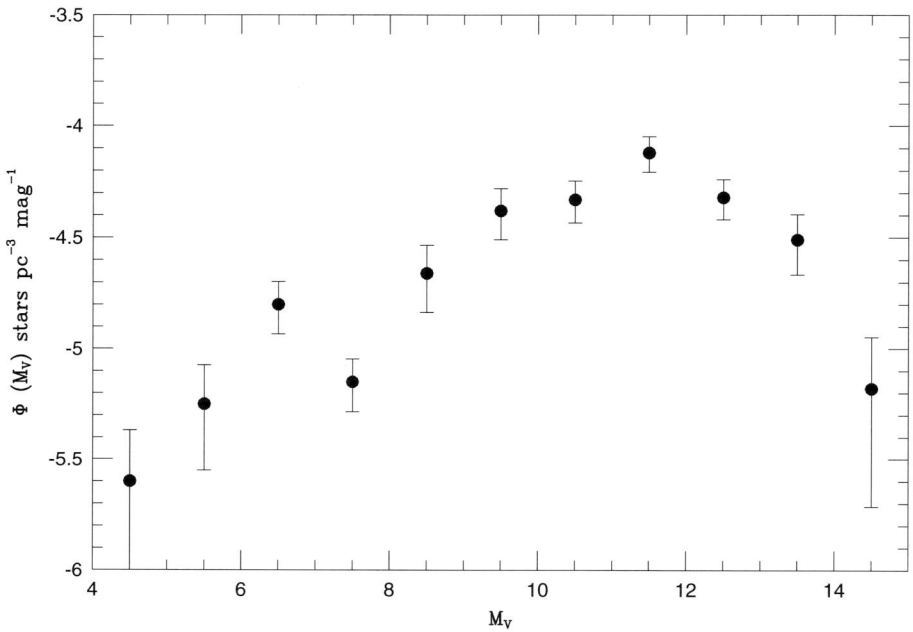

Figure 11.10. The composite halo luminosity function, combining data from the analyses by [B1] to $M_V < 9$, and [D1] at fainter magnitudes.

11.5   LUMINOSITY FUNCTIONS IN GLOBULAR CLUSTERS

Globular clusters provide an alternative means of exploring the halo luminosity function. Cluster studies offer the advantage of dealing with a coeval population of known metal-abundance, and since the cluster population spans a broad abundance range, one can hope to identify significant metallicity-dependent trends. There are, however, compensating complications. First, internal dynamical evolution is expected to lead to mass segregation, while external gravitational interactions can strip the outermost stars from the cluster. Extensive simulations [R3], [K1] show that internal evolution least affects the mass distribution at intermediate radii near the half-light radius, r_h (the radius which encompasses half the total flux). As a result, luminosity functions measured at intermediate radii are reasonably close approximations to the global luminosity function in most globular clusters, and, as discussed further in the following section, dynamical models can be used to estimate the appropriate corrections to the derived mass functions.

A more important consideration is the necessity of obtaining accurate star-counts at faint magnitudes and high surface source-densities. With distances of 3–10 kpc for the nearer cluster systems, early-type M subdwarfs have apparent magnitudes of $I \sim 21$, while the hydrogen-burning limit lies at apparent magnitudes fainter than $I = 24$ and $V = 26.5$. The star density to $I = 22$ magnitude in even the outer regions of these clusters can exceed 2,000 stars per square arcminute, corresponding to an average star-to-star separation of ~ 1 arcsec. As a result, image crowding and confusion limits both the photometric and star-counting accuracy attainable with ground-based, seeing-limited observations.

Despite these substantial difficulties, considerable effort was devoted in the late 1980s and early 1990s to obtaining deep ground-based photometry – often in a single passband – of the nearer cluster systems. Many of these observations were undertaken by Richer, Fahlman and collaborators using the Canada–France–Hawaii Telescope [R3], [R4]. These studies employ sophisticated automated software to identify stars by finding local intensity maxima in the CCD frames, and determine magnitudes by fitting the full image profile. Comparably deep exposures of fields offset from the cluster are used to assess the contribution from background field-stars, while the completeness of the final stellar census is estimated by adding artificial stars and determining, as a function of magnitude, the fraction recovered by the analysis software. Summarising the results from these programmes – which extended to fainter than 23rd magnitude in the I-band – [R3] argued that the rising luminosity functions measured in most clusters provided clear evidence for a steeply-rising mass function, perhaps extending into the brown dwarf régime.

This conclusion has not been borne out by higher spatial-resolution HST observations. Diffraction-limited images with a point-spread function of FWHM $\sim 0\rlap{.}''09$ at $\lambda = 8000\,\text{\AA}$ provide significant advantages in analysing observations of crowded fields. The first globular cluster studied in detail was NGC 6397, the second-nearest system to the Sun, observed by Paresce *et al.* [P1]. Their I-band luminosity function shows a maximum at $I \sim 20.75 \pm 0.25$ ($M_I \sim +8.0$) with an almost-symmetric decline in number densities towards fainter magnitudes (Figure 11.11). This is not due to incompleteness. The colour–magnitude diagram (Figure 11.12) shows a strikingly narrow main sequence which peters out well above the limiting magnitude reached by the observations.

The discrepancy between the ground-based and HST observations stems from two sources. First, severe crowding within the field led to an *overestimation* of the necessary corrections for incompleteness. Second, NGC 6397 lies at low Galactic latitude, and, as a result, field stars contribute significantly to the total number counts. This contribution can be assessed accurately from the *HST* colour–magnitude diagram. [R3], however, had observations in only the I passband, and estimated the contamination from observations of a control field, offset from the cluster. The latter observations underestimated the level of contamination, and the two errors combined to give an overestimate of the number of low-luminosity cluster members.

The HST data suffer from some residual contamination, but it has proven possible to use images separated by only 32 months to determine proper motions with sufficient accuracy to isolate cluster members [K2]. Analysis of these observations produces the colour–magnitude diagram plotted in Figure 11.12. The s-shaped morphology was predicted by D'Antona [D2], and stems from the same changes in internal structure that occur in solar-abundance disk dwarfs (see Chapter 3). Moving down the main sequence, the change in slope at $M_V \sim 8$ marks the onset of H_2 association, and the corresponding decrease in the adiabatic gradient, while the

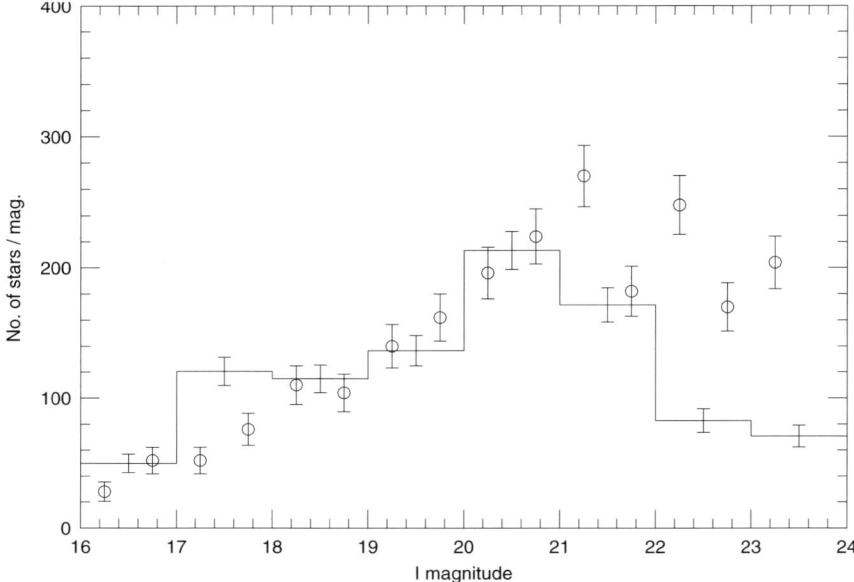

Figure 11.11. Luminosity functions derived for the nearby cluster NGC 6397, from ground-based observations [R3] (open circles), and using HST data [P1] (histogram).

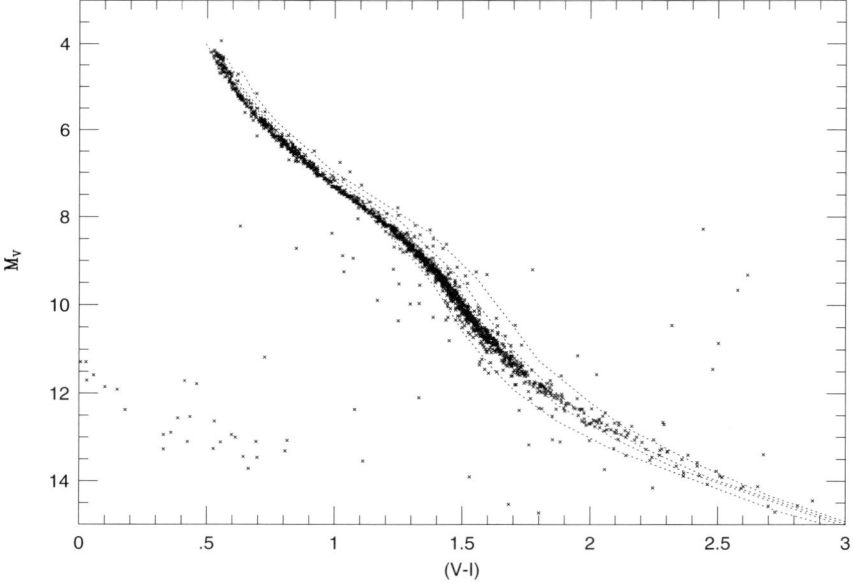

Figure 11.12. The main sequence of NGC 6397, from HST observations [C5], compared with theoretical isochrones for metal-poor stars. The lowest luminosity track has $[m/H] = -2.0$; isochrones for $[m/H] = -1.5$, -1.2 and -1.0 are also shown. (Data courtesy of A. Cool.)

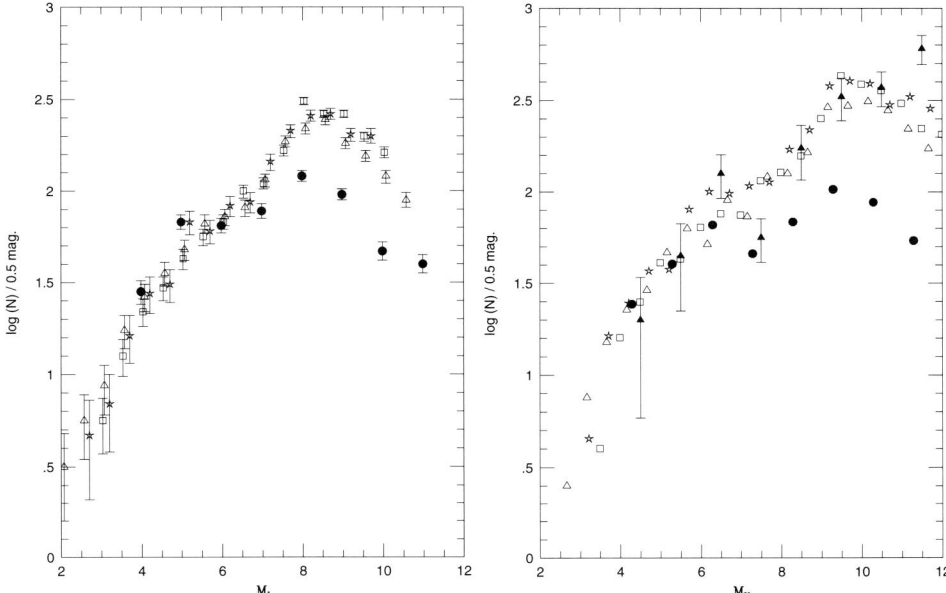

Figure 11.13. (*Left*) $\Phi(M_V)$ and (*right*) $\Phi(M_I)$ for the metal-poor clusters M15 (open squares), M30 (stars), M92 (open triangles) and NGC 6397 (solid points), scaling the number densities to match at $M_V = +5$, $M_I = +6$. The composite field-halo luminosity function from Figure 11.10 is also plotted for comparison in the left panel. (Cluster data from [P2].)

second point of inflection, at $M_V \sim 13$ ($\sim 0.12\,M_\odot$), occurs when pressure, rather than temperature, becomes responsible for the ionisation of the stellar envelope [D3]. Below this point, stars become increasingly degenerate, with the radius remaining nearly constant at $0.1\,R_\odot$. The hydrogen-burning limit in NGC 6397 lies at $M_V \sim 14.5$ and $(V–I) \sim 2.6$, both brighter and bluer than in the disk, reflecting the lower opacities in these metal-poor stars. With an age of more than 11 Gyr, any brown dwarfs in these systems have cooled to temperatures substantially less than 1,000 K.

HST photometric data have now been obtained for more than a dozen clusters, and observations confirm the existence of mass segregation [D8], [M5] at the level predicted by dynamical models. Figure 11.13 plots results for four metal-poor clusters, while Figure 11.14 plots similar data for several intermediate-abundance systems. There is no obvious correlation between the form of the luminosity function and abundance amongst those systems.

While $\Phi(M_I)$ generally peaks at $M_I \sim 8.5$, cluster-to-cluster variations are evident, particularly when deep HST data can be combined with groundbased data at brighter magnitudes, allowing calculation of $\Phi(M_I)$ over the full magnitude range from the turn-off to the hydrogen-burning limit. In particular,

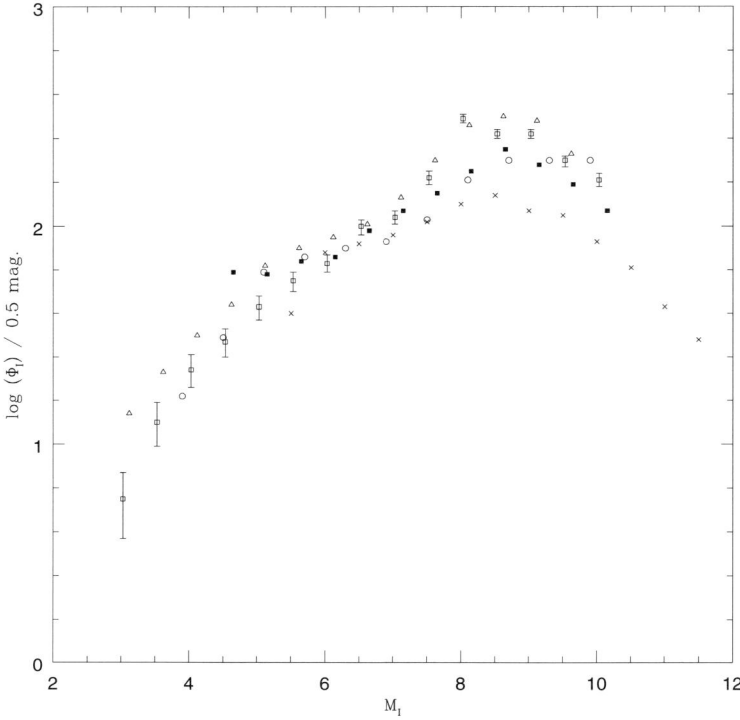

Figure 11.14. Luminosity functions for the metal-rich globular cluster 47 Tuc (solid squares [D7]) and the intermediate abundance clusters Ω Cen (open circles [P4]), M3 (triangles [M1]) and NGC 6752 (crosses [F1]). The luminosity function of the metal-poor cluster M15 (open squares [P2]) is plotted as a reference.

the metal-poor cluster NGC 6397 has a noticeably flatter luminosity function (fewer low-mass stars) than the similar-abundance systems M92, M15 and M30 [P2], while the intermediate-abundance cluster M10 has a steeper luminosity function than do M22 and M55 [P3]. In the former case, the difference probably reflects different dynamical evolution. The orbit followed by NGC 6397 takes the cluster through the Galactic Plane at a distance of only a few kpc from the Galactic centre, and reaches a maximum height of only ∼1.5 kpc above the Plane [D9]. Each passage probably results in tidal shocks, which strip the outermost stars from the cluster. Since NGC 6397 has a relatively short dynamical relaxation time (∼200 Myr), those stars are rapidly replenished from the cluster core, with mass segregation favouring low-mass stars and leading to a systematic depletion. In contrast, the other three metal-poor clusters follow orbits in which the cluster spends relatively little time near the Plane, and those systems have suffered correspondingly less dynamical attrition. Averaging the data for these three metal-poor clusters provides our current best estimate of the form of the luminosity function for $[m/H] = -2$ stellar systems.

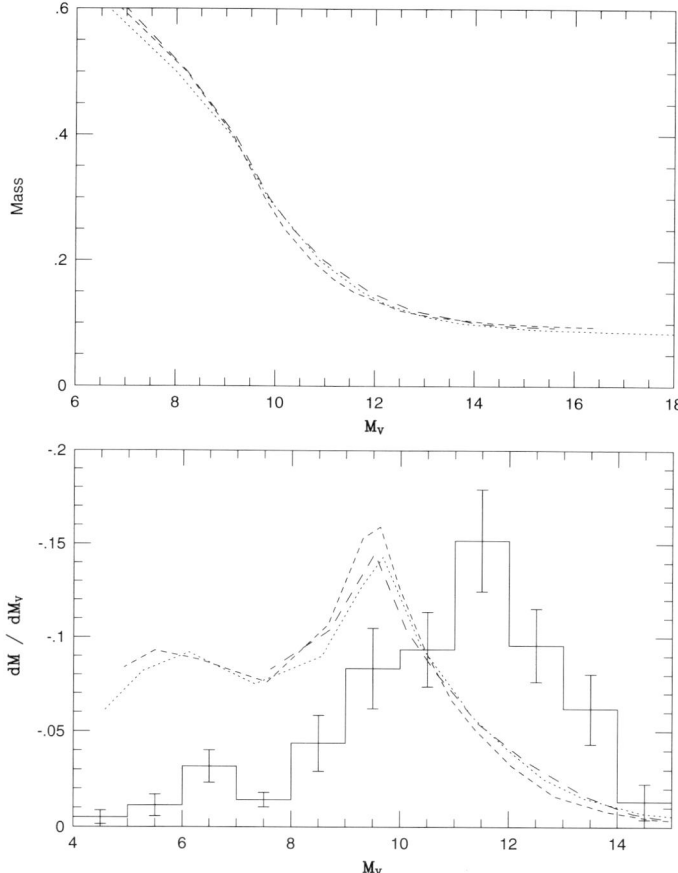

Figure 11.15. The upper panel plots (mass–M_V) relationships predicted by intermediate-abundance models from [D4] ([Fe/H] = -1.3, dotted line), [A2] ($[m/H] = -1.5$, dashed line) and [B2] ($[M.H] = -1.5$, long-dashed line). The results are in good agreement. The lower panel compares the first derivatives from those relationships to $\Phi_{halo}(M_V)$.

11.6   THE MASS FUNCTION OF THE STELLAR HALO

The relationship between mass and luminosity depends on chemical composition. None of the binaries with known orbits listed in Table 8.2 has an abundance significantly below the solar value, so no empirical calibration of this dependence exists. However, several groups have recently computed sets of models of metal-poor dwarfs [D4], [A2], [B2]. Figure 11.15 compares their intermediate-abundance (mass, M_V) relationships, plotting the [Fe/H] = -1.3 results from [D4] with $[m/H] = -1.5$ data from the other groups. The different analyses produce similar results.

As discussed in Chapter 8, rapid changes in slope in the mass–luminosity

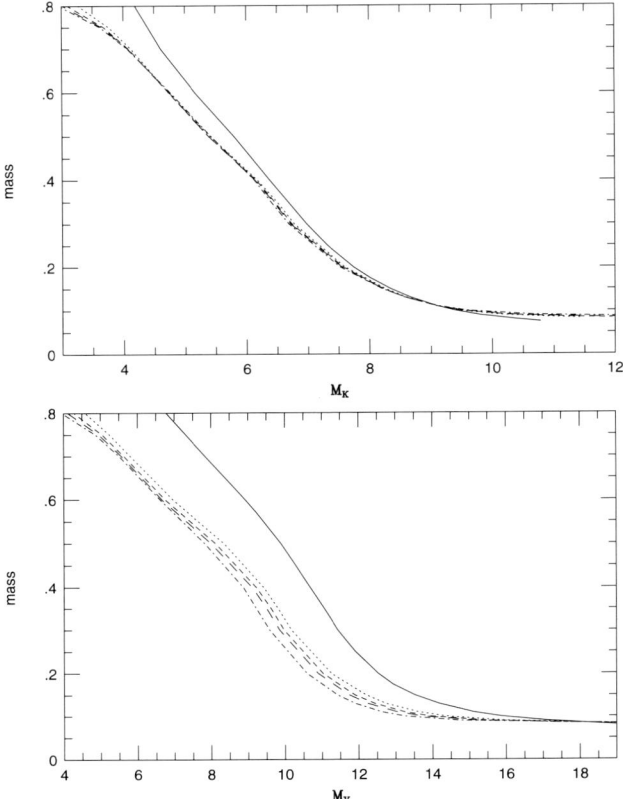

Figure 11.16. The (mass–M_V) and (mass, M_K) relationships predicted by the Lyon models for abundances $[m/H] = -1.0$ (dotted line), -1.5 (short dash), -1.8 (long dash) and -2.0 (dash–dot). The solar abundance relationship for $\tau = 1\,\mathrm{Gyr}$ is shown as the solid line.

relationship can introduce features into the mass function. The lower panel in Figure 11.15 compares the composite luminosity function (from Figure 11.10) against the first derivative of each of the three (mass, M_V) relationships plotted in the upper panel. Again, the three theoretical relationships are in reasonable agreement, but there is no obvious correlation between the maximum in dM/dM_V and the maximum at $M_V \sim 11.5$. Given the good agreement of the three theoretical analyses, further comparisons are restricted to the Lyon models [B2].

11.6.1 $\Psi_{\mathrm{halo}}(M)$ in the field

Field subdwarfs span a substantial range in abundance. Figure 11.16 plots the (mass, M_V) and (mass, M_K) relationships predicted for abundances $[m/H] = -1.0$ to -2.0;

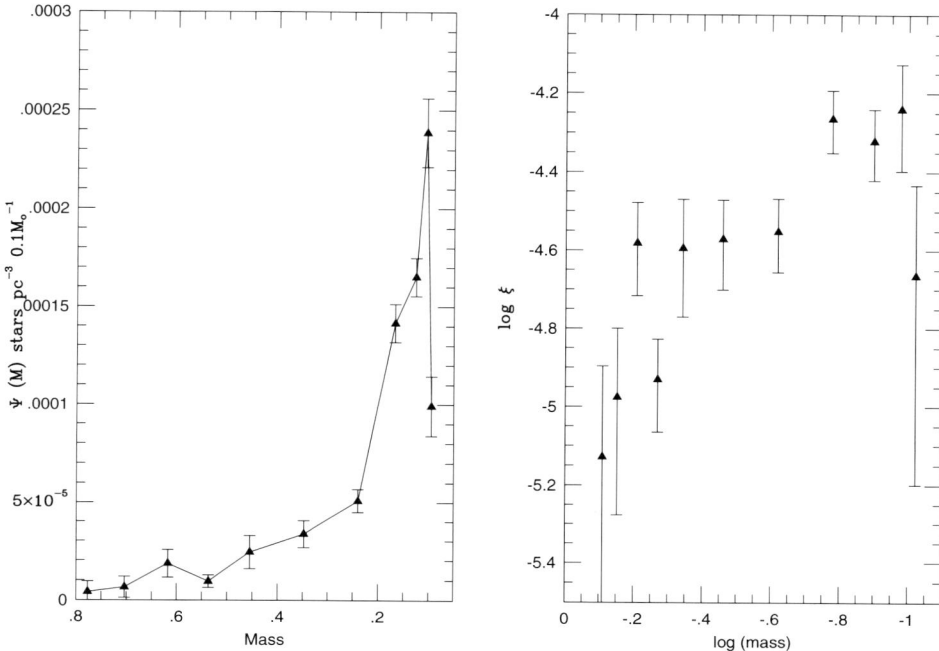

Figure 11.17. The halo mass function derived from field subdwarfs adopting the Lyon $[m/H] = -1.5$ mass–luminosity relationship. The left panel plots the linear function, dN/dM; the right panel plots the logarithmic function, $dN/d\log M$.

1-Gyr solar abundance isochrone is plotted as a reference. The models preserve a consistent morphology, with decreased line-blanketing at lower abundances leading to brighter M_V for a given mass, while the mass limit for hydrogen-burning also increases. The effect is more pronounced at V than at K, as would be expected given the greater prominence of atomic and molecular features at optical wavelengths. Unfortunately, few subdwarfs contributing to the field $\Phi_{\mathrm{halo}}(M_V)$ have infrared photometry, so we are forced to use the (mass, M_V) relationship. As Figure 11.16 shows, this complicates the derivation of $\Psi_{\mathrm{halo}}(M)$ by introducing uncertainties of 20–30% in the inferred mass.

In principle, field subdwarfs could be segregated by abundance to compute $\Psi_{\mathrm{halo}}(M)$ for a range of metallicities; in practice, the total sample is too small, and the available data too inhomogeneous, to permit this approach. Faced with such limitations, we can only estimate $\Psi_{\mathrm{halo}}(M)$ for the field by applying a single mass–luminosity relationship to the composite luminosity function plotted in Figure 11.10. We use the Lyon (mass, M_V) relationship for $[m/H] = -1.5$, corresponding to the modal abundance of field subdwarfs (Chapter 6). The derived mass function is plotted in Figure 11.17, and is substantially steeper than the corresponding relationship for the Galactic disk (Figure 8.8). Indeed, fitting the distribution with a power-law produces $\alpha \sim 1.65$ (where $\alpha = 2.35$ is the Salpeter value). The juncture between

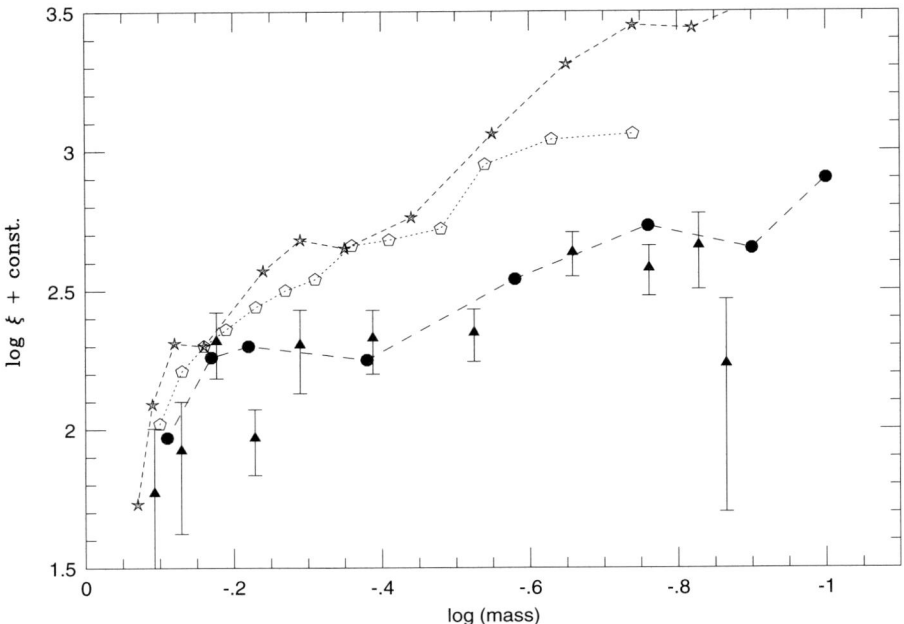

Figure 11.18. Mass functions, $\Psi(M)$, for the globular clusters NGC 6397 (solid points, [P2]) M3 (open pentagons, [M1]) and M30 (stars, [P2]). The solid triangles plot the field-halo mass function from Figure 11.17. The cluster data have not been corrected for dynamical effects.

the [B1] and [D1] luminosity function data occurs at a mass of $\sim 0.45\,M_\odot$, so the sharp increase below $0.2\,M_\odot$ is not due to a mismatch in density zero-points.

11.6.2 $\Psi_{halo}(M)$ in globular clusters

As with the luminosity function, the halo mass function can also be estimated from globular clusters. Since the abundances of clusters are known to relatively high precision, there is no uncertainty involved in selecting the appropriate mass–luminosity relationship. Mass functions have been derived for a number of clusters by [D7], [D8], [P2], [P3], [M1], [P4], [P5]. Fitting the data for each cluster as a single power-law leads to indices ranging from $\alpha \sim 1.6$ (M30) to $\alpha \sim 0.8$ (NGC 6397), although in general the mass functions have a more complex form. Several studies suggest that $\Psi(M)$ may turn over at the lowest masses [P3], [P4]. The [P3] results, for example, indicate $\alpha \sim 1$ for masses between $\sim 0.8\,M_\odot$ (after the turnoff) and $\sim 0.25\,M_\odot$, with $\alpha \sim 0$ at lower masses.

The functions derived in these direct analyses, plotted in Fig 11.18, represent the present-day mass function, $\Psi^{PD}(M)$, at the particular radii sampled within each cluster. As noted in Section 11.5, dynamical evolution can lead to significant differences between that function and both the present-day *global* mass function and the initial mass function. Due allowance can be made for these effects if the dynamical evolution is not too drastic and, as already discussed, models suggest that in most clusters this condition holds at intermediate radii. Dynamical models track the time evolution of mass segregation through a variety of techniques based on either direct N-body simulations [V1] or analytic techniques, such as the Fokker–Planck method [C4]. Stellar two-body interactions can lead to both cluster disruption through evaporation and/or core collapse, in which the central density peaks sharply (such as in M15). Both evaporation and tidal stripping due to shocks as the cluster passes through the Galactic disk lead to the total cluster mass decreasing with time; mass segregation promotes preferential depletion of low-mass stars in the outer regions.

The appropriate dynamical model for a given cluster is selected by comparing the predicted kinematics and density distributions (as a function of stellar mass) against observational determinations. However, deriving these observational parameters is not straightforward. The density distribution for stars of different masses can be calculated from radial number counts of stars of different apparent magnitudes; but since we are observing a two-dimensional projection, some assumptions are required regarding isotropy before the three-dimensional distribution can be inferred. Similarly, while astrometry can, in principle, be used to determine the internal kinematics, the measurements require sub-milliarcsecond accuracy. Most current studies are based on radial velocity observations, which provide only a one-dimensional estimate of σ_V (and possible rotation) and are limited to the brighter cluster members, and therefore stars within a restricted mass range ($\sim 0.8\,M_\odot$). Since the typical velocity dispersion near the centre of a globular cluster is only $\sim 10\,\mathrm{km\,s^{-1}}$, high accuracy observations of large samples of stars are required.

Despite these stringent requirements, observational analyses have been undertaken of a number of clusters. These studies consider multi-mass models, in which the cluster stars are grouped in dynamical units by mass, with typically 20% of the total mass assigned to white dwarf remnants and a few percent to neutron stars. These analyses can be used to estimate both the total mass and the likely form of $\Psi^{\mathrm{PD}}(M)$. As examples, [D10] derives a total mass of $\sim 6.6 \times 10^4\,M_\odot$ for NGC 6397, and a mass function with $\alpha \sim 1.9$ for $M > 0.4\,M_\odot$, $\alpha \sim 2.5$ for $0.1 < M/M_\odot < 0.4$; [D11] estimate a total mass of $\sim 5 \times 10^5\,M_\odot$ for M15, with a mass function of slope $\alpha \sim 1.9$ to $\sim 0.3\,M_\odot$, flattening sharply at lower masses; and [P3] find that relatively small corrections for dynamical effects need to be applied to the slopes derived in their analyses of seven intermediate abundance and metal-poor clusters. It is significant that the strongest correlations derived in the last study are between α and dynamical parameters – the cluster destruction rate and the relaxation time. NGC 6397, with the least steep mass function, has the shortest relaxation time and the highest disruption rate.

11.6.3 Summary

At present, the halo mass function is known with relatively low accuracy. Analyses based on samples of nearby subdwarfs must allow for heterogeneous and poorly-defined abundances, as well as relatively small sample-sizes and kinematic selection effects which underlie $\Phi(M_V)$; determinations from globular cluster data are vulnerable to the choice of distance scale and the effects of dynamical evolution. Both sets of analyses are forced to rely solely on theoretical mass-luminosity relationships, and while the available models are in good agreement, no empirical calibration is available to test for possible systematic errors. However, with all these caveats, current results suggest that the halo mass function is not identical in shape to the disk mass function. The halo mass function exhibits a steeper rise in number density between ~ 0.8 and $0.25\,M_\odot$, but flattens and perhaps turns over at lower masses.

The mass density of the local halo, obtained by integrating the function plotted in Figure 11.17, is $7.8 \times 10^{-5}\,M_\odot\,\mathrm{pc}^{-3}$, or 0.26% of the mass density due to main-sequence disk dwarfs. Stellar remnants, such as white dwarfs and neutron stars, probably contribute a further $2\text{--}3 \times 10^{-5}\,M_\odot\,\mathrm{pc}^{-3}$. If the mass function turns over at lower masses, the contribution from undetected halo brown dwarfs will produce only a marginal increase in this mass density.

11.7 M SUBDWARFS AND THE DISTANCE SCALE

In main-sequence fitting, the distance to a star cluster is estimated by matching the cluster's colour–magnitude diagram against an absolute calibration defined by parallax stars. Traditionally, the latter stars lie on the upper main sequence, with spectral types no later than about K1. Only a small number of F and G subdwarfs have parallaxes determined to an accuracy of better than 10%, even after the completion of the Hipparcos astrometric survey. Since the halo mass function rises toward lower masses, the local volume encloses larger numbers of K and M dwarfs, offering the prospect of a more accurate definition of the main-sequence relationship. Later-type stars have been incorporated in a few distance studies of disk open clusters, but until recently their application to globular clusters was prevented by the necessity for both accurate abundance estimates for field stars and accurate photometry at faint magnitudes in the clusters. With the availability of atmosphere models [A3] to calibrate abundance and deep HST photometry for the clusters, main-sequence fitting distances can now be measured using M subdwarfs.

NGC 6397 provides the best target for distance determination, since observations reach the hydrogen-burning limit. Lying at low Galactic latitude ($l = 338°$, $b = -12°$), the cluster is subject to substantial foreground reddening, but there is good agreement among the various indicators at $E_{B-V} = 0.18 \pm 0.02$. Thus, intrinsic colours can be calculated with a high degree of confidence, while the proper-motion study by King et al. provides effective elimination of all field-star contamination.

NGC 6397 has a metal abundance of [Fe/H] = −1.82 [C2]. esdM subdwarfs are estimated as having abundance $\langle [m/H] \rangle \sim -2.0$, so these stars are well matched as distance calibrators. Figure 11.19 shows the results of fitting the cluster lower main-sequence against 11 esdM subdwarfs with parallaxes measured to a precision of better than 10% [R2]. The significant dispersion in M_V among the field subdwarfs is

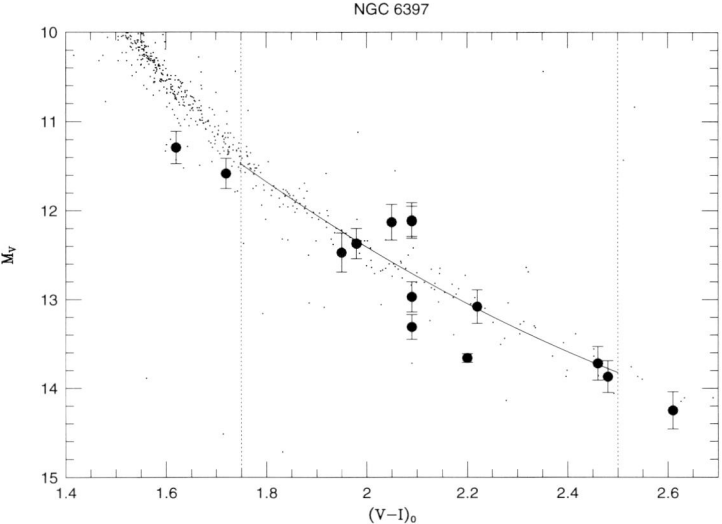

Figure 11.19. Main-sequence fitting, using M subdwarfs. The solid line is the mean relationship defined by the lowest-luminosity stars in NGC 6397; the distance modulus is derived by matching against field subdwarfs within the colour range isolated by the vertical lines. (From [R2], courtesy of the *Astronomical Journal*.)

not surprising, given that these stars must span a range of abundances. The derived distance modulus is $(m–M)_0 = 12.12 \pm 0.15$, which compares favourably with $(m–M)_0 = 12.24 \pm 0.1$ derived by matching the $(M_V, (B–V))$ cluster colour-magnitude diagram against F and G subdwarfs with Hipparcos parallax data. This is a technique in need of more refinement, but the results augur well for the future.

11.8   SUMMARY

As in the Galactic disk, low-mass subdwarfs represent the majority constituent of the halo – by number, if not by mass. While the properties of these stars are less well understood than their higher-temperature counterparts, recent theoretical work, both on the atmospheres and on the internal structure, combined with the availability of better data, have opened the way to more quantitative analysis. Abundances can be estimated, both from the location of stars in the H–R diagram and from molecular bandstrengths, and the abundance distribution appears to be consistent with that derived for more luminous subdwarfs. The overall luminosity function is well determined, although further and more detailed work is required to establish both the overall mass function and the extent of any metallicity dependence. Present results suggest that $\Psi_{\text{halo}}(M)$ is somewhat steeper than the disk function over the range 0.8–0.25 $M_\odot$, but may turn over at low masses and is poorly approximated as a power law. Current results in this area are at best provisional. Despite these uncertainties, however, and even given the current low-accuracy abundance estimates, we can use late-type subdwarfs as calibrators in main-sequence fitting. The initial results from those analyses are in good agreement with the more conventional main-sequence fitting studies based on F and G subdwarfs.

11.9   REFERENCES

A1   Ake, T. B., Greenstein, J. L., 1980, *ApJ*, **240**, 859.

A2   Alexander, D. R., Brocato, E., Cassisi, S., Castellani, V., Ciacio, F., Degl'Innocenti, S., 1997, *A&A*, **317** 90.

A3   Allard, F., Hauschildt, P. H., 1995, *ApJ*, **445**, 433.

B1   Bahcall, J. N., Casertano, S., 1986, *ApJ*, **308**, 347.

B2   Baraffe, I., Chabrier, G., Allard, F., Hauschildt, P., 1997, *A&A*, **327**, 1057.

B3   Beers, T. C., Rossi, S., Norris, J. E., Ryan, S. G., Molaro, P., Rebolo, R., 1998, *Sp. Sci. Rev.*, **84**, 139.

B4   Bessell, M. S., 1982, *Proc. ASA*, **4**, 417.

C1   Carney, B. W., Latham, D. W., Laird, J. B., Aguilar, L. A., 1994, *AJ*, **107**, 2240.

C2   Carretta, E., Gratton, R. G., 1997, *A&AS*, **121**, 95.

C3   Casertano, S., Ratnatunga, K. U., Bahcall, J. N., 1990, *ApJ*, **357**, 435.

C4   Chernoff, D. J., Weinberg, M., 1990, *ApJ*, **351**, 121.

C5   Cool, A. M., Piotto, G., King, I. R., 1996, *ApJ*, **468**, 655.

C6   Cottrell, P. L., 1978, *ApJ*, **223**, 544.

D1   Dahn, C. C., Liebert, J. W., Harris, H., Guetter, H. C., 1995, in *The Bottom of the Main Sequence and Beyond*, ed. C. G. Tinney, Springer, Heidelberg, p. 239.

D2   D'Antona, F., 1990, in *Physical Processes in Fragmentation and Star Formation*, ed. R. Capuzzo Dlocetta, C. Chiosi and A. Di Fazio, Kluwer, Reidel, Dordrecht, p. 367.

D3   D'Antona, F., 1995, in *The Bottom of the Main Sequence and Beyond*, ed. C. G. Tinney, Springer, Heidelberg, p. 17.

D4   D'Antona, F., Mazzitelli, I., 1995, *ApJ*, **456**, 329.

D5   Dawson, P. C., 1986, *ApJ*, **311**, 984.

D6   De Marchi, G., Paresce, F., 1995, *A&A*, **304**, 202.

D7   De Marchi, G., Paresce, F., 1995, *A&A*, **304**, 211.

D8   De Marchi, G., Paresce, F., 1996, *ApJ*, **467**, 568.

D9   Dinescu, D. I., Girard, T. M., van Altena, W. E., 1999, *AJ*, **117**, 1792.

D10 Drukier, G. A., 1995, *ApJS*, **100**, 347.

D11 Dull, J. D., Cohn, H. N., Lugger, P. M., Murphy, B. W., Seitzer, P. O., Callanan, P. J., Rutter, R. G. M., Charles, P. A., 1997, *ApJ*, **481**, 267.

E1   Eggen, O. J., 1983, *ApJS*, **51**, 183.

F1   Ferraro, F. R., Carretta, E., Bragaglia, A., Renzini, A., Ortolani, S., 1997, *MNRAS*, **286**, 1012.

G1   Gizis, J. E., 1997, *AJ*, **113**, 806.

G2   Gizis, J. E., Reid, I. N., 1998, *PASP*, **109**, 1223.

G3   Gizis, J. E., Reid, I. N., 1998, *AJ*, **117**, 508.

G4   Gould, A., Flynn, C., Bahcall, J. N., 1998, *ApJ*, **503**, 798.

H1   Hartwick, F. D. A., Cowley, A. P., Mould, J. R., 1984, *ApJ*, **286**, 269.

J1   Jones, D. H. P., 1973, *MNRAS*, **161**, 19P.

K1   King, I. R., Sosin, C., Cool, A. M., 1995, *ApJ*, **452**, L33.

K2   King, I. R., Anderson, J., Cool, A. M., Piotto, G., 1998, *ApJ*, **492**, L37.

K3   Kirkpatrick, J. D., Henry, T. J., McCarthy, D. W., 1991, *ApJS*, **77**, 417.

L1   Layden, A.C., Hanson, R.B., Hawley, S.L., Klemola, A.R., Hanley, C.J. 1996, *AJ*, **112**, 2110.

M1   Marconi, G. *et al.*, 1998, *MNRAS*, **293**, 479.

M2   Merritt, D., Meylan, G., Mayor, G. 1997, *AJ*, **114**, 1074.

M3   Mould, J. R. 1976, *A&A*, **48**, 443.

M4   Mould, J. R., McElroy, D. B. 1978, *ApJ*, **220**, 935.

M5   Mould, J. R. *et al.*, 1996, *PASP*, **108**, 682.

N1   Norris, J. E., 1986, *ApJS*, **61**, 667.

O1   Ohman, Y., 1936, *Stockholm Obs. Ann.*, 12, No. 3.

O2   Oort, J. H., 1965, in *Galactic Structure*, ed. A. Blaauw and M. Schmidt, University of Chicago Press, p. 455.

P1   Paresce, F., De Marchi, G, Romaniello, M., 1995, *ApJ*, **440**, 216.

P2   Piotto, G., Cool, A. M., King, I. R., 1997, *AJ*, **113**, 1345.

P3   Piotto, G., Zoccali, M., 1999, *A&A*, **345**, 485.

P4   Pulone, L., De Marchi, G., Paresce, F., Allard, F., 1998, *ApJ*, **492**, L41.

P5   Pulone, L., De Marchi, G., Paresce, F., 1999, *A&A*, **342**, 440.

R1   Reid, I. N., 1984, *MNRAS*, **206**, 1.

R2   Reid, I. N., Gizis, J. E., 1998, *AJ*, **116**, 2929.

R3   Richer, H. B., Fahlman, G. G., Buonnano, R., Fusi Pecci, F., Searle, L., Thompson, I., 1991, *ApJ*, **381**, 147.

R4   Richer, H. B., Fahlman, G. G., 1992, *Nature*, **358**, 383.

R5   Richstone, D. O., Graham, F. G., 1984, *ApJ*, **277**, 227.

S1   Schmidt, M., 1975, *ApJ*, **202**, 22.

S2   Sommer-Larsen, J. Zhen, C., 1990, *MNRAS*, **242**, 10.

T1   *The 19th Texas Symposium on Relativistic Astrophysics and Cosmology*, 1998, eds. J. Paul, T. Montmerle and E. Aubourg, CEA, Saclay.

V1   Vesperini, E., Heggie, D. C., 1997, *MNRAS*, **289**, 898.

Epilogue

The first drafts of sections of this book were written in early 1996. At that time it seemed that the main focus would centre squarely on the M dwarfs which numerically dominate the Galactic disk stellar population. Neither of us anticipated the explosion of discoveries of cool, lower-mass brown dwarfs, requiring the addition of two new spectral classes, and, almost simultaneously, the first detections of extrasolar planets. We have, as far as possible, modified and expanded our discussion to keep pace with those discoveries. However, even as we write these final words, new insights are being obtained into the nature and number of L and T dwarfs, and new planetary systems are being added to the catalogue.

Over the next decade we expect to see a consolidation of knowledge in these new areas. As of January 2000, we know of 110 L dwarfs, 16 T dwarfs and over 25 extrasolar planets. The next big step forward should come with the mid-infrared surveys of the Space Infrared Telescope Facility and the detection of room temperature brown dwarfs for eventual follow-up observations by the Next Generation Space Telescope.

Meanwhile, then M dwarfs and subdwarfs in the disk and halo will continue to provide significant challenges to our understanding of the physical properties of cool stars. We can hope that future observations will shed more illumination on such areas as the mechanism responsible for driving magnetic activity and producing spots and flares; the temperature structure and chemical processes in cool atmospheres; and, most significantly, the means by which molecular clouds form distributions of stars of different masses. Understanding those physical processes in our own Galaxy through investigations such as the ones we have described, is essential for a broad range of astrophysical problems, including disentangling the complex behaviour of galaxy formation at high redshifts in the early Universe.

Appendix

The 8-parsec sample

A.1 INTRODUCTION

The Sun's immediate neighbours have attracted particular attention for most of this century. Following van de Kamp [V1], [V2], listings are generally limited to stars within 5 parsecs of the Sun. However, the more extensive observations currently available – particularly speckle imaging and spectroscopic follow-up observations, justify extending the distance limit to 8 parsecs. The following tables collect together photometric and spectroscopic data for these stars.

The starting point for this compilation was the preliminary version of the *Third Catalogue of Nearby Stars* (CNS3) [G1], supplemented by spectroscopic observations from Reid *et al.* [R1] and Hawley *et al.* [H1], and astrometry for the brighter stars from the Hipparcos catalogue. The distance estimates are a weighted combination of trigonometric parallaxes and spectroscopic parallaxes, assigning an uncertainty of 30% to the latter. Imposing a distance limit of 8 parsecs produces a sample of 150 stars (including eight white dwarfs) in 105 systems with declinations north of −30° (75% of the celestial sphere) and 38 stars (three white dwarfs) in 33 systems south of that declination limit.

The spectral types listed in these tables come from three sources: M dwarfs with types M7 and later have been classified by Henry *et al.* [H2]; spectral types for white dwarfs and early-type (<K5) stars are from the CNS3; and data for the remaining stars are from [R1], [H1]. Most of the photometry is taken from either Bessell's [B1] observations or Leggett's [L1] compilation; columns 7 and 11 (headed S_O and S_{IR}) in Table A.1 identify additional sources of optical and near-infrared data, respectively. All of the stars have V-band data, but many lack observations at near-infrared wavelengths. We have used three techniques to estimate K-band absolute magnitudes for stars without direct JHK photometry: first, we have used data from [L1] to delineate a (V–I)/(V–K) relationship, which can be approximated as

$$(V\text{–}K) = 1.58(V\text{–}I) + 0.82, \qquad \sigma_{V\text{–}K} = 0.15$$

to estimate (V–K) for stars with measured (V–I) colours redder than 1.0 magnitudes;

second, we have adopted the (B–V)/(V–K) relationship defined by stars from the ESO JHK list of standards [B2],

$$(V–K) = 2.22(B–V) + 0.10, \qquad \sigma_{V–K} = 0.12$$

for bluer stars with (B–V) data; and, finally, for stars with no colour information, the K-band absolute magnitudes are derived from the approximate relationships

$$M_K \sim 0.75M_V - 1.5, \qquad 6 < M_V < 10$$

$$M_K \sim 0.50M_V + 1.0, \qquad 10 < M_V < 18.$$

A.2   REFERENCES

B1   Bessell, M. S., 1991, *A&AS*, **83**, 357.

B2   Bouchet, P., Manfroid, J., Schmider, F. X., 1991, *A&AS*, **91**, 409.

G1   Gliese, W., Jahreiss, H., 1991, *Third Catalogue of Nearby Stars*, CNS3.

H1   Hawley, S. L., Gizis, J. E., Reid, I. N., 1996, *AJ*, **112**, 2799.

H2   Henry, T. J., Kirkpatrick, J. D., Simons, D. A., 1995, *AJ*, **108**, 1437

L1   Leggett, S. K., 1992, *ApJS*, **82**, 351.

R1   Reid, I. N., Hawley, S. L., Gizis, J. E., 1995, *AJ*, **110**, 1838.

V1   van de Kamp, P., 1953, *PASP*, **65**, 73.

V2   van de Kamp, P., 1977, *Vistas in Astronomy*, **20**, 501.

Table A.1. The northern 8-parsec sample.

Name	M_V	(U–B)	(B–V)	(V–R)	(V–I)	S_O	M_K	(J–H)	(H–K)	S_{IR}	Sp.	r	ϵ_r	M_{bol}	Comment
Sun	4.79	0.18	0.65				3.30^b				G2 V			4.67	
GJ 1002	15.43	1.86	1.99	1.60	3.60	1	9.10	0.55	0.33	1	M5.5	4.64^G	0.07	11.91	G158-27
GJ 1005 A	12.84	1.26	1.71	1.39	2.77	2	7.90	0.64	0.18	3	M4	5.60^G	0.24	10.50	
GJ 1005 B	15.64			1.55	3.51	2	9.90	0.24	0.77	3	M4	5.60	0.24	12.26	
Gl 15 A	10.32	1.24	1.55		2.14	1	6.30	0.57	0.22	1	M1	3.57^H	0.01	8.72	
Gl 15 B	13.31	1.40	1.79	1.24	2.82	1	8.21	0.54	0.26	1	M3.5	3.57	0.01	10.90	
Gl 33	6.36	0.58	0.88	0.51	0.96	4	4.21	0.48	0.09	5	K2 V	7.46^H	0.05	6.09	
Gl 34 A	4.58	0.02	0.57	0.40	0.67	7	3.09	0.31	0.04	5	G3 V	5.95^H	0.02	4.53	η Cas
Gl 34 B	8.64		1.39			6	4.91	0.58	0.11	3	K7 V	5.95	0.02	7.6	
Gl 35	14.22	0.02	0.55	0.26	0.49	4					DZ7	4.33^H	0.06		VMa 2
Gl 53 A	5.78	0.09	0.69			6	4.12^a				G5 VI	7.55^H	0.03	5.62	μ Cas
Gl 53 B	11.6					6	6.70^b					7.55	0.02	9.7	
Gl 54.1	14.17	1.33	1.86	1.37	3.15	1	8.54	0.55	0.29	1	M4.5	3.77^G	0.04	11.32	
Gl 65 A	15.32	1.09	1.85	1.65	3.69	1	8.78	0.60	0.28	3	M5.5	2.62^G	0.03	11.66	
Gl 65 B	15.98						9.18	0.68	0.28	3	M6	2.62	0.03	12.09	UV Cet
Gl 68	5.86	0.49	0.83	0.35	0.86	5	3.85	0.37	0.21	5	K1 V	7.47^H	0.05	5.64	
Gl 71	5.68	0.21	0.72	0.43	0.81	4	3.85	0.36	0.12	7	G8 Vp	3.65^H	0.01	5.52	τ Cet
Gl 83.1	14.02	1.35	1.83	1.37	3.12	1	8.40	0.55	0.29	1	M4.5	4.50^G	0.06	11.21	
Gl 105 A	6.52	0.80	0.97	0.57	1.07	4	4.15	0.51	0.06	8	K3 V	7.21^H	0.05	6.17	
Gl 105 B	12.38	1.09	1.61	1.19	2.79	1	7.33	0.55	0.22	1	M4	7.21	0.05	10.01	
Gl 105 C	17.55				4.60	9	9.45^c					7.21	0.05	12.35	
Gl 109	11.19	1.11	1.56	1.10	2.45	4	6.50	0.53	0.27		M3	7.55^H	0.14	9.24	
LP771-95	11.92^d		1.69	1.39	2.55	6	6.9^c				M3.5	7.86^H	1.97	9.85	
LP771-96	12.5					6	7.1^b				M3	7.86	1.97	10.1	
LP771-96A	14.5					6	7.5^b				M4	7.86	1.97	10.3	
Gl 144	6.20	0.58	0.88	0.50	0.94	4	4.12	0.42	0.08	8	K2 V	3.22^H	0.01	5.83	ε Eri
Gl 166 A	5.92	0.44	0.82	0.47	0.88	4	3.92	0.47	0.07		K1 Ve	5.04^H	0.02	5.79	40 Eri
Gl 166 B	11.01	−0.68	0.03			6					DA4	5.04	0.02		

(continued)

Table A.1 (*cont.*)

Name	M_V	(U–B)	(B–V)	(V–R)	(V–I)	S_O	M_K	(J–H)	(H–K)	S_{IR}	Sp.	r	ϵ_r	M_{bol}	Comment
Gl 166 C	12.68	1.03	1.63	1.24	2.87	1	7.44	0.53	0.24	1	M4.5	5.04	0.02	10.20	
Gl 169.1A	12.39	1.21	1.64	1.21	2.81	1	7.07	0.65	0.26	1	M4	5.48^H	0.03	9.99	Stein2051
Gl 169.1B	13.74	−0.52	0.31			6					DC5	5.48	0.03		
LHS 1723	12.81			1.30	2.99	14	7.27^c				M4	7.51^S	2.25	10.18	
Gl 185 A	8.97	1.15	1.41			6	5.49	0.60	0.17	3	K7	7.95^G	0.44	7.97	
Gl 185 B	11.21					6	6.56	0.69	0.34	3		7.95	0.45	9.9	
Gl 205	9.17	1.21	1.47	0.97	2.07	1	5.09	0.68	0.19	1	M1.5	5.69^H	0.04	7.66	
Gl 213	12.72	1.22	1.60	1.23	2.82	1	7.56	0.52	0.26	1	M4	5.79^H	0.13	10.31	
Gl 223.2	15.40	0.77	1.77	0.97	1.87	4					DZ9	6.46^G	0.17		
GJ 1087	14.59	−0.14	0.60			6					DAP9	7.99^G	0.37		
LHS 1805	12.30	1.60	1.18	2.72	1	2	7.15^b				M3.5	7.63^G	0.17	10.1	
G 99-49	12.66	1.10	1.68			6	7.33^b				M4	5.42^G	0.18	10.3	
Gl 229 A	9.32	1.19	1.50	0.96	2.01	1	5.35	0.63	0.20	1	M0.5	5.77^H	0.04	7.86	
Gl 229 B	28::				6.9:	10	15.6	−0.10	−0.10	10	T	5.77	0.04	17.7	
Gl 234 A	13.01	1.23	1.72	1.31	3.02	1	7.41	0.63	0.26	3	M4.5	4.13^G	0.03	10.33	Ross 614
Gl 234 B	16.5						9.17	0.68	0.41	3		4.13	0.03	12.2	
Gl 244 A	1.46	−0.04	0.00	−0.01	−0.02	4	1.54	0.00	−0.02	11	A1 V	2.64^H	0.01	1.28	Sirius
Gl 244 B	11.33		−0.03			6					DA2	2.64	0.01		Sirius B
Gl 251	11.30	1.20	1.55	1.09	2.50	1	6.57	0.60	0.25	1	M3	5.52^H	0.06	9.29	
Gl 268 A	13.03^d	1.19	1.70	1.33	3.04	1	7.49	0.60	0.28	3	M4.5	6.36^H	0.13	10.33	
Gl 268 B	13.4						7.7	0.60	0.29	3		6.36	0.13	10.75	
Gl 273	11.87	1.12	1.57	1.15	2.70	1	6.89	0.54	0.25	1	M3.5	3.95^H	0.02	9.62	
Gl 280 A	2.65	0.03	0.42	0.25	0.49	4	1.63	0.17	0.04	7	F5 IV	3.50^H	0.01	2.67	Procyon
Gl 280 B	13.0:					6					DA	3.50	0.01		
Gl 285	12.28	0.97	1.61	1.26	2.95	1	6.86	0.60	0.26	1	M4.5	5.93^H	0.09	9.08	YZ CMi
Gl 299	13.65	1.19	1.77	1.25	2.92	1	8.46	0.47	0.28	1	M4.5	6.85^G	0.12	11.11	
Gl 300	13.14		1.58	1.22	2.90	1	7.75	0.64	0.25	1	M4	6.10^G	0.35	10.63	
GJ 1111	16.99	2.11	2.05	2.00	4.26	1	9.46	0.58	0.36	1	M6.5	3.63^G	0.04	12.39	G51-15
GJ 1116 A	15.46		1.84	1.67	3.78	1	8.83	0.57	0.28	3	M5.5	5.24^G	0.07	11.66	

(continued)

Name															
GJ 1116 B	16.31		1.93			6	9.22					5.24	0.07	12.1	
G 41-14 A[e]	12.18	1.21	1.67	1.24	2.84	15	6.87[c]	0.50	0.403	3	M3.5	7.85[S]	1.66	9.75	
G 41-14 B	12.18						6.87[c]			3	M3.5	7.85	1.66	9.75	
Gl 338 A	8.67	1.20	1.41	0.85	1.74	6	5.11	0.61	0.21	1	K7	6.23[H]	0.16	7.48	
Gl 338 B	8.70	1.17	1.42				5.13	0.62	0.19	1	K7	6.23	0.16	7.70	
Gl 380	8.15	1.28	1.38	0.85	1.62	1	4.77	0.62	0.13	1	K2 V	4.87[H]	0.02	7.08	
Gl 382	9.80		1.50	0.99	2.17	1	5.61	0.64	0.20	1	M1.5	7.81[H]	0.09	8.30	
Gl 388[h]	10.88	1.06	1.53	1.09	2.51	1	6.17	0.64	0.21	1	M3	4.87[G]	0.07	8.86	AD Leo
Gl 393	10.36	1.23	1.51	1.01	2.24	1	6.05	0.56	0.26	1	M2	7.23[H]	0.11	8.65	
LHS 292	17.32		2.10	2.06	4.40	1	9.68	0.58	0.36	1	M6.5	4.53[G]	0.07	12.47	
Gl 402	12.46	1.06	1.64	1.21	2.79	1	7.23	0.61	0.27	1	M4	6.88[G]	0.23	10.09	
Gl 406	16.54	1.59	1.99	1.88	4.06	1	9.17	0.62	0.36	1	M6	2.41[G]	0.01	12.28	Wolf 359
Gl 408	10.92	1.22	1.54	1.05	2.39	1	6.44	0.59	0.24	1	M2.5	6.62[H]	0.07	9.04	
Gl 411	10.43	1.14	1.51	1.01	2.15	1	6.32	0.54	0.20	1	M2	2.55[H]	0.01	8.82	Lalande 21185
Gl 412 A	10.34	1.18	1.54	1.00	2.02	1	6.34	0.58	0.21	1	M0.5	4.83[H]	0.03	8.87	
Gl 412 B	15.98		2.08	1.66	3.77	1	9.43	0.52	0.29	1	M6	4.83	0.03	12.2	WX UMa
Gl 445	12.17	1.08	1.58	1.12	2.64	1	7.28	0.51	0.25	1	M3.5	5.39[H]	0.04	10.00	
Gl 447	13.50	1.34	1.76	1.30	2.98	1	8.01	0.58	0.28	1	M4	3.34[G]	0.02	10.88	Ross 128
GJ 1156	14.76	1.01	1.88	1.51	3.46	16	8.54	0.52	0.31	3	M5	6.64[G]	0.13	11.46	
Gl 473 A	14.85	1.18	1.81	1.54	3.54	1	8.45	0.61	0.38	3	M5	4.35[G]	0.08	11.43	
Gl 473 B	15.10						8.50	0.53	0.26	3	M7	4.35	0.08	11.45	
Gl 514	9.63	1.22	1.49	0.96	2.01	1	5.64	0.61	0.19	1	M0.5	7.63[H]	0.08	8.17	
Gl 526	9.80	1.11	1.43	0.96	2.04	1	5.79	0.59	0.18	1	M1.5	5.43[H]	0.04	8.31	
Gl 555	12.39	1.20	1.63	1.24	2.86	1	7.09	0.59	0.26	1	M4	6.12[H]	0.10	9.92	
Gl 566 A	5.57	0.22	0.73			6	3.85[a]	0.60	0.40		G8 V	6.70[H]	0.03	5.42	ξ Boö
Gl 566 B	7.84	1.15	1.16			6	4.38[b]	0.57	0.11		K4 V	6.70	0.03	7.28	
LHS 3003	18.08			2.17	4.52		9.96			3	M6.5	6.22[G]	0.23	13.02	
Gl 570 A	6.78	1.06	1.10	0.70	1.34	1	4.24	0.61	0.21		K5 V	5.91[H]	0.06	6.29	
Gl 570 B	9.23	1.22	1.48	1.00	2.12	4	5.07	0.72	0.29		M1	5.91	0.06	7.7	
Gl 570 C	11.5						6.50					5.9	10.06	9.5	
Gl 570 D						1	16.7	−0.1	−0.1	20	T	5.91	0.06	17.5	

Table A.1 (*cont.*)

Name	M_V	(U–B)	(B–V)	(V–R)	(V–I)	S_O	M_K	(J–H)	(H–K)	S_{IR}	Sp.	r	ϵ_r	M_{bol}	Comment
Gl 581	11.58	1.2	1.60	1.10	2.51	1	6.87	0.61	0.26	1	M3	6.27^H	0.09	9.56	
Gl 625	11.02	1.28	1.60	0.99	2.21	1	6.76	0.57	0.23	1	M1.5	6.58^H	0.02	9.35	
Gl 628	11.93	1.12	1.57	1.17	2.68	1	6.95	0.54	0.27	1	M3.5	4.26^H	0.03	9.70	
Gl 643	12.70	1.35	1.69	1.19	2.73	1	7.67	0.55	0.25	1	M3.5	6.50^H	0.17	10.41	Wolf 629
Gl 644 A	10.9	1.13	1.57	1.10	2.46	1	6.25	0.64	0.25	1	M3	5.74^H	0.13	8.90	Wolf 630
Gl 644 B	11.0:					6	6.4					5.74	0.13	9.0	
Gl 644 C	18.01		2.20	2.15	4.56		9.76	0.58	0.37	1	M7	5.74	0.13	12.87	VB 8
Gl 644 D	12.2:					6	8.6:					5.74	0.13	9.7	
G 203-47 A^f	12.48	0.89	1.46	1.20	2.81	16	7.2^c				M3.5	7.25^H	0.17	10.08	
G 203-47 B	>15						>8				WD	7.25	0.17	11.6	
Gl 661 A	10.95	1.01	1.49	1.10	2.51	1	6.37	0.54	0.21	3	M3.5	6.32^H	0.13	8.93	
Gl 661 B	11.40						6.75	0.49	0.25	3		6.32	0.13	9.4	
Gl 663 A	6.18	0.49	0.86	0.49	0.93	4	4.2^b				K1 V	5.99^H	0.04	5.93	
Gl 663 B	6.22		0.86			6	4.2^b				K1 V	5.99	0.04	5.97	
Gl 664	7.44	1.08	1.16	0.70	1.28	4	4.60^c				K5 V	5.97^H	0.04	6.88	
Gl 673	8.10	1.27	1.36	0.85	1.61	1	4.74	0.61	0.13	1	K5	7.72^H	0.06	7.04	
Gl 687	10.94	1.04	1.50	1.09	2.50	1	6.24	0.57	0.22	1	M3	4.53^H	0.02	8.93	
GJ 1221	15.21	−0.30	0.40			6					DXP9	6.14^G	0.11		
Gl 699	13.26	1.26	1.73	1.21	2.78	1	8.22	0.50	0.26	1	M4	1.82^H	0.01	10.90	Barnard's Star
Gl 701	9.91		1.50	0.98	2.06	1	5.88	0.60	0.22	1	M1	7.80^H	0.09	8.40	
Gl 702 A	5.68	0.51	0.86	0.51	0.96	4	3.68^b				K0 V	5.09^H	0.04	5.43	
Gl 702 B	7.47		1.15			6	4.8^b				K5 V	5.09	0.04	6.92	
GJ 1224	14.18	1.79	1.39	3.18	1	7	8.58	0.43	0.42	3	M4.5	7.70^G	0.21	11.28	
LHS 3376	14.14	1.80	1.40	3.12	1	3	8.39^c				M4.5	7.54^G	0.28	11.33	
LP 229-17	12.14	1.42	1.16	2.68	1	5	7.09^c				M3.5	7.18^S	2.15	9.91	
Gl 721	0.59	−0.01	0.00	−0.01	0.00	4	0.55	0.00	0.00	7	A0 V	7.76^H	0.03	0.44	Vega
Gl 725 A	11.14	1.11	1.54	1.07	2.46	1	6.68	0.53	0.23	1	M3	3.57^H	0.03	9.18	
Gl 725 B	11.92	1.14	1.59	1.12	2.55	1	7.21	0.52	0.23	1	M3.5	3.57	0.03	9.84	
Gl 729	13.11		1.72	1.23	2.78	1	8.03	0.54	0.28	1	M3.5	2.97^H	0.02	10.74	

Gl 752 A	10.28	1.16	1.50	1.03	2.32	1	5.82	0.62	0.22	1	M2.5	5.87^H	0.05	8.48	
Gl 752 B	18.65		2.20	2.10	4.33	13	9.95	0.66	0.44	1	M8	5.87	0.05	12.93	VB 10
Gl 764	5.87	0.37	0.78	0.45	0.87	7	3.97	0.39	0.10	7	K0 V	5.77^H	0.02	5.69	σ Dra
Gl 768	2.22	0.08	0.22	0.14	0.27	4	1.71	0.08	0.03	7	A7 IV	5.14^H	0.02	2.25	Altair
GJ 1245 A	15.05		1.90	1.59	3.60	1	8.53	0.52	0.37	1	M5.5	4.70^G	0.10	11.53	
GJ 1245 B	15.65		1.98	1.66	3.73	1	9.09	0.50	0.39	1	M5.5	4.70	0.10	11.93	
GJ 1245 C	18.09						10.09	0.66	0.40	3		4.70	0.10	12.79	
Gl 793	11.05		1.56	1.06	2.43	1	6.45	0.60	0.23	1	M2.5	7.96^H	0.07	9.10	
LP 816-60	12.71		1.65			19	8.0^c					5.49^H	0.11	10.7	
Gl 809	9.27	1.24	1.49	0.92	1.99	1	5.41	0.65	0.19	1	M0.5	7.04^H	0.04	7.83	
Gl 820 A	7.49	1.11	1.19	0.67	1.27	4	4.68	0.50	0.19	5	K5	3.50^H	0.01	6.91	61 Cyg
Gl 820 B	8.31	1.22	1.37		1.62	1	4.98	0.65	0.14	1	K5	3.50	0.01	7.26	
Gl 829 A	11.87	1.31	1.61	1.11	2.58	1	6.34	0.58	0.25	1	M3.5	6.74^H	0.08	9.08	
Gl 829 B	11.87	1.31	1.61	1.11	2.58	1	6.34	0.58	0.25	1	M3.5	6.74^H	0.08	9.08	
Gl 831 A	12.60		1.67	1.31	2.97	1	7.01	0.59	0.28	1	M4.5	7.76^G	0.28	9.99	
Gl 831 B	15.76						8.95	0.88	0.18	3		7.94	0.29	11.8	
LHS 3799	13.88		1.84	1.41	3.22	16	7.98^c	0.47	0.30	3	M4.5	7.48^G	0.31	10.93	
Gl 860 A	11.88	1.23	1.66	1.19	2.68	1	7.03	0.62	0.07	3	M3	4.01^H	0.05	9.66	
Gl 860 B	13.28		1.80				7.63			3	M4	4.01	0.05	10.7	
Gl 866 A	15.4^d	1.54	1.97	1.68	3.70	1	8.8	0.63	0.34	3	M5.5	3.38^G	0.04	11.7	
Gl 866 B	15.94				4.20		8.92	0.53	0.34	3		3.38	0.04	11.8	
Gl 866 C^g	15.4						8.8					3.38	0.04	11.7	
Gl 873	11.75	1.10	1.58	1.19	2.69	1	6.79	0.59	0.23	1	M3.5	5.05^H	0.05	9.50	EV Lac
Gl 876	11.82	1.14	1.56	1.20	2.74	1	6.70	0.60	0.26	1	M4	4.70^H	0.05	9.54	
Gl 880	9.47	1.19	1.48	0.99	2.11	1	5.34	0.63	0.21	1	M1.5	6.88^H	0.06	7.93	
Gl 881	1.72	0.06	0.09	0.05	0.08	4	1.55	0.00	0.02	7	A3 V	7.69^H	0.05	1.75	Fomalhaut
Gl 892	6.48	0.88	1.00	0.57	1.05	7	4.10	0.48	0.08	8	K3 V	6.53^H	0.03	6.10	
Gl 896 A	11.43	1.06	1.71	1.22	2.84	1	6.36	0.55	0.30	3	M3.5	6.25^H	0.11	9.00	EQ Peg
Gl 896 B	13.42		1.65				7.60	0.79	0.13	3	M4.5	6.25	0.11	10.8	
Gl 896 C	13.4						7.60			3	M4.5	6.25	0.11	10.8	
GJ 1286	15.37		1.95	1.59	3.63	18	9.01	0.61	0.28	3	M5.5	7.24^G	0.18	11.81	

(continued)

Table A.1 (*cont.*)

Name	M_V	(U–B)	(B–V)	(V–R)	(V–I)	S_O	M_K	(J–H)	(H–K)	S_{IR}	Sp.	r	ϵ_r	M_{bol}	Comment
Gl 905	14.78	1.48	1.90	1.52	3.45	1	8.44	0.63	0.32	1	M5	3.17[G]	0.02	11.49	
Gl 908	10.11	1.09	1.46	0.95	2.03	1	6.18	0.52	0.22	1	M1	5.97[H]	0.05	8.63	

[a] M_K inferred from (B–V)/(V–K) relationship.

[b] M_K inferred from M_V/M_K relationship.

[c] M_K inferred from (V–I)/(V–K) relationship.

[d] M_V allows for contribution of secondary star to joint photometry.

[e] Identified as SB2 in echelle observations with Palomar 60-inch (Gizis, Reid and Hawley). Based on the relative line-strengths, we assume equal components.

[f] Identified as SB1 in echelle observations with Palomar 60-inch (Gizis, Reid and Hawley). $\Delta R > 1.5$ magnitudes assumed.

[g] Noted as triple by Henry, echelle observations with Palomar 60-inch (Gizis, Reid and Hawley) show that the system includes an SB2 binary with a period inconsistent with the astrometric orbit. Based on the approximately equal line-strengths, we assume that the brighter star of the two resolved by speckle interferometry is the spectroscopic binary.

[h] Balega et al. (*Sov. Astr. Lett.*, **17**, 226) claim detection of a companion at $\sim0.''1$ separation through H-band speckle observations, but Henry found no evidence for a companion brighter than $M_K = 11$ to similar limits.

All photometry transformed to the Cousins BVRI and CIT JHK systems. Sources:

1. Leggett, 1992, *ApJS*, **82**, 351; 2. Reid, 1982, *MNRAS*, **201**, 73; 3. Henry, 1991, PhD thesis; 4. Bessell, 1988, *A&AS*, **83**, 357; 5. Johnson et al., 1968, *ApJ*, **152**, 465; 6. Jahreiss and Gliese, 1991, CNS3; 7. Johnson et al., 1966, *Comm. LPL*, **4**, 99 (H estimated); 8. Glass, 1975, *MNRAS*, **171**, 19P; 9. Golimowski et al., 1995, *ApJ*, L101; 10. Matthews et al., 1996, *AJ*, **112**, 1678; 11. Glass, 1974, *MNASSA*, **31**, 81; 12. Weis, 1996, *AJ*, **112**, 1759; 13. Bessell, 1991, *AJ*, **91**, 616; 14. Gliese, 1982, *A&AS*, **47**, 471; 15. Weis, 1991, *AJ*, **102**, 1975; 16. Weis, 1987, *AJ*, **93**, 451; 17. Weis, 1984, *ApJS*, **55**, 289; 18. Weis and Upgren, 1982, *PASP*, **94**, 821; 19. ESA, 1997, *Hipparcos Catalogue*; 20. Burgasser et al., 1999, *ApJL*, **525**.

Distances: [H], from *Hipparcos* parallaxes; [G] from ground-based parallaxes; [S] from TiO spectroscopic parallaxes.

Table A.2. The southern 8-parsec sample

Name	M_V	(U–B)	(B–V)	(V–R)	(V–I)	M_K	(J–H)	(H–K)	Sp.	r	ϵ_r	Comments
Gl 1	10.36	0.96	1.45	0.97	2.12	6.33	0.51	0.22	M1.5	4.36[H]	0.02	1
Gl 19	3.45	0.10	0.62	0.35	0.69	1.97	0.32	0.08	G2 IV	7.48[H]	0.02	2, A
LHS 1513	12.07	1.22	1.51	1.39	2.13	7.03			M3.5	7.73[S]	1.16	
Gl 139	5.25	0.22	0.71	0.41	0.80	3.61	0.36	0.07	G5 V	6.06[H]	0.02	1,3
GJ 1061	15.13	1.52	1.90			8.57			M5.5	3.80[S]	0.57	C
Gl 191	10.90	1.21	1.56	0.95	1.95	7.11	0.49	0.21	K7	3.91[H]	0.01	4, Kapteyn's Star
Gl 293	14.88	–0.17	0.66						DQ9	6.91[G]	0.93	
LHS 288	15.66		1.82	1.51	3.61	8.83				4.49[G]	0.27	A
Gl 440	13.20	–0.63	0.19						DQ6	4.62[H]	0.04	
Gl 480.1	12.82	1.39	1.73	1.17	2.61	7.7			M3.5	7.66[H]	0.23	
LHS 337	13.33	1.25	1.69	1.91	2.98	7.66			M4.5	7.62[S]	1.14	2, A
Gl 551	15.49	1.37	1.85	1.63	3.62	8.76	0.58	0.34	M5.5	1.29[H]	0.01	Proxima Cen
Gl 559 A	4.38	0.23	0.64	0.36	0.69	2.86			G2 V	1.35[H]	0.01	B, α Cen
Gl 559 B	5.71	0.63	0.84	0.47	0.88	3.75			K0 V	1.35	0.01	4, B
Gl 588	10.45	1.11	1.51	1.07	2.39	5.92	0.61	0.25	M2.5	5.93[H]	0.05	
HIC 82725	13.26									4.93[H]	0.71	
Gl 667 A	7.02	0.82	1.04	0.64	1.20	4.61			K3 V	7.25[G]	0.31	B
Gl 667 B	7.90					7.80			K5 V			B
Gl 667 C	10.75	1.15	1.57	0.90	2.12	6.54	0.55	0.24	M1.5			
Gl 674	11.09	1.21	1.55	1.08	2.41	6.60	0.56	0.24	M2.5	4.54[H]	0.03	4
Gl 682	12.43	1.20	1.64	1.25	2.82	7.16	0.61	0.28	M3.5	5.05[H]	0.06	
GJ 2130A	11.53		1.46							6.18[H]	0.43	
GJ 2130B	12.43		1.43							6.18	0.43	
Gl 693	11.93		1.63	1.16	2.56	7.07			M2	5.81[H]	0.09	A
Gl 754	13.38	1.22	1.66	1.29	2.96	7.88			M4.5	5.90[S]	0.43	A
Gl 780	4.62	0.45	0.76	0.41	0.76	3.00	0.32	0.10	G8 V	6.11[H]	0.02	1, δ Pav
Gl 783 A	6.41	0.46	0.87	0.52	1.00	4.38			K3 V	6.05[H]	0.03	B

(continued)

Table A.2 (*cont.*)

Name	M_V	(U–B)	(B–V)	(V–R)	(V–I)	M_K	(J–H)	(H–K)	Sp.	r	ϵ_r	Comments
Gl 783 B	12.55								M3.5			C
Gl 784	9.01	1.22	1.41	0.91	1.83	5.32	0.62	0.18	K7	6.20^H	0.04	
Gl 825	8.71	1.22	1.42	0.90	1.76	5.11	0.63	0.17	K7	3.51^H	0.01	
Gl 832	10.19	1.20	1.48	1.01	2.18	5.93	0.55	0.22	M1.5	4.94^H	0.03	
Gl 845	6.89	0.99	1.05	0.62	1.14	4.46			K5 V	3.63^H	0.01	
LHS 3836	11.95					6.97			M3.5	7.93^S	1.19	C
Gl 879	7.07	1.02	1.10	0.66	1.21	4.53			K5 V	7.64^H	0.05	
Gl 887	9.77	1.18	1.50	0.98	2.02	6.10	0.56	0.22	M0.5	3.29^H	0.003	5, A
HIC 114176	13.13									6.76^H	0.63	
Gl 915	13.59	-0.87	0.07						DA5	7.80^S	0.55	
LP 944-20	20.18				4.50	11.10			M8	4.97^G	0.02	

Sources of photometry listed under Comments:

Photometry from Bessell [B1], Leggett [L1] and the CNS3 [G1], except:

A. M_K estimated from (V–I)/(V–K) relationship.
B. M_K estimated from (B–V)/(V–K) relationship.
C. M_K estimated from M_V/M_K relationship.
1. JHK from Glass, 1974, *MNASSA*, **31**, 81.
2. RI from Eggen, 1986, *AJ*, **92**, 910.
3. (U–B) from Johnson *et al.*, 1966, *Comm. LPL*, **4**, 99.
4. (U–B) from Eggen, 1987, *AJ*, **93**, 379 and references within.
5. (U–B) from Cousins, 1971, SAAO Circ. No. 1.
6. LP 944-20 is also known as BRI0037-3535.

Index

chemical signatures, 411
direct imaging, 397
Doppler techniques, 401
formation, 410
mass distribution, 406
photometric detection, 397
radial velocity surveys, 402
searches, 395
Extreme Ultraviolet Explorer (EUVE), 175

Fermi exclusion principle, 96
Fermi–Dirac statistics, 85, 96
flare stars, 65
flares, 65, 69, 163, 179, 183
flat-field exposure, 22
flux, 12, 21
flux tubes, 164
Fornax dwarf ellipticals, 210
free–free process, 138

G 29-38, 345
γ Vir, 307
G96−45 (GJ 1081), 396
Galactic disk, 234, 262
  abundance distribution, 242
  age, 243
  density distribution, 238
  kinematics, 235
  luminosity function, 295
  radial density law, 239
  star-formation history, 244
  vertical density law, 240
Galactic halo, 60, 214, 421
  abundance distribution, 221
  age, 227
  density distribution, 218
  formation, 231
  kinematics, 220
Galactic structure, 209
gas polytropes, 87
GD 165B, 43, 44, 348, 376
Gemini telescope, 13
Georgia State University, 413
GG Tau, 118, 319, 320
GJ 1081 (G96-45), 396
GJ 1111, 58
Gl 96-3, 376, 377
Gl 166C, 62

Gl 169.1A (Stein 2051A), 74, 396
Gl 191 (Kapteyn's Star), 61, 422
Gl 229B, 48, 58, 74, 352, 359, 360, 365, 368
Gl 278C (YY Gem), 54, 97, 114, 154
Gl 285 (Ross 882, YZ CMi), 50, 65
Gl 388 (AD Leo), 70, 71, 170, 172, 178, 184, 189, 396
Gl 406, 49, 151
Gl 411 (Lalande 21185), 49, 151, 396, 397
Gl 412B (Lalande 21258B, WX UMa), 65
Gl 417, 377
Gl 570, 74, 366, 377
Gl 584, 376
Gl 630.1A (CM Dra), 54, 90, 97, 114, 310, 398, 399
Gl 643, 344
Gl 644C (VB 8), 54, 74, 344
Gl 65AB (L726-8, UV Cet), 50, 65, 70, 97
Gl 699 (Barnard's Star), 54, 397
Gl 702 (70 Oph), 396
Gl 729, 170
Gl 752B (VB 10), 43, 65, 97, 137, 151, 200, 376
Gl 820 (61 Cyg), 9, 252, 396
Gl 860AB (Krüger 60AB), 97
Gl 866AB, 54, 309
Gl 873, 50
Gl 876, 122, 407, 411
Gl 884, 54
Gl 890, 181, 182, 195
globular clusters, 211, 434
GM Aur, 118
gravitational collapse, 117
gravitational contraction, 84
gravitational lensing, 311
Great Debate, 217

H^- ion, 132
$H\alpha$ emission, 197
halo subdwarfs, 421
  classification, 423
  luminosity function, 429
  photometric properties, 422
  proper-motion samples, 432
  searches, 423
  star-count analyses, 434
Harvard classification system, 39
Haute-Provence Observatory, 405
Hayashi tracks, 118